Algebra II

An Incremental Development

Algebra II

An Incremental Development

JOHN H. SAXON, JR.

SAXON PUBLISHERS, INC.

Algebra II: An Incremental Development

Teacher's Edition

Copyright © 1984 by Saxon Publishers, Inc.

Printed in the United States of America
ISBN: 0-939798-11-5
Fourth Printing, May 1987

Saxon Publishers, Inc.
1002 Lincoln Green
Norman, Oklahoma 73069

To the Student

Algebra is an abstract study of the way numbers behave and interrelate. In Algebra 1, we found that algebra is not difficult—it is just different—and concepts that were confusing when first encountered became familiar concepts after they had been practiced for a period of weeks or months—until finally they were understood. Then further study of the same concepts caused additional understanding as totally unexpected ramifications appeared. And, as we mastered these new abstractions, our understanding of seemingly unrelated concepts became clearer.

Thus, algebra does not consist of unconnected topics that can be filed in separate compartments, studied once, mastered, and then neglected. Algebra is like a big ball made of pieces of string that have been tied together. Many pieces touch directly, but the others are all an integral part of the ball, and all must be rolled along together if understanding is to be achieved.

A total assimilation of the fundamentals of algebra is the key that will unlock the doors of higher mathematics and the doors to chemistry, physics, engineering, and other mathematically based disciplines. In addition, it will also unlock the doors to the understanding of psychology, sociology, and other nonmathematical disciplines in which research depends heavily on mathematical statistics. Thus, we see that algebraic ability is necessary in almost any field of endeavor.

One must be able to apply algebraic fundamentals automatically if these fundamentals are to be useful. There is insufficient time to relearn basics every time a basic principle must be applied, and familiarity or a slight acquaintance with a basic principle does not suffice for its use. Testing has indicated that many students have only a tenuous grasp of basics at this point even though some topics have been practiced for almost a full year.

Thus, in this book we go back to the beginning—to signed numbers, and then quickly review all of the topics of Algebra 1 and practice these topics as we weave in more advanced concepts. We will treat these concepts as skills that can be learned through practice. The applicability of some of these skills, such as completing the square, deriving the quadratic formula, simplification of radicals, and complex numbers, might not be apparent at this time, but the benefits of having mastered these skills will become evident as your education continues. Problems have been selected in various skill areas and these problems will be practiced again and again in the problem sets. **It is wise to strive for speed and accuracy when working these review problems. If you feel that you have mastered a type of problem, don't skip it when it appears again. If you have really mastered the concept, the problem should not be troublesome; you should be able to do the problem quickly and accurately. If you have not mastered the concept, you need the practice that working the problem will provide. With the exception of the problems designated as optional problems, every problem in every problem set should be worked.** Don't skip problems. Master musicians practice fundamental musical skills every day. All experts practice fundamentals as often as possible. To attain and maintain proficiency in algebra, it is necessary to practice fundamental algebraic skills constantly as new concepts are being investigated. And, as in the last book, you are encouraged to be diligent and to work at developing defense mechanisms whose use will protect you against your seemingly uncanny ability to invent ways to make mistakes.

One last word. There is no requirement to like algebra. I am not especially fond of algebra—and I wrote the book—but I do love the ability to pass through doors that knowledge of algebra has unlocked for me. I did not know what was behind the doors when I began. Some things I found there were not appealing while others were fascinating. I enjoyed being an Air Force test pilot. A degree in engineering was a requirement to be admitted to test pilot school. My knowledge of mathematics enabled me to obtain this degree. At the time I began my study of mathematics, I had no idea that I would want to be a test pilot or would ever need to use mathematics in any way.

I encourage you to work hard because you will never know when the knowledge will be useful, and if you don't have the knowledge when the need arises, it will be too late. Don't make the mistake of deciding now that you will never need to know algebra—or chemistry—or physics—or history—or English—or a foreign language. You do not know now, for you cannot peer into the future with accuracy—you can only make an uneducated guess.

To the Teacher

One year of algebra does not give students the mastery of algebraic concepts that is required by the detailed developments of the advanced topics of algebra such as logarithms, the complex plane, probability, and conic sections. The introduction of these concepts before students are prepared has been one of the major causes for the mass exodus of students from advanced mathematics. These students are not prepared for nor do they have the maturity for a full appreciation of the proofs of Euclidian geometry. Certainly, they need to conquer and totally understand the geometric concepts of area, volume, and surface area, and concomitantly need to understand unit conversions within systems and from system to system. And most of all, they need more gentle practice in the fundamental topics of algebra—equations—graphs—basic techniques of solving word problems—percent—ratio—exponents. For too long, we have allowed these topics to frighten and intimidate rather than comfort and soothe. When the students have become proficient in these concepts, they will have acquired the basic skills that will permit survival in and enjoyment of advanced mathematics and science. Students avoid advanced science courses because they are not prepared algebraically. They avoid chemistry because they have difficulty with simple ratio and percent problems and scientific notation. This book will prepare the student totally for chemistry without neglecting the abstractions of algebra. There is no reason we cannot have both. This book is an attempt to pull the pendulum back to the center.

I believe that the use of this book will significantly increase enrollment in third-year mathematics and in chemistry because students will be prepared—and they will know that they are prepared. Further, since this book concentrates on fundamentals rather than on esoterica, I believe that its use will cause an immediate significant increase in scores on college entrance examinations.

This is a final course in beginning algebra for all students. The contents should not be modified so that it is a comprehensive course for the gifted and a less comprehensive course for the less gifted since the applications of the fundamentals are the same for both groups. Instead, the time allotted should be varied, with three or even four semesters being programmed for completion if necessary. It is vital that students

work all the problems in every problem set. In this book the concentration is on review rather than on the new topic. This shift in emphasis allows the learning process to be spread out rather than being concentrated. This spreading out has been shown to lead to greater comprehension and to improved retention by the student. Adding additional problems of the new kind will normally necessitate lessening the emphasis on the review topics and will result in improved short-term learning but will cause the long-term retention to be vitiated.

One of the major factors in the decline in mathematics test scores has been the so-called mastery learning in which proficiency in a skill, once attained and demonstrated, is allowed to fade away through lack of practice. Piano virtuosos practice the scales every day, even though "mastery" of this skill could have been proclaimed for most of them by the time they were six years old. Then, where should final concept or unit tests be given? The answer is never. **Every test should be a final examination and should test every concept that has been presented previously.** After students have learned a topic through long-time practice, they should not be denied the opportunity to demonstrate what they have accomplished. They have worked hard and learned, and they have the reward coming.

The abstractions of algebra are applicable to all our thoughts—to the happy and to the frivolous as well as to the mundane and lifeless. Algebraic applications need not be restricted to computing the price of avocados at the grocery store, borrowing money from several banks at varying rates of interest, and computing a diameter in the machine shop. Attempts to restrict word problems to these and similar applications have restricted the dimensions of algebraic thinking and have made algebra books unnecessarily dry and unstimulating. The historical references and unusual words in the word problems in this book were not included in an attempt to teach history or vocabulary but to demonstrate that the study of algebra can be just as interesting and entertaining as the study of history or English or any other subject.

Acknowledgments

None of my books would have been possible without the able assistance of my typist, Detia Roe, who is a graduate physicist. I thank her again. I thank Frank Wang for his help in proofreading, checking accuracy, and for his other contributions. Lastly, I thank my mentor, publisher Bob Worth, for his assistance in getting my book published.

                                                                John Saxon

Norman, Oklahoma
April 1983

Contents

Additional Topics

Appendix

Basic

110 Lessons

Course

LESSON 1 *Absolute value • Properties and Definitions*

1.A
absolute value

A number is an idea. A numerical expression (often called a numeral) is a single symbol or a collection of symbols that designates a particular number. We say that the number designated is the **value** of the expression. All of the following numerical expressions designate the number positive three, and we say that each of these expressions has a value of positive three.

$$ 3 \qquad \frac{7+8}{5} \qquad 2+1 \qquad \frac{12}{4} \qquad \frac{75}{25} \qquad \frac{16}{2}-5 $$

We have agreed that a positive number can be designated by a numeral preceded by a plus sign or by a numeral without a sign. Thus we can designate positive three by writing either

$$ +3 \qquad \text{or} \qquad 3 $$

The number zero is neither positive nor negative and can be designated with the single symbol

$$ 0 $$

Every other real number is either positive or negative and can be thought of as having two qualities or parts. One of the parts is designated by the plus or minus sign, and the other part is designated by the numerical part of the numeral. The two numerals

$$ +3 \qquad \text{and} \qquad -3 $$

designate a positive number and a negative number. The signs of the numerals are different, but the numerical part of each is

$$ 3 $$

We say that this part of the numeral designates the **absolute value** of the number. It is difficult to find a definition of absolute value that is acceptable to everyone. Many people object to saying that the absolute value is the same thing as the "bigness" of a number because "bigness" might be confused with the concept of "greater than" that is used to order numbers. Some explain absolute value by saying that all nonzero real numbers can be paired, each with its opposite, and that the absolute value of either is the positive member of the pair. Thus

$$ +3 \qquad \text{and} \qquad -3 $$

are a pair of opposites, and both have an absolute value of 3. Other people prefer to define the absolute value of a number as the number that describes the distance of the graph of the number from the origin. If we use this definition, we see that the graphs of $+3$ and -3 are both 3 units from the origin, and thus both numbers have an absolute

value of 3,

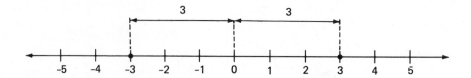

Many feel that words should not be used to define absolute value because absolute value can be defined exactly by using only symbols and using two vertical lines to indicate absolute value. This definition is in three parts. Unfortunately, the third part is confusing to some.

(a) If $x > 0$ $|x| = x$

(b) If $x = 0$ $|x| = x$

(c) If $x < 0$ $|x| = -x$

Part (c) does not say that the absolute value of x is a negative number. It says that if x is a negative number (all numbers less than zero are negative), the absolute value of x is the opposite of x. Since -15 is a negative number, its absolute value is its opposite, which is $+15$.

$$|-15| = -(-15) = 15$$

In the same way, if we designate the absolute value of an algebraic expression such as

$$|x + 2|$$

and x has a value such that $x + 2$ is a negative number, then the absolute value of the expression would be the negative of the expression.

If $x + 2 < 0$,

$$|x + 2| = -(x + 2)$$

To demonstrate, we will give x a value of -5 and then we will have

$$|-5 + 2| = |-3| = -(-3) = +3$$

No matter how we think of absolute value, we must remember that the absolute value of zero is zero and that the absolute value of every other real number is a positive number.

$$|-5| = 5 \qquad |5| = 5 \qquad |-2.5| = 2.5$$

In this book, we will sometimes use the word number when the word numeral would be more accurate. We do this because overemphasizing the distinction between the two words can be counterproductive.

1.B
properties and definitions

Understanding algebra is easier if we make an effort to remember the difference between properties and definitions. A **property** describes the way something is. We can't change properties. We are stuck with properties because they are what they are. For instance,

$$3 + 2 = 5 \qquad \text{and} \qquad 2 + 3 = 5$$

The order of addition of two real numbers does not change the answer. We can understand this property better if we use dots rather than numerals.

Here we have represented the number 5 with five dots. Now, on the left below we separate the dots to show what we mean by 3 + 2, and on the right we show 2 + 3. The answer is 5 in both cases because there are a total of 5 dots regardless of the way in which they are arranged. We call this property the commutative property of real numbers in addition.

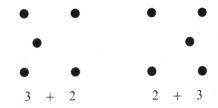

$$3 \quad + \quad 2 \qquad\qquad 2 \quad + \quad 3$$

Definitions are different because they are things that we have agreed on. For instance,

$$3^2$$

means

$$3 \text{ times } 3$$

It didn't have to mean that. We could have used 3_2 to mean

$$3 \text{ times } 3$$

but we didn't. We note that the order of operations is also a definition. When we write

$$3 + 4 \cdot 5$$

we could mean to multiply first or to add first. Since we cannot have two different answers to the same problem, it is necessary to agree on the meaning of the notation. We have agreed to do multiplication before algebraic addition, and so this expression represents the number $+23$.

Also, when we wish to write the negative of *three squared*, we write

$$-3^2$$

and when we wish to indicate that the quantity -3 is to be squared, we write

$$(-3)^2$$

These are definitions of what we mean when we write

$$-3^2 \quad \text{ and } \quad (-3)^2$$

and there is nothing to understand. We have defined these notations to have the meanings shown.

The first problem set consists of review problems in simplifying expressions containing signed numbers. When these expressions are simplified, one should try to remember which steps are because of properties and which steps are because of definitions.

**problem
set 1**

Simplify:

1. $-3 - 2^2 - 2^3 - 2^4 - 2$ 2. $-3 - (-2)^3 - 3^2 - (-2)$

3. $-|-2| - |-4 - 2| + |8|$ 4. $-6 - |-2 - 3| - (-2)^3 - 2$

5. $-[-2(-3 - 2) - (-2 - 3)]$

6. $-2[-2 - 3(-2 - 2)][-2(-4) - 3]$ 7. $-2[-3 - (-4) - 2^2][-|2|]$

8. $-2^2 - 2^3[-2 + 3(-2)] - |-2^3|$ 9. $3(-2 + 5) - 2^2(2 - 3) - |-2|$

10. $-3 - 2^3 - 4^2 - |-2 - 3(2)|$ 11. $-|-3(2) - 3| - 2^2$

12. $-3 - (-2 + 7)(-2 + 4) - (-3 + 2) - 3^2$

13. $-\{-2[(-3 + 7) - (-2)][-3(-2 + 1)]\}$

14. $3^2 - 3^3 + 3^4 - (-3)^3 - 3$ 15. $-|-2 - 3| - (-2)^3 - 2$

16. $-2[(-3 + 1) - (-2 - 2)(-1 + 3)]$

17. $-(-4)^2 - 4|-2| - 2^3 + |-11 - 4|$

18. $6 - \{[3^2 - 8 + (-2)][-(4 - 6)(-3)^2 + 2]2\}$

19. $-[-(-2)] - |-4 - 3|2^2 - 4$ 20. $-8 - 3^2 - (-2)^2 - 3(-2) + 2$

21. $[-|-3|][(2 - 7)(-3 - 2) + (-2)^2]$

22. $\dfrac{-|-4| - (-3) + 7 - 6(4 - |7 - 11|)}{7 - |(3)(-2)|}$

23. $-(-3 - 2)(-7 - |-3 - 2|) - (-3)^2$ 24. $-\{-[-5(-3 + 2)7]\}$

25. $[-3][|-2 - 7 - 2| - (-3)^2 - (-2)]$ 26. $\dfrac{-4 - (-3) + 7 - 6(2)}{7 - (3)(-2)}$

27. $3 - 5 - 2^2 - 4^2(-1)(-3 - |-2 - 5| - 3)$

28. $-8 + (-3)(-2)^2 + (-7) - 2(-4 - 2)$

29. $6(-3)[-(5 - 4)(6 - 2)3]$

30. $3 - (-3) - 7(4 - 2) + 7(-2)(-4)^3$

LESSON 2   *Negative exponents • Product and power theorems for exponents*

2.A
**negative
exponents**

Negative exponents cannot be "understood" because they are the result of a definition and thus there is nothing to understand. We define 2 to the third power as follows:

$$2^3 = 2 \cdot 2 \cdot 2$$

We have agreed that 2^3 means 2 times 2 times 2. In a similar fashion, we define 2 to the negative third power to mean one over 2 to the third power

$$2^{-3} = \frac{1}{2^3}$$

Thus, we have two ways to write the same thing. We give the formal definition of negative exponents as follows:

DEFINITION OF x^{-n}

If n is any real number and x is any real number that is not zero,

$$x^{-n} = \frac{1}{x^n}$$

This definition tells us that when we write an exponential in reciprocal form, the sign of the exponent must be changed. If the exponent is negative, it is positive in reciprocal form; and if it is positive, it is negative in reciprocal form. In the definition we say that x cannot be zero because division by zero is not permissible, and we must guard against indicating that it is permissible.

example 2.A.1   Simplify: (a) $\dfrac{1}{3^{-2}}$ (b) 3^{-3} (c) $(-3)^{-3}$

solution   (a) $\dfrac{1}{3^{-2}} = 3^2 = \mathbf{9}$ (b) $3^{-3} = \dfrac{1}{3^3} = \dfrac{\mathbf{1}}{\mathbf{27}}$ (c) $(-3)^{-3} = \dfrac{1}{(-3)^3} = -\dfrac{\mathbf{1}}{\mathbf{27}}$

2.B
product theorem for exponents

Remember that x^2 means x times x

$$x^2 = x \cdot x$$

and x^3 means x times x times x

$$x^3 = x \cdot x \cdot x$$

Using these definitions, we can find the answer to x^2 times x^3 as follows:

$$x^2 \cdot x^3 \qquad \text{means} \qquad x \cdot x \quad \text{times} \quad x \cdot x \cdot x \quad \text{which equals } x^5$$

This demonstrates the product theorem for exponents, which we state formally in the following box.

PRODUCT THEOREM FOR EXPONENTS

If m and n and x are real numbers and $x \neq 0$,

$$x^m \cdot x^n = x^{m+n}$$

This theorem holds for all real number exponents.

example 2.B.1   Simplify: $x^2 y x^{-5} y^{-4} x^5 x^0$

solution   We simplify by adding the exponents of like bases and get

$$x^2 y^{-3}$$

example 2.B.2   Simplify: $\dfrac{y y^{-3} x^4 y^5 x^{-10}}{y^{-6} x^{-3} y^{10} x^2}$

solution   First we simplify above and below. Then we decide to write the answer with all exponents in the numerator.

$$\frac{y^3 x^{-6}}{y^4 x^{-1}} = y^{-1} x^{-5}$$

2.C
power theorem for exponents

We can use the product theorem to expand $(x^2)^3$ as

$$(x^2)^3 = x^2 \cdot x^2 \cdot x^2 = x^6$$

This procedure generalizes to the power theorem for exponents.

POWER THEOREM FOR EXPONENTS

If m and n and x are real numbers,

$$(x^m)^n = x^{mn}$$

This theorem can be extended to any number of exponential factors.

EXTENSION OF THE POWER THEOREM

If the variables are real numbers,

$$(x^m y^a z^b k^c \cdots)^n = x^{mn} y^{an} z^{bn} k^{cn} \cdots$$

---

example 2.C.1   Simplify: $\dfrac{x(x^{-3})^2 y(xy^{-2})^{-3}}{(y^2)^3 y^{-3}(x^2)^3}$

solution   First we will use the power theorem in both the numerator and the denominator and get

$$\frac{xx^{-6} yx^{-3} y^6}{y^6 y^{-3} x^6}$$

Now we simplify both the numerator and the denominator, and as the last step, we decide to write all exponentials with positive exponents.

$$\frac{x^{-8} y^7}{y^3 x^6} = \frac{y^4}{x^{14}}$$

problem set 2

Simplify. Write answers with all exponentials in the numerator.

1. $\dfrac{(x^0)^2 xy}{(xy^{-2})^2}$

2. $\dfrac{(x^2 y)^0 xy}{x^2 (y^{-2})^3}$

3. $\dfrac{xx^2 (x^0 y^{-1})^2}{x^2 x^{-5}(y^2)^5}$

4. $\dfrac{m^2 p^0 (m^{-2} p)^2}{m^{-2} p^{-1}(m^{-3} p^2)^3}$

5. $\dfrac{k^2 a^0 a(k^2 a)^{-2}}{(k^{-3})^0 a^0 ka^{-4}}$

6. $\dfrac{(a^2 b^0)^2 ab^{-2}}{a^2 b^{-2}(ab^{-3})^2}$

Simplify. Write answers with positive exponents.

7. $\dfrac{(xm^{-1})^{-3} x^2 m^2}{(x^0 y^2)^{-2} xy}$

8. $\dfrac{(a^2 b^{-3})^2 (b^2)^0}{(a^{-2} b)^{-3} b^2}$

9. $\dfrac{(c^2 d)^{-3} c^{-5}}{(c^2 d^0)^{-2} d^3}$

10. $\dfrac{(m^2 n^{-5})^{-2} m(n^0)^2}{(m^2 n^{-2})^{-3} m^2}$

11. $\dfrac{(x^{-2} y^5)^3 (x^2)^0 y}{xy^{-3} x^{-2}}$

12. $\dfrac{(b^2 c^{-2})^{-3} c^{-3}}{(b^2 c^0 b^{-2})^4}$

Simplify. Write answers with negative exponents.

13. $\dfrac{(abc)^{-3}c^2b}{a^{-4}bc^2a}$ **14.** $\dfrac{kL^2k^{-2}}{(k^0L)^2L^{-3}k}$ **15.** $\dfrac{(p^2rr^{-4})^{-2}r^0}{r^2pp^{-3}}$

16. $\dfrac{s^2ym^{-3}}{(s^0t^2)^{-3}m^{-3}st}$ **17.** $\dfrac{(x^{-3}yz^{-3})^2xy^0}{(xy^0z^{-2})^{-3}xy}$ **18.** $\dfrac{x^{-3}y^2xy^4}{(x^{-2}y)^3y^{-3}x}$

Simplify:

19. -3^{-2} **20.** $-(-2)^3$

21. $\dfrac{1}{-2^{-3}}$ **22.** $-3^2 - [-2^0 - (3 - 2) - 2]$

23. $-2\{[-3 - 2(-2)][-2 - 3(-2)]\}$ **24.** $-2[(-3 + 7 - 2) - 4(-6)]$

25. $2\{-3^0[(-5 - 2)(-3) - 2]\}$ **26.** $-3[6 - 2^0(2 - 6)2 - 2^2]$

27. $-3[2^2 - 3(-2) - 2^2 - 3]$ **28.** $-3[4^0 - 7(2 - 3) - 2^2]$

29. $-|-2 - 3| - (-5) - 3^3$ **30.** $-|-3^2 - 2| - 2^0 - (-3)$

LESSON 3   *Evaluation of expressions • Adding like terms*

3.A
evaluation of expressions

In the first lesson, we noted that a numerical expression is a meaningful arrangement of numerals and symbols that designate operations. Thus, each of the following can be called a numerical expression.

$$4 \qquad 2 + 2 \qquad \frac{7 + 3 + 14}{6} \qquad \sqrt{16} \qquad 48 \div 12$$

Every numerical expression represents a single number. We say that this number is the **value** of the expression. The value of each of the above expressions is 4, for each one is a different way to designate the number 4.

     The value of the expression

$$x + 4$$

depends on the number we use as a replacement for x. If we replace x with -32, the expression will have a value of -28.

$$(-32) + 4 = -28$$

When we replace the variables in an expression with selected numbers and simplify, we say that we have **evaluated** the expression.

example 3.A.1   Evaluate $x^2y - y$ if $x = -2$ and $y = -4$.

solution   We replace x with -2 and y with -4.

$$(-2)^2(-4) - (-4) = (+4)(-4) + 4 = -16 + 4 = -12$$

example 3.A.2   Evaluate $a(-b - a) - ab$ if $a = -2$ and $b = 4$.

solution   We replace a with -2 and b with 4.

$$-2(-4 + 2) - (-2)(4) = -2(-2) - (-8) = 4 + 8 = 12$$

3.B
adding like terms

Like terms are terms whose literal components represent the same number regardless of the numbers used to replace the variables. Thus,

$$3xyz \quad \text{and} \quad -2zyx$$

are like terms because xyz and zyx have the same value regardless of the replacement values of the variables. We demonstrate this by replacing x with 2, y with 3, and z with 4.

xyz	zyx
(2)(3)(4)	(4)(3)(2)
(6)(4)	(12)(2)
24	24

We add like terms by adding the coefficients of the terms as we show in the following examples.

example 3.B.1 Simplify by adding like terms: $3xy - 2x + 4 - 6yx + 3x$.

solution We add like terms and get

$$-3yx + x + 4$$

example 3.B.2 Simplify by adding like terms:

$$\frac{3a^{-2}b}{c} - \frac{4b}{a^2c} + 7a^{-2}c^{-1}b - \frac{4ba^2}{c}$$

solution If we write the terms in the same form, we can see which terms are like terms. This time we choose to write the terms with all exponents positive, and we get

$$\frac{3b}{a^2c} - \frac{4b}{a^2c} + \frac{7b}{a^2c} - \frac{4ba^2}{c}$$

We see that the first three terms are like terms and can be added by adding the numerical coefficients. We do this and get

$$\frac{6b}{a^2c} - \frac{4ba^2}{c}$$

problem set 3

Evaluate:

*1. $x^2y - y$ \quad if $x = -2$ and $y = -4$

*2. $a(-b - a) - ab$ \quad if $a = -2$ and $b = 4$

3. $x - |x|y^2 - xy$ \quad if $x = -2$ and $y = -3$

4. $(a - b) - a(-b)$ \quad if $a = -5$ and $b = 3$

5. $-a(a - ax)(x - a)$ \quad if $a = -2$ and $x = 4$

6. $a^2 - y^3(a - y^2)y$ \quad if $a = -2$ and $y = -3$

7. $-p^2 - p(a - p^2)$ \quad if $a = 4$ and $p = -3$

8. $a^2 - y^3(a - y^2)y^2$ \quad if $a = -2$ and $y = 3$

9. $a^2 - a(x - ax)$ \quad if $a = 1$ and $x = 2$

Simplify by adding like terms:

*10. $\dfrac{3a^{-2}b}{c} - \dfrac{4b}{a^2c} + 7a^{-2}c^{-1}b - \dfrac{4ba^2}{c}$

11. $\dfrac{p^2x^4}{m^5} - \dfrac{2p^4m^5}{x^{-4}} - \dfrac{3p^2x^2}{x^{-2}m^5} + \dfrac{7p^2m^{-5}m^{10}x^4}{p^{-2}}$

12. $-\dfrac{m^4x^5}{k^5} + \dfrac{2m^2x^5}{k^5m^{-2}} - \dfrac{3m^3x^2k}{m^{-4}x^3k^4}$

13. $-2x^5y^4 + \dfrac{3xy^3}{x^{-4}y^{-1}} + \dfrac{4x^3y^2}{x^2y}$

14. $2xy^2m - \dfrac{3x^2y^2m^4}{m^3x} + \dfrac{2x^2ym^3}{mx^2}$

Simplify:

15. $\dfrac{(x^0y^2)^{-3}y^{-2}p^0}{(x^2)^{-4}(y^2)^0(p^3)^{-2}}$

16. $\dfrac{(mxy^2p)^2p^{-2}x^2}{(p^0xmy^3)^{-2}xp^{-2}}$

17. $\dfrac{xx^2(x^{-2})^{-2}(mpx^2)^{-4}}{(x^0)^2(x^2)^0(x^2mp^{-2})^3}$

18. $\dfrac{m^2(x^0y^{-2})^2p^4}{p^{-2}(x^0)^{-5}m^5}$

19. $\dfrac{p^2x^{-4}k^5(p^2k)^{-2}}{(p^2x^{-3})^{-2}}$

20. $\dfrac{x^2xx^0(x^{-2})^2}{xx^3x^{-14}(x^{-2})^{-3}}$

21. $\dfrac{(x^2y^{-2}p^0)^{-3}p^2}{x^2(x^{-4})^0(p^{-2}y^5)^{-2}}$

22. $\dfrac{(x^{-2}y^2)^5(x^0y^{-2})^{-4}}{(x^{-4}yy^2)^2(p^{-4})^0}$

Simplify:

23. $-3^{-2} + \dfrac{1}{2^{-3}}$

24. $-(-2)^3 - \dfrac{1}{(-3)^{-2}}$

25. $-3[-2 - 2 - (-3)][-2 - 3]$

26. $-2(-3 + 7^0) - |(-2 - 3)|$

27. $|-2| - 3^2 - (-3)^3 - 2$

28. $-4(-2 + 7) - (-4 - 6^0)$

29. $-2\{[(-3 - 2)(-2)][-2 - 3]\}$

30. $-[(-3)(-2) - (-3)(-2 + 4)]$

LESSON 4   *Distributive property • Solution of equations • Change sides—change signs*

4.A
distributive property

The distributive property is a property of real numbers that permits two approaches to the simplification of expressions such as

$$3(4 + 5)$$

We find that we can get the same answer with either approach.

3(4 + 5)	3(4 + 5)
3(9)	$3 \cdot 4 + 3 \cdot 5$
27	12 + 15

On the left we added 4 and 5 to get 9 and then multiplied by 3. On the right we multiplied first and then we added. We got the same answer each time. We note that on the right we *distributed* the multiplication over the addition.

example 4.A.1   Expand: $\dfrac{4a^2}{b}\left(\dfrac{b^{-2}}{a^4}-\dfrac{3ba}{a^2}\right)$

solution   **Since letters stand for unspecified numbers, all the rules for numbers also hold for letters.** Two multiplications are indicated. We do them both and then simplify.

$$\frac{4a^2b^{-2}}{ba^4}-\frac{12a^2ba}{ba^2}=\frac{4}{b^3a^2}-12a$$

example 4.A.2   Expand: $\dfrac{a^{-3}b^0}{c}\left(\dfrac{a^2bc}{c^2}-\dfrac{3a^{-2}}{b^{-2}}\right)$

solution   The distributive property permits two multiplications. Then we simplify.

$$\frac{a^{-3}a^2bc}{c^3}-\frac{3a^{-3}a^{-2}}{cb^{-2}}=\frac{a^{-1}b}{c^2}-\frac{3a^{-5}}{cb^{-2}}$$

4.B
solution of equations

We remember that the two rules for solving equations are the addition rule and the multiplication/division rule. These rules are extensions of the additive property of equality and the multiplicative property of equality, and these rules apply to both true equations and false equations.

   (a) $4+3=7$ (true)   (b) $4+3=5$ (false)

We can add the same number to both sides of an equation without changing the truth or falsity of the equation. We will demonstrate this by adding -5 to both sides of equations (a) and (b).

(a)			(b)		
	$4+3=7$	true		$4+3=5$	false
	$4+3-5=7-5$	added -5		$4+3-5=5-5$	added -5
	$2=2$	still true		$2=0$	still false

We can use the additive property of equality to prove that the same number can be added to both sides of an equation without changing the solution set of the equation. To demonstrate, we will use the equation

$$x+4=6$$

The number 2 is the solution to this equation. If we add -5 to both sides of the equation, we get

$$x+4-5=6-5$$

or $$x-1=1$$

We did not change the solution by adding -5 to both sides, as 2 is also the solution to the new equation. **Equations that have the same solution sets are called *equivalent equations.*** Thus, the new equation and the original equation are equivalent equations.

A similar explanation could be used for the multiplication/division rule. We will forego this explanation and state the two rules as follows.

ADDITION RULE FOR EQUATIONS

The same quantity can be added to both sides of an equation without changing the solution set of the equation.

MULTIPLICATION/DIVISION RULE FOR EQUATIONS

Every term on both sides of an equation can be multiplied/divided by the same nonzero quantity without changing the solution set of the equation.

We remember that we always use the addition rule before we use the multiplication/division rule. This is because the solution of an equation undoes a normal order of operations problem. To demonstrate, we will begin with 4, then multiply by 3, and then add -2 to get 10.

$$3(4) - 2 = 10$$

Now, to undo what we have done and get back to 4, we must undo the addition of -2 first and then undo the multiplication. To demonstrate this procedure, we replace 4 with x and get the equation

$$3x - 2 = 10$$

Now we solve to find that x equals 4.

$$
\begin{array}{ll}
3x - 2 = 10 & \text{replace 4 with } x \\
\underline{+ 2 \quad +2} & \text{add } +2 \text{ to both sides} \\
3x \quad\;\; = 12 & \\
\dfrac{3x}{3} = \dfrac{12}{3} & \text{divided by 3} \\
x = 4 &
\end{array}
$$

We remember that the five steps for solving equations whose variable is x are:

1. Eliminate parentheses.
2. Add like terms on both sides.
3. Eliminate x on one side or the other.
4. Eliminate the constant term on the x side.
5. Eliminate the coefficient of x.

We will use these steps to solve the equations in the next two examples.

example 4.B.1   Solve: $12 - (2x + 5) = -2 + (x - 3)$

solution   As the first step, we eliminate the parentheses, remembering that if the parentheses are preceded by a minus sign, we must change all signs therein.

$$12 - 2x - 5 = -2 + x - 3$$

Now we simplify on both sides of the equation

$$7 - 2x = x - 5$$

Next we eliminate the x term on the left side by adding $+2x$ to both sides.

$$
\begin{array}{rl}
7 - 2x = & x - 5 \\
+2x & +2x \\
\hline
7 \quad = & 3x - 5
\end{array}
$$

Now we eliminate the -5 on the right by adding $+5$ to both sides.

$$
\begin{array}{rl}
7 = & 3x - 5 \\
+5 & +5 \\
\hline
12 = & 3x
\end{array}
$$

Then we complete the solution by dividing both sides by 3.

$$\mathbf{4 = x} \qquad \text{divided by 3}$$

The same procedure is used when the numbers in the equation are fractions or mixed numbers.

example 4.B.2 Solve: $3\left(\dfrac{5}{6} - \dfrac{5}{3}x\right) = -\left(-\dfrac{1}{2} + x\right)$

solution As the first step, we eliminate the parentheses. Then we solve.

$$
\frac{5}{2} - 5x = \frac{1}{2} - x
$$

$$
\begin{array}{ll}
\underline{\quad + 5x \qquad\qquad + 5x \quad} & \text{add } 5x \text{ to both sides} \\[4pt]
\dfrac{5}{2} \quad = \quad \dfrac{1}{2} + 4x & \\[6pt]
\underline{-\dfrac{1}{2} \qquad\qquad -\dfrac{1}{2} \quad} & \text{add } -\dfrac{1}{2} \text{ to both sides} \\[6pt]
2 = 4x & \\[6pt]
\dfrac{1}{2} = x & \text{divided both sides by 4}
\end{array}
$$

4.C
**change sides—
change signs**

It is important to understand why we do things in algebra, but it is also important not to let the emphasis on understanding interfere with our ability to do. The use of the addition rule for equations is a case in point. We can use this rule to eliminate a term from one side of an equation by adding the opposite of the term to both sides of the equation. For example, if we wish to solve the equation

$$y + 2x = 4$$

for y, we add $-2x$ to both sides of the equation.

$$
\begin{array}{ll}
y + 2x = 4 & \text{equation} \\
\underline{\quad - 2x \qquad - 2x \quad} & \text{add } -2x \text{ to both sides} \\
y \quad = 4 - 2x &
\end{array}
$$

We were able to eliminate the $2x$ term from the left side of the equation, but when we did, the same term appeared on the right side of the equation with its sign changed. **This happens every time we use the addition rule. The term will disappear on one side of the equation and will appear on the other side with its sign changed.** Many people use this thought process. Rather than mentally adding the same quantity to both sides, they simply pick up a term, carry it across the equals sign, and change the sign of the term. This leads to the adage

Change sides—change signs

Authors of algebra books published in the late 1800's called this process **transposition.** If we use transposition to solve the last equation for y, we transpose the $+2x$ to the right side where it becomes $-2x$.

$$y + 2x = 4 \longrightarrow y = 4 - 2x$$

example 4.C.1 Use the rule change sides—change signs to solve for x: $x - 2 = 7$.

solution We move the -2 from the left side to the right side and change its sign.

$$x - 2 = 7 \longrightarrow x = 7 + 2 \longrightarrow \mathbf{x = 9}$$

example 4.C.2 $p - 3x + 4 = 7y$: solve for p.

solution We move $-3x + 4$ from the left side to the right side, where it becomes $+3x - 4$.

$$p - 3x + 4 = 7y \longrightarrow \mathbf{p = 7y + 3x - 4}$$

example 4.C.3 $3y - 2x + 5 = 0$: solve for y.

solution We move $-2x + 5$ to the right side and change both signs. Then we complete the solution by dividing by 3.

$$3y - 2x + 5 = 0 \qquad \text{equation}$$
$$3y = 2x - 5 \qquad \text{changed sides and changed signs}$$
$$y = \frac{2}{3}x - \frac{5}{3} \qquad \text{divided}$$

problem set 4 Solve:

*1. $12 - (2x + 5) = -2 + (x - 3)$

2. $15(4 - 5b) = 16(4 - 6b) + 10$

*3. $3\left(\frac{5}{6} - \frac{5}{3}x\right) = -\left(-\frac{1}{2} + x\right)$

4. $3\frac{1}{3}x - \frac{5}{6} = -\frac{2}{3}$

5. $3(-2x - 3) - 2^2 = -(-3x - 5) - 2$

6. $-2(2x - 3) - 2^3 - 3 = -x - (-4)$

7. $4\frac{1}{3}x - \frac{1}{2} = 3\frac{2}{5}$

8. $-\frac{3}{5}x + \frac{2}{7} = 4\frac{3}{8}$

Expand:

*9. $\dfrac{4a^2}{b}\left(\dfrac{b^{-2}}{a^4} - \dfrac{3ba}{a^2}\right)$

*10. $\dfrac{a^{-3}b^0}{c}\left(\dfrac{a^2bc}{c^2} - \dfrac{3a^{-2}}{b^{-2}}\right)$

11. $\dfrac{xy^2}{x^0x^{-3}}\left(\dfrac{xy^{-2}}{x(y^2)^0} - \dfrac{3y^{-2}}{x^4}\right)$

12. $\dfrac{ay^{-4}}{p}\left(\dfrac{p^{-2}}{ay^2} - \dfrac{3a^{-1}y}{p^{-2}}\right)$

Simplify. Write all exponentials in the numerator.

13. $\dfrac{x^0x^{-2}(y^2)^{-3}y}{y^2(yx^{-2})y}$

14. $\dfrac{(3x^2)^{-2}y^0y^5}{(9y)^{-2}yy^2x^{-3}}$

15. $\dfrac{(2yx^{-2})^{-2}yx^2}{(x^2)^0y^{-3}x^2}$

16. $\dfrac{2(x^{-2})^{-2}yx^2y^{-3}}{x^0xx^2x^{-5}(x^2)^3}$ **17.** $\dfrac{(x^2y2x)^{-2}y}{(x^{-4})^0xxy^2}$ **18.** $\dfrac{3x^2xy^2x^{-4}}{(x^2y)^{-2}(-2)^{-2}}$

Simplify by adding like terms:

19. $\dfrac{2x^2xyx}{x^2y^{-1}} - \dfrac{3x^2y^4}{yy} + \dfrac{7xx^{-3}y^{-2}}{x^{-4}y^{-4}}$ **20.** $\dfrac{3x}{y} - 7x^2x^{-1}y^{-1} + 2y^2y^{-1}x^{-1}$

21. $\dfrac{2ay^2}{x} + \dfrac{5a^2x^{-1}}{ay^{-2}} + \dfrac{2xy^2}{ay}$

Evaluate:

22. $a^2(a - ab)$ if $a = -2$ and $b = 3$

23. $x^0yx(xy - x^2)$ if $x = -3, y = -1$

24. $a^{-2}b(a - b)(b - a)$ if $a = -1, b = -2$

25. $ab(a^2 - b)a - b$ if $a = -3, b = -1$

Simplify:

26. $-3^0 - 2^0 - 2^0(-2 - 3^2) - (-2 + 7) - |-2 - 3|$

27. $-2\{2[(-3 - 2^2) - (-3 + 7)] - 2\}$

28. $-3^0 - (-2 - 3 - 2^0)(-3) - (-2 - 4) + (-6)$

29. $-2^{-2}(-16)$

30. $-(-2^{-3}) - \dfrac{1}{(-2)^{-2}}$

LESSON 5   *Word problems • Fractional parts of a number*

5.A
word problems

Word problems that contain one statement of equality can usually be solved by writing one equation and using one unknown (variable). Word problems that contain two statements of equality can usually be solved by writing two equations and using two unknowns. Three statements of equality require three equations and three unknowns, etc. **In general, to obtain a unique solution, the number of equations must equal or exceed the number of unknowns. If the number of equations exceeds the number of unknowns, at least one of the equations is redundant.**

We will begin with problems that can be solved by writing one equation in one unknown. **If the equation tells us how much two quantities differ, then one of the quantities must be increased or decreased as required so that a statement of equality can be written.**

example 5.A.1   Twice a number is decreased by 7, and this sum is multiplied by 3. The result is 9 less than 10 times the number. What is the number?

solution   **In this kind of problem, we can prevent the most common mistake if we begin by writing an equation that we know is untrue.**

$$(2N - 7)3 = 10N \qquad \text{untrue}$$

We know that the left side is 9 less than the right side. We can make the sides equal by adding 9 to the left side or by adding -9 to the right side. We choose the second option and get

$$(2N - 7)3 = 10N - 9 \qquad \text{added } -9 \text{ to the right side}$$

Now we solve to find that the number is -3.

$$6N - 21 = 10N - 9 \qquad \text{multiplied}$$
$$-12 = 4N \qquad \text{added } -6N + 9 \text{ to both sides}$$
$$\mathbf{-3 = N} \qquad \text{divided by 4}$$

example 5.A.2   The number of ducks on the pond was doubled when the new flock landed. Next 7 more ducks came. The resulting number of ducks was 13 less than 3 times the original number. How many ducks were there to begin with?

solution   Again we begin with an equation that is untrue.

$$2N_D + 7 = 3N_D \qquad \text{untrue}$$

We can make this a true equation by adding $+13$ to the left side or by adding -13 to the right side. We decide to add $+13$ to the left side. Then we solve

$$2N_D + 7 + 13 = 3N_D \qquad \text{added 13 to the left side}$$
$$2N_D + 20 = 3N_D \qquad \text{simplified}$$
$$\mathbf{20 = N_D} \qquad \text{added } -2N_D \text{ to both sides}$$

example 5.A.3   The sum of -7 and 6 times a number is multiplied by 5. The result is 332 less than 3 times the number. What is the number?

solution   Again we begin by writing an equation that is untrue.

$$(6N - 7)5 = 3N \qquad \text{untrue}$$

The left side is 332 less than the right side, so we add 332 to the left side to make the sides equal.

$$(6N - 7)5 + 332 = 3N$$

Now we solve to find that N equals -11.

$$30N - 35 + 332 = 3N \qquad \text{multiplied}$$
$$30N + 297 = 3N \qquad \text{added}$$
$$27N = -297 \qquad \text{rearranged}$$
$$\mathbf{N = -11} \qquad \text{divided}$$

5.B
fractional parts of a number

When we multiply a number by a fraction, we say that we have taken a fractional part of the number. For instance, if we multiply $\frac{3}{8}$ by 40, we get 15.

$$\frac{3}{8} \times 40 = 15$$

We say this with words by saying that three-eighths of 40 is 15. We see that 40 associates with the word **of** and 15 associates with the word **is**. Thus, the general form of the equation is

$$(F) \times (\text{of}) = (\text{is})$$

example 5.B.1   One-fifth of the clowns had red noses. If 30 clowns had red noses, how many clowns were there in all?

solution   We can write the statement of the problem as

$$\frac{1}{5} \text{ of the clowns is } 30$$

Now we write the equation

$$(F) \times (\text{of}) = (\text{is})$$

and replace F with $\frac{1}{5}$, *of* with C, and *is* with 30. Then we solve.

$$\frac{1}{5} \cdot C = 30 \qquad \text{equation}$$

$$\frac{5}{1} \cdot \frac{1}{5} \cdot C = \frac{5}{1} \cdot 30 \qquad \text{multiplied both sides by } \frac{5}{1}$$

$$C = 150 \qquad \text{solution}$$

example 5.B.2   Seven-eighths of the Tartar horde rode horses. If 140,000 were in the horde, how many did not ride horses?

solution   If seven-eighths rode, then one-eighth did not ride.

$$(F) \times (\text{of}) = (\text{is})$$

and we have

$$\frac{1}{8} \times (140{,}000) = NR$$

$$17{,}500 = NR$$

problem set 5

*1.   Twice a number is decreased by 7, and this sum is multiplied by 3. The result is 9 less than 10 times the number. What is the number?

*2.   The number of ducks on the pond was doubled when the new flock landed. Next 7 more ducks came. The resulting number of ducks was 13 less than 3 times the original number. How many ducks were there to begin with?

*3.   The sum of -7 and 6 times a number is multiplied by 5. The result is 332 less than 3 times the number. What is the number?

*4.   One-fifth of the clowns had red noses. If 30 clowns had red noses, how many clowns were there in all?

*5.   Seven-eighths of the Tartar horde rode horses. If 140,000 were in the horde, how many did not ride horses?

Solve:

6. $-3x^0(2x - 3) - (-2^0) - 2 = 5(x - 3^0)2$

7. $-2^2(-2 - x) - x^0(3 - 2) = -2(x + 3)$

8. $3\frac{1}{2}x + 2\frac{1}{4} = -\frac{1}{8}$ 9. $\frac{1}{2}(6 - 8x) + \frac{3}{4}(8x - 12) = 4x + 6$

10. $-3 - 3^0 - 3^2(2x - 5) - (-2x - 3) = -x^0(x - 3)$

11. $-2^3 - \dfrac{1}{-2^{-2}}(x + 2) - 3x = -2^0(-2x^0 - 4)$

12. $-3[x - 2 - 3(2)] + 2[x - 3(x - 2)] = 7(x - 5)$

Expand:

13. $\dfrac{x^{-2}}{y}\left(2x^2y - \dfrac{3x^{-3}y}{y^{-2}}\right)$ **14.** $\dfrac{2ab}{c^2}\left(\dfrac{c^2a^{-1}}{b} - \dfrac{3ac}{b}\right)$ **15.** $-\dfrac{ax^2}{b}\left(\dfrac{bax^3}{a^2} - 3ax\right)$

Simplify:

16. $\dfrac{a^0a^2(ab^{-2})^{-2}a}{ba(a^2b^3)^{-3}b^2}$

17. $\dfrac{(xm^{-2})^0x^0m^0}{xx^2m^0(2x)^{-2}}$

18. $\dfrac{4c^2dc^{-3}(2cd^{-2})^{-2}}{c^0c^{-3}(c^{-2}d)^2}$

19. $\dfrac{p^2m^5(p^{-3})(2p)^{-3}}{m^6(m^{-2})^2mp^3}$

Simplify by adding like terms:

20. $\dfrac{x^2xy}{y^{-2}} - \dfrac{3x^5}{xxy^{-3}} + \dfrac{7x^7}{y^3x^4}$

21. $-\dfrac{3a^2x^4}{x} + \dfrac{2aax^2}{x} - \dfrac{5x^3}{a^{-2}}$

Evaluate:

22. $a^0x(a - ax^2)$ if $a = -3$ and $x = -2$

23. $mx - m(m - mx^2)$ if $m = -2$ and $x = -1$

24. $a^2 - b(a - b)$ if $a = -\dfrac{1}{2}$ and $b = \dfrac{1}{4}$

25. $a - ba(a^2 - b)$ if $a = -\dfrac{1}{2}$ and $b = -\dfrac{1}{4}$

Simplify:

26. $-2(-2 - 3^2) - 2[-2(-3)]$

27. $-3^2 - (-3)^3 - \dfrac{1}{-2^2}$

28. $-3^0[-2^0 - 2^2 - 2^3(-2 - 3)]$

29. $-3[(-2^0 + 5) - (-3 + 7) - |-2|]$

30. $\dfrac{1}{-4^{-2}}$

LESSON 6 *Equations with decimal numbers •*
Consecutive integer word problems

6.A
equations with decimal numbers

When equations contain decimal numbers, it is sometimes helpful to multiply every term by a power of 10 that will turn all numbers into integers.

example 6.A.1 Solve: $.003x + .4 = 2.05$

solution If we begin by multiplying every term by 1000, we get

$$3x + 400 = 2050$$

which is an equation that contains only the variable and integers. Then we solve the equation.

$$3x = 1650 \qquad \text{added } -400 \text{ to both sides}$$

$$x = 550 \qquad \text{divided both sides by 3}$$

Many people use the words **decimal fraction** to describe numbers that have internal decimal points. This is because numbers such as

$$2.0413$$

can be written in fractional form. We can write 2.0413 as a fraction as

$$\frac{20{,}413}{10{,}000}$$

Thus, the general equation for a fractional part of a number can also be used for problems that involve a decimal part of a number.

example 6.A.2   The students found that .015 of the teachers were either brave or completely fearless. If 300 teachers fell into one of these categories, how many teachers were there in all?

solution   We use the equation for a fractional part of a number and replace F with WD for what decimal.

$$(WD) \times (\text{of}) = (\text{is})$$

We replace WD with .015, *of* with T, and *is* with 300.

$$.015T = 300$$

We finish by dividing both sides by .015.

$$T = 20{,}000$$

example 6.A.3   An analysis of the old woman's utterances showed that .932 were vaticinal. If she spoke 2000 times during the period in question, how many utterances were not vaticinal?

solution   If .932 were vaticinal, the decimal fraction that was not vaticinal was

$$1 - .932 = .068$$

So we can write

$$(WD)(\text{of}) = (\text{is})$$

$$(.068)(2000) = NV$$

$$136 = NV$$

So 136 utterances were nonvaticinal in nature.

6.B
consecutive integer word problems

In algebra, we study problems whose mastery will provide the skills necessary to solve real problems that may be encountered in higher mathematics or in mathematically based disciplines such as chemistry or physics. Problems about consecutive integers are of this type. They help us remember which numbers are integers and allow us to practice our word problem skills.

We remember that we designate an unspecified integer with the letter N and greater consecutive integers with $N + 1$, $N + 2$, etc.

Consecutive integers N, $N + 1$, $N + 2$, etc.

Consecutive odd integers are 2 units apart, and consecutive even integers are also 2 units apart. Thus, we can designate both of them with the same notation.

Consecutive odd integers N, $N + 2$, $N + 4$, $N + 6$, etc.

Consecutive even integers N, $N + 2$, $N + 4$, $N + 6$, etc.

example 6.B.1   Find three consecutive even integers such that 5 times the sum of the first and the third is 16 greater than 9 times the second.

solution   We designate the consecutive even integers as

$$N \quad N + 2 \quad N + 4$$

and write the necessary equation and solve.

$$
\begin{array}{ll}
5(N + N + 4) - 16 = 9(N + 2) & \text{equation} \\
10N + 20 - 16 = 9N + 18 & \text{multiplied} \\
10N + 4 = 9N + 18 & \text{simplified} \\
N = 14 & \text{added } -9N - 4
\end{array}
$$

So the desired integers are **14, 16, and 18.**

example 6.B.2   Find four consecutive integers such that 5 times the sum of the first and the fourth is 1 greater than 8 times the third.

solution   We designate the consecutive integers as

$$N \quad N + 1 \quad N + 2 \quad N + 3$$

Now we write the equation and solve

$$
\begin{array}{ll}
5(N + N + 3) - 1 = 8(N + 2) & \text{equation} \\
10N + 15 - 1 = 8N + 16 & \text{multiplied} \\
10N + 14 = 8N + 16 & \text{simplified} \\
2N = 2 & \text{added } -8N - 14 \\
N = 1 & \text{divided by 2}
\end{array}
$$

Thus, the desired integers are **1, 2, 3, and 4.**

problem set 6

*1.   The students found that .015 of the teachers were either brave or completely fearless. If 300 teachers fell into one of these categories, how many teachers were there in all?

*2.   An analysis of the old woman's utterances showed that .932 were vaticinal. If she spoke 2000 times during the period in question, how many utterances were not vaticinal?

3.   A number is multiplied by -3 and then this product is decreased by 7. The result is 4 less than twice the opposite of the number. What is the number?

4.   When Cleopatra called for barge workers, $2\frac{1}{2}$ times the number needed showed up. If 175 showed up, how many barge workers did she need?

*5.   Find three consecutive even integers such that 5 times the sum of the first and the third is 16 greater than 9 times the second.

*6.   Find four consecutive integers such that 5 times the sum of the first and the fourth is 1 greater than 8 times the third.

Solve:

*7. $.003x + .4 = 2.05$ 8. $3\frac{2}{5}x + 1\frac{1}{4} = 7\frac{1}{3}$

9. $-3(x - 2 + 1) - (-2)^2 - 3(x - 2) = 5x^0(2 - x) - 2x$

10. $-3 - 2^2 - 2(x - 3) = 2[(x - 5)(2 - 5)]$

11. $4(x + 3) - 2^0(-x - 3) = 2x - 4(x^0 - x) - 3^2$

Expand:

12. $\dfrac{xy}{p}\left(\dfrac{-3p^{-1}}{xy}+\dfrac{2p}{x^{-1}y}\right)$ **13.** $-\dfrac{x^0k}{p}\left(\dfrac{k^0p}{x}-2p\right)$

Simplify:

14. $\dfrac{(2x^{-2}y^0)^{-2}yx^{-2}}{xxxy^2(y^{-2})^2}$ **15.** $\dfrac{(3y^{-2})^2yxy^{-2}y}{y^0xx^2x^3(x^{-1})^2}$

16. $\dfrac{a^0bc^0(a^{-1}b^{-1})^2}{ab(ab^0)abc}$ **17.** $\dfrac{(2x^2)^{-3}(xy^0)^{-2}}{2xx^0x^1xxy^2}$

Simplify by adding like terms:

18. $-2xy+\dfrac{5x^0xy^{-1}}{y^{-2}}-\dfrac{5xx^{-1}x^2}{(x^{-1})^{-1}}$ **19.** $-\dfrac{3x^2xy^2}{y^4}+\dfrac{2xxx}{y^{-2}}-\dfrac{3xy}{x^{-2}y^{-1}}$

Evaluate:

20. $xy-x^2y-y$ if $x=-2$ and $y=-4$

21. a^2b-ab^2-b if $a=-4$ and $b=-2$

22. $a^{-2}b-a(a-b)$ if $a=-\dfrac{1}{2}$ and $b=\dfrac{1}{4}$

23. $a(a-ab^2)$ if $a=-\dfrac{1}{3}$ and $b=-\dfrac{1}{6}$

24. $m^2p(mp-p^2)$ if $m=-\dfrac{1}{4}$ and $p=\dfrac{1}{5}$

Simplify:

25. $-3^0[-3^2-2(-2-3)][-2^0]$ **26.** $-3-(-3)^2+(-3)(-6)$

27. $-2[(-2-2)-2(-2^0-1)]$ **28.** $-3^2+(-3)^2-4^2-|-2-2|$

29. $-3^{-2}-\dfrac{2}{-2^{-3}}-2^0$ **30.** $-(-2)^{-3}-3^{-2}-3$

LESSON 7 *Percent*

7.A

percent The Latin word for *by* is *per* and the Latin word for 100 is *centum*. Thus the word **percent** literally means "by the hundred." The following percent equation (b) is exactly the same equation as the fractional part of a number equation (a) except that the denominator of the fraction is 100.

$$\text{(a)}\quad (WF)\times(\text{of})=(\text{is})\qquad \text{(b)}\quad \left(\dfrac{P}{100}\right)\times(\text{of})=(\text{is})$$

There are two other forms of the percent equation that are often used.

$$\text{(c)}\quad \dfrac{P}{100}=\dfrac{\text{is}}{\text{of}}\qquad\text{and}\qquad \text{(d)}\quad (\text{Rate})\times(\text{of})=(\text{is})$$

We call (c) the ratio form of the percent equation. In form (d) the rate is the percent divided by 100. If the percent were 20 percent, then the rate would be .2, which is 20 divided by 100. **Any of the three percent equations can be used. They are not different equations, but are three different forms of the same equation.**

There are two types of percent problems. In one type, the original quantity is divided into two parts and the final percent is less than 100. In the second type, the original quantity increases and the final percent is greater than 100. It is helpful to be able to draw diagrams that give us a picture of the problem.

example 7.A.1   Eighteen is 20 percent of what number? Work the problem and then draw the completed diagram.

solution   We will use the fractional form of the percent equation.

$$\frac{P}{100} \times (\text{of}) = (\text{is}) \quad \longrightarrow \quad \frac{20}{100} \times WN = 18$$

Now we multiply both sides by $\frac{100}{20}$ to solve.

$$\frac{100}{20} \cdot \frac{20}{100} \, WN = \frac{100}{20} \cdot 18 \quad \longrightarrow \quad WN = 90$$

If one part of 90 is 18 for 20 percent, the other part must be 72 for 80 percent. The diagram is as shown here.

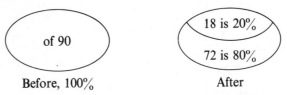

Before, 100%   After

example 7.A.2   Fifteen hundred is what percent of 250? Work the problem and then draw the completed diagram.

solution   Again we choose to use the fractional form of the equation.

$$\frac{WP}{100} \times 250 = 1500$$

To solve, we multiply both sides by $\frac{100}{250}$.

$$\frac{100}{250} \cdot \frac{WP}{100} \cdot 250 = 1500 \cdot \frac{100}{250} \quad \longrightarrow \quad WP = 600 \text{ percent}$$

The diagram shows 250 increased to 1500, which is 600 percent.

Before, 100%   After

problem set 7   *1.   Eighteen is 20 percent of what number? Draw the diagram.

*2.   Fifteen hundred is what percent of 250? Draw the diagram.

3.   Twenty percent of what number is 460? Draw the diagram.

4.   What percent of 20 is 680? Draw the diagram.

5. Three hundred eighty is 1900 percent of what number? Draw the diagram.

6. Find three consecutive odd integers such that 7 times the sum of the first and the third is 120 less than 10 times the opposite of the second.

7. Find three consecutive even integers such that 6 times the sum of the first and third is 8 less than 14 times the second.

8. If twice the opposite of a number is increased by 5, the result is the opposite of the number. What is the number?

Solve:

9. $-3p(-2 - 3) + p - 2^2 = -(2p + 4) - p^0$

10. $.005x - .07 = .02x + .0032$

11. $3\frac{2}{5}k + 2\frac{1}{3} = -\frac{3}{5}$

12. $2\frac{1}{5} + 3\frac{1}{8} + 2\frac{1}{2}x = 4\frac{3}{20}$

13. $3x - 2 - 2^0(x - 3) - 2^0 + 2^2 = 5(-x - 2) + 3^0$

Expand:

14. $\dfrac{2xyp}{y^{-2}}\left(\dfrac{x^{-1}}{y^3p} - \dfrac{3x}{yyp^2}\right)$

15. $\dfrac{4x^{-2}y}{k}\left(\dfrac{2kx^2}{y} - \dfrac{3xy}{k}\right)$

Simplify:

16. $\dfrac{(x^{-2}yp)^{-2}yx^0}{(xy)^{-5}(xp^{-2})^5 y^0}$

17. $\dfrac{(2x^2y^3)^{-3}y}{(4xy)^{-2}(x^{-2}y)^3 y}$

18. $\dfrac{xx^{-2}y(x^{-3})^2xy^0}{(2xy)^{-2}x^2(y^{-3})^2}$

Simplify by adding like terms:

19. $-\dfrac{3x^2xy}{p} + \dfrac{7xyp^{-1}}{x^{-2}} - \dfrac{2xxxp^{-1}}{y^{-1}}$

20. $-4xp^2 + \dfrac{3xxp^4}{p^2x^2} - \dfrac{2xp}{p^{-1}}$

Evaluate:

21. $-a(a - b)$ if $a = -\dfrac{1}{2}, b = \dfrac{1}{3}$

22. $-xy(-x^2 - y)$ if $x = -\dfrac{1}{2}, y = \dfrac{1}{4}$

23. $x^2 - x(x - y)$ if $x = -\dfrac{1}{2}, y = -\dfrac{1}{4}$

24. $x^3 - x(xy - y)$ if $x = -2, y = -4$

25. $x - a(a - xa)$ if $x = 2, a = -\dfrac{1}{2}$

Simplify:

26. $-2\{[-2^0 - 3(-2)] - [-2(-3 - 2)(-2)]\}$

27. $-2^0 - 2 - 2^2 - (-2)^3 - 2(-2 - 2) - 2$

28. $3^0(-2 - 3)(-2 + 5)(-2) - (-3 + 7)(-4^0 - 3^0)$

29. $-5(-2^0) - 1(-3 - 2) - (-2 + 7)(4^0 - 3^0)$

30. $2[(-2^0 - 1)(-2^0 - 15^0) - (-2)^2 - 3^0] - 2$

LESSON 8 *Polynomials • Graphing linear equations • Intercept-slope method*

8.A
polynomials

It is convenient to have a word to describe the simplest kind of algebraic expression. These expressions have coefficients that are real numbers and variables that have whole numbers as exponents. No fractional exponents or negative exponents are allowed. The following are examples of these very simple expressions.

$$-4 \qquad \frac{1}{2} \qquad 2x \qquad -3x^2 \qquad -5x^2 + 6x + 2$$

Unfortunately, we use an intimidating word to designate these simple expressions—**polynomial**. It would have been helpful had we called them **"simplenomials"** instead, but we didn't. But we can think "simplenomial" when we hear the word polynomial.

A polynomial in one variable has a real number for a coefficient and has one of the numbers 0, 1, 2, 3, . . . , etc. as the exponent of the variable. Thus, all of the following are polynomials. They are also called **monomials** because they have only one term.

$$\text{(a)} \quad -4 \qquad \text{(b)} \quad 2x^2 \qquad \text{(c)} \quad 3x^{14} \qquad \text{(d)} \quad .004x^5 \qquad \text{(e)} \quad \sqrt{2}x$$

The first one, (a), can be thought of as $-4x^0$, and since x^0 equals 1, this expression fits the definition of a polynomial. The rest of the expressions have real number coefficients and whole number exponents so they are all polynomials.

Polynomials of two terms are also called **binomials** and polynomials of three terms are also called **trinomials.**

$$\text{(f)} \quad x + 2 \qquad \text{(g)} \quad x^4 + 2x \qquad \text{(h)} \quad 2x^2 + 3x + 2$$

Thus, (f) and (g) are binomials, and (h) is a trinomial.

The degree of a polynomial is the same as the degree of the highest-degree term of the polynomial. Thus (f) is a first-degree polynomial because the exponent of x is 1. The polynomial (g) is a fourth-degree polynomial because the exponent of x^4 is 4. Using the same reasoning, the polynomial (h) is a second-degree polynomial because the greatest exponent is 2.

An equation that contains only polynomial terms is called a **polynomial equation.** The degree of a polynomial equation is the same as the degree of the highest-degree term in the equation. Thus, the equations

$$2x + 3y = 6 \qquad 3x - 2y = 0 \qquad -3x = 2y + 4$$

are all first-degree polynomial equations. If we use two number lines to form a coordinate plane, we can graph the set of ordered pairs of x and y that satisfy one of these equations. **The graph of a first-degree polynomial equation in two unknowns is a straight line.**

8.B
graphing linear equations

To find two or more ordered pairs of x and y that satisfy the equation of a line, we often use five steps.

1. Solve the equation for y.
2. Make a table and select convenient values of x.
3. Use these values of x in the equation to find the matching values of y.
4. Complete the table.
5. Graph the ordered pairs and draw the line.

example 8.B.1   Graph the equation $2x + 3y = 6$.

solution   We will use the five steps listed above.

1. First we solve the equation for y.

$$2x + 3y = 6 \longrightarrow 3y = -2x + 6 \longrightarrow y = -\frac{2}{3}x + 2$$

2. Next we make the table and select 0, 6 and -6 as values for x.

x	0	6	-6
y			

3. Now we find the matching values of y.

When $x = 0$:   When $x = 6$:   When $x = -6$:

$$y = -\frac{2}{3}(0) + 2 \qquad y = -\frac{2}{3}(6) + 2 \qquad y = -\frac{2}{3}(-6) + 2$$

$$y = 2 \qquad\qquad y = -2 \qquad\qquad y = 6$$

4. Next we complete the table.

x	0	6	-6
y	2	-2	6

5. And then we graph the points and draw the line.

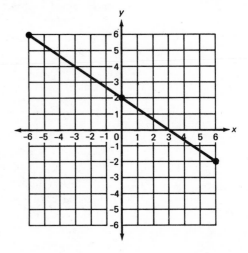

8.C

intercept-slope method   The method of graphing a line shown in the example above is exact and will always work. However, the method is time-consuming, and there is a quicker way to graph a line that is just as accurate.

Recall that two points are all that is needed to graph a line. We can often use the y intercept as one of the points and use the slope to find another point. Since we use the intercept first, we call this method the **intercept-slope method.** To demonstrate we will graph the same equation again.

example 8.C.1   Use the intercept-slope method to graph the equation $2x + 3y = 6$.

solution   The first step is the same. We solve the equation for y.

$$2x + 3y = 6 \longrightarrow 3y = -2x + 6 \longrightarrow y = -\frac{2}{3}x + 2$$

The equation has two numbers. The first number is $-\frac{2}{3}$ and is the slope. The second

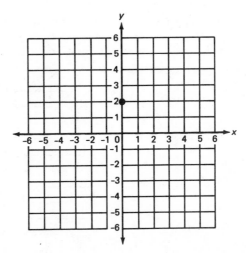

number is 2 and is the y intercept. This is the value of y when x equals zero. We graph the intercept, which is (0, 2). Next we write the slope, $-\frac{2}{3}$ as either (a) $\frac{-2}{+3}$ or (b) $\frac{+2}{-3}$. We remember that the slope is the rise over the run. Thus, from the point we have graphed, to find a second point we can (a) take a rise of -2 and a run of $+3$, or (b) take a rise of $+2$ and a run of -3. Both ways are shown below: (a) on the left and (b) on the right.

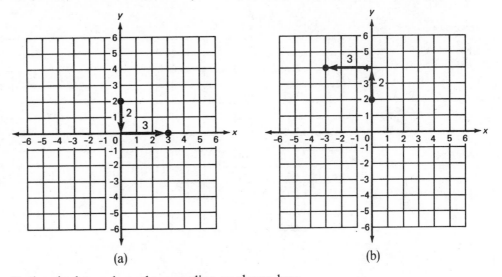

Both pairs let us draw the same line, as shown here.

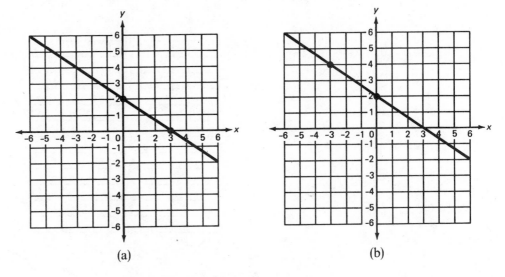

example 8.C.2   Use the intercept-slope method to graph the line $y = -3x - 3$.

solution   The intercept is $y = -3$. First graph this point.

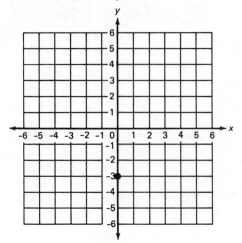

The slope can be written as (a) $\frac{-3}{+1}$ or as (b) $\frac{+3}{-1}$. Thus, from the intercept we can (a) take a rise of -3 and a run of $+1$, or (b) take a rise of $+3$ and a run of -1.

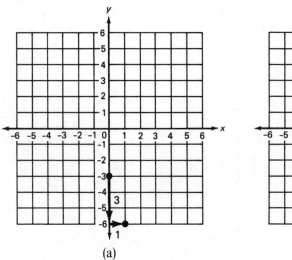
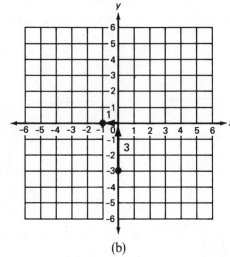

(a)   (b)

All three points lie on the line, as shown here.

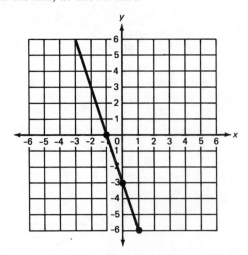

problem set 8

1. When the tournament began, only .36 of the knights wore new armor. If 828 knights wore new armor, how many knights participated in the tournament?

2. Sir Lancelot found four consecutive even integers such that 10 times the sum of the first and fourth was 24 greater than 9 times the sum of the second and fourth. What were his integers?

3. When the Danish invaders multiplied their secret number by 3 and then added -7, the result was 72 less than twice the opposite of the number. What was the secret number of the Danes?

4. Only seven-sixteenths of the men necessary to defend the castle answered the call to arms. If 420 answered the call, how many were required to defend the castle?

5. The defenders thought of consecutive odd integers while waiting for the fusilade from the trebuchet. Their integers were three in number and were such that 5 times the sum of the first and third was 108 greater than twice the opposite of the second. What were the integers?

6. Eighty-six is 20 percent of what number? Draw a diagram of the problem.

7. What number is 340 percent of 56? Draw a diagram of the problem.

*8. Graph $2x + 3y = 6$ on a rectangular coordinate system.

*9. Graph $y = -3x - 3$ on a rectangular coordinate system.

Solve:

10. $.003x + .02x - .03 = .177$

11. $2\frac{1}{3}x + 1\frac{3}{5} = 7\frac{2}{5}$

12. $-4^0 - 2^2 - (-2)^3 - 3 - (2 - 2x) - 4 = -3(-2 + 2x)$

13. $-\frac{1}{2} + 2\frac{3}{8} - 7\frac{1}{4} + 3\frac{1}{2}x = 4\frac{1}{16}$

Expand:

14. $\dfrac{x^{-2}y}{p}\left(\dfrac{-3x^2p}{y} - \dfrac{4xy^2}{p^2}\right)$

15. $\dfrac{-3^0x^0}{p^0}\left(-3x + \dfrac{5xy}{p^{-2}}\right)$

Simplify:

16. $\dfrac{(-2x^2y^{-2})^2x}{2x^4x^0xx^2y}$

17. $\dfrac{(4x)^{-2}y^0(y^{-2})^2y}{32^{-1}x^2(yx^0)^{-3}}$

18. $\dfrac{5x^{-2}(y^2x^3)^{-3}}{2^{-2}xyx^2(x^{-2}y)}$

Simplify by adding like terms:

19. $3x^{-2}y + 5x^2y^{-1} - \dfrac{3}{xxy^{-1}}$

20. $\dfrac{2xxy^{-3}y}{xy} + \dfrac{2y^{-1}y^{-2}}{x^{-1}} - \dfrac{7xy^2}{y}$

Evaluate:

21. $ax - a(x^2)$ \quad if $a = -\dfrac{1}{2}, x = \dfrac{1}{4}$

22. $ab^2(a - ab)$ \quad if $a = \dfrac{1}{2}, b = -\dfrac{1}{3}$

23. $a^2 - a^3b$ \quad if $a = -\dfrac{1}{2}, b = \dfrac{1}{5}$

24. $m^2 - (m - am)$ \quad if $m = \dfrac{1}{3}, a = -\dfrac{1}{5}$

25. $mx - (m^2 - x)$ \quad if $m = -\dfrac{1}{3}, x = \dfrac{1}{2}$

Simplify

26. $-3[(-2^0 - 4) - (-2)(-3)] - [(-6^0 - 2) - 2^2(-3)]$

27. $-3 - 3^0 - 3^{-2} + \dfrac{1}{9} - 3^0(-3 - 3)$ **28.** $-|-2^0| - 2^{-2} - (-2)^{-2}$

29. $(-1)^{-3} - 1^{-2} - 1^2 - (-1)^3$ **30.** $2^0 - 2^{-2} - 2^{-3} - 2$

LESSON 9 *Percent word problems*

9.A
percent word problems

The Latin word per means by and the Latin word centum means one hundred. We combine these words to form the English word **percent** which means by the one hundred. To compute a given percent of a number, we can first divide the number into 100 parts. Then 30 percent of the number means 30 of these parts and 193 percent of the number means 193 of these parts, etc. To demonstrate let us begin with the number

$$242$$

Now if we divide 242 into 100 parts we find that each part equals 2.42.

$$\frac{242}{100} = 2.42$$

Thus, 30 percent of 242 means 30 of these parts or 30 times 2.42.

(a) $30 \times 2.42 = \textbf{72.6}$

Likewise, 193 percent of 242 means 193 of these parts or 193 times 2.42.

(b) $193 \times 2.42 = \textbf{467.06}$

If we use the percent equations, we get the same answers.

(a) $\dfrac{30}{100} \times 242 = \textbf{72.6}$ (b) $\dfrac{193}{100} \times 242 = \textbf{467.06}$

Percent word problems fall into several different categories, and it is most helpful if a diagram of the problem is drawn as the first step. A diagram allows the visualization of the problem and will help prevent mistakes. Some students believe that drawing diagrams is childish, for they can work the problems without diagrams. The author believes that making preventable mistakes is childish and that drawing pictures that will prevent these mistakes is an indication of maturity. Check his opinion with an engineer or a graduate physicist or graduate mathematician before making up your mind. There is no excuse for making errors that can be prevented by drawing a picture of the problem!

example 9.A.1   The wood nymphs and the maids gamboled and frolicked before the banquet began. If 70 percent of those present were wood nymphs and 120 maids were present, how many wood nymphs came to the banquet?

solution   We use the following diagrams to visualize the problem

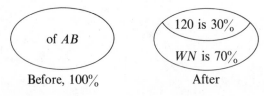

of *AB*

Before, 100%

120 is 30%

WN is 70%

After

The diagram shows *AB* for *A*ll at the *B*anquet and *WN* for *W*ood *N*ymphs. Since 70 percent were wood nymphs, then 30 percent were maids. We see that 120 is 30 percent of all at the banquet.

$$\frac{30}{100} \times AB = 120$$

To solve we multiply both sides by $\frac{100}{30}$.

$$\frac{100}{30} \cdot \frac{30}{100} \cdot AB = 120 \cdot \frac{100}{30} \longrightarrow AB = 400$$

Since the guests totaled 400 and 120 were maids, then there must have been **280** wood nymphs.

example 9.A.2 The harvest was cornucopian, as it was 120 percent greater than last year. If the yield was 140,800 bushels, how many bushels were harvested last year?

solution Again we find that a diagram is helpful.

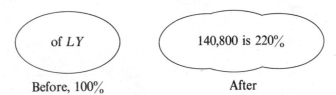

Before, 100%   After

We see that we began with 100 percent last year and that a 120 percent increase means 220 percent this year.

$$\frac{220}{100} \times LY = 140{,}800$$

We complete the solution by multiplying both sides by $\frac{100}{220}$.

$$\frac{100}{220} \cdot \frac{220}{100} \cdot LY = 140{,}800 \cdot \frac{100}{220}$$

$$LY = \textbf{64,000 bushels}$$

problem set 9

*1. The wood nymphs and the maids gamboled and frolicked before the banquet began. If 70 percent of those present were wood nymphs and 120 maids were present, how many wood nymphs came to the banquet?

*2. The harvest was cornucopian, as it was 120 percent greater than last year. If the yield was 140,800 bushels, how many bushels were harvested last year?

3. Twenty percent of the income was used to pay for raw materials. If $78,000 was spent for other purposes, what was the total income?

4. Find three consecutive integers such that the product of -5 and the sum of the first two is 43 less than twice the second.

5. When the war tocsin sounded, 84 percent of the soldiers staggered to their feet. If 40,000 did not get up, how many soldiers were present?

6. Diomedes peered into the darkness and saw 1400 Trojans. If he could not see seven-eighths of the Trojans, how many Trojans were there?

7. Graph $y - 2x + 3 = 0$ on a rectangular coordinate system.

8. Graph $3y + 6 = -x$ on a rectangular coordinate system.

Solve:

9. $.02 - .003x + x = 5.005$

10. $-3\frac{1}{5}x + 7\frac{1}{10} = 4\frac{2}{9}$

11. $-2[(2 - 3)x + 7(2^0 - 1)] = -3(x - 2)$

12. $3\frac{1}{2}x - 2\frac{1}{4} = \frac{7}{4}$

13. $-2^0(2x - 3) - 4x^0 = 2x - 3^0$

Expand:

14. $\dfrac{x^0 y^2}{p^{-2}}\left(\dfrac{p^2}{y^2} - \dfrac{y^{-2}}{p^{-2}}\right)$

15. $\dfrac{ak^{-2}}{a^{-3}}\left(\dfrac{2k^4}{a^4} - 3k\right)$

Simplify:

16. $\dfrac{(2x^2 ya)^{-3} ya^3}{x^2 y(ay)^{-2} y}$

17. $\dfrac{(3a)^{-2} a^2 y^{-2}}{a^3 y^3 a^0}$

18. $\dfrac{(-2xyz)^{-3}}{(x^2 z^{-3})^{-3}}$

Simplify by adding like terms:

19. $3x - \dfrac{2xy^2}{y} + \dfrac{4xx^{-2}}{(x^2)^{-1}}$

20. $\dfrac{2xy}{p} - \dfrac{5xxx}{(x^{-2})^{-1} y^{-1}} + \dfrac{3xp^{-1}}{y^{-1}}$

Evaluate:

21. $-a^2 b - a$ if $a = -\dfrac{1}{2}, b = \dfrac{1}{4}$

22. $a(a - ab)$ if $a = -\dfrac{1}{2}, b = -\dfrac{1}{8}$

23. $a(a - b)(ab - b)$ if $a = -2$ and $b = 3$

24. $a^2(x - ax^2)$ if $a = -2, x = -4$

25. $a^2 - a^0 - ax^0$ if $a = -2, x = -3$

Simplify:

26. $-2(-3 - 2^0) - 2^0(-2^2 - 2)$

27. $-3 - 3^0 - \dfrac{1}{-3^3} - (-3)^2 - 3$

28. $-3[(-5 + 2)(-2) - (3^0 - 2) - 2]$

29. $-2^0(-2 - 3^0) - (-2)^3 - |-3|$

30. $-\dfrac{1}{-2^{-3}} + \dfrac{1}{-(-2)^{-3}} - 3^2$

LESSON 10 *Pythagorean theorem*

10.A
Pythagorean theorem

Thus far, we have discussed properties and definitions. Properties are the way things are because they just are. Definitions are things we have agreed on. For instance,

PROPERTIES

(a) $3 + 2 = 2 + 3$

(b) If $a = b$, then $a + c = b + c$.

DEFINITIONS

(c) $x^{-2} = \dfrac{1}{x^2}$

(d) $4 + 3 \cdot 2 = 10$

Both properties and definitions can be called **rules.** Another kind of rule is a **theorem.** A theorem is just like a property because theorems tell us the way things are. The difference is that theorems can be proved by using properties and definitions. **The Pythagorean theorem states that the area of the square drawn on the hypotenuse of a right triangle equals the sum of the areas of the squares drawn on the other two sides.** On the left, we show a right triangle whose sides are 3, 4, and 5. The area of the square on the hypotenuse is 25 square units, which equals the sum of the areas of the squares drawn on the other two sides, because 9 plus 16 equals 25. In the center, we show a right triangle with sides a, b, and c; and on the right, we have drawn the squares on the sides of this triangle.

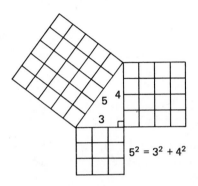
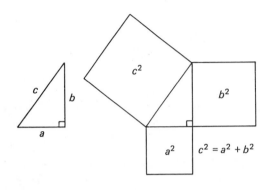

We see that the resulting algebraic equation is

$$c^2 = a^2 + b^2$$

where c is the length of the hypotenuse. This theorem is proved in geometry. For now, we will use it just as if it were a property.

example 10.A.1   Use the Pythagorean theorem to find side a.

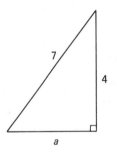

solution   Since $c^2 = a^2 + b^2$, we write

$$7^2 = 4^2 + a^2 \qquad \text{Pythagorean theorem}$$
$$49 = 16 + a^2 \qquad \text{squared 7 and 4}$$
$$33 = a^2 \qquad \text{added } -16$$
$$\sqrt{33} = a \qquad \text{solved}$$

example 10.A.2   Use the Pythagorean theorem to find the distance between the points $(4, 2)$ and $(-3, 4)$.

solution   We could use the distance formula, which is an algebraic statement of the Pythagorean theorem. However, the problem can be worked with fewer mistakes and more understanding by graphing the points, drawing the triangle, and using the theorem.

$D^2 = 2^2 + 7^2$ Pythagorean theorem

$D^2 = 4 + 49$ squared 2 and 7

$D^2 = 53$ added

$\mathbf{D = \sqrt{53}}$ solved

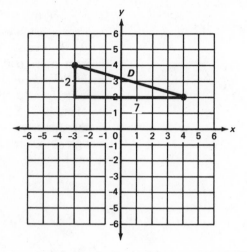

problem set 10

1. Twenty percent of the people at the fair were in a festive mood. If 1400 were not in a festive mood, how many attended the fair?

2. When Julius crossed the Rubicon, he had with him $3\frac{1}{4}$ times as many soldiers as he needed to conquer Rome. If he had 26,000 soldiers with him, how many were needed to conquer Rome?

3. When the Theban legion refused to obey the orders, it was decreed that the Legion should be decimated—every tenth man killed. If 590 men were killed, how many men did the Theban legion have after it was decimated?

4. Find three consecutive odd integers such that 4 times the first is 8 less than 3 times the sum of the last two.

5. When Aegisthus found that his secret had been discovered, he upped the ante by 160 percent. If the ante was now 10,400 minas,† what was the original ante?

6. Atreus could see 4200 Argives marching toward the lion's gate. If he could see 14 percent of the Argives, how many were hidden from view?

7. Graph $2y = 3x + 2$ on a rectangular coordinate system.

8. Graph $y = -3$ on a rectangular coordinate system.

*9. Find side a.

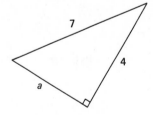

*10. Find the distance between (4, 2) and $(-3, 4)$.

Solve:

11. $\dfrac{3}{4}x - \dfrac{1}{5}x = 2\dfrac{3}{4}$

12. $-5.2 + 3y = .2(y + 2)$

13. $4x(2 - 3^0) + (-2)(x - 5) = -(3x + 2)$

---

† In Greek currency before the fourth century B.C. it took 60 minas to equal 1 talent.

Expand:

14. $\dfrac{4xy}{m^{-2}}\left(\dfrac{3y^{-1}}{m^2x} - \dfrac{2x}{ym}\right)$ **15.** $\dfrac{2x^0y}{p}\left(\dfrac{2p}{y} - \dfrac{3xy}{p}\right)$

Simplify:

16. $\dfrac{(3x^{-2})^{-2}xy}{3^{-3}x^{-2}(yx^0)^{-3}}$ **17.** $\dfrac{(2a^{-2})^{-2}(a^0)^2xa^2}{a^2xxx^2(xa)^2}$ **18.** $\dfrac{(x^2p^2)^{-3}x^0p^2}{x^{-2}px^0(xp)^{-3}}$

Simplify by adding like terms:

19. $3x + \dfrac{2x^2x^{-3}}{x^{-2}y^0} - x^0$ **20.** $\dfrac{5x^2y}{z} - \dfrac{3z^{-1}y}{x^{-2}} + \dfrac{7xxy^2z}{yz^2}$

Evaluate:

21. $x^2 - xa$ if $x = -\dfrac{1}{3}, a = \dfrac{1}{2}$ **22.** $x(x^2 - a)$ if $x = -\dfrac{1}{3}, a = \dfrac{1}{2}$

23. $b(ab - b)$ if $a = -\dfrac{1}{3}, b = -\dfrac{1}{2}$

24. $ab - a^2b^2 - b$ if $a = -\dfrac{1}{2}, b = -\dfrac{1}{2}$

25. $a^2(b - ab)b$ if $a = -\dfrac{1}{2}, b = -\dfrac{1}{2}$

Simplify:

26. $-(-2 - 3 - |-3|)(-5 + 3)$ **27.** $-\dfrac{1}{2^{-3}} - \dfrac{1}{-2^{-3}} - (-3 - 2^0) - 2$

28. $-|-2| - |-2^0| - 3^2 - (-3)^3$ **29.** $-[(-2 - 3)(-2^0)(-2) - 3^0(-2)]$

30. $-\dfrac{1}{2} - \left(\dfrac{1}{2}\right)^2 - \left(-\dfrac{1}{2}\right)^3 - \dfrac{1}{2}$

LESSON 11 *Addition of fractions*

11.A
addition of fractions

To add fractions whose denominators are equal, we add the numerators algebraically and record their sum over a single denominator.

$$\frac{1}{11} + \frac{5}{11} - \frac{4}{11} = \frac{1 + 5 - 4}{11} = \frac{2}{11}$$

In this example, each of the denominators is 11, and the sum of the numerators is 2. The same procedure is used when the fractions contain letters.

$$\frac{a}{b} + \frac{2}{b} + \frac{a + 3}{b} = \frac{2a + 5}{b}$$

In this example, each of the denominators is b, and the sum of the numerators is $2a + 5$.

When the denominators are not equal, the form of one or more of the fractions must be changed so that the denominators will be equal. One-half and three-fourths cannot be added because the denominators are not the same.

$$\frac{1}{2} + \frac{3}{4}$$

To make the denominators the same, we will change $\frac{1}{2}$ to $\frac{2}{4}$ by multiplying $\frac{1}{2}$ by $\frac{2}{2}$.

$$\frac{1}{2}\left(\frac{2}{2}\right) + \frac{3}{4} = \frac{2}{4} + \frac{3}{4} = \frac{5}{4}$$

If we wish to add the algebraic fractions

$$\frac{a}{b} + \frac{c}{x} + \frac{d+e}{4}$$

we need to change the forms of the fractions so that the denominators are equal.

$$\frac{a}{b}\left(\frac{4x}{4x}\right) + \frac{c}{x}\left(\frac{4b}{4b}\right) + \frac{(d+e)}{4}\left(\frac{bx}{bx}\right) = \frac{4ax + 4cb + bx(d+e)}{4bx}$$

We used a different multiplier for each fraction but did not change the value of any of the fractions because each of the multipliers had a value of 1. The multipliers that we used were

$$\frac{2}{2} = 1 \qquad \frac{4x}{4x} = 1 \qquad \frac{4b}{4b} = 1 \qquad \frac{bx}{bx} = 1$$

The fact that the denominator and the numerator can be multiplied by the same nonzero quantity without altering the value of the fraction is often called the fundamental principle of fractions or the fundamental theorem of rational expressions. We will call it the **denominator-numerator-same-quantity theorem** because this name helps us remember what the theorem is.

DENOMINATOR-NUMERATOR–SAME-QUANTITY THEOREM

$$\frac{a}{b} = \frac{ac}{bc} \qquad \text{because} \qquad \frac{c}{c} = 1 \qquad (b, c \neq 0)$$

The denominator and the numerator of a fraction may be multiplied by the same nonzero quantity without changing the value of the fraction.

A three-step procedure can be used to add fractions whose denominators are different as we will show in the next two examples.

example 11.A.1   Add $\dfrac{k}{2a} + \dfrac{bc}{ax^2} - \dfrac{m}{ax^3}$.

solution   The least common multiple of the denominators of these fractions is $2ax^3$. Thus, each new denominator will be $2ax^3$.

$$\frac{}{2ax^3} + \frac{}{2ax^3} - \frac{}{2ax^3}$$

We see that the original denominator of the first fraction has been multiplied by x^3. Thus, the numerator k must also be multiplied by x^3.

$$\frac{kx^3}{2ax^3} + \frac{}{2ax^3} - \frac{}{2ax^3}$$

The second denominator has been multiplied by $2x$, so the numerator bc must also be multiplied by $2x$.

$$\frac{kx^3}{2ax^3} + \frac{2xbc}{2ax^3} - \frac{}{2ax^3}$$

The multiplier in the last denominator is 2, so we must also multiply m by 2.

$$\frac{kx^3}{2ax^3} + \frac{2xbc}{2ax^3} - \frac{2m}{2ax^3}$$

Now the denominators are the same, so the numerators are added and their sum is recorded over a single denominator.

$$\frac{kx^3 + 2xbc - 2m}{2ax^3}$$

example 11.A.2   Add $\dfrac{m}{k} - b + \dfrac{cx}{ak^2}$.

solution   The new denominators will be ak^2.

$$\frac{}{ak^2} - \frac{}{ak^2} + \frac{}{ak^2}$$

Next we multiply each original numerator by the same quantity used as a multiplier for its denominator.

$$\frac{akm}{ak^2} - \frac{bak^2}{ak^2} + \frac{cx}{ak^2}$$

As the last step, we record the sum of the numerators over a single denominator.

$$\frac{akm - bak^2 + cx}{ak^2}$$

problem set 11

1. Thirteen percent of the people believed in lycanthropes. If 5220 did not believe in lycanthropes, how many believed?

2. Find four consecutive even integers such that the product of -2 and the sum of the first and the fourth is 20 less than the opposite of the third.

3. The recomputed price was $5599. If this was 120% greater than the original price, by how much had the original price been increased?

4. Thirty percent of the people refused to work and just sat around. If 1400 people worked, how many just sat around?

5. If Peter picked 240 percent more plums than Roger picked and if Peter picked 6800, how many did both boys pick together?

6. Gilbreda listed four consecutive odd integers as she watched steam spew out. Her integers were such that the product of -4 and the sum of the first and the fourth was 10 greater than 10 times the opposite of the third. What were the first 4 integers on her list?

Add:

*7. $\dfrac{m}{k} - b + \dfrac{cx}{ak^2}$

*8. $\dfrac{k}{2a} + \dfrac{bc}{ax^2} - \dfrac{m}{ax^3}$

9. $\dfrac{p}{ak} - c + \dfrac{3a}{4k}$

10. $\dfrac{m^2}{p} - \dfrac{3p}{cx} - \dfrac{5}{4c^2x}$

11. Find side x.

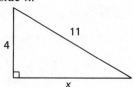

12. Find the distance between $(-3, -3)$ and $(5, 2)$.

13. Graph (a) $4x + 3y - 6 = 0$ and (b) $x = -3$ on the same rectangular coordinate system.

Solve:

14. $3\dfrac{1}{2} - 2\dfrac{1}{3}x = 3\dfrac{1}{4}$

15. $.03x - x + 2 = -.91$

16. $3x(2 - 3^0) - 7^0 = -2x(3 - 7^0) + 2$

Expand:

17. $\dfrac{a^{-2}}{y}\left(3a^2y - \dfrac{2y}{a^2}\right)$

18. $\dfrac{2x^0yp}{k}\left(\dfrac{3k}{yp} - \dfrac{2yk}{p}\right)$

Simplify:

19. $\dfrac{(2a^{-2})^{-2}ax^2}{a^0(x^2a)^{-3}ax}$

20. $\dfrac{2p^2a^{-2}ap^0p^4}{(2pa^{-2})^{-3}ap^0}$

21. $\dfrac{xym^2m^{-4}xm}{(2x^2y)^{-3}xy^0x^{-3}y}$

Simplify by adding like terms:

22. $\dfrac{xp^{-3}}{y} - \dfrac{3y^{-1}}{x^{-1}p^3} + \dfrac{2x}{pppy}$

23. $-3ka + \dfrac{3k^2a^2}{ka} - \dfrac{5a^0k}{a^{-1}}$

Evaluate:

24. $-a - ax(a - x)$ if $a = -\dfrac{1}{2}$, $x = \dfrac{3}{2}$

25. $-a^2(b - a)$ if $a = -\dfrac{1}{2}$, $b = \dfrac{3}{2}$

26. $a^3 - (a - ab)$ if $a = -\dfrac{1}{2}$, $b = \dfrac{5}{4}$

27. $ab(a^2 - b)$ if $a = -\dfrac{1}{2}$, $b = -\dfrac{3}{5}$

Simplify:

28. $-2(-3 - 2^0 - 2)(-2 + 5)(-2)$

29. $-\dfrac{1}{-2^0} - \dfrac{1}{-2^2} - \dfrac{1}{-2^{-2}}$

30. $|-3^0| - |-2 - 3| + (-2^0)(-2 - 5)$

LESSON 12   *Equation of a line*

12.A
equation of a line

All first-degree equations in x and y such as $2x + 3y - 6 = 0$ are called **linear equations,** and the graph of the ordered pairs of x and y that satisfy one of these equations is a straight line. There is an infinite number of ordered pairs of x and y that satisfy this equation. Two of them are $(-3, 4)$ and $(3, 0)$. We have graphed these points and drawn the line in the figure.

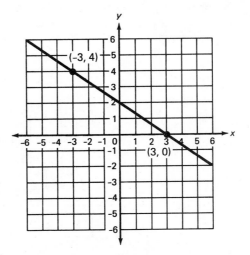

If we write the equation $2x + 3y - 6 = 0$ in the form

$$y = mx + b$$

called the **slope-intercept form,** we get

$$y = -\frac{2}{3}x + 2$$

The letter b in the general equation has been replaced with the number 2 in this equation. This number is the y value of the equation when x equals zero, and on the graph is the y coordinate of the point where the line crosses the y axis.

   The letter m in the general equation has been replaced with the number $-\frac{2}{3}$ in this equation. This number is the ratio of the change in the y coordinate to the change in the x coordinate as we move along the line from one point to another. If we move from $(-3, 4)$ to $(3, 0)$, y changes from 4 to 0, a change of -4, and x changes from -3 to 3, a change of $+6$ so the slope is -4 over $+6$.

$$\text{slope} = \frac{-4}{+6} = -\frac{2}{3}$$

The sign of the slope can be determined visually by using the little man and his car as a mnemonic device. **He always comes from the left side, as we show here.**

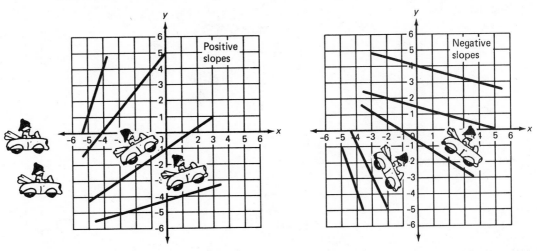

We can find the magnitude of a slope visually by connecting two points on the line with horizontal and vertical lines to form a right triangle as shown here. **The magnitude (absolute value) of the slope is the absolute value of the rise over the absolute value of the run,** or 4 over 6, which equals 2 over 3.

$$|\text{Slope}| = |m| = \frac{|\text{rise}|}{|\text{run}|} = \frac{4}{6} = \frac{2}{3}$$

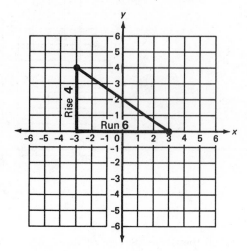

Thus, the line $y = -\frac{2}{3}x + 2$

1. has a negative slope

2. whose magnitude is $\frac{2}{3}$

3. and has an intercept of $+2$

all of which can be verified by looking at the graph. Test your understanding by covering the solution to the next problem until you have worked it. Remember that the equations of vertical and horizontal lines are special cases. The equation of a vertical line has the form $x = \pm k$ and the equation of a horizontal line has the form $y = \pm k$.

example 12.A.1   Write the equations of the four lines shown.

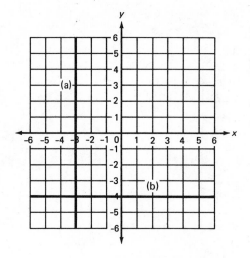

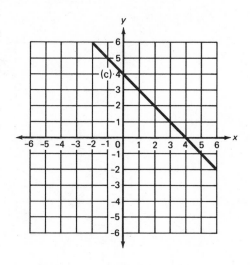

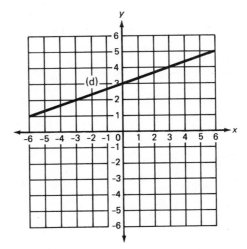

solution   (a) $x = -3$ (b) $y = -4$ (c) $y = -x + 4$ (d) $y = \frac{1}{3}x + 3$

problem set 12

1. Forty percent of the vases were monochromatic and the rest were variegated. If 2400 were variegated, how many were monochromatic?

2. Find four consecutive integers such that twice the sum of the first, second, and fourth is 40 less than 3 times the opposite of the third.

3. For a given performance, $4\frac{1}{4}$ times as many tickets were sold as there were seats. If 5100 tickets were sold, how many could be seated in the auditorium?

4. Wilbur thought of a number. He calculated that $2\frac{1}{5}$ of his number was equal to 1. What was his number?

5. Five times the opposite of a number was increased by 25. This was exactly 90 greater than 8 times the number. What was the number?

6. The little train had completed 30 percent of the journey. If 6300 miles still remained, what was the total length of the journey?

Add:

7. $m + \dfrac{x}{c} + \dfrac{c}{x^2 b}$

8. $\dfrac{a}{b} - \dfrac{3b}{a^2} - \dfrac{2}{abc}$

9. $\dfrac{1}{k} + k + \dfrac{k^2}{m}$

10. $1 + \dfrac{a}{b}$

11. Find side k.

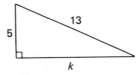

12. Find the distance between $(-2, 7)$ and $(-8, -2)$.

13. Graph (a) $3y + x - 9 = 0$ and (b) $x = 2$ on a rectangular coordinate system.

14. Find the equation of lines (a) and (b).

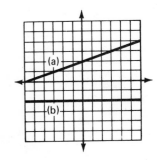

Solve:

15. $8\frac{1}{4} + 2\frac{1}{2}x = \frac{1}{8}$ **16.** $.001 + .02x - .1 = .002x$

17. $-3(-2 - 2^0x) - (-2) - 2(-2 - 3x) = -2(x + 4)$

Expand:

18. $\dfrac{a^{-3}x}{y^{-3}}\left(\dfrac{xxx^{-2}}{y^{-2}yy} - 3\right)$ **19.** $\dfrac{x^0x^{-2}}{y}\left(x^2y - \dfrac{2x^2y}{x^4}\right)$

Simplify:

20. $\dfrac{x^{-2}(2x^{-3})y^2y^0}{x^{-3}yx^2y^{-7}}$ **21.** $\dfrac{a^0ba^2a^{-1}a}{(a^2b^{-2})^{-3}}$

Simplify by adding like terms:

22. $\dfrac{a}{x} - \dfrac{3a^2y^0x^{-1}}{a} + \dfrac{4x^{-1}}{aa^{-2}}$ **23.** $abc - \dfrac{5a^2c^3}{ab^{-1}c^2} - \dfrac{3}{a^{-1}b^{-1}c^{-1}}$

Evaluate:

24. $a^2(a^0 - ab)$ if $a = -\dfrac{1}{2}, b = \dfrac{3}{4}$ **25.** $x(x^2 - x^2y)$ if $x = -\dfrac{1}{2}, y = \dfrac{1}{2}$

26. $-a^2(a - ab^2)$ if $a = -2, b = -1$

27. $-ab(a - a^2b - b)$ if $a = 2, b = -3$

Simplify:

28. $-2^0(-2^0 - 3^0 - |-2|) - (-2)(-3)$ **29.** $-\dfrac{1}{3^{-2}} - 3^2 - 3$

30. $3 - |-2 - 3 + 5| - 2^0 - 2$

LESSON 13 *Substitution*

13.A

substitution   **In mathematics, equal quantities can always be substituted for one another.** We identify this property by calling it the **substitution axiom.** The substitution axiom is stated in different ways by different authors. Three statements of this axiom that are frequently used are given here.

SUBSTITUTION AXIOM

1. Changing the numeral by which a number is named in an expression does not change the value of the expression.
2. For any numbers a and b, if $a = b$, then a and b may be substituted for each other.
3. If $a = b$, then a may replace b or b may replace a in any statement without changing the truth or falsity of the statement.

The definition given in 1 seems to apply only to individual expressions. Definition 2 is general enough but not sufficiently specific. Definition 3 seems to apply only to statements and not to individual expressions. **We will use the definition below to state formally and exactly the thought that if two expressions have equal value, it is permissible to use either expression.**

SUBSTITUTION AXIOM

If two expressions a and b are of equal value, $a = b$, then a may replace b or b may replace a in another expression without changing the value of the expression. Also a may replace b or b may replace a in any statement without changing the truth or falsity of the statement. Also a may replace b or b may replace a in any equation or inequality without changing the solution set of the equation or inequality.

Thus the substitution axiom applies to expressions, equations and inequalities. We have been using this axiom in evaluation problems when we have replaced the variables with numbers. Now we will use the axiom to solve a system of first-degree linear equations in two unknowns.

example 13.A.1   Use substitution to solve: $\begin{cases} x = y + 5 \\ 3x + 2y = 5 \end{cases}$

solution   We will replace x in the lower equation with $y + 5$ and then solve for y.

$$3(y + 5) + 2y = 5 \qquad \text{substituted}$$
$$3y + 15 + 2y = 5 \qquad \text{multiplied}$$
$$5y = -10 \qquad \text{simplified}$$
$$\boldsymbol{y = -2} \qquad \text{divided}$$

Now we replace y with -2 in the top equation and find that x equals 3.

$$x = y + 5 \qquad \text{top equation}$$
$$x = (-2) + 5 \qquad \text{substituted}$$
$$\boldsymbol{x = 3} \qquad \text{simplified}$$

Thus the solution is the ordered pair **(3, −2)**.

example 13.A.2   Use substitution to solve: $\begin{cases} 3x - y = 11 \\ 2x + 3y = -11 \end{cases}$

solution   We solve the top equation for y and get

$$y = 3x - 11$$

Then we substitute $3x - 11$ for y in the bottom equation and solve.

$$2x + 3y = -11 \qquad \text{bottom equation}$$
$$2x + 3(3x - 11) = -11 \qquad \text{substituted}$$
$$2x + 9x - 33 = -11 \qquad \text{multiplied}$$
$$11x = 22 \qquad \text{simplified}$$
$$\boldsymbol{x = 2} \qquad \text{divided}$$

Now we replace x with 2 in the bottom equation and solve for y.

$$2(2) + 3y = -11 \qquad \text{replaced } x \text{ with 2}$$
$$4 + 3y = -11 \qquad \text{multiplied}$$
$$3y = -15 \qquad \text{simplified}$$
$$\mathbf{y = -5} \qquad \text{divided}$$

Thus our solution is $\mathbf{(2, -5)}$.

example 13.A.3   Use substitution to solve: $\begin{cases} x + y = 20 \\ 5x + 10y = 150 \end{cases}$

solution   We solve the top equation for x.

$$x + y = 20 \quad \longrightarrow \quad x = 20 - y$$

Now we substitute $20 - y$ for x in the bottom equation.

$$5(20 - y) + 10y = 150 \qquad \text{replaced } x \text{ with } 20 - y$$
$$100 - 5y + 10y = 150 \qquad \text{multiplied}$$
$$5y = 50 \qquad \text{simplified}$$
$$\mathbf{y = 10} \qquad \text{divided by 5}$$

And since $x + y = 20$,

$$x = 20 - 10 \quad \longrightarrow \quad \mathbf{x = 10}$$

Thus our solution is $\mathbf{(10, 10)}$.

problem set 13

1. Sixty percent of the boats had blue sails. If 300 boats did not have blue sails, how many boats did have blue sails?

2. Three-sixteenths of all the citizens in Rome thought Caligula was sane. If 93,750 believed he was sane, how many citizens were there in Rome?

3. Find four consecutive even integers such that 3 times the sum of the first and the fourth is 14 greater than 5 times the third.

4. A number was increased by 14 and this sum was tripled. This result was 67 greater than twice the opposite of the number. What was the number?

Solve by using substitution:

*5. $\begin{cases} x = y + 5 \\ 3x + 2y = 5 \end{cases}$

*6. $\begin{cases} 3x - y = 11 \\ 2x + 3y = -11 \end{cases}$

*7. $\begin{cases} x + y = 20 \\ 5x + 10y = 150 \end{cases}$

8. $\begin{cases} x + y = 20 \\ 25x + 10y = 395 \end{cases}$

Add:

9. $p + \dfrac{k}{c} + \dfrac{m}{pc^2}$

10. $4 + \dfrac{2}{a}$

11. $\dfrac{a^2}{k} + k + \dfrac{k}{4}$

12. $m^2 + \dfrac{m}{p} + \dfrac{m}{ap^2}$

13. Find side p.

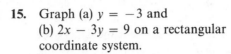

14. Find the distance between $(-2, -2)$ and $(4, -6)$.

15. Graph (a) $y = -3$ and (b) $2x - 3y = 9$ on a rectangular coordinate system.

16. Find the equations of lines (a) and (b).

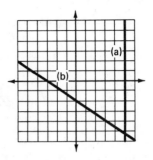

Solve:

17. $3\frac{1}{2}x + 4\frac{1}{3} = 7\frac{2}{9}$

18. $.03 + .03x = .003$

19. $-3^0 - 3^2 - (-2x - |2|) = 7x$

Expand:

20. $\dfrac{x^2a}{3}\left(\dfrac{9a^{-1}}{x^2} - \dfrac{2xa^2}{xa}\right)$

21. $\dfrac{-2a^0p}{m}\left(\dfrac{mp}{a^{-2}} - \dfrac{2a^{-4}a}{p^2m}\right)$

Simplify:

22. $\dfrac{xa^2(x^0a^{-2})^4}{(2x^{-2})^{-2}}$

23. $\dfrac{m^2pxx^{-4}(x^{-2})^2}{(3p^{-2})^{-2}xpx}$

Simplify by adding like terms:

24. $xa - \dfrac{3x^2a^3}{xa^2} + \dfrac{2x}{a^{-1}}$

25. $\dfrac{amp^{-1}}{m^{-1}} - \dfrac{3a^2m^2}{pa} + \dfrac{5pa}{m^2}$

Evaluate:

26. $-x^3 - x^2 - x(a - x)$ if $x = -\dfrac{1}{2}$ and a $= 2$

27. $a^2 - ax(x - ax)$ if $x = -3$ and $a = 4$

Simplify:

28. $-2 - 2^0(-3 - |-2|) - 3^0$

29. $\dfrac{1}{2^{-3}} - \dfrac{2}{-2^{-2}} - \dfrac{1}{(-2)^{-2}} - 2^0$

30. $3^0 - 2(-2) - |-2 - 4^0 - 3|$

LESSON 14 *Equation of a line through two points and with a given slope*

14.A
equation
of a line

The easiest equations to solve are first-degree equations and second-degree equations, and it is fortunate that these are the equations most commonly encountered in everyday life and in science courses. Cubic equations (third-degree) and quartic equations (fourth-degree), such as

$$3x^3 + 2x^2 - 5x + 6 = 0 \quad \text{and} \quad 7x^4 + 2x + 6 = 0$$

are much more difficult to solve. Happily, they seldom occur in real-life problems.

Linear equations in two unknowns are also easy equations to understand, and they are useful equations that occur often. We will concentrate on mastering these equations to prepare for chemistry, physics, and advanced courses in mathematics. To make our task easier, we will always use y to represent the dependent variable and x to represent the independent variable. Also we will always try to use the slope-intercept form of the equation.

$$y = mx + b$$

Other forms of this equation are often used, but we will avoid them until we master the slope-intercept form.

We use the letter m to represent the slope and the letter b to represent the intercept. **When we are given the exact coordinates of two points on a line, we can find the exact value of the slope and the exact value of the intercept.**

example 14.A.1   Find the exact equation of the line that passes through $(-3, 2)$ and $(3, -3)$.

solution   We will graph the line to find the slope.

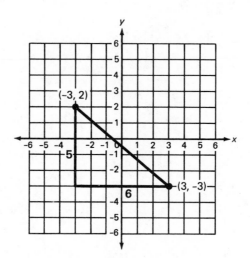

We see that the line has a negative slope whose magnitude is $\frac{5}{6}$. Since the slope is $-\frac{5}{6}$, we can write

$$y = -\frac{5}{6}x + b$$

This slope is exact because we were given the exact coordinates of both points. Now we can use this exact slope and the coordinates of one of the points to find the exact value of the intercept. We can use $(-3, 2)$ or $(3, -3)$ for x and y and solve algebraically for b. To demonstrate that both sets of coordinates will yield the same value for b, we will do the problem twice.

USING $(-3, 2)$:

$$2 = -\frac{5}{6}(-3) + b$$

$$\frac{12}{6} = \frac{15}{6} + b$$

$$-\frac{3}{6} = b$$

$$-\frac{1}{2} = b$$

USING $(3, -3)$:

$$-3 = -\frac{5}{6}(3) + b$$

$$\frac{-18}{6} = \frac{-15}{6} + b$$

$$-\frac{3}{6} = b$$

$$-\frac{1}{2} = b$$

Note that either way we get a value of $-\frac{1}{2}$ for b. Also, note the trick of writing 2 as $\frac{12}{6}$ and writing -3 as $-\frac{18}{6}$ to make the solution easier.

Now since $m = -\frac{5}{6}$ and $b = -\frac{1}{2}$, we can write the equation of the line as

$$y = -\frac{5}{6}x - \frac{1}{2}$$

The slope is exact and the intercept is exact; so, this equation is an exact equation, not an estimate.

example 14.A.2   Find the exact equation of the line that passes through $(-4, 7)$ and that has a slope of $-\frac{3}{5}$.

solution   This time we don't have to graph the points since we are told that the slope is $-\frac{3}{5}$. Thus, we can write

$$y = -\frac{3}{5}x + b$$

Now we can find the exact intercept algebraically by using -4 for x and 7 for y in the equation. Note in the third line how we write 7 as $\frac{35}{5}$ to make the equation easier to solve.

$$y = -\frac{3}{5}x + b \qquad \text{equation}$$

$$7 = -\frac{3}{5}(-4) + b \qquad -4 \text{ for } x \text{ and 7 for } y$$

$$\frac{35}{5} = \frac{12}{5} + b \qquad \text{simplified}$$

$$\frac{23}{5} = b \qquad \text{solved for } b$$

Thus our equation is

$$y = -\frac{3}{5}x + \frac{23}{5}$$

problem set 14

1. When the piper increased his volume, the number of the rats increased 160%. If he ended up with 6578 rats, how many did he have before the volume was increased?

2. To pass the time at Loch Leven, Mary counted sheep. One day she counted 250 percent more than ever before. If the highest previous total had been 4900, how many sheep did she count this time?

3. The number was doubled and then the product was increased by 7. This sum was multiplied by -3, and the result was 9 greater than 3 times the opposite of the number. What was the number?

4. Find three consecutive odd integers such that 4 times the sum of the last two is 2 greater than 10 times the first.

Solve by using substitution:

5. $\begin{cases} 3x - 3y = 21 \\ 2x - y = 12 \end{cases}$

6. $\begin{cases} 4x - y = 22 \\ 2x + 3y = 4 \end{cases}$

7. $\begin{cases} x + y = 28 \\ 5x + 10y = 230 \end{cases}$

8. $\begin{cases} x + y = 22 \\ 100x + 25y = 2050 \end{cases}$

Add:

9. $x + \dfrac{x^2}{y} - \dfrac{3x}{cy^2}$

10. $\dfrac{m}{x} + 4$

11. $1 + \dfrac{a}{x}$

12. $4 + \dfrac{c}{x} - cxy$

13. Find side k.

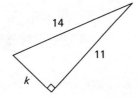

14. Find the distance between $(-3, 2)$ and $(5, 2)$.

15. Graph (a) $x = -4$ and (b) $3y + 2x = 6$ on a rectangular coordinate system.

16. Find the equations of lines (a) and (b).

*17. Find the equation of the line that passes through $(-3, 2)$ and $(3, -3)$.

*18. Find the equation of the line that passes through $(-4, 7)$ and that has a slope of $-\frac{3}{5}$.

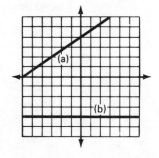

Solve:

19. $2\dfrac{1}{4}x - \dfrac{3}{5} = -1\dfrac{1}{20}$

20. $.005x - .05 = .5$

21. $-2^0 - 3^2 = -2(x - 3^0) - 4(2x - 5)$

22. Expand: $\dfrac{-3^{-2}x}{y}\left(\dfrac{9y^0 x}{-x} - \dfrac{3x}{y}\right)$

23. Simplify: $\dfrac{(-2x)^{-2}xy^0y^2(x)^{-2}}{(x^2y)^{-2}xyxy^{-2}}$

24. $mp^2\left(m^{-1}p^{-2} - \dfrac{4m}{p^2}\right)$

25. Simplify by adding like terms: $-\dfrac{3xy}{m} + \dfrac{7x^2x^{-1}m^{-1}}{y} - \dfrac{8m^{-1}x^{-1}}{x^{-2}y^{-1}}$

Evaluate:

26. $x^2 - y - xy^2$ if $x = -\dfrac{1}{3}, y = \dfrac{1}{2}$

27. $a^2x - a(xa - a)$ if $a = -1, x = -3$

Simplify:

28. $-5^0(-2 - 3^0) - 2^2 - \dfrac{1}{(-2)^{-2}}$ **29.** $4^0(-2 - 3) - 2^0(-2) - 3^0$

30. $5 - |-2 - 3| - 4^0 - 2|-2^0| - 3$

LESSON 15   *Elimination*

15.A
elimination

The addition rule for equations tells us that the same quantity can be added to both sides of an equation without changing the solution to the equation. If we look at the equation

$$x + 2 = 7$$

we know that the solution is 5. We can use the addition rule to get this solution if we add -2 to both sides.

$$
\begin{array}{rr}
x + 2 = & 7 \\
-2 & -2 \\
\hline
x \quad\; = & 5
\end{array}
$$

We can use a similar process to help us solve systems of equations. If we have the system of equations

$$\begin{cases} 2x + y = 11 \\ 3x - 2y = 6 \end{cases}$$

we can eliminate the variable y if we multiply the top equation by 2 and add the equations. Since we eliminate one variable by adding, this method is called the **elimination method** and is sometimes called the **addition method.**

$$
\begin{array}{lllll}
2x + y = 11 & \longrightarrow & (2)^\dagger & \longrightarrow & 4x + 2y = 22 \\
3x - 2y = 6 & \longrightarrow & (1) & \longrightarrow & 3x - 2y = 6 \\
& & & & \hline \\
& & & & 7x \quad\;\;\; = 28 \\
& & & & \mathbf{x = 4}
\end{array}
$$

To use the addition method we had to assume that $3x - 2y$ equaled 6. Since we made this assumption, we can say that we were adding equal quantities to both sides of the top equation.

Now we will use 4 for x in the bottom equation and solve for y.

$$3(4) - 2y = 6 \quad \longrightarrow \quad 12 - 2y = 6 \quad \longrightarrow \quad 6 = 2y \quad \longrightarrow \quad \mathbf{y = 3}$$

---

† The notation $\rightarrow (2) \rightarrow$ has no mathematical meaning. In this book we use this notation to help us remember the number by which we have multiplied.

Since we made an assumption, we must check the values $x = 4$ and $y = 3$ in both of the original equations

TOP EQUATION	BOTTOM EQUATION
$2x + y = 11$	$3x - 2y = 6$
$2(4) + (3) = 11$	$3(4) - 2(3) = 6$
$8 + 3 = 11$	$12 - 6 = 6$
$11 = 11$ Check	$6 = 6$ Check

example 15.A.1   Use elimination to solve: $\begin{cases} 3x + 2y = 23 \\ -2x + 3y = 2 \end{cases}$

solution   We will multiply the top equation by 2 and the bottom equation by 3 and then add the equations.

$$3x + 2y = 23 \longrightarrow (2) \longrightarrow 6x + 4y = 46$$
$$-2x + 3y = 2 \longrightarrow (3) \longrightarrow \underline{-6x + 9y = 6}$$
$$13y = 52 \longrightarrow \mathbf{y = 4}$$

Now we replace y with 4 in the top equation and solve for x.

$$3x + 2(4) = 23 \longrightarrow 3x + 8 = 23 \longrightarrow 3x = 15 \longrightarrow \mathbf{x = 5}$$

Thus, our solution is the ordered pair (5, 4). Now to check:

TOP EQUATION	BOTTOM EQUATION
$3x + 2y = 23$	$-2x + 3y = 2$
$3(5) + 2(4) = 23$	$-2(5) + 3(4) = 2$
$15 + 8 = 23$	$-10 + 12 = 2$
$23 = 23$ Check	$2 = 2$ Check

problem set 15

1. Twenty percent of those interviewed were students. If in his lifetime the announcer interviewed 1400 students, how many people did he interview in all?

2. Sammy slammed everything he could reach. If he saw 1000 and slammed 200, what percent of what he saw did he slam?

3. The pharmacist calculated to make the elixir. She took 5 times the opposite of a number and added it to -7. If the result was 35 less than twice the number, what was the number?

4. Find three consecutive even integers such that -4 times the sum of the first and the third is 12 greater than the product of 7 and the opposite of the second.

Use elimination to solve:

*5. $\begin{cases} 2x + y = 11 \\ 3x - 2y = 6 \end{cases}$ ***6.** $\begin{cases} 3x + 2y = 23 \\ -2x + 3y = 2 \end{cases}$

Use substitution to solve:

7. $\begin{cases} 3x + y = 16 \\ 2x - 3y = -4 \end{cases}$ **8.** $\begin{cases} x + 3y = -9 \\ 5x - 2y = 23 \end{cases}$

Add:

9. $y + \dfrac{x}{a^2} - \dfrac{mx}{3y^2}$

10. $4 - \dfrac{3a}{x}$

11. $\dfrac{7}{p} - 3$

12. $c + \dfrac{c^2}{x} + ac^2$

13. Find side m.

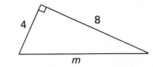

14. Find the distance between $(-4, 2)$ and $(3, -4)$.

15. Graph (a) $y = -3$, (b) $3x - 4y = 8$ on a rectangular coordinate system.

16. Find the equations of lines (a) and (b).

17. Find the equation of the line through $(-2, -2)$ and $(4, 3)$.

18. Find the equation of the line through $(-2, 5)$ that has a slope of $-\frac{1}{7}$.

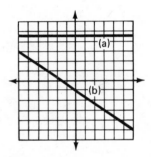

Expand:

19. $\dfrac{-3x^{-1}y^2}{p}\left(\dfrac{-px}{y^2} - \dfrac{3x^2y}{p^{-3}}\right)$

Solve:

20. $3\dfrac{2}{5}x - \dfrac{4}{5} = 7\dfrac{3}{10}$

21. $.07x - .02 = .4$

22. $-3x(-2 - 3^0) = -2^0(-x - 3x^0)$

Simplify:

23. $\dfrac{x^2y^{-2}x^0(y^{-1})}{x^2y^2(x^0)^2}$

24. $\dfrac{px^2yx^2py^{-3}}{(p^0x^{-2})^2}$

Simplify by adding like terms:

25. $\dfrac{8x^2yy}{m^{-2}x^2} + \dfrac{3y^2}{m^{-2}} - \dfrac{5m^2x^{-2}}{x^{-2}y^{-2}}$

Evaluate:

26. $axy - a^2x$ if $a = -\dfrac{1}{2}, x = 3, y = -\dfrac{1}{3}$

27. $a^0(a^0x - ax)$ if $a = -2$ and $x = -3$

Simplify:

28. $-2^0(2^0 - 3^2) - (-2 - 3 - (-2))$

29. $3^0 - \dfrac{3}{3^{-2}} + 2 - 5^0$

30. $3 - |-2 - 3| - |-2| - |-2^0|$

LESSON 16 *Multiplication of polynomials • Division of polynomials*

16.A
multiplication of polynomials

When algebraic expressions are multiplied, each term of one expression is multiplied by every term of the other expression. Then, like terms in the product are added. Some people prefer to use a vertical format, while others prefer a horizontal format. Neither format is more correct than the other, and many people use both formats, sometimes using one and sometimes using the other. The format is not important—the procedure is important. The simplest kind of algebraic expression is a polynomial, so we will demonstrate by multiplying polynomials.

example 16.A.1 Multiply $(2x + 2)(3x^2 - 5x + 2)$.

solution We decide to use the vertical format. People tend to put the longer expression on the top; but this time, just to be different, we will put the longer expression on the bottom.

$$
\begin{array}{r}
2x \ + \ 2 \\
3x^2 \ - \ 5x \ + \ 2 \\
\hline
6x^3 \ + \ 6x^2 \\
- \ 10x^2 \ - \ 10x \\
+ \ 4x \ + \ 4 \\
\hline
\mathbf{6x^3 \ - \ 4x^2 \ - \ 6x \ + \ 4}
\end{array}
$$

example 16.A.2 Multiply $(3x^2 - 2x - 2)(x - 2)$.

solution We will use the horizontal format this time. Either expression can be first. We decide to use the same order as given in the problem.

$$(3x^2 - 2x - 2)(x - 2) = 3x^3 - 6x^2 - 2x^2 + 4x - 2x + 4$$
$$= \mathbf{3x^3 - 8x^2 + 2x + 4}$$

16.B
division of polynomials

When we divide polynomials, we use a procedure that is very similar to that used for long division with real numbers. To demonstrate, we will divide 49 by 12.

$$
\begin{array}{r}
4 \\
12 \overline{)\ 49} \\
48 \\
\hline
1
\end{array}
\qquad \text{so} \qquad \frac{49}{12} = \mathbf{4\frac{1}{12}}
$$

Note that the fraction is formed by writing the remainder over the divisor.

example 16.B.1 Divide $2 + 5x - 2x^2 + 5x^3$ by $-4 + x$.

solution It is customary to begin by writing both expressions in descending powers of the variable. This step is not absolutely necessary, but it helps to keep like terms below each other when doing the division.

$$\begin{array}{r}
5x^2 + 18x \ + 77 \\
x - 4 \overline{\smash{\big)}\ 5x^3 - \ 2x^2 + \ 5x + \quad 2} \\
\underline{5x^3 - 20x^2} \\
18x^2 + \ 5x \\
\underline{18x^2 - 72x} \\
77x + \quad 2 \\
\underline{77x - 308} \\
310
\end{array}$$

Thus, $\dfrac{5x^3 - 2x^2 + 5x + 2}{x - 4}$ equals $5x^2 + 18x + 77 + \dfrac{310}{x - 4}$.

Check: We can check our division by writing the first three terms with denominators of $x - 4$ and adding.

$$\frac{5x^2(x - 4)}{x - 4} + \frac{18x(x - 4)}{x - 4} + \frac{77(x - 4)}{x - 4} + \frac{310}{x - 4}$$

$$= \frac{5x^3 - 20x^2 + 18x^2 - 72x + 77x - 308 + 310}{x - 4}$$

$$= \frac{5x^3 - 2x^2 + 5x + 2}{x - 4} \qquad \text{Check}$$

example 16.B.2   Divide $-6 + x^3$ by $-2 + x$.

solution   Again we rewrite each expression in descending powers of the variable. Also, we insert $0x^2$ and $0x$ into the cubic expression to help us keep the proper spacing.

$$\begin{array}{r}
x^2 + 2x \ + 4 \\
x - 2 \overline{\smash{\big)}\ x^3 + 0x^2 + 0x - 6} \\
\underline{x^3 - 2x^2} \\
2x^2 + 0x \\
\underline{2x^2 - 4x} \\
4x - 6 \\
\underline{4x - 8} \\
2
\end{array}$$

Thus, $\dfrac{x^3 - 6}{x - 2}$ equals $x^2 + 2x + 4 + \dfrac{2}{x - 2}$.

We will check our work by adding:

$$x^2\left(\frac{x - 2}{x - 2}\right) + 2x\left(\frac{x - 2}{x - 2}\right) + 4\left(\frac{x - 2}{x - 2}\right) + \frac{2}{x - 2}$$

$$= \frac{x^3 - 2x^2 + 2x^2 - 4x + 4x - 8 + 2}{x - 2} = \frac{x^3 - 6}{x - 2} \qquad \text{Check}$$

problem set 16

1. Two and one-seventh times the acceptable number had crawled into the space provided. If 900 had crawled in, what was the acceptable number?

2. When the smoke cleared, only .016 of the microbes had disassociated. If 420,000 microbes were in the vial, how many had disassociated?

3. Only 14 percent of the reception guests were uninvited. If 903 of those present had been invited, how many attended the reception?

4. Of the angry crowd 40 percent had been mollified, and now it was necessary to placate the rest. If 8600 had been mollified, how many had to be placated?

Use elimination to solve:

5. $\begin{cases} 4x - 3y = -1 \\ 2x + 5y = 19 \end{cases}$

6. $\begin{cases} 7x - 2y = 13 \\ 4x + 7y = 40 \end{cases}$

Use substitution to solve:

7. $\begin{cases} 3x + y = 2 \\ 2x - 5y = 7 \end{cases}$

8. $\begin{cases} x - 3y = 4 \\ 4x - 7y = 16 \end{cases}$

*9. Multiply: $(2x + 2)(3x^2 - 5x + 2)$

Divide and check:

*10. $2 + 5x - 2x^2 + 5x^3$ by $-4 + x$ *11. $-6 + x^3$ by $-2 + x$

Add:

12. $2 + \dfrac{a}{2x^2}$

13. $\dfrac{4}{cx} + c - \dfrac{3}{4c^2x}$

14. $-3x + \dfrac{2x^2}{yp}$

15. Find p.

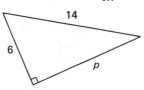

16. Find the distance between $(-3, -5)$ and $(-1, 2)$.

17. Graph (a) $x = -2\frac{1}{2}$ and, (b) $x - 3y = 6$ on a rectangular coordinate system.

18. Find the equations of lines (a) and (b).

19. Find the equation of the line through $(-2, -4)$ and $(5, -6)$.

20. Find the equation of the line through $(-4, 1)$ and that has a slope of $\frac{3}{5}$.

Solve:

21. $3\dfrac{1}{2}x - \dfrac{1}{5} = 3\dfrac{3}{20}$

22. $.05m - .05 = .5$

23. $-4(-x - 2) - 3(-2 - 4) = -4^0(x - 2)$

24. Expand: $\dfrac{3^{-2}x^{-2}y}{z}\left(\dfrac{18x^2z}{y} - \dfrac{3xyz^2}{x}\right)$

25. Simplify: $\dfrac{3^{-2}x^0(x^{-2}y^4)^{-3}}{xxy^2y^0(y^3)^{-4}}$

26. Simplify by adding like terms: $\dfrac{3x^2xxy}{y^{-4}} - \dfrac{2x^2}{(x^{-1})^2y^{-5}} - \dfrac{7xxyx^2}{(y^{-2})^2}$

27. Evaluate $p^2 - xp - x(p - x)$ if $x = -2$ and $p = \dfrac{1}{2}$.

Simplify:

28. $-2^0(-2 - 3) - (-2)^0 - 2[-1 - (-3 - 7) - 2(-5 + 3)]$

29. $2^0 - \dfrac{27}{3^2} - 3^1 - 3^0 - |-3^0 - 2| - (-3)$

30. $-2 - (-3)^{-3}$

LESSON 17   *Subscripted variables*

In first-degree equations in two unknowns, we often use the letter x to represent the independent variable and the letter y to represent the dependent variable. Thus, if we are given the linear equation

$$2x + 3y = 6$$

and are asked to graph the equation, we would first solve the equation for y and get

$$y = -\frac{2}{3}x + 2$$

This is the familiar slope-intercept form of the equation, and we can now graph the equation by choosing values for x (the independent variable) and seeing which values of y (the dependent variable) that this equation pairs with the values of x. When we graph an equation, we always use the horizontal number line for x and the vertical number line for y. Thus, the graph of the equation is as follows:

x	0	-6	6
y	2	6	-2

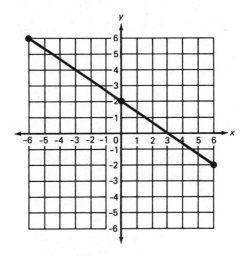

We note that the intercept is $+2$ and that the sign of the slope is negative. If we visualize a triangle, we see that the slope is $-\frac{2}{3}$.

When we work word problems, the use of x and y as variables is often not helpful because it is difficult to remember what these letters stand for. But if we use subscripted variables, there is no difficulty in recognizing the variable. The large letter gives the general description, and the small letter supplies more specific information. To use subscripted variables to say that the total number of nickels and dimes was 40, we could write

$$N_N + N_D = 40$$

Here N_N means number of nickels and N_D means number of dimes.

example 17.A.1   Solve the system: $\begin{cases} N_N + N_D = 40 \\ 5N_N + 10N_D = 250 \end{cases}$

solution We decide to use elimination, and we multiply the top equation by -5 and then add.

$$-5N_N - 5N_D = -200$$
$$5N_N + 10N_D = 250$$
$$\overline{5N_D = 50}$$

$$N_D = 10$$

and since $N_N + N_D = 40$, we must find that $N_N = 30$.

example 17.A.2 Solve the following system of equations.

$$R_M T_M + R_W T_W = 260 \qquad R_M = 40 \qquad R_W = 60 \qquad T_M + T_W = 5$$

solution These equations come from a word problem in which a man walked part of a distance and rode a motorcycle the rest of the distance. Thus, R_M and T_M stand for the rate and time on the motorcycle, and R_W and T_W stand for the rate and time walking. To solve, we substitute 40 for R_M and 60 for R_W and get

$$40T_M + 60T_W = 260$$

Now we rearrange the last equation.

$$T_M + T_W = 5 \quad\longrightarrow\quad T_M = 5 - T_W$$

Next we substitute $5 - T_W$ for T_M and then solve.

$$40(5 - T_W) + 60T_W = 260 \qquad \text{substituted}$$
$$200 - 40T_W + 60T_W = 260 \qquad \text{multiplied}$$
$$20T_W = 60 \qquad \text{simplified}$$
$$T_W = 3 \qquad \text{divided}$$

Now since $T_M + T_W = 5$, we conclude that $T_M = 2$.

**problem
set 17**

1. Only $\frac{2}{5}$ of the students doubted. If 600 were nondoubters, how many students were there in all?

2. Only $\frac{7}{13}$ of the teachers believed. If 210 teachers believed, how many teachers were there?

3. The number of skeptics increased 260 percent overnight. If 400 were skeptical at sunset, how many were skeptical at sunrise?

4. Find three consecutive integers such that 7 times the sum of the last two is 109 greater than 10 times the first.

*5. Use elimination: $\begin{cases} N_N + N_D = 40 \\ 5N_N + 10N_D = 250 \end{cases}$

6. Use substitution: $\begin{cases} N_p + N_D = 50 \\ N_p + 10N_D = 140 \end{cases}$

7. Multiply: $(2x + 3)(2x^2 + 2x + 2)$

8. Divide $3x^3 - 2$ by $x + 1$ and then check.

Solve for all unknown variables:

*9. $R_M T_M + R_W T_W = 260$, $R_M = 40$, $R_W = 60$, $T_M + T_W = 5$

10. $R_E T_E = R_W T_W$, $R_E = 200$, $R_W = 250$, $9 - T_E = T_W$

11. $R_M T_M = R_R T_R$, $R_M = 8$, $R_R = 2$, $5 - T_M = T_R$

Add:

12. $4x + \dfrac{3x}{a}$ **13.** $4 + \dfrac{p}{x}$ **14.** $-\dfrac{2x}{y} - cx + \dfrac{7x^2y}{np^2}$

15. Find side x: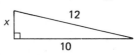

16. Find the distance between $(2, 4)$ and $(6, -2)$.

17. Graph: (a) $y = -3$
(b) $3x - 5y = 10$

18. Find the equations of lines (a) and (b).

19. Find the equation of the line through $(4, 2)$ and $(6, -3)$.

20. Find the equation of the line through $(-3, 5)$ that has a slope of $-\frac{2}{7}$.

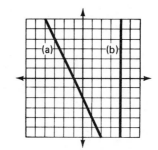

Solve:

21. $2\dfrac{1}{5}x - 3\dfrac{1}{4} = \dfrac{7}{20}$ **22.** $.003 - .03 + .3x = 3.3$

23. $3^0(2x - 5) + (-x - 5) = -3(x^0 - 2)$

24. Expand: $\dfrac{x^0y^{-2}x}{x^3y}\left(\dfrac{x^2y}{m} - \dfrac{3x^4y^2}{m^{-2}}\right)$ **25.** Simplify: $\dfrac{2^{-3}x^0(x^2)}{x^{-3}xy^{-3}y}$

26. Simplify by adding like terms: $\dfrac{xy}{y^{-2}} - \dfrac{3x^4y^4}{x^3y} + \dfrac{7xy^{-2}}{xy^{-3}}$

27. Evaluate $xy - x(x - y^0)$ if $x = 2$ and $y = -\dfrac{1}{2}$.

Simplify:

28. $-(-3 - 2) + 4(-2) + \dfrac{1}{-2^{-3}}$

29. $-|-2 - 7| - |-2 - 4| - 3|-2 + 7|$

30. $-2^{-3} - (-2)^{-3}$

LESSON 18 *Ratio word problems*

18.A
ratio word problems

A **ratio** is a fraction. If we write the fraction

$$\frac{3}{4}$$

we can say that we have written the ratio of 3 to 4. All of the following ratios designate the same number and thus are equal ratios.

$$\frac{3}{4} \qquad \frac{6}{8} \qquad \frac{300}{400} \qquad \frac{15}{20} \qquad \frac{27}{36} \qquad \frac{111}{148}$$

An equation or other statement which indicates that two ratios are equal is called a **proportion.** Thus, we say that

$$\frac{3}{4} = \frac{15}{20}$$

is a proportion. We note that cross products of equal ratios are equal, as we show in the following example.

$$\frac{3}{4} =\!\!< \frac{15}{20} \qquad \begin{array}{c} 4 \times 15 \\ \\ 3 \times 20 \end{array} \qquad \begin{array}{c} 4 \times 15 = 3 \times 20 \\ \\ 60 = 60 \qquad \text{True} \end{array}$$

We can solve proportions that contain an unknown by setting the cross products equal and then dividing to complete the solution. To solve

$$\frac{4}{3} = \frac{5}{k}$$

we first set the cross products equal and get

$$4k = 15$$

We finish by dividing by 4.

$$\frac{4k}{4} = \frac{15}{4} \longrightarrow \boldsymbol{k = \frac{15}{4}}$$

When we set the cross products equal, we say that we have *cross multiplied.*

In some ratio word problems the word ratio is used but in other ratio word problems the word ratio is not used and it is necessary to realize that a constant ratio is implied.

example 18.A.1 The ratio of Arabians to mixed breeds in the herd was 2 to 17. If there were 380 horses in the herd, how many were Arabians?

solution We are given 2 Arabians and 17 mixed breeds for a total of 19. Thus we can write

$$\begin{array}{cl} 2 & \text{Arabians} \\ 17 & \text{mixed breeds} \\ 19 & \text{total} \end{array}$$

Now by looking at what we have written, we see that three proportions are indicated.

$$\text{(a)} \quad \frac{2}{17} = \frac{A}{M} \qquad\qquad \text{(b)} \quad \frac{2}{19} = \frac{A}{T} \qquad\qquad \text{(c)} \quad \frac{17}{19} = \frac{M}{T}$$

We are given a total of 380 and are asked for the number of Arabians, so we will use proportion (b) and replace T with 380 and solve.

$$\frac{2}{19} = \frac{A}{380} \longrightarrow 2 \cdot 380 = 19A \longrightarrow 760 = 19A \longrightarrow A = \mathbf{40}$$

example 18.A.2 It took 600 kilograms of sulfur to make 3000 kilograms of the new compound. How many kilograms of other materials would be required to make 4000 kilograms of the new compound?

solution This wording is typical of real-world ratio problems. The word ratio is not used in the statement of the problem, but the statement implies that the ratio of kilograms of sulfur to kilograms of the compound is constant. We will call the rest *NS* for "not sulfur." If 600 kilograms (kg) was sulfur, then 2400 kg must have been "not sulfur."

$$600 \quad S$$
$$2400 \quad NS$$
$$3000 \quad P$$

Thus, the three implied equations are:

(a) $\dfrac{600}{2400} = \dfrac{S}{NS}$ (b) $\dfrac{600}{3000} = \dfrac{S}{P}$ (c) $\dfrac{2400}{3000} = \dfrac{NS}{P}$

Since we are given 4000 kg of the compound and are asked for the amount of "not sulfur," we will use equation (c).

$$\frac{2400}{3000} = \frac{NS}{4000} \longrightarrow (2400)(4000) = 3000NS \longrightarrow \mathbf{3200\ kg = NS}$$

problem set 18

*1. The ratio of Arabians to mixed breeds in the herd was 2 to 17. If there were 380 horses in the herd, how many were Arabians?

*2. It took 600 kilograms of sulfur to make 3000 kilograms of the new compound. How many kilograms of other materials would be required to make 4000 kilograms of the new compound?

3. The law of the land displeased 27 percent of the natives. If 54,000 natives were displeased, how many natives were there?

4. The percentage of nonagenarians in the population was only .004 percent. If there were 40 nonagenarians in the village, what was the total population?

5. Use elimination: $\begin{cases} N_D + N_Q = 200 \\ 10N_D + 25N_Q = 2750 \end{cases}$

6. Use substitution: $\begin{cases} N_P + N_D = 30 \\ N_P + 10N_D = 291 \end{cases}$

7. Multiply: $(2x + 4)(3x^2 - 2x - 10)$

8. Divide $5x^3 - 1$ by $x - 2$ and check.

Solve for all unknown variables:

9. $R_F T_F = R_S T_S,\ T_S = 6,\ T_F = 5,\ R_F - 16 = R_S$

10. $R_M T_M = R_R T_R,\ R_M = 8,\ R_R = 2,\ T_R = 5 - T_M$

11. $R_G T_G + R_B T_B = 100,\ R_G = 4,\ R_B = 10,\ T_B = T_G + 3$

Add:

12. $7xyz + \dfrac{1}{xyz}$

13. $1 + \dfrac{a}{x}$

14. $-\dfrac{3x}{y} - c + \dfrac{7c}{xy^3}$

15. Find side *m*.

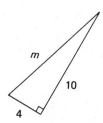

16. Find the distance between $(-3, 5)$ and $(4, -2)$.

17. Graph: (a) $y = 2$ (b) $y = 2x$

18. Find the equations of lines (a) and (b).

19. Find the equation of the line through $(-3, 5)$ and $(4, -2)$.

20. Find the equation of the line that has a slope of $\frac{5}{3}$ and that passes through $(4, -2)$.

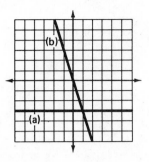

Solve:

21. $4\frac{1}{6}x - 2\frac{1}{12} = -\frac{5}{24}$

22. $1.04x - .02x + 2.2 = 6.28$

23. $-2[(x - 2) - 4x - 3] = -(-4 - 2x)$

24. Expand: $\dfrac{3x^0 y^{-2}}{z^2}\left(\dfrac{2xyz^{-1}}{p} - \dfrac{y^2}{(3x)^{-2}}\right)$

25. Simplify: $\dfrac{a^0 xy^2(a^2)^{-2}(xy^{-2})^2}{ax^0(y^{-2})^2}$

26. Simplify by adding like terms: $-\dfrac{2x^2}{y^2} - \dfrac{5y^{-2}p^0}{x^{-2}} + \dfrac{7xxy}{y^3}$

27. Evaluate $ax - a(a - x)$ if $a = -\dfrac{1}{2}$ and $x = \dfrac{1}{4}$.

Simplify:

28. $-2^0(-3^0 - 2) - |-2 + 5| - 7(-2 - 5^0)$

29. $-2|-2 - 5| + (-3)| - 2(-2) - 3| + 7$

30. $-\dfrac{1}{(-2)^{-3}} + \dfrac{3}{-3^{-2}}$

LESSON 19 *Value word problems*

19.A
value word problems

Value word problems are a genre of problems in which one or more statements in the problem are about the value of items. The total value of one kind is the value of one item times the number of items of that kind. For example, every nickel has a value of 5 cents; thus, N_N nickels would have a value of $5N_N$ cents. In a like manner, the total value of N_D dimes would be $10N_D$ cents. We will begin value problems with problems that contain two statements of equality that lead to two equations in two unknowns. Either substitution or elimination can be used to solve these equations.

example 19.A.1   Karamagu had 50 nickels and dimes whose value was $4. How many of each kind of coin did he have?

solution   If we use dollars as the basic value in the problem, we get the two equations

$$N_N + N_D = 50$$

$$.05N_N + .1N_D = 4$$

We can avoid the decimal fractions at the outset if we use pennies as our basic value in the problem. If we do this, our equations are

$$N_N + N_D = 50$$

$$5N_N + 10N_D = 400$$

We will solve the top equation for N_N and use substitution.

$5(50 - N_D) + 10N_D = 400$	substituted $(50 - N_D)$ for N_N
$250 - 5N_D + 10N_D = 400$	multiplied
$250 + 5N_D = 400$	simplified
$5N_D = 150$	added -250 to both sides
$N_D = 30$	divided

Since $N_N + N_D = 50$, it follows that $N_N = 20$.

example 19.A.2   The fishmonger sold codfish for 6 pence each and mussels for 1 pence each. If Harriet bought a total of 26 items and spent 86 pence, how many codfish did she buy?

solution   The two equations are

$$N_C + N_M = 26$$

$$6N_C + 1N_M = 86$$

We will use elimination. We will multiply the top equation by -1 and then add the equations.

$$-N_C - N_M = -26$$
$$\underline{6N_C + N_M = \quad 86}$$
$$5N_C \qquad = \quad 60$$
$$N_C \qquad = 12$$

Thus $N_M = 26 - 12 = 14$.

**problem
set 19**

*1.   Karamagu had 50 nickels and dimes whose value was $4. How many of each kind of coin did he have?

*2.   The fishmonger sold codfish for 6 pence each and mussels for 1 pence each. If Harriet bought a total of 26 items and spent 86 pence, how many codfish did she buy?

3.   Sulfur is mixed with other chemicals to make sulfuric acid. If it takes 16 tons of of sulfur to make 49 tons of sulfuric acid, how many tons of other constituents are needed to make 294 tons of acid? (Hint: Sulfuric acid is the total.)

4.   Nineteen percent of the nitric acid was used in the experiment. If 1134 liters remained, how much nitric acid had been available in the beginning? How much had been used?

5.   A number was multiplied by -7 and this product was increased by -7. This sum was doubled and the result was 4 greater than 5 times the opposite of the number. What was the number?

6.   Use elimination: $\begin{cases} 5x + 25y = -160 \\ -3x + 2y = -23 \end{cases}$

7. Multiply: $(x^2 - 2)(x^3 - 2x^2 - 2x + 4)$

8. Divide $-3x^3 - 2$ by $-2 + x$ and then check.

Solve for all unknown variables:

9. $R_H T_H + R_S T_S = 180$, $R_H = 70$, $R_S = 20$, $T_H = T_S$

10. $R_F T_F = R_S T_S$, $T_S = 6$, $T_F = 5$, $R_F - 10 = R_S$

11. $R_M T_M = R_R T_R$, $R_M = 8$, $R_R = 2$, $T_R = 5 - T_M$

Add:

12. $4 + \dfrac{3x^2}{7y^2 z}$

13. $\dfrac{a}{2x^2} - \dfrac{b}{x^2 y} - c$

14. $\dfrac{1}{m^2 p} + m$

15. Find side z:

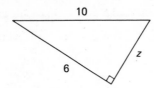

16. Find the distance between $(-3, -5)$ and $(2, 4)$.

17. Graph: (a) $x = -3$
    (b) $5x - 3y = 9$

18. Find the equations of lines (a) and (b).

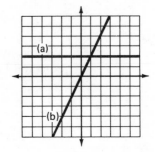

19. Find the equation of the line through $(-3, -5)$ and $(2, 4)$.

20. Find the equation of the line that has a slope of $\frac{2}{7}$ and that passes through $(-3, -5)$.

Solve:

21. $3\frac{1}{4}p - \dfrac{7}{3} = -\dfrac{5}{12}$

22. $.07x - .7 = -7.7$

23. $-[(-2 - 6)(-2) - 2] = -2[(x - 2)2 - (2x - 3)3]$

24. Expand: $-\dfrac{3x^0 y^2}{p^{-2}}\left(\dfrac{x}{p^2 y^2} - \dfrac{3xy}{x^2 p}\right)$

25. Simplify: $\dfrac{x^2 x^0 (xx^{-3})}{x(y^0 y^2)^{-2}}$

26. Simplify by adding like terms: $\dfrac{2x^2 a}{y} - \dfrac{3xy^{-2}a}{y^{-1}x^{-1}} + \dfrac{4x^0 y^{-1}}{x^{-2}a^{-1}}$

27. Evaluate $m - m^2 y(m - y)$ if $m = -\dfrac{1}{3}$ and $y = \dfrac{1}{6}$.

Simplify:

28. $-4^0[(-5 + 2) - |-3 + 7| - 3(-1 - (-2)^0)]$

29. $-1|-1 - 3| + |-2| - 2(-3) + 4$

30. $-\dfrac{-1}{-(-2)^{-2}} + \dfrac{3}{-(-2)^{-2}}$

LESSON 20 *Simplification of radicals •*
Line parallel to a given line

20.A
simplification of radicals

Handheld calculators can often be used to demonstrate properties that would otherwise be difficult to demonstrate. The square root of 10 equals the square root of 2 times the square root of 5.

$$\sqrt{10} = \sqrt{2}\sqrt{5}$$

If we use our calculator to find decimal approximations of these numbers, we get

$$\sqrt{10} = 3.1622777 \quad \text{and} \quad \sqrt{2}\sqrt{5} = (1.4142136)(2.236068)$$

and if we use the calculator to multiply the two numbers on the right, we get a product of

$$3.1622778$$

This number differs from our approximation of $\sqrt{10}$ on the left above by only 1 unit in the seventh decimal place. This exercise helps some to believe that $\sqrt{10}$ really does equal $\sqrt{2}\sqrt{5}$.

The rule that explains this can be proved and is really a theorem that we call the **product-of-square-roots theorem.** The proof of this theorem is so straightforward, however, that some people often use the word property for this rule and call it the product property of radicals. This rule is also applicable to higher-order roots such as cube roots, fourth roots, etc. For now, we restrict its use to square roots and state the rule formally here.

PRODUCT-OF-SQUARE-ROOTS THEOREM

If *m* and *n* are nonnegative real numbers, then

$$\sqrt{m}\sqrt{n} = \sqrt{mn} \quad \text{and} \quad \sqrt{mn} = \sqrt{m}\sqrt{n}$$

This theorem can be generalized to the product of any number of factors and we say that the square root of any product may be written as the product of the square roots of the factors of the product. For example,

$$\sqrt{2 \cdot 5 \cdot 5} \quad \text{can be written as} \quad \sqrt{2}\sqrt{5}\sqrt{5}$$

and

$$\sqrt{3 \cdot 3 \cdot 3 \cdot 5} \quad \text{can be written as} \quad \sqrt{3}\sqrt{3}\sqrt{3}\sqrt{5}$$

We will use this theorem in the following problems to help us simplify radical expressions.

example 20.A.1

Simplify: $3\sqrt{50} - 5\sqrt{200}$

solution

As the first step we write each radical as a product of prime factors.

$$3\sqrt{5 \cdot 5 \cdot 2} - 5\sqrt{2 \cdot 2 \cdot 2 \cdot 5 \cdot 5}$$

Now we use the product-of-square-roots theorem to write

$$3\sqrt{5}\sqrt{5}\sqrt{2} - 5\sqrt{2}\sqrt{2}\sqrt{2}\sqrt{5}\sqrt{5}$$

We finish by remembering that $\sqrt{5}\sqrt{5} = 5$ and $\sqrt{2}\sqrt{2} = 2$. We get

$$15\sqrt{2} - 50\sqrt{2} = -35\sqrt{2}$$

example 20.A.2   Simplify: $3\sqrt{2} \cdot 4\sqrt{12} \cdot 2\sqrt{3}$

solution   We multiply and get

$$24\sqrt{72}$$

Now we simplify $\sqrt{72}$ and multiply by 24.

$$24\sqrt{2}\sqrt{2}\sqrt{2}\sqrt{3}\sqrt{3} = 144\sqrt{2}$$

example 20.A.3   Simplify: $4\sqrt{3}(2\sqrt{3} - \sqrt{6})$

solution   Two multiplications are indicated. We do these and get

$$(4\sqrt{3})(2\sqrt{3}) + (4\sqrt{3})(-\sqrt{6}) = 24 - 12\sqrt{2}$$

20.B
line parallel to a given line

Parallel lines are lines that have the same slopes but have different intercepts. If we look at the equations

$$y = \frac{1}{2}x + 2 \qquad y = \frac{1}{2}x - 1 \qquad y = \frac{1}{2}x - 4$$

we see that they all have a slope of $+\frac{1}{2}$, but that each line has a different intercept.

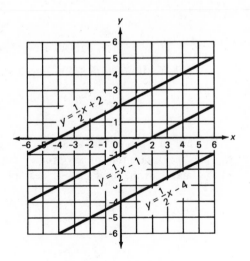

If we are asked to find the equation of a line that is parallel to one of these lines, its slope must also be $+\frac{1}{2}$. The only thing we need to find is the intercept.

example 20.B.1   Find the equation of the line that is parallel to the line $2y - x = 2$ and that passes through the point $(3, -1)$.

solution   If we write the equation of the given line in slope-intercept form, we get

$$y = \frac{1}{2}x + 1$$

The slope of the new line must also be $\frac{1}{2}$ if it is to be parallel, to this line, so we have

$$y = \frac{1}{2}x + b$$

Now we use the coordinates 3 and -1 for x and y and solve algebraically for b.

$$-1 = \frac{1}{2}(3) + b \qquad \text{substituted}$$

$$-\frac{2}{2} = \frac{3}{2} + b \qquad \text{simplified}$$

$$-\frac{5}{2} = b \qquad \text{solved}$$

Thus the intercept of the new line is $-\frac{5}{2}$ and the equation of the new line is

$$y = \frac{1}{2}x - \frac{5}{2}$$

problem set 20

1. The formula required that 20 kilograms of carbon be used to get 160 kilograms of the compound. How many kilograms of other components was needed to make 640 kilograms of the compound?

2. Fewer than half of the performers were virtuosos. In fact only $\frac{3}{10}$ were in this category. If 28 were not virtuosos, how many performers were there? How many virtuosos were there?

3. The expensive ones cost $7 each while the worthless ones sold for only $2 each. Monongahela spent $111 and bought three more expensive ones than worthless ones. How many of each kind did she buy?

4. Only 40 percent of the combustibles burned. If 240 tons burned, how many tons did not burn?

5. Find four consecutive even integers such that -4 times the sum of the first and fourth is 6 greater than the opposite of the sum of the second and third.

6. Use substitution: $\begin{cases} y - 2x = 8 \\ 2y + 2x = 40 \end{cases}$

7. Divide $-2x^3 - x + 2$ by $-1 + x$ and check.

8. Solve for the unknown variables: $R_G T_G + R_B T_B = 100$, $R_G = 4$, $R_B = 10$, $T_B = T_G + 3$

Simplify:

*9. $3\sqrt{50} - 5\sqrt{200}$

*10. $3\sqrt{2} \cdot 4\sqrt{12} \cdot 2\sqrt{3}$

*11. $4\sqrt{3}(2\sqrt{3} - \sqrt{6})$

12. $5\sqrt{5}(2\sqrt{5} - 3\sqrt{10})$

Add:

13. $7 + \dfrac{x}{y}$

14. $\dfrac{m^2}{x^2 a} + \dfrac{5}{ax} - \dfrac{m}{a}$

15. $\dfrac{a}{x^2} - a - \dfrac{3x}{2a^4}$

16. Find side z.

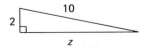

17. Graph: (a) $y = -2$
    (b) $y = -2x - 2$

18. Find the equations of lines (a) and (b).

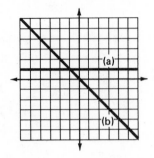

*19. Find the equation of the line that is parallel to the line $2y - x = 2$ and that passes through the point $(3, -1)$.

20. Find the equation of the line that has a slope of $-\frac{1}{12}$ and that passes through $(2, 5)$.

Solve:

21. $3\frac{1}{12} - \frac{1}{6}x = 2\frac{1}{24}$

22. $-.04x - x - .2x = 6.2$

23. $-2[(-3)(-2 - x) + 2(x - 3)] = -2x$

24. Expand: $\dfrac{-x^0 y}{y^2 y^{-2}}\left(\dfrac{x}{y} - \dfrac{3xy^2}{y^3}\right)$

25. Simplify: $\dfrac{(-2x^0)^{-3}x^2 y y^0 y^3}{(xy^{-2})^2(-2x^{-1})^{-2}}$

26. Simplify by adding like terms: $-\dfrac{3x}{a} + \dfrac{(a^0)a}{aax^{-1}} - \dfrac{5x^{-1}}{a}$

27. Evaluate $k - kx(k^2 - x)$ if $k = -\dfrac{1}{4}$ and $x = \dfrac{1}{8}$.

Simplify:

28. $-2^0[(-7 + 3) - |-2 + 9| - 2(-2^0 - (-5))]$

29. $-2^0|-3 - 7| + |(-5^0)| - 2(-5) + 2$

30. $-\dfrac{1}{-(-3)^{-3}} + \dfrac{2}{-(-3)^{-3}}$

LESSON 21   *Scientific notation • Two statements of equality*

21.A
scientific notation

We like to use the decimal system because base 10 numerals have a special advantage. When we multiply a number by a power of 10, the sole effect is to move the decimal point the number of places left or right that equals the power of 10. Multiplying by a positive integral power of 10 moves the decimal point to the right. To demonstrate, we will multiply 412.036 by 10^2 and by 10^4.

$$412.036 \times 10^2 = 41{,}203.6$$

$$412.036 \times 10^4 = 4{,}120{,}360.$$

Multiplying by 10 to a negative integral power moves the decimal point to the left. If we multiply the same number by 10^{-2} and 10^{-4}, we move the decimal point two places to the left and four places to the left as we show here.

$$412.036 \times 10^{-2} = 4.12036$$

$$412.036 \times 10^{-4} = .0412036$$

We find this property of decimal numerals very useful when we deal with very large and very small numbers. **We can put the decimal point anywhere we please as long as we follow the numeral with**

$$\times 10^{\pm b}$$

and use this notation to tell where the decimal point really is. Thus

$$304.162 \qquad .304162 \times 10^3 \qquad 304{,}162 \times 10^{-3}$$

all represent the same number. In the last two notations, the decimal point is incorrectly placed, but that is all right because the notations

$$\times 10^3 \qquad \text{and} \qquad \times 10^{-3}$$

tell us where it should be placed.

example 21.A.1 Simplify: $\dfrac{(.0003 \times 10^{-6})(4000)}{(.006 \times 10^{15})(2000 \times 10^4)}$

solution We begin by writing all four numbers in scientific notation. Then we multiply and divide as indicated.

$$\frac{(3 \times 10^{-10})(4 \times 10^3)}{(6 \times 10^{12})(2 \times 10^7)} = \frac{3 \cdot 4}{6 \cdot 2} \times \frac{10^{-7}}{10^{19}} = \mathbf{1 \times 10^{-26}}$$

21.B
two statements of equality

Thus far, our study of problems that require two equations for their solution has been restricted to coin problems about nickels and dimes and to similar problems about the values of items that are not coins. Now we will begin our investigation of other types of problems that contain two statements of equality. The experience gained with value problems should make these new problems easy to understand.

example 21.B.1 The ratio of two numbers is 3 to 4 and their sum is 84. What are the numbers?

solution The two equations are

$$\text{(a)} \quad \frac{N}{D} = \frac{3}{4} \quad \text{and} \quad \text{(b)} \quad N + D = 84$$

We will cross multiply in (a), and then solve (b) for D and substitute into (a).

$$\text{(a)} \quad \frac{N}{D} = \frac{3}{4} \quad \longrightarrow \quad 4N = 3D \qquad \text{(b)} \quad N + D = 84 \quad \longrightarrow \quad D = 84 - N$$

$$\begin{array}{ll} 4N = 3(84 - N) & \text{substituted} \\ 4N = 252 - 3N & \text{multiplied} \\ 7N = 252 & \text{simplified} \\ N = \mathbf{36} & \text{divided} \end{array}$$

Since $N + D = 84$, then $D = 84 - 36 = \mathbf{48}$.

example 21.B.2 The sum of two numbers is 128 and their difference is 44. What are the numbers?

solution A little thought in choosing the variables is often helpful. Here we have two numbers. One is greater than the other, so we will use S to represent the small number and L to represent the large number. The equations are

$$\text{(a)} \quad L + S = 128 \quad \text{and} \quad \text{(b)} \quad L - S = 44$$

We will solve the equations by using elimination.

$$
\begin{array}{ll}
\text{(a)} & L + S = 128 \\
\text{(b)} & \underline{L - S = \;\;44} \\
& 2L \qquad = 172
\end{array}
$$

$$L = 86$$

and thus

$$S = 128 - 86 \quad \longrightarrow \quad S = 42$$

problem set 21

*1. The ratio of two numbers is 3 to 4 and their sum is 84. What are the numbers?

*2. The sum of two numbers is 128 and their difference is 44. What are the numbers?

3. It took 900 kg of acetelene to make 2400 kg of the compound. How many kilograms of other components was required to make 3600 kg of the compound?

4. Twenty percent of the nitrogen combined with the other elements. If 740 kg of nitrogen did not combine, how much did combine?

5. The nickels and dimes had a value of $5.75. If there were 70 coins in all, how many were nickels?

6. Find three consecutive integers such that -5 times the sum of the first and third is 24 greater than the opposite of the second times 4.

7. Use elimination: $\begin{cases} 8y - 3x = 22 \\ 2y + 4x = 34 \end{cases}$

8. Multiply: $(4x^2 - 2x + 2)(-3 + 2x)$

9. Solve for the unknown variables: $R_K T_K = R_N T_N$, $R_K = 6$, $R_N = 3$, $T_N - T_K = 8$

Simplify:

10. $3\sqrt{200} - 5\sqrt{18} + 7\sqrt{50}$

11. $2\sqrt{3} \cdot 4\sqrt{2} \cdot 5\sqrt{6}$

12. $2\sqrt{2}(6\sqrt{6} - 3\sqrt{2})$

Add:

13. $4x + \dfrac{1}{p}$

14. $\dfrac{m^2}{a^2 x^2} - \dfrac{3}{ax} - \dfrac{m}{x}$

*15. Simplify: $\dfrac{(.0003 \times 10^{-6})(4000)}{(.006 \times 10^{15})(2000 \times 10^4)}$

16. Find the distance between $(-3, -5)$ and $(4, -5)$.

17. Graph: (a) $x = 4$, and
(b) $4x - 3y = 12$

18. Find the equations of lines (a) and (b).

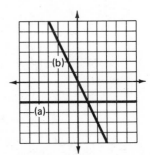

19. Find the equation of the line through $(-3, -5)$ and $(4, -5)$.

20. Find the equation of the line that passes through the point $(2, 2)$ and is parallel to $y = -\frac{3}{7}x + 4$.

Solve:

21. $5\frac{1}{3}x - \frac{3}{4} = \frac{7}{8}$ **22.** $.03(x - 4) = .02(x + 6)$

23. $-[(-4 - 1)(-3) - 6^0] = -2[(y - 4)3 - (2y - 5)]$

24. Expand: $\dfrac{5y^0p^{-2}}{x^2}\left(-\dfrac{2p^2}{x^2} - \dfrac{p^2}{x}\right)$ **25.** Simplify: $\dfrac{x^2(y^{-2})^2(y^0)^2}{(2x^2y^3)^{-2}}$

26. Simplify by adding like terms: $\dfrac{3x^2a}{x} - 5xa + \dfrac{7x^{-2}}{x^{-3}a^{-1}}$

27. Evaluate $x(x - ax)x$ if $x = -\dfrac{1}{2}$ and $a = \dfrac{1}{3}$.

Simplify:

28. $-7^0[(-2 + 3) - |-4 + 3|]$

29. $-3^0|-2 - 5| + |(-2^0)| - 3(-4) + 2$

30. $-\dfrac{1}{-2^{-2}} - \dfrac{1}{-(-2)^0}$

LESSON 22 *Uniform motion problems–equal distances*

In this lesson we begin our study of the solution of uniform motion word problems. It is very probable that we will never encounter one of these problems in a science course or in everyday life. We study these problems because solving them will allow us to develop problem solving skills that will be useful in solving many problems that will be encountered in science courses and in everyday life.

   Uniform motion problems involve statements about people or things that move at a constant velocity (speed)[†] or at an average velocity. The uniform motion problems on which we will concentrate will contain four statements about things that are equal or that differ by a specified amount. Each of these statements can be turned into an equation. We find that these equations can be most easily written if we use four variables. In this book, we will use subscripted variables, but this is not necessary. It is often entertaining to use inventiveness and originality when choosing variables. For example, we could use either

$$R_W \qquad \text{or} \qquad R_{RW} \qquad \text{or} \qquad RRW$$

to stand for the rate that Ruby walked.

   To solve uniform motion problems, we will write equations about rate or velocity, equations about time, and equations about distances traveled. Since the distance equations are the most difficult to write, we will consider these equations to be the key equations and we will write the distance equation first. We will always draw a diagram

---

[†] In science courses a distinction is made between the words speed and velocity. In this book the words are synonymous.

to help us write the distance equation. If both people or things travel the same distance, the diagrams will have one of the two forms shown here.

$$D_1 = D_2$$

$$\text{so } R_1 T_1 = R_2 T_2$$

The distance equation is shown on the right. We always replace D_1 with $R_1 T_1$ and D_2 with $R_2 T_2$ since rate times time equals distance.

$$R_1 T_1 = D_1 \quad \text{and} \quad R_2 T_2 = D_2$$

example 22.A.1   Roger made the trip on Sunday and Judy made the same trip on Monday. Roger traveled at 12 miles per hour. Judy traveled at 20 miles per hour, so her time was 2 hours less than Roger's time. How far did they travel?

solution   We begin by drawing the distance diagram and writing the distance equation.

$$D_R = D_J \quad \text{so} \quad R_R T_R = R_J T_J$$

Next we write the rate equations and the time equation:

$$R_R = 12 \quad R_J = 20 \quad T_J + 2 = T_R$$

To solve, in the distance equation we replace R_R with 12, R_J with 20, and T_R with $T_J + 2$.

$$12(T_J + 2) = 20T_J \quad \text{substituted}$$
$$12T_J + 24 = 20T_J \quad \text{multiplied}$$
$$24 = 8T_J \quad \text{simplified}$$
$$\mathbf{3 = T_J} \quad \text{divided}$$

Now since Judy's time was 3 hours, Roger's was 3 hours + 2 hours = 5 hours. To find the distances traveled, we multiply R_J by T_J and R_R by T_R.

$$D_J = R_J T_J \qquad\qquad D_R = R_R T_R$$
$$D_J = (20)(3) = \mathbf{60 \text{ miles}} \qquad D_R = (12)(5) = \mathbf{60 \text{ miles}}$$

example 22.A.2   Br'er Rabbit hopped off toward the briar patch at 10 kilometers per hour (kph) at 10 a.m. At noon Br'er Wolf began the chase from the same starting point. If he caught Br'er Rabbit at 2 p.m., how fast did he run?

solution   They traveled the same distances. The distance diagram and distance equation are as follows:

$$D_R = D_W \quad \text{so} \quad R_R T_R = R_W T_W$$

One rate and both times were given.

$$R_R = 10 \quad T_R = 4 \quad T_W = 2$$

To solve, we substitute the last three equations into the first.

$$(10)(4) = R_W(2) \qquad \text{substituted}$$
$$40 = 2R_W \qquad \text{multiplied}$$
$$\mathbf{20 \text{ kph} = R_W} \qquad \text{divided}$$

problem set 22

*1. Roger made the trip on Sunday and Judy made the same trip on Monday. Roger traveled at 12 miles per hour. Judy traveled at 20 miles per hour, so her time was 2 hours less than Roger's time. How far did they travel?

*2. Br'er Rabbit hopped off toward the briar patch at 10 kilometers per hour (kph) at 10 a.m. At noon Br'er Wolf began the chase from the same starting point. If he caught Br'er Rabbit at 2 p.m., how fast did he run?

3. The ratio of two numbers is 7 to 5. The sum of the numbers is 144. What are the numbers?

4. The federal tax was $500 more than the state tax. If the sum of the taxes was $6900, what was the amount of the federal tax?

5. Huckleberries were $5 a peck while whortleberries cost $13 a peck. Hortense spent $109 for a total of 9 pecks. How many pecks of whortleberries did she buy?

6. The chemist found that 30 grams of iodine was required to make 600 grams of the solution. How many grams of other things was required to make 5000 grams of the solution?

7. Use substitution: $\begin{cases} 5x + y = 24 \\ 7x - 2y = 20 \end{cases}$

8. Divide $x^3 - 4x + 2$ by $-1 + x$ and check.

Simplify:

9. $3\sqrt{2} \cdot 4\sqrt{3} \cdot 2\sqrt{6}\sqrt{2}$

10. $2\sqrt{27} - 3\sqrt{75}$

11. $3\sqrt{2}(2\sqrt{2} - \sqrt{6})$

12. $2\sqrt{3}(5\sqrt{3} - 2\sqrt{6})$

Add:

13. $2 + \dfrac{1}{x}$

14. $\dfrac{5x^2}{y} + p^2 - \dfrac{3x}{py}$

15. Simplify: $\dfrac{(.0035 \times 10^{-4})(200 \times 10^6)}{(700 \times 10^5)(.00005)}$

16. Graph: (a) $x = -5$, and (b) $2x - y = 4$

17. Find the equations of lines (a) and (b).

18. Find the equation of the line that passes through $(6, 0)$ and $(-3, -3)$.

19. Find the distance between $(6, 0)$ and $(-3, -3)$.

20. Find the equation of the line that has a slope of $\frac{2}{5}$ and passes through $(3, -5)$.

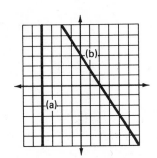

Solve:

21. $3\frac{2}{5}x - 4\frac{1}{10}x = 2\frac{1}{4}$ **22.** $.02(p - 2) = .03(2p - 6)$

23. $-[(-3 - 6)(-1^0) - 6^0] = -4[(x - 3)2]$

24. Expand: $\dfrac{xy^{-2}}{z^0 p}\left(\dfrac{py^2}{x} - \dfrac{3xy^{-4}}{py}\right)$ **25.** Simplify: $\dfrac{(x^{-2}yp)^{-3}(x^0 yp)^2}{(2x^2)^{-2}}$

26. Simplify by adding like terms: $-\dfrac{3x^2 y}{xx} + \dfrac{2x^{-2}x^4}{y^{-1}x^2} - \dfrac{5xy^2}{xy}$

27. Evaluate $ya(y - a)y$ if $y = -\dfrac{1}{2}$ and $a = \dfrac{1}{5}$

Simplify:

28. $-2^0[(5 - 7 - 2) - |-2 - 7| - 2^0]$

29. $-4^0|3 - 7| + |(-2^0)| - 2(-2) + 2$ **30.** $\dfrac{-3}{-3^{-2}} - \dfrac{2}{-2^{-3}}$

LESSON 23 *Graphical solutions*

23.A
graphical solutions

We have been solving two equations in two unknowns by using either substitution or elimination. These equations can also be solved by graphing. To do this, we graph both of the equations and visually determine the coordinates of the point where the lines cross. The answer we get is an approximation because we must estimate the crossing point. The advantage of the graphical method is that we can see what we are doing. If a more exact answer is required, it can be obtained by using either substitution or elimination.

example 23.A.1   Solve this system by graphing. Check the solution by using either substitution or elimination.

$$\begin{cases} 3y - 2x = 6 & \text{(a)} \\ y + x = -1 & \text{(b)} \end{cases}$$

solution   To graph, we first solve each equation for y to get the slope-intercept form. Then we graph the equations.

(a) $y = \dfrac{2}{3}x + 2$

(b) $y = -x - 1$

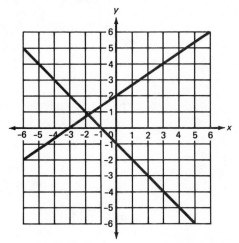

The lines appear to cross at $(-1.8, .75)$. We will check this by using the original equations and using elimination.

$$
\begin{array}{lll}
3y - 2x = 6 & \longrightarrow (1) \longrightarrow & 3y - 2x = 6 \\
y + x = -1 & \longrightarrow (2) \longrightarrow & \underline{2y + 2x = -2} \\
& & 5y = 4 \longrightarrow y = \dfrac{4}{5}
\end{array}
$$

We use $y = \frac{4}{5}$ in equation (b) to solve for x.

$$\left(\frac{4}{5}\right) + x = -1 \qquad\qquad \text{substitution}$$

$$x = -1 - \frac{4}{5} \qquad\qquad \text{added } -\frac{4}{5} \text{ to both sides}$$

$$x = -\frac{9}{5} \qquad\qquad\quad \text{simplified}$$

Thus the exact solution is $\left(-\dfrac{9}{5}, \dfrac{4}{5}\right)$.

example 23.A.2   Solve this system by graphing. Check the solution by using either substitution or elimination.

$$\begin{cases} y - 2x = 2 & \text{(a)} \\ y = -3 & \text{(b)} \end{cases}$$

solution   We rewrite equation (a) in slope-intercept form and graph both lines.

(a) $y = 2x + 2$

(b) $y = -3$

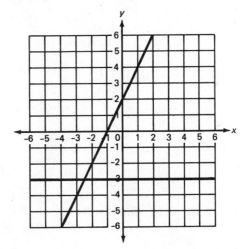

This time the graphical solution appears to be $(-2.6, -3)$. We will use substitution to check. We use -3 for y in equation (a).

$$(-3) = 2x + 2 \qquad\qquad \text{substitution}$$

$$-5 = 2x \qquad\qquad\quad \text{added } -2 \text{ to both sides}$$

$$-\frac{5}{2} = x \qquad\qquad\quad \text{divided}$$

Thus the exact solution is $\left(-\dfrac{5}{2}, -3\right)$.

problem set 23

1. At the sound of the explosion Mary began to run north at 600 feet per minute. Jim regained consciousness 4 minutes later and began to run after Mary at 800 feet per minute. How long did it take Jim to catch Mary?

2. The fast freight made the trip in 10 hours while the slow freight took 12 hours for the same trip. How long was the trip if the fast freight was 10 kph faster than the slow freight?

3. The fraction had a value of $\frac{11}{12}$. The sum of the numerator and the denominator was 230. What was the fraction?

4. The vivandière sold viands and sandwiches. If she sold 300 total and 50 more viands than sandwiches, how many of each did she sell?

5. The value of the quarters and nickels was $5. If there were 40 more nickels than quarters, how many coins of each type were there?

6. Twenty percent of the compound was copper sulfate. If there was 400 tons of the compound in the warehouse, how much was not copper sulfate?

7. Use elimination: $\begin{cases} 5x + 2y = 70 \\ 3x - 2y = 10 \end{cases}$

8. Multiply: $(3x^3 - 2x)(2x^2 - x - 4)$

Simplify:

9. $4\sqrt{3} \cdot 3\sqrt{12} \cdot 2\sqrt{3}$

10. $3\sqrt{75} - 4\sqrt{48}$

11. $4\sqrt{3}(5\sqrt{2} - 2\sqrt{3})$

12. $2\sqrt{5}(5\sqrt{5} - 3\sqrt{15})$

Add:

13. $3xy^2m + \dfrac{4}{x}$

14. $\dfrac{5x^2}{pm} - 4 + \dfrac{c}{p^2m}$

15. Simplify: $\dfrac{(.00003)(.006 \times 10^{-6})}{(1800 \times 10^{15})(100,000)}$

Solve by graphing. Then get an exact solution by using substitution or elimination.

*16. $\begin{cases} 3y - 2x = 6 \\ y + x = -1 \end{cases}$

*17. $\begin{cases} y - 2x = 2 \\ y = -3 \end{cases}$

18. Find the equations of lines (a) and (b).

19. Find the equation of the line that passes through $(0, 0)$ and $(4, 2)$.

20. Find the distance between $(0, 0)$ and $(4, 2)$.

21. Find the equation of the line that has a slope of $-\frac{3}{8}$ and passes through $(4, 4)$.

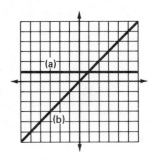

Solve:

22. $3\frac{1}{4}x + 5\frac{1}{3} = 2\frac{1}{6}$

23. $.3(2p - 4) = .1(p + 3)$

24. $-[(-4 - 6)(-4^0) - 2] = -2^0[(x - 4)]$

25. Expand: $xy^{-2}\left(\dfrac{x^0y^2}{x} - \dfrac{3x^0y^2}{x^2}\right)$

26. Simplify: $\dfrac{(-3xy)^{-2}x^2y^2}{(x^{-3})^2yx^2}$

27. Simplify by adding like terms: $\dfrac{3m}{x} - \dfrac{2x^{-1}}{m^0 m^{-1}} + \dfrac{5x^2 m^2}{x^3 m}$

28. Evaluate $xy - (x - y)$ if $x = -\dfrac{1}{2}$ and $y = 2$.

Simplify:

29. $-3^0[(-2 - 3 + 8) - |-3 - 5| - 5^0]$

30. $\dfrac{-2^0}{-2^2} - \dfrac{(-2)^0}{(-2)^{-3}}$

LESSON 24 *Fractional equations*

24.A
fractional
equations

We have noted that the solution of equations that contain decimal numbers can be simplified if we first multiply every term in the equation by a judiciously selected power of 10. For instance, the solution of this equation

$$.03x + .02 = .2$$

can be made much easier if we first multiply every term by 10^2 (or 100 if you prefer). If we do this we get

$$3x + 2 = 20 \qquad \text{multiplied by } 100$$
$$3x = 18 \qquad \text{added } -2$$
$$x = 6 \qquad \text{divided}$$

There is also a procedure that can be used to facilitate the solution of equations that contain fractions. **We can eliminate the fractions by multiplying every numerator by the least common multiple (LCM) of the denominators. Then the denominators will divide into the LCMs, leaving the equation with integral coefficients.**

example 24.A.1 Solve: $4\dfrac{1}{2}x - \dfrac{3}{5} = -1\dfrac{1}{4}$

solution As the first step we will write the mixed numbers as improper fractions.

$$\frac{9}{2}x - \frac{3}{5} = -\frac{5}{4}$$

Next we multiply every numerator by 20, the LCM of the denominators.

$$20 \cdot \frac{9}{2}x - 20 \cdot \frac{3}{5} = 20\left(-\frac{5}{4}\right)$$

Now we divide each denominator into the LCM and simplify to get an equation with integral coefficients.

$$90x - 12 = -25 \qquad \text{simplified}$$
$$90x = -13 \qquad \text{added } +12$$
$$x = -\frac{13}{90} \qquad \text{divided}$$

example 24.A.2   Solve: $\dfrac{4x + 2}{5} - \dfrac{3}{2} = \dfrac{1}{3}$

solution   We begin by multiplying every numerator by 30, the LCM of the denominators.

$$30\,\frac{(4x + 2)}{5} - 30 \cdot \frac{3}{2} = 30 \cdot \frac{1}{3}$$

Now we divide, simplify and solve.

$$24x + 12 - 45 = 10 \qquad \text{divided}$$
$$24x - 33 = 10 \qquad \text{simplified}$$
$$24x = 43 \qquad \text{added } +33$$
$$x = \frac{43}{24} \qquad \text{divided}$$

example 24.A.3   Solve: $\dfrac{3x}{2} + \dfrac{8 - 4x}{7} = 3$

We can eliminate the denominators if we multiply every numerator by 14.

$$14 \cdot \frac{3x}{2} + 14\,\frac{(8 - 4x)}{7} = 14(3)$$

Now we divide, simplify and solve

$$21x + 16 - 8x = 42 \qquad \text{divided}$$
$$13x + 16 = 42 \qquad \text{simplified}$$
$$13x = 26 \qquad \text{added } -16$$
$$x = 2 \qquad \text{divided}$$

problem set 24

1. The fast freight left at noon and got there at 6 p.m. The next day the slow freight made the same trip in 8 hours. What was the speed of the slow freight if the speed of the fast freight was 60 miles per hour?

2. Henry can ride his horse at 4 miles per hour and get to the battlefield on time. If he stops for 1 hour to make a speech to his troops, he must ride at 5 miles per hour to get to the battlefield on time. How far is it to the battlefield?

3. The fraction had a value of $\frac{5}{7}$. Amy R found that the sum of the numerator and the denominator was 120. What was the fraction?

4. Charles and Nelle picked 173 quarts of berries. How many did each pick if Charles picked 11 more quarts than Nelle picked?

5. Raisins were $700 for a measure while plums cost $900 a measure. Jenai and Marvin bought 50 measures and spent $41,000. How many measures of raisins did they buy?

6. It took 700 kilograms of potassium to make 49,000 kilograms of the new fertilizer. How many kilograms of other components did Gerd have to use to make 4200 kilograms of the new fertilizer?

7. Use substitution: $\begin{cases} 7x + 9y = 119 \\ 2x + y = 23 \end{cases}$

8. Divide $3x^3 - 3$ by $-2 - x$ and check.

Simplify:

9. $4\sqrt{3} \cdot 5\sqrt{2} \cdot 6\sqrt{12}$

10. $4\sqrt{63} - 3\sqrt{28}$

11. $3\sqrt{2}(5\sqrt{2} - 6\sqrt{12})$

12. $2\sqrt{2}(5\sqrt{10} - 3\sqrt{2})$

Add:

13. $4m^2yp + \dfrac{6}{m^2y}$

14. $\dfrac{k^2}{2p} + c - \dfrac{4}{p^2c}$

15. Simplify: $\dfrac{(.0007 \times 10^{-23})(4000 \times 10^6)}{(.00004)(7,000,000)}$

16. Solve by graphing. Then get an exact solution by using substitution or elimination.
$$\begin{cases} 3x + 2y = 12 \\ 5x - 4y = 8 \end{cases}$$

17. Find the equations of lines (a) and (b).

18. Find the equation of the line that passes through $(-2, 5)$ and $(-6, -3)$.

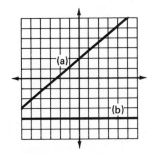

19. Find side k.

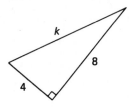

20. Find the equation of the line through $(5, -3)$ that has a slope of $\frac{2}{9}$.

Solve:

*21. $\dfrac{4x + 2}{5} - \dfrac{3}{2} = \dfrac{1}{3}$

*22. $\dfrac{3x}{2} + \dfrac{8 - 4x}{7} = 3$

*23. $4\dfrac{1}{2}x - \dfrac{3}{5} = -1\dfrac{1}{4}$

24. $.07 - .003x + .2 = 1.02$

25. $-2[x - (-2 - 4^0) - 3] + [(x - 2)(-3)] = -x$

26. Expand: $\dfrac{xy^{-2}}{p^2}\left(\dfrac{p^2y^2}{x} + \dfrac{5x^2y^3}{p^{-2}}\right)$

27. Simplify: $\dfrac{4x^2y^{-2}x}{(-2x^0)^{-3}}$

28. Simplify by adding like terms: $\dfrac{3p^2x^2}{xy} - \dfrac{5ppy^{-1}}{x^{-1}} - \dfrac{5p^2x^2x^2}{x^{-3}y}$

29. Simplify: $-2^0[(-1 - 3^0 - 2^2) - |-2^0 - 5|]$

30. Evaluate $xa(ax - a)$ if $a = -\dfrac{1}{3}$ and $x = \dfrac{1}{2}$.

LESSON 25 *Monomial factoring · Cancellation*

25.A
monomial factoring

The Latin word for *maker* or *doer* is the word *factor*. From this came the Old French word *facteur*, and this led to the Middle English word *factour*. In modern English we use the same spelling as did the ancient Romans, and in mathematics the meaning is very close to the Latin meaning of maker or doer. If we multiply 3 and 2, we get 6.

$$3 \cdot 2 = 6$$

We say that 3 and 2 are **factors** of 6 because we can make 6 by multiplying these two numbers. If we can make an expression by multiplying two or more other expressions, we say that each of the expressions that is multiplied is a factor of the final expression.

> A factor is one of two or more expressions that is multiplied to form a product.

Thus, since we multiply $4x$ and $ax + y$ to get $4ax^2 + 4xy$,

$$4x(ax + y) = 4ax^2 + 4xy$$

we say that $4x$ and $ax + y$ are factors of $4ax^2 + 4xy$. **When we write a sum as the product of factors, we say that we are factoring.**

example 25.A.1 Factor $4ax^2 + 4xy$.

solution We have been given a sum and asked to write this expression as a product. We begin by recording an empty set of parentheses.

$$(\qquad)$$

Now in front of the parentheses, we write the greatest common factor of the original terms.

$$4x(\qquad)$$

Now we decide what must be recorded inside the parentheses so that when we multiply by $4x$ the result will be $4ax^2 + 4xy$, our original expression. The correct entry is $ax + y$.

$$\mathbf{4x(ax + y)}$$

We say that we have factored the expression by factoring out $4x$.

example 25.A.2 Factor $4x^2p^2k - 6k^2p^4x$.

solution We begin by writing a set of parentheses preceded by the greatest common factor of both terms.

$$2xp^2k(\qquad)$$

Next we decide the proper entry for the parentheses so that the product will be our original expression.

$$\mathbf{2xp^2k(2x - 3kp^2)}$$

example 25.A.3 Factor $4x^2y - 2xy + 10xy^2$.

solution   Again we begin with empty parentheses preceded by the greatest common factor of all the terms.

$$2xy(\quad)$$

We finish by finding the proper entry for the parentheses.

$$2xy(2x - 1 + 5y)$$

25.B
cancellation

We say that multiplication and division are inverse operations because they undo each other. For example, if we begin with 7 and then multiply and divide by 2

$$\frac{7 \cdot 2}{2} = 7$$

the result is 7, the number with which we began. This procedure can sometimes be used to simplify expressions that appear rather formidable.

$$\frac{(4.062)(3.0176)}{4.062}$$

Here we can see that the result will be 3.0176 because multiplication and division by 4.062 undo each other. We can draw a line through each of these numbers and say that we have canceled.

$$\frac{(4.062)(3.0176)}{(4.062)} = 3.0176$$

The only time cancellation is possible is when both the numerator and denominator are products of factors and each contains one or more common factors. Thus we may cancel the 4's in the following expression

$$\frac{4(x + 2)}{4} = \frac{4(x + 2)}{4(1)} = x + 2$$

because the 4 in the denominator can be thought of as the product of 4 and 1. No cancellation is possible here

$$\frac{x + 4}{4}$$

because addition and division are not inverse operations. We cannot cancel

$$\frac{4x + 4}{4} \qquad \text{incorrect}$$

in the present form; but if we factor out a 4 in the numerator, cancellation is possible.

$$\frac{4x + 4}{4} = \frac{4(x + 1)}{4} = x + 1$$

example 25.B.1   Cancel if possible: (a) $\dfrac{6x^3 + x^2}{x^2}$ (b) $\dfrac{6a^2 + x^2}{x^2}$

solution   We can factor and cancel in (a), but no cancellation is possible in (b).

$$\text{(a)} \quad \frac{6x^3 + x^2}{x^2} = \frac{x^2(6x + 1)}{x^2} = 6x + 1$$

problem set 25

1. The Mary Sue stayed on the Grand Banks for 38 days and salted down 14,440 pounds of codfish. How long would she have had to stay to salt down 36,100 pounds?

2. Brown Bear made the trip in 40 hours. Flying Fish took only 30 hours to make the trip because his speed exceeded that of Brown Bear by 6 kilometers per hour. How long was the trip?   ·

3. There were 5 times as many boys as girls at the party. Also, the number of boys was 100 less than 15 times the number of girls. How many boys and girls came to the party?

4. The day was a concatenation of disasters. If the ratio of minor disasters to major disasters was 5 to 2 and there were 980 disasters in all, how many were minor?

5. The machine broke open and the quarters and half-dollars fell to the floor. The mysophobe grimaced but still retrieved them because there were 200 coins whose value was $75. How many of each kind were there?

6. Seventy percent of the compound was sodium chloride. If 660 grams of other chemicals was used, what was the total weight of the compound?

7. Divide $x^3 - 2$ by $x - 5$ and then check.

Factor:

*8. $4x^2p^2k - 6k^2p^4x$

*9. $4x^2y - 2xy + 10xy^2$

10. $x^2y^3m^5 + 12x^3ym^4 - 3x^2y^2m^2$

11. $16m^2p^3y - 8y^4mp^3 + 4m^2p^2y^2$

12. $x^3y^2z^3 + x^2yz^2 - 3x^3yz$

13. $p^5x^3 + p^4x^2 - p^3x$

Simplify:

14. $2\sqrt{3} \cdot 3\sqrt{6} \cdot 5\sqrt{12}$

15. $6\sqrt{18} + 5\sqrt{8} - 3\sqrt{50}$

16. $2\sqrt{5}(3\sqrt{15} - 2\sqrt{5})$

17. $9\sqrt{6}(3\sqrt{6} - 2\sqrt{3})$

Add:

18. $a + \dfrac{a}{b}$

19. $\dfrac{ax^2}{m^2p} - c + \dfrac{2}{m}$

20. Simplify: $\dfrac{(38{,}000 \times 10^3)(300 \times 10^{-4})}{.00019 \times 10^{-5}}$

21. Solve by graphing and then get an exact solution by using either substitution or elimination.
$$\begin{cases} 3x - 2y = 10 \\ y = -\dfrac{1}{2} \end{cases}$$

22. Find the equation of the line that passes through the point (3, 5) and is parallel to the line $y = \dfrac{1}{6}x - 2$.

Solve:

23. $\dfrac{x+1}{4} - \dfrac{3}{2} = \dfrac{2x-9}{10}$

24. $\dfrac{4x-8}{5} + \dfrac{2x-4}{2} = 9$

25. $\dfrac{n+3}{6} - \dfrac{1}{3} = \dfrac{2n-2}{5}$

26. $\dfrac{5x+3}{2} - \dfrac{3}{4} = \dfrac{5}{2}$

*27. Cancel if possible: $\dfrac{6x^3 + x^2}{x^2}$

28. Expand: $\dfrac{x^2 y^{-2}}{z^2}\left(\dfrac{z^2}{y^2(2x^{-2})^{-1}} - \dfrac{4x^2 y^0}{z^{-2}}\right)$

29. Simplify: $-3^0[(-2 - 4 - 2^2 - 2^0) - |-3 - 2|]$

30. Evaluate $km(m^2 k - k)$ if $k = \dfrac{1}{3}$ and $m = -\dfrac{1}{4}$.

LESSON 26   *Trinomial factoring*

26.A
trinomial factoring

When we factor, we undo a multiplication. Thus to get rules for factoring trinomials, we look at the pattern that develops when we multiply binomials.

(a)
$$\begin{array}{r} x + 2 \\ x - 5 \\ \hline x^2 + 2x \\ -5x - 10 \\ \hline x^2 - 3x - 10 \end{array}$$

(b)
$$\begin{array}{r} b + 4 \\ b - 1 \\ \hline b^2 + 4b \\ -b - 4 \\ \hline b^2 + 3b - 4 \end{array}$$

(c)
$$\begin{array}{r} x - 3 \\ x - 4 \\ \hline x^2 - 3x \\ -4x + 12 \\ \hline x^2 - 7x + 12 \end{array}$$

In each of the multiplications shown here, we note that in the trinomial products,

(a) The first term is the product of the first terms of the binomials.
(b) The last term is the product of the last terms of the binomials.
(c) The coefficient of the middle term is the sum of the last terms of the binomials.

Thus, to factor a trinomial whose lead coefficient is 1, we need to find two numbers whose sum and product meet the requirements of (b) and (c).

example 26.A.1   Factor $x^2 + 6 - 5x$.

solution   It is helpful if we begin by writing the trinomial in descending powers of the variable.

$$x^2 - 5x + 6$$

Now we write two sets of parentheses with x as the first entry in each set.

$$(x \quad)(x \quad)$$

The second entries are the two numbers whose sum is -5 and whose product is $+6$. The numbers are -3 and -2. Thus, our answer is

$$(x - 3)(x - 2)$$

example 26.A.2   Factor $-x^2 + 5x + 14$.

solution   When the x^2 term has a negative coefficient, it is helpful to first factor out a negative quantity. Here we will factor out -1.

$$(-1)(x^2 - 5x - 14)$$

Next we factor the trinomial and get

$$(-1)(x - 7)(x + 2)$$

We could leave the answer in this form, but many people like to multiply the (-1) by one of the other factors. If we do this, we can get

$$(-x + 7)(x + 2) \quad \text{or} \quad (x - 7)(-x - 2)$$

either of which is acceptable.

example 26.A.3   Factor $24 + 6x - 3x^2$.

solution   First we rearrange the trinomial in descending powers of the variable.

$$-3x^2 + 6x + 24$$

Next we factor out the common negative factor, which is -3.

$$-3(x^2 - 2x - 8)$$

Then we factor the trinomial

$$(-3)(x - 4)(x + 2)$$

example 26.A.4   Factor $32x^2 + 12x^3 - 2x^4$.

solution   We begin by writing the trinomial in descending powers of the variable.

$$-2x^4 + 12x^3 + 32x^2$$

Let's hope that this trinomial has a common factor, for we don't know how to factor quartic equations yet. The trinomial does have a common factor which is $-2x^2$. Thus we factor out $-2x^2$ and get

$$-2x^2(x^2 - 6x - 16)$$

Then we finish by factoring the trinomial

$$-2x^2(x + 2)(x - 8)$$

problem set 26

1. The Silver Arrow made the trip in 20 hours. The Orange Blossom Special made the same trip in only 8 hours because its speed was 60 mph greater than that of the Silver Arrow. How far was the trip?

2. The number of blues was 50 greater than 5 times the number of reds. Also, the number of reds was 210 less than the number of blues. How many were red and how many were blue?

3. The bag contained used $5 bills and used $10 bills. If there were 1200 bills and their total value was $7000, how many were $5 bills and how many were $10 bills?

4. Fourteen percent were belligerent while the rest were merely eristic. If 4300 were eristic, how many were belligerent?

5. The compound contained 140 grams of calcium and its total weight was 1960 grams. If the total weight had been 2240 grams, how many grams of substances other than calcium was there?

6. Ten percent of the chemicals were nonabsorbent. If 7290 chemicals were absorbent, how many chemicals were being considered?

7. Divide $x^4 - 2$ by $x + 1$ and check.

Factor the common factor:

8. $35x^7y^5m - 7x^5m^2y^2 + 14y^7x^4m^2$ **9.** $6x^2ym^5 - 2x^2ym + 4xym$

10. $4x^2y^4p^6 - 2xp^5y^7 + 8x^4p^5y^5$

Factor completely. Always factor the greatest common factor (GCF) as the first step.

*11. $x^2 + 6 - 5x$ ***12.** $-x^2 + 5x + 14$

*13. $24 + 6x - 3x^2$ ***14.** $32x^2 + 12x^3 - 2x^4$

15. $-2ab + abx + abx^2$

Simplify:

16. $\dfrac{6x^2y - xy}{xy}$ **17.** $3\sqrt{2} \cdot 2\sqrt{6} \cdot 3\sqrt{6}$

18. $-3\sqrt{12} + 5\sqrt{27} - 8\sqrt{25}$ **19.** $3\sqrt{2}(5\sqrt{3} - 2\sqrt{2})$

Add:

20. $1 + \dfrac{x^2y}{m}$ **21.** $\dfrac{a^2m}{x^2} - x^2 - \dfrac{x^2}{c}$

22.   Simplify: $\dfrac{(3000 \times 10^{-14})(.00008)}{(.0002 \times 10^5)(200{,}000)}$

23.   Solve by graphing and then get an exact solution by using either substitution or elimination.
$$\begin{cases} 2y - x = 6 \\ y - 2x = -3 \end{cases}$$

24.   Find the distance between $(4, -2)$ and $(-3, -5)$.

Solve:

25. $\dfrac{2x}{3} + \dfrac{6 - 3x}{4} = 7$ **26.** $\dfrac{4 + x}{2} - \dfrac{1}{3} = 6$

27. $\dfrac{4 - x}{3} + \dfrac{1}{6} = 5$ **28.** $-2x - \dfrac{3^0 - x}{2} + \dfrac{x - 5^0}{7} = 2$

29.   Simplify: $\dfrac{x^2(yz^0)^{-2}}{(-2x^2y)^{-3}}$

30.   Evaluate $-\dfrac{1}{-(-3)^{-2}} + \dfrac{1}{-2^{-3}}(a - ba)$ if $a = -\dfrac{1}{2}$ and $b = 3$.

LESSON 27   *Rational expressions*

27.A
rational expressions   A **ratio** is another name for a fraction. Thus

$$\dfrac{3}{4}$$

can be described by saying three-fourths or by saying the ratio of 3 to 4. Whenever an algebraic expression is written in the form of a fraction, such as

$$\frac{a}{b} \qquad \frac{a^2x}{4y} \qquad \frac{(m+p)^2}{k} \qquad \frac{3xy-4}{7p}$$

we can call it a **rational expression.** We have been adding rational expressions by using the least common multiple of the denominators as the new denominator. **When one of the denominators has a factor that is a sum, then this sum is one of the factors of the least common multiple.**

example 27.A.1   Add: $\dfrac{4}{x} + \dfrac{6}{x+a}$

solution   The least common multiple is always a product of factors. The least common multiple of these denominators is $x(x+a)$.

$$\frac{}{x(x+a)} + \frac{}{x(x+a)}$$

Next we use the denominator-numerator–same-quantity theorem to help determine the new numerators. Then we add.

$$\frac{4(x+a)}{x(x+a)} + \frac{6x}{x(x+a)} = \frac{10x+4a}{x(x+a)}$$

example 27.A.2   Add: $\dfrac{x+2}{x+4} + 6 + \dfrac{2}{x^2}$

solution   The least common multiple is $x^2(x+4)$. We will use the same steps we used in the last example. First we write the new denominators.

$$\frac{}{x^2(x+4)} + \frac{}{x^2(x+4)} + \frac{}{x^2(x+4)}$$

Now we determine the new numerators.

$$\frac{x^2(x+2)}{x^2(x+4)} + \frac{6x^2(x+4)}{x^2(x+4)} + \frac{2(x+4)}{x^2(x+4)}$$

Next we expand the numerators,

$$\frac{x^3 + 2x^2 + 6x^3 + 24x^2 + 2x + 8}{x^2(x+4)}$$

and then simplify as the last step.

$$\frac{7x^3 + 26x^2 + 2x + 8}{x^2(x+4)}$$

example 27.A.3   Add: $\dfrac{x+2}{x^2+4x+3} - \dfrac{1}{x(x+1)}$

solution   Algebra books often have contrived problems which must be recognized or the solution is difficult. This is one of those problems. The denominator of the first term can be factored. If we do the factoring, the rest of the problem is straightforward.

$$\frac{x + 2}{(x + 3)(x + 1)} - \frac{1}{x(x + 1)} \qquad \text{factored}$$

$$= \frac{}{x(x + 3)(x + 1)} - \frac{}{x(x + 3)(x + 1)} \qquad \text{new denominators}$$

$$= \frac{x(x + 2)}{x(x + 3)(x + 1)} - \frac{(x + 3)}{x(x + 3)(x + 1)} \qquad \text{new numerators}$$

$$= \frac{x^2 + 2x - x - 3}{x(x + 3)(x + 1)} \qquad \text{added}$$

$$= \frac{x^2 + x - 3}{x(x + 3)(x + 1)} \qquad \text{simplified}$$

problem set 27

1. The hiking club hiked to Robbers Cave State Park at 4 mph. They got a ride back to town in a truck that went 20 mph. If the round trip took 18 hours, how far was it from town to the park?

2. Jojo set out on a hike. After walking for some time at 5 kph, he caught a ride back home in a truck that traveled at 20 kph. If the round trip took 10 hours, how far did he walk?

3. There were 6 more girls than twice the number of boys. There were 36 boys and girls in all. How many boys and how many girls were there?

4. The general expression for consecutive multiples of 7 is $7N$, $7(N + 1)$, $7(N + 2)$, etc., where N is some unspecified integer. Find three consecutive multiples of 7 such that the product of -3 and the sum of the first and third is 21 less than 5 times the opposite of the second.

5. To form the compound, 600 grams of barium was mixed with 2400 grams of other chemicals. If 9000 grams of compound was needed, how much barium was required?

6. Silver iodide made up 70 percent of the total. If the total weighed 2000 grams, how much was not silver iodide?

7. Divide $x^3 - 6$ by $x + 2$. Then check by adding.

Factor the greatest common factor:

8. $9m^2x^4p^2 + 3x^2p^6m^4 - 6x^4m^3p^2$ 　　9. $mx^4y - mx^2y^3 - 4mx^2y$

10. $a^2x^3p - 4a^3x^3p - a^2x^4p$

Factor completely. Always look for a common monomial factor.

11. $4ax + ax^2 - 5a$ 　　12. $8x^2 - x^3 - 15x$

13. $24ax - 5ax^2 - ax^3$ 　　14. $-ax^4 + 4ax^3 + 5ax^2$

15. $56p - 15px + px^2$

Simplify:

16. $\dfrac{4xa + 4x}{4x}$ 　　17. $3\sqrt{2} - 2\sqrt{3} \cdot 3\sqrt{12}$

18. $-3\sqrt{20} + 2\sqrt{125} + 5\sqrt{45}$ 　　19. $2\sqrt{3}(3\sqrt{2} - 3\sqrt{3})$

Add:

*20. $\dfrac{4}{x} + \dfrac{6}{x + a}$ 　　*21. $\dfrac{x + 2}{x + 4} + 6 + \dfrac{2}{x^2}$

*22. $\dfrac{x + 2}{x^2 + 4x + 3} - \dfrac{1}{x(x + 1)}$

23. Simplify: $\dfrac{(.00056 \times 10^4)(7 \times 10^3)}{(.00049 \times 10^{16})(.00002 \times 10^{-5})}$

24. Solve by graphing and then get an exact solution by using either substitution or elimination.
$$\begin{cases} 3x + 2y = 12 \\ 8x - 2y = 10 \end{cases}$$

25. Find the equation of the line that passes through $(-3, -3)$ and is parallel to the line $y = -\frac{3}{8}x + 17$.

Solve:

26. $\dfrac{6 + x}{5} + \dfrac{1}{3} = 8$ 27. $\dfrac{4 - x}{6} - \dfrac{1}{2} = 11$ 28. $\dfrac{3x - 2}{3} + 4 = \dfrac{1}{6}$

29. Simplify by adding like terms: $\dfrac{3a^{-2}y}{b} + 7b^{-1}ya^{-2} - \dfrac{5b^{-1}}{a^2 y^{-1}}$

30. Evaluate $-3^0(-3 - 2^2) - 4(-2)ax - a$ if $a = -2$ and $x = 4$.

LESSON 28 *Complex fractions · Rationalizing the denominator*

28.A
complex fractions

Possibly the most useful rule in algebra is the denominator–numerator–same-quantity theorem.

> **DENOMINATOR–NUMERATOR–SAME-QUANTITY THEOREM**
>
> The denominator and the numerator of a fraction may be multiplied by the same nonzero quantity without changing the value of the fraction.

This rule allows us to change the form of any fraction to a form that is more convenient. We have used this rule to help us add abstract fractions. To review the use of this procedure, we will add two fractions.

$$\frac{a}{b} + \frac{c}{d}$$

We will multiply the first fraction by d over d and multiply the second fraction by b over b. Then we can add the fractions.

$$\frac{a}{b}\left(\frac{d}{d}\right) + \frac{c}{d}\left(\frac{b}{b}\right) = \frac{ad + cb}{bd}$$

We also use the denominator-numerator–same-quantity theorem to help us simplify fractions of fractions.

$$\frac{\dfrac{a}{b}}{\dfrac{c}{d}}$$

If we multiply the denominator of this fraction by its reciprocal, which is $\frac{d}{c}$, the resulting denominator will be 1. If we do this we must also multiply the numerator by the same quantity, $\frac{d}{c}$, so that the value of the expression will not be changed.

$$\frac{\dfrac{a}{b}\cdot\dfrac{d}{c}}{\dfrac{c}{d}\cdot\dfrac{d}{c}} = \frac{\dfrac{ad}{bc}}{1} = \boldsymbol{\frac{ad}{bc}}$$

The denominator-numerator–same-quantity theorem can be used anywhere—and at any time—on a term in an equation or on a term that is not in an equation. In everyday language we can say,

> **Anytime, anywhere, the denominator and the numerator of an expression can be multiplied by the same quantity (not zero) without changing the value of expression. Only the form of the expression is changed.**

example 28.A.1   Simplify: $\dfrac{\dfrac{a}{b}}{\dfrac{x+y}{b}}$

solution   We will multiply the denominator and the numerator by b over $x + y$.

$$\frac{\dfrac{a}{b}}{\dfrac{x+y}{b}} \cdot \frac{\dfrac{b}{x+y}}{\dfrac{b}{x+y}} = \boldsymbol{\frac{a}{x+y}}$$

example 28.A.2   Simplify: $\dfrac{\dfrac{a}{a+b}}{\dfrac{c}{a+b}}$

solution   We will multiply both the top and bottom by $a + b$ over c.

$$\frac{\dfrac{a}{a+b}}{\dfrac{c}{a+b}} \cdot \frac{\dfrac{a+b}{c}}{\dfrac{a+b}{c}} = \boldsymbol{\frac{a}{c}}$$

28.B
rationalizing the denominator

Often we encounter expressions such as

$$\frac{4}{\sqrt{7}}$$

that have a radical in the denominator. Some people like to change the form of these expressions so that the radical does not appear in the denominator. We remember that we can always change the form of an expression by multiplying both the denominator and the numerator by the same quantity. For this example, we choose to multiply by $\sqrt{7}$ over $\sqrt{7}$.

$$\frac{4}{\sqrt{7}} \frac{\sqrt{7}}{\sqrt{7}} = \frac{4\sqrt{7}}{7}$$

This new expression has the same value as the original expression, but the denominator is a rational number. This procedure is called **rationalizing the denominator.** The instructions for one of these problems will use the one word simplify.

> **An expression that contains square root radicals is in simplified form when no radicand has a perfect-square factor and no radicals are in the denominator.**

example 28.B.1   Simplify: $\dfrac{3}{2\sqrt{5}}$

solution   We can eliminate the radical in the denominator by multiplying by $\sqrt{5}$ over $\sqrt{5}$. Of course, we still can't get rid of $\sqrt{5}$ completely, for it will now appear in the numerator.

$$\frac{3}{2\sqrt{5}} \frac{\sqrt{5}}{\sqrt{5}} = \frac{3\sqrt{5}}{10}$$

example 28.B.2   Simplify: $\dfrac{2}{3\sqrt{12}}$

solution   We will multiply by $\sqrt{12}$ over $\sqrt{12}$ and then simplify the result.

$$\frac{2}{3\sqrt{12}} \frac{\sqrt{12}}{\sqrt{12}} = \frac{2\sqrt{12}}{36} = \frac{2(2\sqrt{3})}{36} = \frac{\sqrt{3}}{9}$$

problem set 28

1. Bronson roared off on his motorcycle at 60 mph. Then, much to his chagrin, he ran out of petrol. He pushed the motorcycle all the way back at 3 mph. If the entire trip took 21 hours, how far did he push the motorcycle?

2. The general expression for multiples of 3 is $3N$, $3(N + 1)$, $3(N + 2)$, etc., where N is some unspecified integer. Find four consecutive multiples of 3 such that 5 times the sum of the first and fourth is 6 less than 13 times the third.

3. The number of girls in the class was 1 less than 3 times the number of boys. There were 15 students in all. How many were boys, and how many were girls?

4. The class treasury contained $30 in nickels and dimes. If there were 500 coins, how many coins of each type were there?

5. Arthur found that for every 2000 peasants, only 100 had seen a Dane. If there were 150,000 peasants in the kingdom, how many had never seen a Dane?

6. Sixteen percent of the mixture was arsenic and the rest was silicon. Mendeleev had 7350 kilograms of silicon. How much arsenic did he have? What did the entire mixture weigh?

7. Divide $x^3 - 7$ by $x - 5$ and then check.

Factor the greatest common factor:

8. $2x^2y - 8x^4y^4$

9. $4x^2y^3p^3 - 16x^2y^3p - x^4y^3p^4$

Factor completely. Always factor the GCF as the first step:

10. $-35xy + 2x^2y + x^3y$

11. $-8a - 7ax + ax^2$

12. $2m^2 + 3xm^2 + m^2x^2$

13. $-a^2 - a^2x^2 - 2xa^2$

Simplify:

14. $\dfrac{4x^2 + x}{x}$

15. $4\sqrt{27} - 3\sqrt{48} + 2\sqrt{75}$

16. $3\sqrt{5}(\sqrt{15} - 2\sqrt{5})$

17. $\dfrac{(.00077 \times 10^{-3})(40 \times 10^6)}{(.00011 \times 10^5)(140{,}000)}$

*18. $\dfrac{\dfrac{a}{b}}{\dfrac{x + y}{b}}$

*19. $\dfrac{\dfrac{a}{a + b}}{\dfrac{c}{a + b}}$

*20. $\dfrac{3}{2\sqrt{5}}$

*21. $\dfrac{2}{3\sqrt{12}}$

Add:

22. $\dfrac{4a}{a + x} + \dfrac{6}{a}$

23. $\dfrac{2x}{x^2 + 2x + 1} + \dfrac{3}{x + 1}$

24. Solve by graphing and then get an exact solution by using either substitution or elimination.
$$\begin{cases} 5x + 2y = 6 \\ y = \dfrac{1}{2}x \end{cases}$$

25. Find the equations of lines (a) and (b).

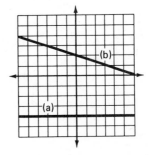

Solve:

26. $\dfrac{2x}{3} + \dfrac{x - 3}{5} = 1$

27. $4\dfrac{1}{3} - 2\dfrac{1}{5}x = -\dfrac{3}{10}$

28. Find the distance between $(-2, 5)$ and $(-6, -3)$.

29. Expand: $\dfrac{3x^{-2}y^2}{z}\left(\dfrac{x^2z}{3y^2} - \dfrac{4x^{-2}y}{z^{-3}}\right)$

30. Evaluate $x^0 - xy^0(x - y)$ if $x = \dfrac{1}{3}$ and $y = -\dfrac{1}{4}$.

LESSON 29 *Uniform motion problems, $D_1 + D_2 = k \cdot$ Polygons and angles*

29.A
uniform
motion
problems

We remember that the key equation in a uniform motion problem is the equation that describes the distances that have been traveled. In all the problems thus far, the persons or objects have traveled equal distances, so the distance diagrams for these problems have been similar to one of the two diagrams shown here.

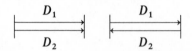

$$D_1 = D_2 \quad \text{so} \quad R_1 T_1 = R_2 T_2$$

In the left-hand diagram both objects began at the same point and traveled in the same direction. In the right-hand diagram the objects began at different points and traveled in opposite directions, but again the distances traveled were equal. Thus, the same equation is applicable to both diagrams.

In some problems, the problem tells us that the sum of one distance and another distance equals a certain number. If the number is 460, the distance diagram would look like one of the following.

$$\begin{array}{ccc} D_1 \quad D_2 & D_1 \quad D_2 & D_1 \quad D_2 \\ \overline{460} & \overline{460} & \overline{460} \end{array}$$

In each diagram we see that the sum of distance 1 and distance 2 equals 460, so the same equation is applicable to all three diagrams.

$$D_1 + D_2 = 460 \quad \text{so} \quad R_1 T_1 + R_2 T_2 = 460$$

example 29.A.1   Napoleon walked part of the 60 miles to the site of the battle and rode the rest of the way on a caisson. He walked at 3 mph and rode at 9 mph. If the total time of the trip was 8 hours, how long did he walk?

solution   His distance walking plus his distance riding equaled 60 miles. This leads to the following distance diagram and distance equation.

$$\begin{array}{ccc} D_W \quad D_R & & \\ \overline{} & D_W + D_R = 60 & \text{so} \quad R_W T_W + R_R T_R = 60 \end{array}$$

We have four unknowns and only one equation. Thus we need three more equations. We reread the problem and find that two are rate equations and one is a time equation.

$$R_W = 3 \quad R_R = 9 \quad T_W + T_R = 8$$

We finish by substituting into the distance equation.

$$\begin{array}{ll} 3(8 - T_R) + 9T_R = 60 & \text{substituted} \\ 24 - 3T_R + 9T_R = 60 & \text{multiplied} \\ 24 + 6T_R = 60 & \text{added} \\ 6T_R = 36 & \text{simplified} \\ T_R = 6 & \text{divided} \end{array}$$

Since he rode for 6 hours and his total time was 8 hours, he must have walked 2 hours.

$$T_W = \text{2 hours}$$

example 29.A.2   Edward Longshanks and Queen Eleanor were 54 miles apart at dawn. Edward began the journey to the meeting place at 8 a.m. at 3 mph; 2 hours later, the queen set out to meet him. If they met at 4 p.m., how fast did the queen travel?

solution   Between them they covered 54 miles, so the following distance diagram and distance equation apply.

$$D_L + D_Q = 54 \qquad \text{so} \qquad R_L T_L + R_Q T_Q = 54$$

We have one equation and four unknowns. Thus we need three more equations. Two are time equations, and one is a rate equation.

$$T_L = 8 \qquad T_Q = 6 \qquad R_L = 3$$

We substitute these into the distance equation and solve.

$$(3)(8) + (R_Q)(6) = 54 \qquad \text{substituted}$$
$$24 + 6R_Q = 54 \qquad \text{multiplied}$$
$$6R_Q = 30 \qquad \text{added } -24$$
$$R_Q = 5 \qquad \text{divided}$$

Thus, Queen Eleanor traveled at 5 mph, which was a fast speed for the roads of thirteenth-century England.

example 29.A.3   At noon, Rocketman whizzed off toward Rocketland; 1 hour later, Moonfa whizzed off in the opposite direction at a speed 200 kph less than that of Rocketman. If they were 11,800 kilometers (km) apart at 5 p.m., how fast did each travel?

solution   Rocketman and Moonfa began at the same point and traveled in opposite directions. Together they covered 11,800 km. Thus the distance diagram and distance equation are as follows.

$$D_R + D_M = 11{,}800 \qquad \text{so} \qquad R_R T_R + R_M T_M = 11{,}800$$

We need three more equations. Two are time equations and one is a rate equation.

$$T_R = 5 \qquad T_M = 4 \qquad R_M = R_R - 200$$

Now we substitute these equations into the distance equation and solve.

$$(R_M + 200)(5) + R_M(4) = 11{,}800 \qquad \text{substituted}$$
$$5R_M + 1000 + 4R_M = 11{,}800 \qquad \text{multiplied}$$
$$9R_M = 10{,}800 \qquad \text{simplified}$$
$$R_M = \text{1200 kph} \qquad \text{divided}$$

Since the rate of Rocketman was 200 kilometers per hour greater than that of Moonfa, Rocketman's rate was

$$R_R = 1400 \text{ kph}$$

29.B
polygons

Definitions often change. The definition of a polygon is a good example. The word is formed from the Greek roots *poly*, which means more than one or many, and *gonon*, which means angle. Thus, polygon literally means "more than one angle." In 1571 Diggs said that "Polygona are such figures that haue moe than foure sides." In 1656 Blount said that a polygon was a geometrical figure that "hath many corners."

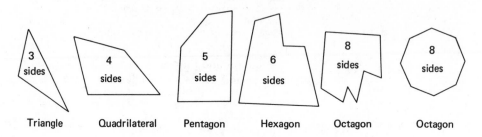

All of the figures shown here fit Blount's definition of a polygon, but the two on the left do not have enough sides for Diggs's definition. Modern authors tend to define polygons as closed geometric figures whose sides are straight lines. We name polygons by using Greek prefixes that denote the number of sides the figure has, as shown in the accompanying figure. *Tri* means 3, *quad* means 4, etc. If the sides of a polygon are all the same length and all of the angles are the same, we say that the polygon is a **regular polygon.** Thus the octagon shown on the extreme right is a regular octagon.

29.C
angles

The definition of the word **angle** has also changed. It comes from the Greek word meaning "to bend," and European authors tend to define an angle to be the opening between two half-lines or rays that have a common endpoint. If the rays are perpendicular, the angle is called a **right angle.** If the angle is smaller than a right angle, it is called an **acute angle;** and if it is greater than a right angle, it is called an **obtuse angle.**

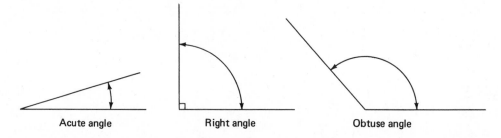

Often, as shown in the center figure, a small square is drawn at the point of intersection of perpendicular lines to indicate that the lines are perpendicular and that the angle is a right angle. If a right angle is divided into 90 equal parts, we say that each of the parts is 1 **degree.** Thus, a right angle is also called a **90-degree angle.** We often use a small raised circle instead of writing the word degree. We would write ninety degrees in this notation by writing 90°.

Here we show two intersecting perpendicular lines and note that four 90° angles are formed.

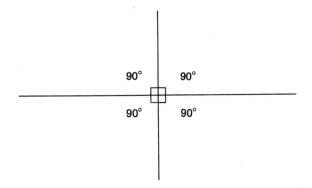

Some American authors say that an angle is generated when a ray is rotated about its endpoint from an initial position to a terminal position. When they do this, they often neglect to say whether the angle is the opening or is the conjunction of the initial and terminal rays.

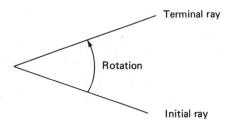

From this diagram we see that an angle consists of an opening that is bounded by two rays. Most American authors define the angle to be the intersecting rays themselves instead of the opening between the rays. This definition is especially helpful when discussing the properties of linear geometric figures. When we do this and consider the angle to be the rays themselves, the magnitude of the opening is called the measure of the angle. To use this terminology, we would speak of an angle whose measure is ninety degrees instead of saying a ninety degree angle. In this book we will think of the angle as being the opening and save the other definition for geometry.

> An angle is the opening between two rays that have a common endpoint.

If we use this definition, we see that a straight line can be thought of as defining a 180° angle.

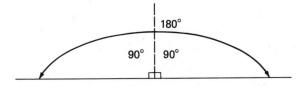

problem set 29   *1.   Napoleon walked part of the 60 miles to the site of the battle and rode the rest of the way on a caisson. He walked at 3 mph and rode at 9 mph. If the total time of the trip was 8 hours, how long did he walk?

*2. Edward Longshanks and Queen Eleanor were 54 miles apart at dawn. Edward began the journey to the meeting place at 8 a.m. at 3 mph. Two hours later, the queen set out to meet him. If they met at 4 p.m., how fast did the queen travel?

*3. At noon, Rocketman whizzed off toward Rocketland. One hour later, Moonfa whizzed off in the opposite direction at a speed 200 kph less than that of Rocketman. If they were 11,800 kilometers apart at 5 p.m, how fast did each travel?

4. The number of roses was 15 greater than twice the number of prunes. If the roses and prunes totaled 255, how many roses were there?

5. For every 130 squirrels in the forest, there were 156,000 good places to hide. If there was a total of 3250 squirrels in the forest, how many good places to hide were there?

6. In a contiguous forest, 17 percent of the places to hide contained xenophobes. If 116,200 places to hide did not contain xenophobes, how many hiding places were there in the contiguous forest?

7. Divide $x^3 + 3x^2 + 7x + 5$ by $x + 1$ and then check.

Factor the greatest common factor:

8. $16x^3y^2z^3 - 8x^2y^2z^2$

9. $2x^2yp^4 - 6x^3yp^3 - 2x^2yp^2$

Factor completely. Always factor the GCF as the first step.

10. $12a^2x + a^2x^2 + 35a^2$

11. $15p - px^2 - 2px$

12. $-2m^2x - m^2 - m^2x^2$

13. $x^2k + 3kx - 40k$

Simplify:

14. $\dfrac{x^2 + ax^2}{x^2}$

15. $2\sqrt{75} - 5\sqrt{48} + 2\sqrt{12}$

16. $2\sqrt{3}(3\sqrt{6} - 4\sqrt{3})$

17. $\dfrac{(.00052 \times 10^{-4})(5000 \times 10^7)}{(.0026 \times 10^{21})(10,000 \times 10^{-42})}$

18. $\dfrac{\dfrac{m}{x}}{\dfrac{m + x}{x}}$

19. $\dfrac{\dfrac{a}{m + x}}{\dfrac{b}{m + x}}$

20. $\dfrac{3}{5\sqrt{12}}$

21. $\dfrac{14}{3\sqrt{75}}$

Add:

22. $\dfrac{4x}{x + 4} + \dfrac{6}{x + 2}$

23. $\dfrac{3m}{m^2 + 3m + 2} - \dfrac{5m}{m + 1}$

24. Solve by graphing. Then get an exact solution by using either substitution or elimination.
$$\begin{cases} y - x = 3 \\ y + 2x = 6 \end{cases}$$

25. Find the equation of the line that passes through $(2, -3)$ and is parallel to $y = -\dfrac{3}{8}x + 2$.

Solve:

26. $\dfrac{3x}{2} - \dfrac{5}{7} = \dfrac{x+2}{3}$ **27.** $2\dfrac{1}{5} - \dfrac{1}{10}x = \dfrac{3}{15}$

28. $\dfrac{5x+3}{4} = -\dfrac{3}{7}$ **29.** Simplify: $\dfrac{(x^2 y^{-2} z)^{-3} x^0}{(x^0 y^{-3} z^2)^3}$

30. Evaluate $x^2 - y^2(x - y)$ if $x = \dfrac{1}{2}$ and $y = \dfrac{1}{3}$.

LESSON 30   *Area*

30.A
area
of a rectangle

If we have a table that is 6 feet long and 3 feet wide, we see that we can divide the top into 18 squares that measure 1 foot on each side.

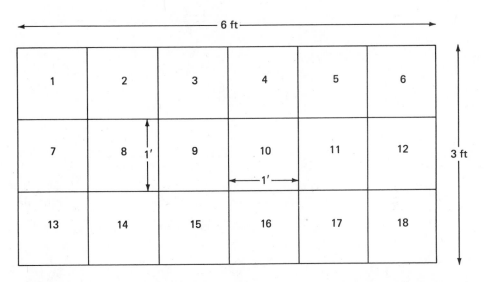

We say that each of the squares shown in the figure has an area of 1 square foot; and, thus, the table top has a total area of 18 square feet. Rather than write out the words square feet, we often use exponential notation with the abbreviation ft as the base and say that the area is

$$18 \text{ ft}^2$$

If we had 18 floor tiles that measured 1 foot on a side, we could use the tiles to completely cover the table top. Thus, we see that we can associate the abstraction of an area of 18 ft² with 18 floor tiles which we can see and feel. Associating abstractions with things we can see and feel is often helpful, and many find it helpful to think of **floor tiles** when they hear the word *area.* Thus when we hear

18 square feet	we can think	18 floor tiles each 1 foot square
21 square inches	we can think	21 floor tiles each 1 inch square
33 square meters	we can think	33 floor tiles each 1 meter square

From the last diagram, we see that we can calculate the area of a rectangle by multiplying the length times the width. Thus, for the table top the area is

$$\text{Area} = \text{length} \times \text{width}$$

$$\text{Area} = 6 \text{ feet} \times 3 \text{ feet}$$

To multiply, we will treat the words feet just as if they were numbers. We mentally rearrange and multiply 6 by 3 to get 18 and feet by feet to get feet squared, which we abbreviate as ft^2.

$$\text{Area} = 18 \text{ ft}^2$$

example 30.A.1   (a)   How many 1-inch-square floor tiles will it take to cover a rectangular area that is 4 feet long and 20 inches wide? (b)   What is the perimeter of the rectangle?

solution   (a)   To find the area in square inches, both the length and width must be in inches. So first, we convert 4 feet to 48 inches.

$$\text{Area} = L \times W \quad \longrightarrow \quad \text{Area} = 48 \text{ in} \times 20 \text{ in} = \mathbf{960 \text{ in}^2}$$

←———————————————— 48 inches ————————————————→

20 inches

Thus, it would take 960 1-inch-square tiles to cover this rectangle.
   (b)   The perimeter is the distance around the rectangle, which is

$$48 \text{ in} + 48 \text{ in} + 20 \text{ in} + 20 \text{ in} = \mathbf{136 \text{ in}}$$

30.B
area of a triangle

A point where two sides of a triangle meet is called a **vertex of the triangle.** The plural of vertex is **vertices,** so the three corner points of a triangle are the vertices of the triangle. Any of the three sides of a triangle can be designated as the base of the triangle. Once the base has been designated, we define the **altitude** or **height** of the triangle to be the perpendicular distance from this base (or its extension) to the other vertex of the triangle. In some triangles, the altitude is one of the sides of the triangle. In other triangles, the altitude will be drawn either inside the triangle or outside the triangle. Here we show an example of each type.

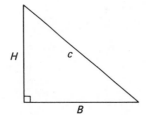
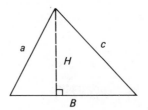
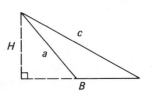

In all three triangles, we will call side B the base, and we have designated the altitude (or height) of the triangle with the letter H. The triangle on the left is a right triangle, and the altitude is the side H. In the center triangle, the perpendicular distance from B is designated with the dotted line H. In the triangle on the right, it was necessary to extend the base so that the altitude H could be drawn.

The area of a triangle is found by dividing the product of the base and the altitude by 2. We remember that any side may be selected as the base. However, in the problems in this book, the choice may be restricted, as the value of only one altitude may be designated.

example 30.B.1 Find the areas of triangles (a), (b), and (c). Lengths are in feet. (d) What is the perimeter of triangle (c)?

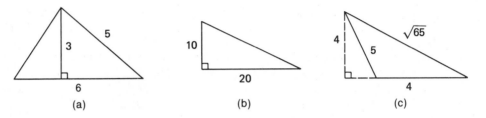

(a) (b) (c)

solution To find the areas, we divide the product of the base and the altitude by 2.

(a) $\quad A = \dfrac{B \times H}{2} \quad \longrightarrow \quad A = \dfrac{6 \text{ ft} \times 3 \text{ ft}}{2} = \textbf{9 ft}^2$

(b) $\quad A = \dfrac{B \times H}{2} \quad \longrightarrow \quad A = \dfrac{10 \text{ ft} \times 20 \text{ ft}}{2} = \textbf{100 ft}^2$

(c) $\quad A = \dfrac{B \times H}{2} \quad \longrightarrow \quad A = \dfrac{4 \text{ ft} \times 4 \text{ ft}}{2} = \textbf{8 ft}^2$

(d) The perimeter of (c) is $5 \text{ ft} + \sqrt{65} \text{ ft} + 4 \text{ ft} = \textbf{(9} + \sqrt{\textbf{65}}\textbf{) ft}$

30.C
circles

A circle is often defined as a set of points in a given plane, each point being equidistant from a designated point, which is called the **center** of the circle.

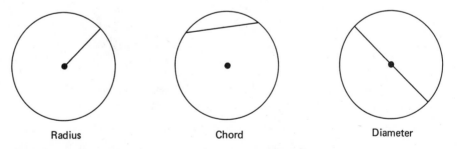

Radius Chord Diameter

A **radius** of a circle is a line segment that connects the center of the circle with a point on the circle. The word radius is the Latin word for the spoke of a wheel. The plural of radius is **radii,** and we can see from the circle on the left that all radii of a given circle are of equal length.

A **chord** is a line segment that connects two points on a circle and a **diameter** is a chord that passes through the center of the circle. In the center figure, we show a chord, and in the right figure we show a diameter. We note that a diameter of a given circle is twice as long as a radius of the same circle or

$$D = 2R$$

Since early times, learned people have been intrigued by the relationship between the circumference of a circle (perimeter of a circle) and the length of the diameter of the same circle. Here we show three circles.

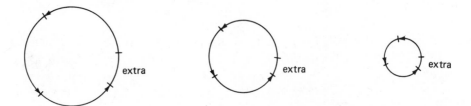

On each one we have laid out lengths that equal diameters. In each case, we see that we can lay out three diameters and have a little extra length left over. We see that three diameters are not enough to reach all the way around, and four diameters would be too much. A close approximation of the number of times the diameter can be laid out is 3.14, and we say that

$$1 \text{ circumference} \approx 3.14 \text{ diameters}$$

This is an approximation, for the exact number is not 3.14 but is an irrational number whose decimal representation would require an infinite number of digits. We represent this number with the symbol

$$\pi$$

and call it *pi* (pronounced "pie"). Thus the circumference of a given circle is π times the diameter of the circle.

$$C = \pi D$$

The area of a circle equals the product of π and the square of the radius.

$$A = \pi R^2$$

example 30.C.1   A circle has a radius of 4 ft. (a) What is the circumference of the circle, and (b) what is the area of the circle?

4 feet

solution   (a)   The circumference is π times the diameter. If the radius is 4 ft, the diameter must be 8 ft. We will use 3.14 as an approximation for π.

$$C = \pi D \quad \longrightarrow \quad C = 3.14D \quad \longrightarrow \quad C = (3.14)(8 \text{ ft}) = \textbf{25.12 ft}$$

(b)   The area of a circle equals π times the square of the radius.

$$A = \pi R^2 \quad \longrightarrow \quad A = 3.14(4 \text{ ft})(4 \text{ ft}) \quad \longrightarrow \quad A = (3.14)(16 \text{ ft}^2) = \textbf{50.24 ft}^2$$

problem set 30

1.   When 240 grams of barium was mixed with 40 grams of sulfite, the desired reaction occurred. If a total of 3360 grams of barium and sulfite was to be used, how much should be sulfite?

2. Fourteen percent of the mass was consumed in the reaction. If 430 grams remained, how much was the initial mass?

3. At 10 a.m., Little Flower trotted off in one direction at 6 mph. At noon, Laughing Boy loped off in the opposite direction. If they were 68 miles apart at 4 p.m., how fast did Laughing Boy lope?

4. The daisies proliferated until 5 times the number of daisies equaled twice the number of prunes. If the daisies and prunes totaled 35, how many of each were there?

5. Yellow Basket found that she had 15 dimes and quarters and that their total value was $2.25. How many of each kind of coin did she have?

6. Find the (a) area and (b) perimeter of a rectangular plot of land whose length is 40 feet and whose width is 120 inches.

7. Find the area of this triangle. The lengths are in feet.

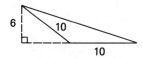

8. Find (a) the circumference and (b) the area of a circle whose radius is 5 feet. Use 3.14 for π.

9. Divide $x^3 + 2$ by $x + 1$ and then check.

Factor completely. Always factor the GCF as the first step.

10. $35a^2 + 2a^2x - a^2x^2$

11. $30 + 3x^2 - 21x$

12. $-x^2ab - 25ab + 10axb$

13. $14a^2b^2 + 9a^2xb^2 + x^2a^2b^2$

Simplify:

14. $\dfrac{p - 4px}{p}$

15. $5\sqrt{18} - 10\sqrt{50} + 3\sqrt{72}$

16. $3\sqrt{12}(4\sqrt{2} - 2\sqrt{3})$

17. $\dfrac{(.00035)(5000 \times 10^{42})}{.00025 \times 10^{-4}}$

18. $\dfrac{\dfrac{x}{y}}{\dfrac{x + y}{y}}$

19. $\dfrac{\dfrac{a}{a + b}}{\dfrac{p}{a + b}}$

20. $\dfrac{2}{3\sqrt{6}}$

21. $\dfrac{2}{5\sqrt{18}}$

Add:

22. $\dfrac{4a}{a + 4} + \dfrac{a + 2}{2a}$

23. $\dfrac{4x}{x^2 + 5x + 6} + \dfrac{2}{x + 2}$

24. Solve by graphing and then get an exact solution by using either substitution or elimination.
$$\begin{cases} 3x + 2y = 8 \\ 2x + 3y = 6 \end{cases}$$

25. Find the equation of the line that passes through $(-2, 4)$ and $(1, 3)$.

Solve:

26. $\dfrac{2x}{3} - \dfrac{2}{5} = \dfrac{x - 5}{2}$

27. $3\dfrac{2}{3} + \dfrac{1}{5}x = \dfrac{1}{15}$

28. $\dfrac{2x - 3}{2} = -\dfrac{2}{5}$

29. Simplify by adding like terms: $\dfrac{4a^2x^2y}{p} - \dfrac{3p^{-1}x^4y}{a^{-2}x^2} + \dfrac{2xa}{a^{-1}y^{-1}p} - \dfrac{2a^2yx^2}{p}$

30. Evaluate $x^2 - y(x - y)$ if $x = -\dfrac{1}{2}$ and $y = \dfrac{1}{4}$.

LESSON 31 *Negative reciprocals • Perpendicular lines • Combination geometric figures*

31.A
negative reciprocals

We remember that the reciprocal of a number is designated when we write the number in inverted form. Thus

$\dfrac{2}{5}$	is the reciprocal of	$\dfrac{5}{2}$
$\dfrac{5}{2}$	is the reciprocal of	$\dfrac{2}{5}$
$-\dfrac{1}{4}$	is the reciprocal of	-4
-4	is the reciprocal of	$-\dfrac{1}{4}$
3	is the reciprocal of	$\dfrac{1}{3}$
$\dfrac{1}{3}$	is the reciprocal of	3

We note that the sign of a number is the same as the sign of its reciprocal. This is not true for numbers that are negative reciprocals because each of these numbers is the inverted form of the other and also the signs of the numbers are different. Thus,

$-\dfrac{2}{5}$	is the negative reciprocal of	$\dfrac{5}{2}$
$\dfrac{5}{2}$	is the negative reciprocal of	$-\dfrac{2}{5}$
$-\dfrac{1}{4}$	is the negative reciprocal of	4
4	is the negative reciprocal of	$-\dfrac{1}{4}$
-3	is the negative reciprocal of	$\dfrac{1}{3}$
$\dfrac{1}{3}$	is the negative reciprocal of	-3

31.B
perpendicular lines

We recall that lines that are parallel have equal slopes and different intercepts. When a linear equation is written in slope-intercept form, the coefficient of the x term designates the slope of the line, and the constant term designates the y-intercept. Note that the coefficient of the x term in both of the following equations is $-\frac{1}{2}$, and that the intercepts are $+2$ and -3.

$$y = -\frac{1}{2}x + 2 \qquad y = -\frac{1}{2}x - 3$$

So the slope of both of these lines is $-\frac{1}{2}$, and the lines cross the y axis at $+2$ and -3, respectively. The graphs of these lines are shown in the figure on the left.

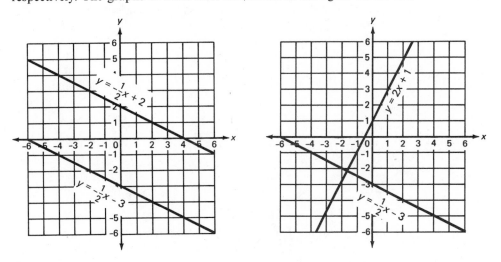

The slopes of lines that are perpendicular are negative reciprocals of each other. On the right we show the graphs of the equations

$$y = -\frac{1}{2}x - 3 \qquad \text{and} \qquad y = 2x + 1$$

and note that the lines appear to be perpendicular, and that the slopes are $-\frac{1}{2}$ and $+2$, numbers that are negative reciprocals. We can use this negative reciprocal relationship to help us write the equation of a line that is perpendicular to a given line and that passes through a designated point.

example 31.B.1

Write the equation of the line that is perpendicular to $y = \frac{2}{5}x - 3$ and that passes through the point $(3, 2)$.

solution

The new line is to be perpendicular to a line whose slope is $\frac{2}{5}$. Thus, the slope of the new line must be $-\frac{5}{2}$. This gives us

$$y = -\frac{5}{2}x + b$$

To finish the equation, we must find the value of the intercept b. To do this, we will use 3 for x and 2 for y and solve algebraically for b.

$$2 = -\frac{5}{2}(3) + b \qquad \text{substituted}$$

$$\frac{4}{2} = -\frac{15}{2} + b \qquad \text{simplified}$$

$$\frac{19}{2} = b \qquad \text{solved}$$

Thus the intercept is $\frac{19}{2}$, and the full equation of the perpendicular line is

$$y = -\frac{5}{2}x + \frac{19}{2}$$

31.C
combination geometric figures

The areas of real-life figures are seldom just areas of rectangles or triangles or circles. Often they are a combination of several of these geometric forms.

example 31.C.1   Find the perimeter and the area of this figure. The dimensions are given in feet.

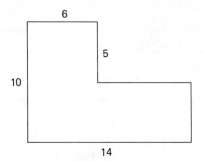

solution   We assume that all lines are perpendicular. If we do this, we can deduce that the missing lengths are 5 ft and 8 ft. We write these numbers and find the perimeter.

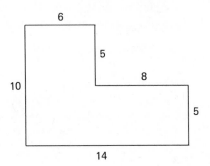

Perimeter = 6 ft + 5 ft + 8 ft + 5 ft + 14 ft + 10 ft = **48 ft**

Next we divide the area into convenient rectangles and find the area of each rectangle. Then we add these areas to find the total area.

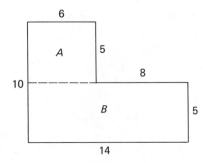

$$A = 5\text{ ft} \times 6\text{ ft} = 30\text{ ft}^2$$
$$B = 14\text{ ft} \times 5\text{ ft} = 70\text{ ft}^2$$
$$\text{Total area} = \mathbf{100\text{ ft}^2}$$

example 31.C.2 Find the perimeter of this figure. How many floor tiles 1-inch square would it take to cover the figure? The dimensions are given in inches. Assume that the tiles can be cut or otherwise reshaped to make them fit exactly.

solution The circumference of a circle is π times the diameter; and since the radius is 4 in, the diameter is 8 in. We will use 3.14 as an approximation of π. The circumference of a half circle is half the circumference of a circle, so

$$\frac{\text{Circumference}}{2} = \frac{(3.14)(\text{diameter})}{2}$$

$$= \frac{(3.14)(8)}{2} = 12.56 \text{ in}$$

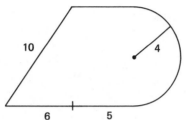

Now we can calculate the perimeter of the figure as

$$10 \text{ in} + 6 \text{ in} + 5 \text{ in} + 5 \text{ in} + 12.56 \text{ in} = \textbf{38.56 in}$$

The number of floor tiles required equals the sum of the areas of the triangle, the rectangle, and half circle.

$$\text{Area of triangle} = \frac{8 \text{ in} \times 6 \text{ in}}{2} = 24 \text{ in}^2$$

$$\text{Area of rectangle} = 5 \text{ in} \times 8 \text{ in} = 40 \text{ in}^2$$

$$\text{Area of half circle} = \frac{(3.14)(4 \text{ in})^2}{2} = 25.12 \text{ in}^2$$

$$\overline{\text{Total area} = \textbf{89.12 in}^2}$$

Thus, 89.12 floor tiles that are 1-inch square would be required to cover the figure.

problem set 31

1. The general expression for consecutive multiples of 5 is $5N$, $5(N + 1)$, $5(N + 2)$, etc., where N is some unspecified integer. Find four consecutive multiples of 5 such that 6 times the first is 40 greater than 2 times the sum of the second and the fourth.

2. The bus headed north at 40 mph at 10 a.m. At noon the Orange Blossom Special headed south from the same station at 70 mph. What time was it when the train and the bus were 960 miles apart?

3. Susie jogged to the farm at 6 mph and rode back home in a truck traveling at 30 mph. How far was it to the farm if the entire trip took 12 hours?

4. Pansy plants were $4 a crate and tomato plants were $6 a crate. Sowega bought 70 crates for Patsy and spent $360. How many crates of each kind did he buy?

5. Ninety percent of the nitrogen combined. If 1200 kilograms did not combine, what was the total weight of the nitrogen?

6. Forty grams of potassium combined with 1400 grams of other elements to form the compound. If 4320 grams of the compound was needed, how many grams of potassium was required?

*7. Write the equation of the line that is perpendicular to $y = \frac{2}{5}x - 3$ and passes through the point $(3, 2)$.

*8. Find both the area and the perimeter of this figure. Dimensions are in feet.

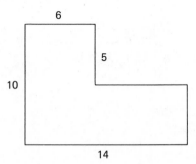

*9. Find the perimeter of this figure. How many floor tiles 1-inch square would it take to cover the figure? The dimensions are given in inches. Assume that the tiles can be reshaped as necessary.

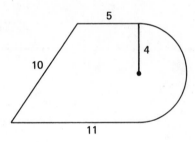

Simplify:

10. $\dfrac{(.0006 \times 10^{-42})(2000 \times 10^{-4})}{.004 \times 10^{-13}}$

11. $\dfrac{\dfrac{a}{b}}{\dfrac{a+b}{b}}$

12. $\dfrac{\dfrac{4}{x+y}}{\dfrac{m}{x+y}}$

13. $\dfrac{3}{2\sqrt{5}}$

14. $\dfrac{7}{3\sqrt{2}}$

15. $5\sqrt{75} - 3\sqrt{300} + 2\sqrt{27}$

16. $3\sqrt{2}(5\sqrt{2} - 4\sqrt{6})$

17. $\dfrac{x + 4x^2}{x}$

Add:

18. $\dfrac{x}{x+2} + \dfrac{3+x}{x^2+4x+4}$

19. $\dfrac{x}{x-3} + \dfrac{2x}{x^2-3x}$

Factor completely. Always factor the GCF as the first step.

20. $-x^3 + 5x^2 - 6x$

21. $2ax^3 - 18ax^2 + 40ax$

22. $-3pax + pax^2 + 2pa$

23. $-10mc + 3mxc + mx^2c$

Solve:

24. $\dfrac{2x+3}{6} - \dfrac{x}{2} = 1$

25. $\dfrac{3x+2}{3} - \dfrac{2}{5} = \dfrac{x+2}{6}$

26. Solve by graphing and then get an exact solution by using either substitution or elimination.
$$\begin{cases} 2x + 3y = 18 \\ -12x + 6y = -18 \end{cases}$$

27.   Divide: $x^4 - 2$ by $x + 1$

28.   Use elimination to solve.
$$\begin{cases} 6x + 4y = 11 \\ 2x - 3y = -5 \end{cases}$$

29.   Solve for m.

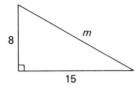

30.   Solve: $-2(-x^0 - 3) + 4(-x - 5^0) = -3^0(2x - 5)$

LESSON 32 *Quotient theorem for square roots •*
Area as a difference

32.A
quotient
theorem
for square roots

We remember that the square root of a product can be written as the product of the square roots of its factors. Thus

$$\sqrt{3 \cdot 2} = \sqrt{3}\sqrt{2}$$

A similar rule applies to the square root of a quotient (fraction), for the square root of a quotient can be written as a quotient of square roots. Thus,

$$\sqrt{\frac{3}{2}} = \frac{\sqrt{3}}{\sqrt{2}}$$

Recall that it is customary to rationalize the denominators of expressions that have radicals in the denominator. In this expression, we do this by multiplying by $\sqrt{2}$ over $\sqrt{2}$. This changes the denominator from the irrational number $\sqrt{2}$ to the rational number 2.

$$\frac{\sqrt{3}}{\sqrt{2}} \frac{\sqrt{2}}{\sqrt{2}} = \frac{\sqrt{6}}{2}$$

The expression $\dfrac{\sqrt{6}}{2}$ has the same value as $\sqrt{\dfrac{3}{2}}$ but is in a different form.

example 32.A.1   Simplify: $\sqrt{\dfrac{3}{7}}$

solution   First we will write the root of the quotient as the quotient of roots.

$$\frac{\sqrt{3}}{\sqrt{7}}$$

We finish by rationalizing the denominator.

$$\frac{\sqrt{3}}{\sqrt{7}} \frac{\sqrt{7}}{\sqrt{7}} = \frac{\sqrt{21}}{7}$$

example 32.A.2 Simplify: $\sqrt{\dfrac{2}{5}} + \sqrt{\dfrac{5}{2}}$

solution This problem was contrived so that when the terms are rationalized, each term will contain $\sqrt{10}$. Then by using a common denominator, the terms can be added. Problems like this one are good problems for practice with quotients of radicals, so they will appear often in the problem sets. We begin our simplification by writing each term as a quotient of radicals.

$$\frac{\sqrt{2}}{\sqrt{5}} + \frac{\sqrt{5}}{\sqrt{2}}$$

Next we rationalize both denominators.

$$\frac{\sqrt{2}}{\sqrt{5}}\frac{\sqrt{5}}{\sqrt{5}} + \frac{\sqrt{5}}{\sqrt{2}}\frac{\sqrt{2}}{\sqrt{2}} = \frac{\sqrt{10}}{5} + \frac{\sqrt{10}}{2}$$

We finish by using 10 as a common denominator and adding.

$$\frac{\sqrt{10}}{5}\left(\frac{2}{2}\right) + \frac{\sqrt{10}}{2}\left(\frac{5}{5}\right) = \frac{2\sqrt{10}}{10} + \frac{5\sqrt{10}}{10} = \frac{\mathbf{7}\sqrt{\mathbf{10}}}{\mathbf{10}}$$

example 32.A.3 Simplify: $3\sqrt{\dfrac{3}{7}} - 5\sqrt{\dfrac{7}{3}}$

solution We begin by writing each expression as a quotient of radicals.

$$\frac{3\sqrt{3}}{\sqrt{7}} - \frac{5\sqrt{7}}{\sqrt{3}}$$

Next we rationalize both denominators.

$$\frac{3\sqrt{3}}{\sqrt{7}}\frac{\sqrt{7}}{\sqrt{7}} - \frac{5\sqrt{7}}{\sqrt{3}}\frac{\sqrt{3}}{\sqrt{3}} = \frac{3\sqrt{21}}{7} - \frac{5\sqrt{21}}{3}$$

We finish by changing each expression so that the denominators are 21. Then we add.

$$\frac{3\sqrt{21}}{7}\left(\frac{3}{3}\right) - \frac{5\sqrt{21}}{3}\left(\frac{7}{7}\right) = \frac{9\sqrt{21}}{21} - \frac{35\sqrt{21}}{21} = -\frac{\mathbf{26}\sqrt{\mathbf{21}}}{\mathbf{21}}$$

example 32.A.4 Simplify: $2\sqrt{\dfrac{2}{7}} - 5\sqrt{\dfrac{7}{2}}$

solution First we write each term as a quotient of radicals.

$$\frac{2\sqrt{2}}{\sqrt{7}} - \frac{5\sqrt{7}}{\sqrt{2}}$$

Now we rationalize both denominators.

$$\frac{2\sqrt{2}}{\sqrt{7}}\frac{\sqrt{7}}{\sqrt{7}} - \frac{5\sqrt{7}}{\sqrt{2}}\frac{\sqrt{2}}{\sqrt{2}} = \frac{2\sqrt{14}}{7} - \frac{5\sqrt{14}}{2}$$

We finish by writing both terms with the same denominators and adding.

$$\frac{2\sqrt{14}}{7}\left(\frac{2}{2}\right) - \frac{5\sqrt{14}}{2}\left(\frac{7}{7}\right) = \frac{4\sqrt{14}}{14} - \frac{35\sqrt{14}}{14} = -\frac{\mathbf{31}\sqrt{\mathbf{14}}}{\mathbf{14}}$$

Many people find this type of problem troublesome. It is not difficult—it is just different. After it has appeared in several problem sets, it will become more familiar.

32.B
area as a difference

There is no one correct way to calculate the area of an irregular geometric figure. The area of the L-shaped figure on the left (a) can be calculated several ways. In (b) the area of *A* is 3 times 5 and the area of *B* is 10 times 2, so the total area is 35 square units. In (c) we find that the area of *C* is 5 times 5 and the area of *D* is 5 times 2, and again the total of the area is 35 square units.

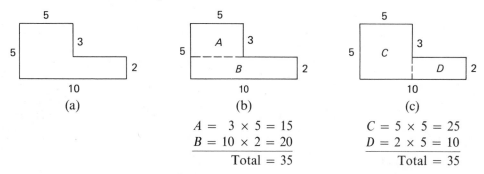

$$A = 3 \times 5 = 15$$
$$B = 10 \times 2 = 20$$
$$\text{Total} = 35$$

$$C = 5 \times 5 = 25$$
$$D = 2 \times 5 = 10$$
$$\text{Total} = 35$$

The areas of some irregular figures can be found by taking the difference of two areas. In the same problem, we can find the area of the outside dimensions (d) to be $5 \times 10 = 50$ and the area of *F* to be $5 \times 3 = 15$.

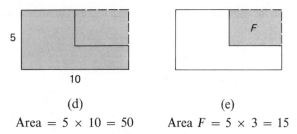

(d)   (e)
Area = $5 \times 10 = 50$ Area $F = 5 \times 3 = 15$

The difference of these areas is 35 square units, the same answer we got when we added areas *A* and *B* and areas *C* and *D*. We will call these two methods of finding areas the **sum method** and the **difference method.**

example 32.B.1 Find the area of the shaded area by using both the sum and difference methods. The dimensions are in feet.

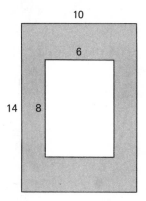

solution　To find the area by adding areas, we divide the figure into four subareas.

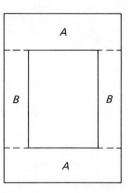

There are two areas labeled *A* and two labeled *B*. We will find the area of each area and then add.

$$\text{Area } A = 3 \text{ ft} \times 10 \text{ ft} = 30 \text{ ft}^2$$
$$\text{Area } A = 3 \text{ ft} \times 10 \text{ ft} = 30 \text{ ft}^2$$
$$\text{Area } B = 2 \text{ ft} \times 8 \text{ ft} = 16 \text{ ft}^2$$
$$\text{Area } B = 2 \text{ ft} \times 8 \text{ ft} = \underline{16 \text{ ft}^2}$$
$$\text{Total} = \mathbf{92 \text{ ft}^2}$$

To use the difference method we subtract the area of the center from the area of the whole.

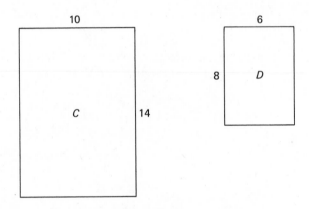

$$\text{Area } C = 10 \text{ ft} \times 14 \text{ ft} = 140 \text{ ft}^2$$
$$\text{Area } D = 6 \text{ ft} \times 8 \text{ ft} = \underline{48 \text{ ft}^2}$$
$$\text{Difference} = \mathbf{92 \text{ ft}^2}$$

We note that to find the area by adding we assumed that the figure was symmetrical to determine the widths of *B* and *A*. When we used the difference of two areas, we did not have to make assumptions.

problem set 32

1. Selby walked for a while at 4 mph and then jogged the rest of the way at 8 mph. If she covered 56 miles in 10 hours, how far did she walk and how far did she jog?

2. Johnny found that the large ones totaled 30 more than 3 times the number of small ones. The ratio of the number of small ones to the number of large ones was 1 to 6. How many were large, and how many were small?

3. Bruce discovered that the wishing well contained $9 in nickels and dimes, and that there were 30 more dimes than nickels. How many coins of each type were there?

4. Sarah knew that forty percent of the mixture was calcium. If 300 kilograms of other elements was used, what was the total weight of the mixture?

5. Four grams of magnesium combined with 20 grams of the other elements to form the compound. If 1440 grams of the compound was required, how many grams of the other elements was required?

6. Hedonism was pandemic as .87 of the students were hedonists. If 1914 students were hedonists, how many students were there in all?

Simplify:

*7. $\sqrt{\dfrac{2}{5}} + \sqrt{\dfrac{5}{2}}$ *8. $3\sqrt{\dfrac{3}{7}} - 5\sqrt{\dfrac{7}{3}}$ *9. $2\sqrt{\dfrac{2}{7}} - 5\sqrt{\dfrac{7}{2}}$

Use both the sum and the difference methods to find the area of the figure in Problem 10 and the area of the shaded area in the figure in Problem 11. Dimensions are in feet.

*10.

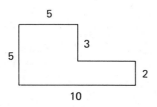

*11.

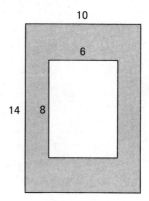

12. Find the equation of the line that passes through the point $(2, 2)$ and that is perpendicular to the line $y = 3x - 5$.

Simplify:

13. $\dfrac{(.0035 \times 10^{15})(.002 \times 10^{17})}{7000 \times 10^{33}}$

14. $\dfrac{\dfrac{x + 4y}{y}}{\dfrac{x + y}{y}}$

15. $\dfrac{xy + 4x^2y^2}{xy}$

16. $3\sqrt{125} - 2\sqrt{45} + 3\sqrt{20}$

17. $4\sqrt{5}(2\sqrt{10} - 3\sqrt{5})$

Add:

18. $\dfrac{x}{x + 3} - \dfrac{2x - 2}{x^2 + 5x + 6}$

19. $\dfrac{m}{m - 5} - \dfrac{2}{m^2 - 5m}$

Factor completely. Always factor the GCF as the first step.

20. $-2x^3 + 8x^2 - 6x$

21. $-14x^3 + 5x^4 + x^5$

22. $7ax + ax^3 - 8ax^2$

23. $-12py + px^2y + 4xpy$

Solve:

24. $\dfrac{x + 2}{5} - \dfrac{3x - 3}{2} = 4$

25. $\dfrac{x}{2} - \dfrac{3x + 2}{4} = 7$

26. Solve by graphing and then get an exact solution by using either substitution or elimination.
$$\begin{cases} -x + 2y = 4 \\ x + y = -2 \end{cases}$$

27. Multiply: $(4x + 2)(x^3 - 2x + 4)$

28. Use substitution to solve:
$$\begin{cases} x + 2y = 5 \\ 3x - y = 7 \end{cases}$$

29. Find the distance between $(-3, -2)$ and $(4, 5)$.

30. Simplify: $-3[(-2^0 - 3^0 - 2)(-3^0) - 2^2] - |3 - 2^0|$

LESSON 33 *Major rules of algebra • Complex fractions*

33.A
major rules of algebra

There are three major rules for algebraic manipulation. Two of these rules require the presence of an equals sign for their use because these are the two rules we use for solving equations.

1. **The same quantity can be added to both sides of an equation without changing the solutions to the equation.**
2. **Every term on both sides of an equation can be multiplied by the same quantity (except zero) without changing the solutions to the equation.**

$$x + 2 = 4 \qquad \frac{x}{3} = 5$$

We can use the first rule to solve the equation on the left and use the second rule to solve the equation on the right

$$\begin{array}{r} x + 2 = 4 \\ -2 \quad -2 \\ \hline x = 2 \end{array} \qquad (3)\frac{x}{3} = (3)(5)$$
$$x = 15$$

These two rules cannot always be used, for, if we have either of the expressions

$$\frac{4}{x + y} + \frac{xy}{x} \qquad \text{or} \qquad \frac{\dfrac{a}{x + y}}{\dfrac{b}{x}}$$

we cannot use the rules for solving equations since neither of these expressions contains an equals sign, and thus neither one is an equation.

When there is no equals sign in the expression, there is only one major rule that we can use, which is

> THE DENOMINATOR–NUMERATOR–SAME-QUANTITY
> THEOREM

3. **The denominator and the numerator of an expression can be multiplied or divided by the same quantity (except zero) without changing the value of the expression. Only the form of the expression is changed.**

This is possibly the most important rule in all algebra and can be used even when there is no equals sign present. **When there is no equals sign present, as in the expression,**

$$\frac{4}{x + y} + \frac{py}{x}$$

we cannot eliminate the denominators! All we can do is make the denominators the same so that the two terms can be added.

$$\frac{4}{x + y} + \frac{py}{x} = \frac{4}{(x + y)}\left(\frac{x}{x}\right) + \frac{py}{x}\left(\frac{x + y}{x + y}\right) = \frac{4x}{x(x + y)} + \frac{py(x + y)}{x(x + y)}$$

Now the terms can be added because the denominators are equal.

$$\frac{4x}{x(x + y)} + \frac{py(x + y)}{x(x + y)} = \frac{4x + py(x + y)}{x(x + y)}$$

Of course, either or both of the expressions in the numerator or denominator that contain parentheses can be multiplied out if we wish. If we do this, we get

$$\frac{4x + pxy + py^2}{x^2 + xy}$$

The numerator-denominator–same-quantity theorem can also be used to simplify the expression

$$\frac{\dfrac{a}{x + y}}{\dfrac{b}{x}}$$

We cannot eliminate the denominators because there is no equals sign in the expression. Therefore the expression is not an equation, and the rules for equations cannot be used. But we can always (even in equations) use the denominator-numerator–same-quantity theorem.

　　　Thus we can simplify this expression by multiplying both the numerator and the denominator by $\frac{x}{b}$, which is the reciprocal of $\frac{b}{x}$.

$$\frac{\dfrac{a}{x + y} \cdot \dfrac{x}{b}}{\dfrac{b}{x} \cdot \dfrac{x}{b}} = \frac{ax}{b(x + y)}$$

We did not eliminate the denominator, but now we have a simpler expression because we have only one fraction instead of one fraction divided by another fraction.

33.B
complex fractions

Fractions of fractions such as the one just simplified are called **complex fractions.** We also use these words to describe expressions such as

$$\frac{\dfrac{x}{y} + \dfrac{4}{yx}}{\dfrac{a}{y}}$$

This expression has a numerator composed of the sum of two fractions and a denominator that has only one fraction. **We will define complex fractions as fractions that contain more than one fraction line.**

example 33.B.1   Simplify: $\dfrac{\dfrac{x}{y} + \dfrac{4}{yx}}{\dfrac{a}{y}}$

solution   **There is no equals sign so we cannot use either of the rules for equations. The only rule that we can use is the denominator-numerator–same-quantity theorem.** We use it first to add the two terms in the numerator.

$$\frac{\dfrac{x}{y} + \dfrac{4}{xy}}{\dfrac{a}{y}} = \frac{\dfrac{x^2}{xy} + \dfrac{4}{xy}}{\dfrac{a}{y}} = \frac{\dfrac{x^2 + 4}{xy}}{\dfrac{a}{y}} .$$

Now we will use the same theorem again to multiply above and below by $\frac{y}{a}$, which is the reciprocal of the denominator $\frac{a}{y}$.

$$\frac{\dfrac{(x^2 + 4)}{xy} \cdot \dfrac{y}{a}}{\dfrac{a}{y} \cdot \dfrac{y}{a}} = \frac{x^2 + 4}{xa}$$

example 33.B.2   Simplify: $\dfrac{\dfrac{a}{x + y} + \dfrac{m}{y}}{\dfrac{x}{a + m}}$

solution   **Again we note that an equals sign is not present. Thus, the only rule we can use is the denominator-numerator–same-quantity theorem.** First, we use it to help us add the two terms in the numerator.

$$\frac{\dfrac{a}{x + y} + \dfrac{m}{y}}{\dfrac{x}{a + m}} = \frac{\dfrac{a}{x + y}\left(\dfrac{y}{y}\right) + \dfrac{m}{y}\left(\dfrac{x + y}{x + y}\right)}{\dfrac{x}{a + m}}$$

$$= \frac{\dfrac{ay + mx + my}{y(x + y)}}{\dfrac{x}{a + m}}$$

Now we finish by using the same theorem again. We multiply above and below by $\frac{a+m}{x}$, which is the reciprocal of $\frac{x}{a+m}$.

$$\frac{\dfrac{(ay + mx + my)}{y(x + y)} \cdot \left(\dfrac{a + m}{x}\right)}{\dfrac{x}{a + m} \cdot \left(\dfrac{a + m}{x}\right)} = \frac{(ay + mx + my)(a + m)}{xy(x + y)}$$

This answer is a little complicated but is closer to real-life answers than the answers to problems that are carefully contrived so that a lot of things can be canceled.

1. Four-seventeenths of the hedonists were also sybarites. If 104 were sybarites, how many hedonists were there in all?

2. Mercury and phosphate were mixed in the ratio of 7 to 2. If 3600 grams of the mixture was required, how much mercury was needed?

3. Twenty percent of the lithium did not combine. If 1620 grams did combine, how much lithium was there in all?

4. There were 10 more reds than 8 times the number of blues. Also the number of reds was 5 less than 11 times the number of blues. How many of each were there?

5. The total distance was 540 miles. Part of the journey was on a motorcycle at 40 mph and part was in a car at 60 mph. What distance was covered by motorcycle if the total time of the journey was 11 hours?

Simplify:

*6. $\dfrac{\dfrac{x}{y}+\dfrac{4}{xy}}{\dfrac{a}{y}}$

*7. $\dfrac{\dfrac{a}{x+y}+\dfrac{m}{y}}{\dfrac{x}{a+m}}$

Use both the sum and the difference methods to find the areas of the following figures. The dimensions are in meters.

8.

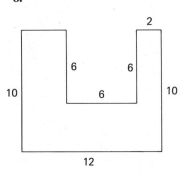

9.

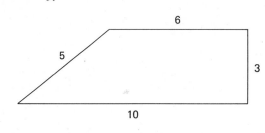

10. Find the equation of the line that is perpendicular to the line $y = -\frac{3}{5}x + 2$ and that passes through the point $(-4, 2)$.

Simplify:

11. $3\sqrt{\dfrac{5}{2}} - 2\sqrt{\dfrac{2}{5}}$

12. $4\sqrt{\dfrac{5}{6}} - 2\sqrt{\dfrac{6}{5}}$

13. Find the perimeter of the figure in Problem 8.

14. How many 1-inch floor tiles would it take to cover the solid line triangle? Dimensions are in inches.

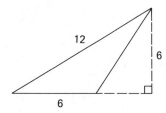

Simplify:

15. $\dfrac{(.0027 \times 10^{15})(500 \times 10^{-20})}{900 \times 10^{14}}$

16. $\dfrac{x + \dfrac{4xy}{x}}{\dfrac{1}{x} - y}$

17. $3\sqrt{18} + 2\sqrt{50} - \sqrt{98}$

18. $\dfrac{4x + 4xy}{4x}$

Add:

19. $\dfrac{a}{x(x + y)} + \dfrac{b}{x^2} + \dfrac{cx + 4}{x + y}$

20. $\dfrac{4}{x + 4} - \dfrac{6x - 2}{x^2 + 2x - 8}$

Factor completely. Always factor the GCF as the first step.

21. $5x^2 + 4x^3 - x^4$ **22.** $10k^2 - 7k^2x + k^2x^2$ **23.** $apx^2 - 20ap - apx$

Solve:

24. $\dfrac{x + 2}{3} - \dfrac{2x - 2}{4} = 5$

25. $\dfrac{3x - 2}{2} - \dfrac{2x + 3}{3} = 4$

26. Solve by graphing and then get an exact solution by using either substitution or elimination.
$$\begin{cases} 2x - y = -5 \\ x + y = 1 \end{cases}$$

27. Divide $2x^4 - x$ by $x - 2$ and check.

28. Evaluate: $-3^0 - x - y^0 - y^2(-2^0 - 3) - |-2 - x|$ if $x = -2$ and $y = -3$.

29. Find the perimeter of a rectangle that measures 2 ft by 4 ft.

30. Find the area of the rectangle of Problem 29.

LESSON 34 *Uniform motion problems:* $D_1 + 40 = D_2$

34.A
uniform motion problems

The distance diagrams for the uniform motion problems encountered thus far have looked like one of the following:

In both diagrams (a), the distance traveled by number 1 equals the distance traveled by number 2, so the distance equation for both diagrams is

$$R_1 T_1 = R_2 T_2$$

In all three diagrams in (b), the distance traveled by number 1 plus the distance traveled by number 2 equals 480, so the distance equation for all three diagrams is

$$R_1 T_1 + R_2 T_2 = 480$$

In some uniform motion problems, the distance traveled by one object exceeds the distance traveled by another object by a specified amount. The diagrams for one of these problems usually look like one of the following:

In the diagram on the left, both A and B began at the same point. For some reason, B traveled 40 more miles (or kilometers or whatever) than A. In the diagram on the right, A started 40 units in front of B, but they ended up at the same place. In both diagrams, the sum of 40 and the distance traveled by A equals the distance traveled by B.

$$\text{Distance } A + 40 = \text{Distance } B$$

Now since rate times time equals distance, we can write the distance equation for both diagrams as

$$R_A T_A + 40 = R_B T_B$$

example 34.A.1   Millicent began the journey at 6 a.m. at 50 kilometers per hour. Beauregard began to chase her at 10 a.m. at 60 kilometers per hour. What time was it when Beauregard got within 40 kilometers of Millicent?

solution   Beauregard and Millicent began at the same point, but Millicent traveled 40 kilometers farther than Beauregard traveled. Thus the distance diagram and distance equation are as follows:

$$R_B T_B + 40 = R_M T_M$$

We have one equation in four unknowns. Thus we need three more equations. They are

$$R_M = 50 \qquad R_B = 60 \qquad T_M = T_B + 4$$

Now we substitute these equivalences into the distance equation and solve for T_B.

$$
\begin{array}{ll}
60T_B + 40 = 50(T_B + 4) & \text{substituted} \\
60T_B + 40 = 50T_B + 200 & \text{multiplied} \\
10T_B = 160 & \text{simplified} \\
T_B = 16 & \text{divided}
\end{array}
$$

Thus, in 16 hours, Beauregard got within 40 kilometers of Millicent. Sixteen hours after 10 a.m. would be **2 o'clock** the next morning.

example 34.A.2   When the sheriff began his pursuit, Robin Hood was already 7 miles out of Nottingham. If the sheriff traveled at 6 miles per hour while Robin Hood's rate was $2\frac{1}{2}$ miles per hour, how long did it take the sheriff to catch up?

solution   The distance diagram, the distance equation, and the rate equations are

$$R_R T_R + 7 = R_S T_S, \qquad R_S = 6, \qquad R_R = 2\tfrac{1}{2}$$

When the problem began, Robin was already 7 miles out of town (how he got there or when is not part of this problem). Thus Robin and the sheriff began traveling at the same time and stopped at the same time, so the time equation is

$$T_S = T_R$$

Next we substitute for R_R, R_S and T_S in the distance equation.

$$\frac{5}{2} T_R + 7 = 6 T_R$$

We substituted $\tfrac{5}{2}$ for R_R, 6 for R_S, and T_R for T_S. We finish by eliminating the denominator by multiplying every term on both sides by 2.

$$5 T_R + 14 = 12 T_R \qquad \text{multiplied by 2}$$
$$14 = 7 T_R \qquad \text{simplified}$$
$$\mathbf{2 = T_R} \qquad \text{divided}$$

Thus the sheriff caught Robin in 2 hours because his traveling time equaled that of Robin.

problem set 34

*1.   Millicent began the journey at 6 a.m. at 50 kilometers per hour. Beauregard began to chase her at 10 a.m. at 60 kilometers per hour. What time was it when Beauregard got within 40 kilometers of Millicent?

*2.   When the sheriff began his pursuit, Robin Hood was already 7 miles out of Nottingham. If the sheriff traveled at 6 miles per hour while Robin Hood's rate was $2\tfrac{1}{2}$ miles per hour, how long did it take the sheriff to catch up?

3.   Kay rode the bicycle into the country at 10 mph, and Yancy pushed it back to town at 3 mph. If the round trip took 13 hours, how far did Kay ride the bicycle into the country?

4.   Thirty percent of the sulfur desiccated. If 42 tons did not desiccate, how much sulfur was there in all?

5.   Fifty grams of sodium bicarbonate was mixed with other compounds to get 150 grams of mixture. If 300 grams of the other compounds was available, how much sodium bicarbonate was needed to make the mixture?

6.   Find three consecutive integers such that 5 times the sum of the first and third is 14 greater than 8 times the second.

Simplify:

7. $\dfrac{\dfrac{y}{ab} - ab}{\dfrac{1}{a} - \dfrac{a}{b}}$

8. $\dfrac{\dfrac{1}{x} - b}{x}$

Use either the sum method or the difference method to find the areas of the shaded parts of the following figures. The dimensions are in feet.

9.

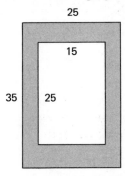

25

15

35   25

10.

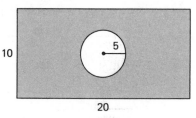

10

5

20

11. Find the equation of the line that is perpendicular to $2x + y = 4$ and that passes through the point $(-2, -1)$.

Simplify:

12. $2\sqrt{\dfrac{2}{9}} - 3\sqrt{\dfrac{9}{2}}$

13. $-3\sqrt{\dfrac{2}{3}} + 2\sqrt{\dfrac{3}{2}}$

14. How many 1-cm-square floor tiles will it take to completely cover the shaded area in the given figure? Dimensions are in centimeters.

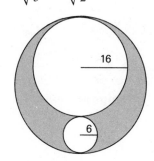

16

6

Simplify:

15. $\dfrac{(.000032 \times 10^4)(700 \times 10^{-14})}{16,000}$

16. $\dfrac{4 + \dfrac{x}{y^2}}{3 - \dfrac{1}{y^2}}$

17. $8\sqrt{27} - 2\sqrt{75} + 2\sqrt{147}$

18. $\dfrac{x^2y - 5x^2y^2}{x^2y}$

Add:

19. $\dfrac{a}{x(x + y)} + \dfrac{bx}{x^2(x + y)} + \dfrac{cx}{x^3}$

20. $\dfrac{x - 4}{x - 3} - \dfrac{2x - 1}{x^2 - 6x + 9}$

Factor completely. Always begin by factoring the GCF.

21. $-4x^2 + 2x^3 + 2x^4$ **22.** $ax^2p - 8pa - 2axp$ **23.** $yx^2 - 4xy + 4y$

Solve:

24. $\dfrac{x - 3}{2} - \dfrac{3x + 4}{2} = 3$ **25.** $\dfrac{x}{3} - \dfrac{2x - 4}{2} = 5$

26. Solve by graphing and then get an exact solution by using either substitution or elimination.
$$\begin{cases} x - 3y = 6 \\ 2x + y = 2 \end{cases}$$

27. Multiply: $(x^2 + x)(x^2 + 2x + 3)$

28. Use substitution to solve: $\begin{cases} 3x - 2y = 2 \\ x - 3y = 4 \end{cases}$

29. Find x.

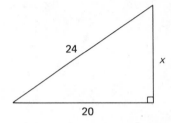

30. Evaluate: $-2^0[-2^0 - 2^2 - (-2)^3 - 2][-2^0 - 2] + x - xy$ if $x = -3$ and $y = -4$.

LESSON 35 · *Volume · Fractional exponents*

35.A
the concept of volume

We have used the words **floor tiles** to help with the abstraction of **area.** When we are asked to find the area of a given geometric figure in square yards, we are really asked to find the number of 1-yard-square floor tiles that it would take to completely cover the figure. If it would require 492.316 floor tiles, we say that the area of the figure is 492.316 square yards. The abstraction of volume can also be reified. We will do this by thinking of **sugar cubes** when we hear the word **volume.** A cube is a six-sided figure whose sides are all squares of equal dimensions.

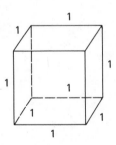

Here we show a sugar cube each of whose edges is 1 inch in length. We say that this cube has a volume of 1 cubic inch, which we can write as 1 in³:

$$1 \text{ cubic inch} = 1 \text{ in}^3$$

We use sugar cubes because sugar cubes are real, while an empty cube is an abstraction. If a particular container will hold exactly 1406.21 of these sugar cubes, we say that the volume of the container is 1406.21 cubic inches, or 1406.21 in³. Thus, when we are told that the volume of a room is 3,200 cubic feet, we are told that it would take 3,200 cubes of sugar (1 ft³ each) to completely fill the room. Of course the sugar cubes might have to be crushed or otherwise reshaped to make them fit exactly.

example 35.A.1   Find the volume of a rectangular box whose base measures 2 ft by 3 ft and whose height is 10 ft.

solution   The base has an area of 6 ft², as we see here. We can set one sugar cube on each of these squares. Thus, we can set 6 sugar cubes on the first layer.

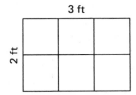 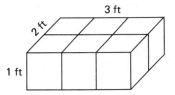

The box is 10 ft tall, so it will have 10 layers, each with 6 cubes, so the total number of cubes will be 60. Thus the volume of the solid is

$$6 \text{ ft}^2 \times 10 \text{ ft} = \textbf{60 ft}^3$$

From this example, we can deduce that the volume of any container that has vertical sides is the area of the base times the height. The area of the base tells us how many cubes can be placed in the first layer, and the height tells us how many layers there are.

example 35.A.2   Find the volume of a solid whose base is as shown and whose height is 5 inches. The dimensions of the base are in inches.

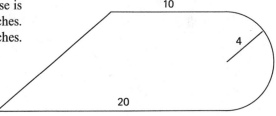

solution   We begin by finding the area of the base.

$$\text{Area of triangle} = \frac{8 \text{ in} \times 10 \text{ in}}{2} = 40 \text{ in}^2$$

$$\text{Area of rectangle} = 8 \text{ in} \times 10 \text{ in} = 80 \text{ in}^2$$

$$\text{Area of half circle} = \frac{(3.14)(4)^2}{2} = 25.12 \text{ in}^2$$

$$\text{Total area} = \textbf{145.12 in}^2$$

Thus 145.12 one-inch cubes can be placed in the first layer. Since the solid will be 5 inches high, the total number of cubes will be

$$5 \times 145.12 \text{ cubes} = 725.6 \text{ cubes}$$

Thus the volume of the solid is **725.6 in³**.

35.B
fractional exponents

There are two ways to write the square root of 2.

$$\sqrt{2} \quad \text{and} \quad 2^{\frac{1}{2}}$$

There is nothing to understand, for this is a definition of what we mean when we use the square root radical sign or the fractional exponent $\frac{1}{2}$. We use the fractional exponent $\frac{1}{3}$ to designate the cube root, the fractional exponent $\frac{1}{4}$ to designate the fourth root, etc.

$$2^{\frac{1}{2}} = \sqrt{2} \qquad 2^{\frac{1}{3}} = \sqrt[3]{2} \qquad 2^{\frac{1}{4}} = \sqrt[4]{2}$$

The rules for exponents are the same for fractional exponents as they are for integral exponents. Thus both the product theorem and power theorem also apply to fractional exponents.

$$x^{\frac{1}{3}} \cdot x^{\frac{4}{3}} = x^{\frac{5}{3}} \quad \text{and} \quad (x^{\frac{1}{3}})^{\frac{4}{3}} = x^{\frac{4}{9}}$$

example 35.B.1 Simplify $4^{-\frac{1}{2}}$.

solution Negative exponents are not operation indicators, so our first step is to write the expression with a positive exponent. Then we simplify.

$$4^{-\frac{1}{2}} = \frac{1}{4^{\frac{1}{2}}} = \frac{1}{2}$$

example 35.B.2 Simplify $-27^{-\frac{1}{3}}$.

solution We can mentally separate the minus sign in front from the rest of the problem. We do this and write the expression with a positive exponent.

$$\underbrace{}_{\text{Minus sign}} \quad - \qquad \frac{1}{27^{\frac{1}{3}}}$$

Now we simplify $27^{\frac{1}{3}}$ and bring the minus sign back and get

$$-\frac{1}{3}$$

example 35.B.3 Simplify $16^{\frac{3}{2}}$.

solution This expression can be simplified in two ways.

$$\text{(a)} \quad (16^3)^{\frac{1}{2}} \qquad \text{(b)} \quad (16^{\frac{1}{2}})^3$$

Both (a) and (b) have the same value as the original expression. Next we simplify within the parentheses and get

$$\text{(a)} \quad (4096)^{\frac{1}{2}} \qquad \text{(b)} \quad (4)^3$$

Both of these have a value of 64, but it is much easier to raise 4 to the third power than it is to take the square root of 4096. For this reason, it is recommended that the fraction always be left inside the parentheses when doing the first step of the simplification. Also we note that it is necessary to learn to recognize that these problems are contrived problems designed to give practice in working with fractional exponents. Not all expressions with fractional exponents can be simplified, for if we have

$$15^{\frac{3}{2}}$$

we can do nothing, for neither

$$(15^{\frac{1}{2}})^3 \qquad \text{nor} \qquad (15^3)^{\frac{1}{2}}$$

can be simplified without using logarithms or a calculator.

example 35.B.4 Simplify $-8^{-\frac{2}{3}}$.

solution As the first step, we mentally separate the minus sign and write the expression with a positive exponent.

$$\underbrace{}_{\text{Minus sign}} \quad - \qquad \frac{1}{8^{\frac{2}{3}}}$$

Next we write the exponent as a product that has the fraction inside the parentheses.

$$- \quad \frac{1}{(8^{\frac{1}{3}})^2}$$

Now we simplify

$$- \quad \frac{1}{2^2}$$

and bring back the minus sign.

$$-\frac{1}{4}$$

problem set 35

1. Matthew was 1200 yards ahead when Lowe began his pursuit. If Lowe ran 3 times as fast as Matthew and overtook him in 30 minutes, how fast did each boy run?

2. Cheryl and Juby trudged at 4 miles per hour until their packs got too heavy. Then they dropped their packs and continued at the brisk pace of 6 mph. If the total trip of 56 miles took 12 hours, how long did they trudge? How long did they walk briskly?

3. The dhow made the trip in 12 hours while the brigantine made the same trip in 4 hours. If the speed of the brigantine was 6 miles per hour greater than the speed of the dhow, what was the distance traveled by each?

4. Blues were $5 each while yellows cost $8 each. Penelope spent $82, and the number of blues she bought was 2 greater than twice the number of yellows. How many of each kind did she buy?

5. Sixty percent of the aluminum fused as it should have. If 40 tons did not fuse, how much aluminum was there in all?

6. Twenty grams of vanadium was melted with other metals to make 40 grams of the alloy. If 400 grams of the other metals was available, how much vanadium should be used?

*7. Find the volume of a solid whose base is as shown and whose height is 5 inches. Dimensions are in inches.

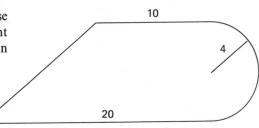

Simplify:

*8. $4^{-\frac{1}{2}}$

9. $27^{-\frac{1}{3}}$

*10. $16^{\frac{3}{2}}$

*11. $-8^{-\frac{2}{3}}$

12. How many 1-foot-square floor tiles would it take to cover this figure? Dimensions are in feet.

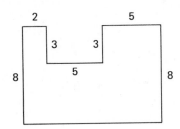

Simplify:

13. $3\sqrt{\dfrac{7}{5}} + 2\sqrt{\dfrac{5}{7}}$ **14.** $2\sqrt{\dfrac{2}{5}} - 9\sqrt{\dfrac{5}{2}}$

15. $\dfrac{\dfrac{a}{b} - 4}{\dfrac{xy}{b}}$ **16.** $\dfrac{\dfrac{x}{x+y} + 6}{\dfrac{4}{x+y}}$

17. Find the perimeter of the figure in Problem 12.

Use either the sum method or the difference method to find the area of the figure in Problem 18 and the shaded area of the figure in Problem 19. Dimensions are in feet.

18. 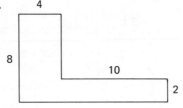   **19.**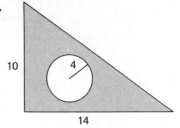

20. Find the equation of the line that passes through the points $(-2, 3)$ and $(4, -5)$.

Add:

21. $\dfrac{4}{x^2(x+y)} + \dfrac{2x-2}{x(x+y)}$ **22.** $\dfrac{3x}{x-2} - \dfrac{2x}{x^2+x-6}$

Factor completely:

23. $35a - ax^2 - 2xa$ **24.** $8x^2 - 2x^3 - x^4$

Simplify:

25. $2\sqrt{3} \cdot \sqrt{12} - 3\sqrt{2} \cdot \sqrt{6} + 4\sqrt{2}(3\sqrt{2} - \sqrt{6})$

26. $\dfrac{(7000 \times 10^{14})(.0002 \times 10^{-11})}{1400 \times 10^{-10}}$

Solve:

27. $\dfrac{x-3}{7} - \dfrac{2x}{4} = 5$ **28.** $.002x = .02 + .04$ **29.** $2\dfrac{1}{3}x - 2x^0 = 3\dfrac{1}{4}$

30. Find the distance between $(-4, 2)$ and $(7, 3)$.

LESSON 36   *Contrived problems • Multiplication of rational expressions • Division of rational expressions*

36.A

problems that teach skills

In beginning algebra courses, we work at developing algebraic skills that can be used later to learn advanced algebraic concepts and skills that can be used in chemistry, physics, and other mathematically based disciplines. There are some skills that we can learn at this level that can be immediately useful. Percent problems, ratio problems, and fractional-part-of-a-number problems fall into this category. We must also learn skills that are not immediately useful. Uniform motion word problems fall into this category. We study these problems because working them will allow us to develop skills with word problems that can be used to solve word problems in more advanced courses. We certainly can find no way to apply immediately our skill of adding abstract expressions such as

$$\frac{a}{b} + \frac{c}{x} + \frac{b}{x + y}$$

but without this skill, many problems that will be encountered in advanced algebra and trigonometry would be difficult, if not impossible. In this lesson, we will investigate multiplication and division of factorable rational expressions that have common factors. These problems have no immediate application but are good problems for practicing factoring and for practicing canceling factors that are sums.

36.B
multiplication of rational expressions

We remember that fractions are multiplied by multiplying the numerators to form the new numerator and by multiplying the denominators to form the new denominator.

$$\frac{3}{5} \cdot \frac{7}{8} = \frac{21}{40} \qquad \frac{x}{y} \cdot \frac{m}{y + k} = \frac{xm}{y^2 + yk}$$

Another name for a fraction is a **ratio,** and we remember that this is the reason that we often call fractional expressions **rational expressions.**

example 36.B.1   Multiply: $\dfrac{x^2 + x - 12}{x^2 - x - 20} \cdot \dfrac{x^2 + 2x - 35}{x^2 + 9x + 14}$

solution   This is not a multiplication problem but is a contrived problem designed to provide practice in factoring and canceling. Thus, we will begin by factoring all four expressions, and then we will cancel the common factors.

$$\frac{\cancel{(x + 4)}(x - 3)}{\cancel{(x - 5)}\cancel{(x + 4)}} \cdot \frac{\cancel{(x + 7)}\cancel{(x - 5)}}{(x + 2)\cancel{(x + 7)}} = \frac{x - 3}{x + 2}$$

36.C
division of rational expressions

We remember that the denominator of a fraction of fractions can be changed to 1 by multiplying it by the reciprocal of the denominator. Thus

$$\frac{\dfrac{x}{y}}{\dfrac{k}{p}}$$

can be simplified by multiplying above and below by $\dfrac{p}{k}$.

$$\frac{\dfrac{x}{y} \cdot \dfrac{p}{k}}{\dfrac{k}{p} \cdot \dfrac{p}{k}} = \frac{\dfrac{xp}{yk}}{1} = \frac{xp}{yk}$$

If the problem had been stated as

$$\frac{x}{y} \div \frac{k}{p}$$

the same result could be obtained by inverting the divisor and then multiplying

$$\frac{x}{y} \div \frac{k}{p} = \frac{x}{y} \cdot \frac{p}{k} = \frac{xp}{yk}$$

This same procedure will be used in the next example.

example 36.C.1   Simplify: $\dfrac{x^2 - 6x + 8}{x^2 + 3x - 28} \div \dfrac{x^2 - 2x - 15}{x^2 + 2x - 35}$

solution   This is also a contrived problem designed to give practice in factoring and canceling. We begin by inverting the divisor and changing the division symbol to a dot that indicates multiplication.

$$\frac{x^2 - 6x + 8}{x^2 + 3x - 28} \cdot \frac{x^2 + 2x - 35}{x^2 - 2x - 15}$$

We finish by factoring all four expressions and canceling, just as we did in the last example.

$$\frac{(x - 2)(x - 4)}{(x + 7)(x - 4)} \cdot \frac{(x + 7)(x - 5)}{(x - 5)(x + 3)} = \frac{x - 2}{x + 3}$$

example 36.C.2   Simplify: $\dfrac{x^2 + x - 6}{x^3 - 2x^2 - 35x} \div \dfrac{x + 3}{x^2 - 7x}$

solution   This time we will factor and invert the divisor in the same step. Then we finish by canceling common factors.

$$\frac{(x + 3)(x - 2)}{x(x + 5)(x - 7)} \cdot \frac{x(x - 7)}{x + 3} = \frac{x - 2}{x + 5}$$

problem set 36

1. The students whose phrasing was pleonastic used 240 percent more words than were necessary. If 400 words were necessary, how many words did these students use?

2. Fats cost 4 cents each while leans cost 21 cents each. Moxley and Rachel bought a total of 30 and spent $2.90. How many of each kind did they buy?

3. The fraction had a value of $\frac{3}{5}$. The sum of the numerator and the denominator was 40. What was the fraction?

4. Don made the trip in 10 hours. Hazel drove 10 miles per hour faster than Don so she made the trip in only 8 hours. How many miles long was the trip?

5. Brett and Julie headed north at 8 a.m. By noon, Brett was 80 miles ahead. What was Brett's speed if Julie's speed was 30 miles per hour?

6. Find four consecutive multiples of 7 such that 4 times the sum of the first and 2 is 15 greater than 3 times the third.

Simplify:

*7. $\dfrac{x^2 + x - 12}{x^2 - x - 20} \cdot \dfrac{x^2 + 2x - 35}{x^2 + 9x + 14}$ *8. $\dfrac{x^2 - 6x + 8}{x^2 + 3x - 28} \div \dfrac{x^2 - 2x - 15}{x^2 + 2x - 35}$

*9. $\dfrac{x^2 + x - 6}{x^3 - 2x^2 - 35x} \div \dfrac{x + 3}{x^2 - 7x}$

10. $\dfrac{1}{-3^{-2}}$

11. $-27^{-\frac{2}{3}}$

12. $\dfrac{1}{81^{-\frac{3}{4}}}$

13. $(-27)^{-\frac{2}{3}}$

14. $\dfrac{\dfrac{1}{x} + \dfrac{4}{y}}{3 + \dfrac{1}{xy}}$

15. $\dfrac{\dfrac{4}{x} - 3}{\dfrac{7}{x} + 2}$

16. $3\sqrt{\dfrac{5}{3}} - 2\sqrt{\dfrac{3}{5}}$

17. $\dfrac{(6000 \times 10^{14})(300 \times 10^{-22})}{.00018 \times 10^{-5}}$

18. $2\sqrt{\dfrac{7}{3}} - 3\sqrt{\dfrac{3}{7}}$

19. $4\sqrt{12}(3\sqrt{2} - 4\sqrt{3})$

20. $2\sqrt{28} - 3\sqrt{63} + 2\sqrt{175}$

Add:

21. $\dfrac{6}{x^2(x + 2)} - \dfrac{3}{x^2 + 3x + 2}$

22. $\dfrac{p}{ax^2} + \dfrac{cx + a}{ax^3} + \dfrac{mx + b}{a^2x^4}$

Solve:

23. $\dfrac{3x - 2}{7} - \dfrac{x}{4} - \dfrac{x - 3}{2} = 1$

24. $3x^0 - 2(x - 3^0) - |-11 - 2| = 4x(2 - 5^0) - 7x$

25. Find the equations of lines A and B.

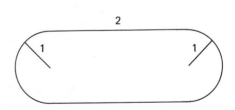

26. The figure shows the base of a container whose sides are perpendicular to the base and are 10 feet high. What is the volume of the container in cubic feet? Dimensions are in feet.

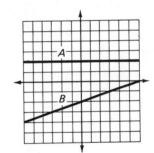

27. Divide $x^4 - 2$ by $x + 1$ and check.

28. Find the distance between $(-3, 7)$ and $(4, -2)$.

Simplify:

29. $\dfrac{x^{-2}y}{zx^4}\left(3x^6y^{-1}z - \dfrac{3x^{-3}y^2}{z^{-1}x^{-4}}\right)$

30. $-3^0(2 - 4^0) - |-3| - 2 - 4^2 - (-2)^3 - 2$

LESSON 37 *Chemical compounds*

37.A
chemical compounds

In chemistry courses, some problems deal with the weight relationships of chemical compounds. Most of these problems are straightforward ratio and percent problems, and it is necessary to know only two things about chemistry to be able to work them.

The first thing is that atoms always combine in the same combinations when they unite to form a specific compound. The chemical formulas for the compounds tell us the number of each kind of atom in a molecule of the compound. For example,

H_2O is the formula for water. Each molecule has two hydrogen (H) atoms and one oxygen (O) atom.

Zn_3N_2 is the formula for zinc nitride. Each molecule has three zinc (Zn) atoms and two nitrogen (N) atoms.

NaOH   is the formula for sodium hydroxide. Each molecule has one sodium (Na) atom, one oxygen (O) atom, and one hydrogen (H) atom.

The second thing we need to know is that every kind of atom has a different weight. We express the weights of atoms in grams. The **gram atomic weight** of oxygen is 16 grams, and the gram atomic weight of hydrogen is 1 gram. We will give the gram atomic weights of the elements in parentheses. In a problem about oxygen and hydrogen, we would note the gram atomic weights as (O, 16; H, 1).

When elements combine to form a compound, we say that the molecules of the compound have a **gram molecular weight** that can be found from the gram atomic weights. Since the formula for water is

$$H_2O$$

we find that the gram molecular weight of water is

$$2(1 \text{ gram}) + 1(16 \text{ grams}) = 18 \text{ grams}$$

example 37.A.1   The chemical formula for water is H_2O. If we have 3600 grams of water, what is the weight of the oxygen? (O, 16; H, 1).

solution   We follow the same procedure we have been using for ratio problems. We note the weight of each element and the total weight. In a molecule of water the weights are

$$\text{Hydrogen:} \quad 2 \times 1 \ = \ 2$$
$$\text{Oxygen:} \quad 1 \times 16 = 16$$
$$\text{Total:} \qquad\qquad = 18$$

Thus the three possible ratios are

(a) $\dfrac{H}{Ox} = \dfrac{2}{16}$ (b) $\dfrac{H}{Total} = \dfrac{2}{18}$ (c) $\dfrac{Ox}{Total} = \dfrac{16}{18}$

The chemical symbol for oxygen is O, but we have used Ox so that O for oxygen won't be confused with 0 for zero. We have been told that the total weight was 3600 grams and were asked for the weight of the oxygen, so we will use (c) and replace total with 3600.

(c) $\dfrac{Ox}{Total} = \dfrac{16}{18} \quad \longrightarrow \quad \dfrac{Ox}{3600} = \dfrac{16}{18}$

To solve, we cross multiply and then divide both sides by 18.

$$18 \cdot \text{Ox} = 16 \cdot 3600 \longrightarrow \frac{18 \cdot \text{Ox}}{18} = \frac{16 \cdot 3600}{18} \longrightarrow \text{Ox} = \textbf{3200 grams}$$

example 37.A.2   The chemical formula for ammonia is NH_3. This tells us that each ammonia molecule contains one nitrogen atom and three hydrogen atoms. If we have 510 grams of ammonia, how much does the nitrogen weigh? (H, 1; N, 14).

solution

Nitrogen: $14 \times 1 = 14$

Hydrogen: $1 \times 3 = 3$

Total: $= 17$

Thus the three ratios are

$$(a) \quad \frac{N}{H} = \frac{14}{3} \qquad (b) \quad \frac{N}{T} = \frac{14}{17} \qquad (c) \quad \frac{H}{T} = \frac{3}{17}$$

We have been told that the total is 510 grams and have been asked for the weight of the nitrogen so we will use (b). We will replace T with 510, cross multiply, and then divide by 17.

$$\frac{N}{T} = \frac{14}{17} \longrightarrow \frac{N}{510} = \frac{14}{17} \longrightarrow 17N = 14 \cdot 510 \longrightarrow \frac{17N}{17} = \frac{14 \cdot 510}{17}$$

$$N = \textbf{420 grams}$$

example 37.A.3   The formula for ammonium chloride is NH_4Cl. This means that in one molecule of ammonium chloride there is one atom of nitrogen, four atoms of hydrogen, and one atom of chlorine. How many grams of chlorine are there in 1060 grams of ammonium chloride? (N, 14; H, 1; Cl, 35)

solution   We begin by finding the molecular weights of each element in a molecule of the compound.

Nitrogen: $1 \times 14 = 14$

Hydrogen: $4 \times 1 = 4$

Chlorine: $1 \times 35 = 35$

Total: $= 53$

We see that the ratio of the weight of the chlorine to the weight of the total solution is 35 to 53.

$$\frac{Cl}{T} = \frac{35}{53}$$

We replace T with 1060 and then solve:

$$\frac{Cl}{1060} = \frac{35}{53} \longrightarrow 53Cl = 35 \cdot 1060 \longrightarrow \frac{\cancel{53}Cl}{\cancel{53}} = \frac{35 \cdot 1060}{53}$$

$$Cl = \textbf{700 grams}$$

*1.   The chemical formula for water is H_2O. If we have 3600 grams of water, what is the weight of the oxygen? (O, 16; H, 1)

*2.   The chemical formula for ammonia is NH_3. This tells us that each ammonia molecule contains one nitrogen atom and three hydrogen atoms. If we have 510 grams of ammonia, how much does the nitrogen weigh? (N, 14; H, 1)

*3. The formula for ammonium chloride is NH_4Cl. This means that in one molecule of ammonium chloride there is one atom of nitrogen, four atoms of hydrogen, and one atom of chlorine. How many grams of chlorine are there in 1060 grams of ammonium chloride? (N, 14; H, 1; Cl, 35)

4. The bus and the train left the same town headed south at 8 a.m. At noon, the bus was 100 miles behind the train. How far did each one travel if the speed of the train was twice the speed of the bus?

5. His imitation turned into a travesty because he used $2\frac{3}{4}$ times as many gestures as were required for a reasonable imitation. If he used 550 gestures, how many were required for a reasonable imitation?

Simplify:

6. $\dfrac{x^2 + x - 6}{x^3 + 7x^2 + 12x} \cdot \dfrac{x^3 + 5x^2 + 4x}{x^2 + 2x - 8}$

7. $\dfrac{ax^3 - ax^2 - 12ax}{x^2 + 7x + 12} \div \dfrac{ax^2 - 4ax}{x^2 + 2x - 8}$

8. $-8^{-\frac{4}{3}}$

9. $(-27)^{-\frac{4}{3}}$

10. $-27^{-\frac{4}{3}}$

11. $\dfrac{-3}{-9^{-\frac{3}{2}}}$

12. Find the area of this figure. The dimensions are in feet.

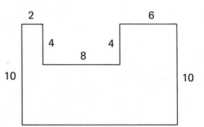

13. Find the volume of a container whose base is the solid line triangle shown and whose sides are 8 feet high. Dimensions are in feet.

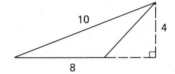

Simplify:

14. $\dfrac{x + \dfrac{1}{x^2}}{x^2 - \dfrac{1}{x^2}}$

15. $\dfrac{a + \dfrac{y}{x}}{a - \dfrac{my}{x}}$

16. $3\sqrt{\dfrac{2}{7}} - 5\sqrt{\dfrac{7}{2}}$

17. $2\sqrt{\dfrac{11}{3}} - 5\sqrt{\dfrac{3}{11}}$

18. Find the equation of the line that passes through $(-4, 2)$ that is perpendicular to the line that goes through $(-4, 6)$ and $(5, 2)$.

19. Find the perimeter of the accompanying figure. Dimensions are in meters.

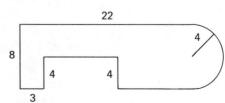

20. Solve by graphing and then get an exact solution by using either substitution or elimination.
$$\begin{cases} 3x - 3y = -6 \\ 3x + y = 6 \end{cases}$$

Add:

21. $\dfrac{x}{x + 5} - \dfrac{3x}{x^2 + 4x - 5}$ 22. $\dfrac{4}{x(x + 2)} + \dfrac{6}{x}$

Solve:

23. $\dfrac{3x + 2}{5} - \dfrac{x - 3}{7} = 2$ 24. $\dfrac{4x}{3} - \dfrac{2x}{4} + x = 5$

Simplify:

25. $3\sqrt{2}(5\sqrt{12} - 2\sqrt{2})$ 26. $4\sqrt{20}(3\sqrt{2} - 2\sqrt{5})$

27. $\dfrac{(x^{-2})^{-1}(y^{-2}x)^{-3}x^0 y}{(x^{-2})^2 x^4 x x^0 y^0 x^2}$ 28. $-2^2 - (-2)^3 - (-2) - 2^0 - 2$

29. $\dfrac{(.00035 \times 10^{-14})(.003 \times 10^5)}{21{,}000 \times 10^{-40}}$

30. Multiply: $\dfrac{4x^{-2}y^{-2}}{z^2}\left(\dfrac{x^2 y}{z^{-2}} - \dfrac{3x^2 y^2 z^2}{p}\right)$

LESSON 38   *Powers of sums • Solving by factoring • Only zero equals zero*

38.A
powers of sums

We know that x^2 means to multiply x by x:

$$x^2 = x \cdot x$$

In the same way, $(x^4 y^{-3} z^2)^2$ means to multiply $x^4 y^{-3} z^2$ by $x^4 y^{-3} z^2$:

$$(x^4 y^{-3} z^2)^2 = (x^4 y^{-3} z^2) \cdot (x^4 y^{-3} z^2) = x^8 y^{-6} z^4$$

The same result can be obtained by using the power theorem for exponents by multiplying the exponents, as shown here:

$$(x^4 y^{-3} z^2)^2 = x^8 y^{-6} z^4$$

We must be careful, however, when the notation indicates that a sum is to be raised to a power because

$$(2x^2 + 3y)^2$$

means that $2x^2 + 3y$ must be multiplied by itself, and the power theorem cannot be used as in the last example. We will use the vertical format to multiply.

$$\begin{array}{r} 2x^2 + 3y \\ 2x^2 + 3y \\ \hline 4x^4 + 6x^2 y \\ 6x^2 y + 9y^2 \\ \hline 4x^4 + 12x^2 y + 9y^2 \end{array}$$

example 38.A.1   Expand $(x + 3)^3$.

solution   This notation indicates that $x + 3$ is to be used as a factor three times.

$$(x + 3)(x + 3)(x + 3)$$

We will use two steps to find the product of the three factors.

$$(x + 3)(x + 3) = x^2 + 3x + 3x + 9 = x^2 + 6x + 9$$

Now we multiply this product by $x + 3$:

$$
\begin{array}{r}
x^2 + 6x + 9 \\
\underline{x \;\; + 3} \\
x^3 + 6x^2 + \;\; 9x \\
\underline{3x^2 + 18x + 27} \\
x^3 + 9x^2 + 27x + 27
\end{array}
$$

We must try to remember that the power theorem can only be used when a product is raised to a power. The power theorem cannot be used when a sum is raised to a power.

38.B
solving by factoring

If a product of two factors equals zero, then one of the factors must be zero. For instance, if we have the notation

$$(4)(\quad) = 0$$

the only possible correct entry for the second parentheses is 0.

$$(4)(0) = 0$$

In the same way, if we have an indicated multiplication of two factors equal to zero,

$$(\quad)(\quad) = 0$$

then either the first factor must equal zero, or the second factor must equal zero. This fact is so important that we give it a name, the **zero factor theorem.** The formal statement of this theorem is as follows:

ZERO FACTOR THEOREM

If p and q are any real numbers and if $p \cdot q = 0$, then either $p = 0$ or $q = 0$, or both.

We use this theorem to help us solve equations by factoring.

example 38.B.1   Solve $-x + x^2 = 6$.

solution   First we rewrite the equation so that the three given terms are on the left side and are in descending powers of the variable.

$$x^2 - x - 6 = 0$$

Next we factor the trinomial and get

$$(x - 3)(x + 2) = 0$$

Now the zero factor theorem tells us that in this case there are just two possibilities. Either $x - 3$ equals zero or $x + 2$ equals zero.

$$\text{If } x - 3 = 0 \qquad \text{If } x + 2 = 0$$
$$\underline{+3 \quad +3} \qquad \underline{-2 \quad -2}$$
$$x \quad = 3 \qquad x \quad = -2$$

We will check both solutions in the original equation.

$$\text{If } x = 3: \qquad\qquad \text{If } x = -2:$$
$$-(3) + (3)^2 = 6 \qquad -(-2) + (-2)^2 = 6$$
$$-3 + 9 = 6 \qquad\qquad 2 + 4 = 6$$
$$6 = 6 \quad \text{Check} \qquad\qquad 6 = 6 \quad \text{Check}$$

The zero factor theorem can be extended to the product of any number of factors by saying that if the product of two or more factors equals zero, then one or more of the factors must equal zero.

EXTENSION OF THE ZERO FACTOR THEOREM

If a, b, c, d, etc. represent real numbers and if

$$a \cdot b \cdot c \cdot d \cdot e \cdot f \cdots = 0$$

then one or more of the factors equals zero.

We will use this fact in the next example.

example 38.B.2   Solve $-35x = -2x^2 - x^3$.

solution   Again the first step is to write the equation in standard form.

$$x^3 + 2x^2 - 35x = 0$$

Next we factor out an x to get

$$x(x^2 + 2x - 35) = 0$$

and then we factor the trinomial.

$$x(x - 5)(x + 7) = 0$$

Now, by the extension of the zero factor theorem, one of these factors must equal 0.

$$\text{If } x = 0 \qquad \text{If } x - 5 = 0 \qquad \text{If } x + 7 = 0$$
$$x = 0 \qquad\qquad x = 5 \qquad\qquad x = -7$$

We finish by checking all three solutions in the original equation.

$$\text{If } x = 0: \qquad \text{If } x = 5: \qquad\qquad \text{If } x = -7:$$
$$-35(0) = -2(0)^2 - (0)^3 \quad -35(5) = -2(5)^2 - (5)^3 \quad -35(-7) = -2(-7)^2 - (-7)^3$$
$$0 = 0 - 0 \qquad -175 = -50 - 125 \qquad 245 = -98 + 343$$
$$0 = 0 \quad \text{Check} \qquad -175 = -175 \quad \text{Check} \qquad 245 = 245 \quad \text{Check}$$

38.C
only zero equals zero

If we solve the following two equations by factoring

$$\text{(a)} \quad x^3 + 2x^2 - 35x = 0 \qquad \text{(b)} \quad 2x^2 + 4x - 70 = 0$$

the results of the factoring are similar.

(a) $x(x - 5)(x + 7) = 0$ (b) $2(x - 5)(x + 7) = 0$

In both cases, we have the product of three factors equal to zero, and it would appear to some that each equation has the three roots. In equation (a), we can set each of the factors equal to zero, solve, and find that the solutions to the equation are

0,   5,   and   −7

Equation (b) is different because although it has three factors, only two of the factors can ever be equal to zero. The factor $(x - 5)$ equals zero if x equals 5 and the factor $(x + 7)$ equals zero if x equals −7, but there is no way that 2 can equal zero. The number 2 equals only the number 2.

Some teachers try to help students avoid a mistake by having them record the solution for (b) as

$$x = 5, \qquad x = -7, \qquad 2 \neq 0$$

problem set 38

1. Verruca counted protrusions. She found that 3 times the number of protrusions was 15 less than −4 times the opposite of the number of protrusions. How many protrusions did she count?

2. Two kilograms of iron was melted with 7 kilograms of other metals to make the alloy. If 1440 kilograms of the alloy was required, how many kilograms of iron should be used?

3. Charles and Matthew knew that the formula for sulfuric acid was H_2SO_4. If they had 196 grams of sulfuric acid, what was the weight of the sulfur? (H, 1; S, 32; O, 16)

4. The ratio of the two numbers was 7 to 2. When Sir Richard and Marion multiplied the denominator by 10, they found that the result was 84 greater than twice the numerator. What were the numbers?

5. Jerry and Milton set out for a ride in the hill country at 30 mph. Their car broke down, and they caught a ride back home in a truck at 20 mph. If they were gone for 10 hours, how far from home did the breakdown occur?

Expand:

*6. $(x + 3)^3$ 7. $(x + 1)^3$

Solve:

*8. $-x + x^2 = 6$ *9. $-35x = -2x^2 - x^3$

*10. $2x^2 + 4x - 70 = 0$

Simplify:

11. $\dfrac{x^2 + 7x + 10}{14x + 9x^2 + x^3} \div \dfrac{x^2 + 2x - 15}{x^3 + 11x^2 + 28x}$

12. $-16^{-\frac{1}{4}}$ 13. $-16^{-\frac{3}{4}}$

14. $8^{\frac{2}{3}}$ 15. $(-8)^{\frac{1}{3}}$

16. How many 1-foot-square floor tiles would it take to cover the area shown. Dimensions are in feet.

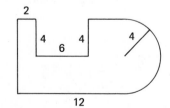

17. If the figure is the base of a solid that is 8 feet high, how many sugar cubes that measure 1 foot on a side will it hold? Dimensions are in feet.

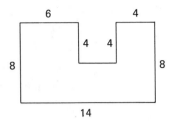

Simplify:

18. $\dfrac{x^2 - \dfrac{a}{x}}{a^2 - \dfrac{a}{x}}$

19. $\dfrac{\dfrac{mp^2}{4} - 5}{4p^2 - \dfrac{p^2}{4}}$

20. $\dfrac{(.00007 \times 10^{-23})(3000 \times 10^{-10})}{700,000 \times 10^{-30}}$

21. $3\sqrt{\dfrac{2}{11}} + 5\sqrt{\dfrac{11}{2}}$

22. $-2\sqrt{\dfrac{11}{3}} + 7\sqrt{\dfrac{3}{11}}$

23. $3\sqrt{24}(2\sqrt{6} - 3\sqrt{12})$

24. Find the equation of the line that goes through the points $(-2, 5)$ and $(3, 4)$.

25. Solve: $\dfrac{5x - 2}{3} - \dfrac{x}{4} = 7$

26. Divide $x^4 - 1$ by $x + 1$.

27. Find the distance between $(-2, 5)$ and $(3, 4)$.

28. Add: $\dfrac{1}{x + 3} + \dfrac{3x}{x + 2} + \dfrac{2x + 1}{x^2 + 5x + 6}$

29. Simplify: $\dfrac{(p^2 y^{-2})^{-3} p^{-2} (y^0)^{-2}}{(p^{-2} p^0 py)^{-3} (yp^{-2})^{-4} p}$

30. $|3^0 - 3^2| - \dfrac{1}{2^{-2}} - 3^0(-3^3 - 3^2)$

LESSON 39 *Difference of two squares • Vertical angles • Similar triangles*

39.A
difference of two squares

If both terms of a two-term algebraic sum are identical except that the sign of the second term is different, we say that each expression is the **conjugate** of the other expression. Thus,

$$-2x - 3y \quad \text{is the conjugate of} \quad -2x + 3y$$
$$-2x + 3y \quad \text{is the conjugate of} \quad -2x - 3y$$
$$3x + 4 \quad \text{is the conjugate of} \quad 3x - 4$$
$$3x - 4 \quad \text{is the conjugate of} \quad 3x + 4$$

When conjugates are multiplied, the product does not have a middle term, as seen here.

$$
\begin{array}{r}
-2x - 3y \\
\underline{-2x + 3y} \\
4x^2 + 6xy \\
\underline{\quad - 6xy - 9y^2} \\
4x^2 \qquad - 9y^2
\end{array}
\qquad
\begin{array}{r}
3x - 4 \\
\underline{3x + 4} \\
9x^2 - 12x \\
\underline{\quad + 12x - 16} \\
9x^2 \qquad - 16
\end{array}
\qquad
\begin{array}{r}
a + b \\
\underline{a - b} \\
a^2 + ab \\
\underline{\quad - ab - b^2} \\
a^2 \qquad - b^2
\end{array}
$$

We see that each of these products of conjugates can be written as the difference of two squared expressions.

$$(2x)^2 - (3y)^2 \qquad (3x)^2 - (4)^2 \qquad (a)^2 - (b)^2$$

These expressions can be factored only by recognizing that the expression is the difference of two squares. There is no procedure to follow.

example 39.A.1   Solve $4x^2 - 9 = 0$ by factoring.

solution   **We recognize that the left side is the difference of two squares.** Thus we factor as follows:

$$(2x - 3)(2x + 3) = 0$$

We finish the solution by using the zero factor theorem.

$$\text{If } 2x - 3 = 0 \qquad \text{If } 2x + 3 = 0$$
$$2x = 3 \qquad\qquad 2x = -3$$
$$x = \frac{3}{2} \qquad\qquad x = -\frac{3}{2}$$

example 39.A.2   Solve $81m^2 - 25 = 0$ by factoring.

solution   **We recognize that the expression is a difference of two squares,** so we factor as

$$(9m + 5)(9m - 5) = 0$$

We complete the solution by setting each factor equal to zero.

$$\text{If } 9m + 5 = 0 \qquad \text{If } 9m - 5 = 0$$
$$9m = -5 \qquad\qquad 9m = 5$$
$$m = -\frac{5}{9} \qquad\qquad m = \frac{5}{9}$$

39.B
triangles   We have said that four right angles are formed at the intersection of two perpendicular lines.

Here we show the intersection of two perpendicular lines and remember that we also call right angles ninety degree (90°) angles. Thus, one degree (1°) is one-ninetieth of a right angle. By using this definition of a 90° angle it is possible to prove that **the sum of the interior angles of any triangle is 180°.**

We note that the sum of the angles in each of these triangles is 180°. Triangle (a) is called a **right triangle** because one angle is a right angle. **Triangles that have all three sides the same length have three 60° angles, as does (b).** We call these triangles **equi-**

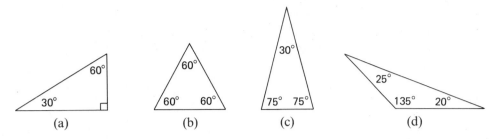

(a)   (b)   (c)   (d)

lateral **triangles.** Triangles that have two sides that are equal also have two angles that are equal, as does (c). We call these triangles **isosceles triangles.**

39.C
vertical angles

Two intersecting straight lines form four angles. The angles that are opposite each other are called **vertical angles** and are equal angles.

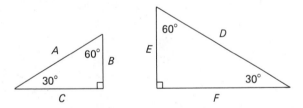

In this figure, angle *A* is the same size as angle *B*, and angle *C* is the same size as angle *D*.

39.D
similar triangles

If two triangles have the same angles, they are similar triangles; and the ratios of corresponding sides are equal. Corresponding sides of similar triangles are the sides that are opposite equal angles.

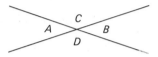

These triangles are similar triangles because the angles in both are 30°, 60°, and 90°. Sides *A* and *D* are corresponding sides because they are the sides opposite the 90° angles. Sides *C* and *F* are corresponding sides because they are the sides that are opposite the 60° angles; and sides *B* and *E* are corresponding sides because they are the sides that are opposite the 30° angles. **We note that two right triangles are similar if an acute angle in one of the triangles is the same size as one of the angles in the other triangle.**

example 39.D.1   Use similar triangles and ratios to find sides *y* and *m*.

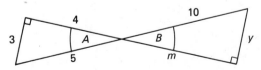

solution   Angles *A* and *B* are equal angles because they are vertical angles. Thus, the triangles are similar triangles because they are right triangles, and angle *A* equals angle *B*. Since ratios of corresponding sides are equal, we can write

$$\frac{3}{5} = \frac{y}{10} \quad \text{and} \quad \frac{4}{5} = \frac{m}{10}$$

These proportions can be solved for y and m.

$$3 \cdot 10 = 5y \qquad 4 \cdot 10 = 5m$$

$$6 = y \qquad\qquad 8 = m$$

example 39.D.2   Use similar triangles and the theorem of Pythagoras to find M, A and B in this figure.

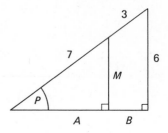

solution   The larger triangle is similar to the smaller triangle because they are both right triangles, and both have angle P as one of the angles. Thus, since ratios of corresponding sides of similar right triangles are equal, we can write

$$\frac{M}{7} = \frac{6}{10}$$

We solve to find that M equals 4.2.

$$10M = 6 \cdot 7 \quad \longrightarrow \quad 10M = 42 \quad \longrightarrow \quad M = 4.2$$

Now we use the Pythagorean theorem to find A.[†]

$$A = \sqrt{7^2 - 4.2^2} \quad \longrightarrow \quad A = \sqrt{31.36} \quad \longrightarrow \quad A = 5.6$$

Now we use similar triangles to find B.

$$\frac{5.6 + B}{6} = \frac{5.6}{4.2} \quad \longrightarrow \quad 23.52 + 4.2B = 33.6 \quad \longrightarrow \quad B = \frac{10.08}{4.2} \quad \longrightarrow \quad B = 2.4$$

problem set 39

1. Some of the freshmen were ingenuous, but $\frac{13}{16}$ had ulterior motives for their actions. If 420 did not have ulterior motives, how many did have ulterior motives?

2. Some of the teacher's proclamations were democratic, but 72 percent were almost ukases. If 9576 were democratic, how many were almost ukases?

3. The chemical formula for carbon dioxide is CO_2. If the reaction produced 528 grams of carbon dioxide, what was the weight of the carbon produced? (C, 12; O, 16)

4. The symbol for strontium is Sr. If the mixture was 14 percent strontium and 688 grams was not strontium, what was the weight of the strontium?

---

† We use a calculator to find the square root.

5. Jimmy and Gary ran from the disaster at 5 mph. Then they reconsidered and walked back at 3 mph. If they were gone for 8 hours, how far did they run from the disaster?

Use similar triangles to find the missing sides of the triangles.

*6.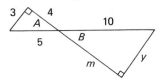

*7.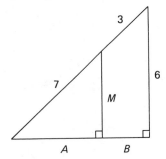

Solve:

8. $x^2 - 9 = 0$

*9. $81m^2 - 25 = 0$

10. $24x = -11x^2 - x^3$

11. Expand: $(x - 1)^3$

Simplify:

12. $\dfrac{x^3 + 6x^2 + 5x}{x^2 + 2x - 15} \div \dfrac{7x + 8x^2 + x^3}{x^2 + 5x - 14}$

13. $\dfrac{2^0}{-4^{-\frac{3}{2}}}$

14. $\dfrac{-3^0}{-27^{-\frac{2}{3}}}$

15. $\dfrac{ax^2 - \dfrac{4}{a}}{\dfrac{x^2}{a} + 6}$

16. $\dfrac{\dfrac{m^2 p}{x} - 6}{m^2 p - \dfrac{4}{x}}$

17. $\dfrac{(3000 \times 10^{-41})(.0008 \times 10^{10})}{2,400,000 \times 10^8}$

18. $2\sqrt{\dfrac{3}{13}} - 5\sqrt{\dfrac{13}{3}}$

19. $5\sqrt{\dfrac{3}{2}} - 2\sqrt{\dfrac{2}{3}}$

20. $5\sqrt{45} - 2\sqrt{75} + 2\sqrt{108}$

21. How many 1-inch-square floor tiles would it take to cover the area shown? Dimensions are in inches.

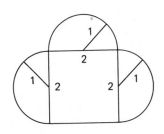

22. If this figure is the base of a solid 10 inches high, how many 1-inch sugar cubes would it hold? Dimensions are in inches.

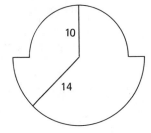

23. Simplify: $3\sqrt{12}(4\sqrt{3} - 3\sqrt{3})$

24. Find the equation of the line that passes through (2, 4) that is perpendicular to the line that passes through (2, 4) and $(-3, -2)$.

Solve:

25. $2\frac{1}{4}x - 3\frac{1}{2} = -\frac{1}{16}$

26. $.002x - .02 = 6.6$

27. Add: $\dfrac{3}{x+1} + \dfrac{2x}{y(x+1)} + \dfrac{3x+2}{x^2+2x+1}$

28. Simplify: $\dfrac{4xy + 4x^2y^2}{4xy}$

29. Expand: $\dfrac{4x^{-2}y^{-2}}{z^2}\left(\dfrac{3x^2y^2z^2}{4} + \dfrac{2x^0y^{-2}}{z^2y^2}\right)$

30. Evaluate: $-2^0 - 3^0(-2 - 5^0) - \dfrac{1}{-2^{-2}} + x^2y - xy$ if $x = -2$ and $y = 3$.

LESSON 40 *Abstract fractional equations*

40.A
fractional equations

We have noted that the easiest way to solve fractional equations is to eliminate the denominators as the first step. Thus to solve this equation

$$\frac{4+x}{5} + \frac{7}{3} = 4$$

we begin by multiplying every numerator by 15, which is the least common multiple of the denominators. This will permit the elimination of the denominators.

$$\frac{15(4+x)}{5} + \frac{15(7)}{3} = 4(15) \qquad \text{multiplied by 15}$$

$$12 + 3x + 35 = 60 \qquad \text{eliminated denominators}$$

$$3x = 13 \qquad \text{simplified}$$

$$x = \frac{13}{3} \qquad \text{divided}$$

We also eliminate the denominators as the first step in the solution of abstract equations, as we will demonstrate in the following examples. **(In the solution of abstract equations, we will assume that no variable or combination of variables in any denominator equals zero.)**

example 40.A.1 Solve for x: $\dfrac{a}{x} + \dfrac{m}{a} = c$

solution We will solve in three steps. **The first step will be to multiply every numerator by the least common multiple of the denominators which is ax. Then we will cancel the denominators.**

Step 1: $ax \cdot \dfrac{a}{x} + ax \cdot \dfrac{m}{a} = c \cdot ax$ multiplied by ax

$$a^2 + mx = cax$$ canceled denominators

Next we move all terms that contain x to the same side of the equation (either side) and factor out the x.

Step 2: $a^2 = cax - mx$ added $-mx$ to both sides

$a^2 = x(ca - m)$ factored out x

We finish by dividing both sides by $(ca - m)$, which is the coefficient of x.

Step 3: $\dfrac{a^2}{ca - m} = \dfrac{x(ca - m)}{ca - m} \longrightarrow \dfrac{a^2}{ca - m} = x$ divided

example 40.A.2   Solve for m: $\dfrac{x}{m} + c = \dfrac{y}{a}$

solution   **Again as the first step, we will eliminate the denominators.** We begin by multiplying every numerator by ma.

Step 1: $(ma)\dfrac{x}{m} + (ma)c = (ma)\dfrac{y}{a}$ multiplied

$$ax + mac = my$$ canceled

Step 2: Next we move all terms that contain an m to the same side of the equation and factor out the m.

$ax = my - mac$ added $-mac$ to both sides

$ax = m(y - ac)$ factored out m

We finish by dividing both sides by $y - ac$, and we get

Step 3: $$\dfrac{ax}{y - ac} = m$$

In step 2, if we had placed all terms that contained m on the left side, our answer would have been

$$m = \dfrac{-ax}{-y + ac}$$

which is the same answer as the one above except that all signs above and below are different. We remember that we can always multiply the denominator and the numerator by the same nonzero quantity. If we use (-1) as the multiplier, we can change this last answer to the first form of the answer.

$$\dfrac{-ax}{-y + ac}\dfrac{(-1)}{(-1)} = \dfrac{ax}{y - ac}$$

Both answers are equally correct and neither is preferable to the other.

example 40.A.3   Solve for p: $\dfrac{6}{p} - ax = \dfrac{m}{y} + k$

solution　Again we begin by canceling the denominators.

$$(py)\frac{6}{p} - py(ax) = (py)\frac{m}{y} + (py)k \qquad \text{multiplied by } py$$

$$6y - pyax = pm + pyk \qquad \text{canceled}$$

Next we place all terms that contain p on the same side and factor out the p.

$$6y = pm + pyk + pyax \qquad \text{added } +pyax$$

$$6y = p(m + yk + yax) \qquad \text{factored}$$

Again we finish by dividing both sides by the coefficient of the variable for which we are solving.

$$\frac{6y}{m + yk + yax} = p$$

**problem
set 40**

1. Many students took a foreign language to increase their vocabularies and improve their grammar. If these people earned 64 percent more than the others, how much did they make if the others earned $1,200,000?

2. The more successful professionals had vocabularies that were 280 percent larger than those of the less successful. If the less successful knew 4800 words, how many words did the more successful know?

3. The formula for iron sulfide is FeS. If the iron (Fe) in a batch of iron sulfide weighed 448 grams, how much did the iron sulfide weigh? (Fe, 56; S, 32)

4. Shoes cost $20 a pair and boots cost $60 a pair. Arlene and Jerry spent $8000 and bought 3 times as many boots as shoes. How many pairs of each did they buy?

5. Amy had a 120-yard head start on Kathy. Kathy ran at 15 yards per second, and Amy ran at 3 yards per second. How far did Kathy have to run to catch Amy?

*6. $\dfrac{a}{x} + \dfrac{m}{a} = c$; find x　　　　　　*7. $\dfrac{x}{m} + c = \dfrac{y}{a}$; find m

*8. $\dfrac{6}{p} - ax = \dfrac{m}{y} + k$; find p　　　　9. $\dfrac{a}{c} - b = \dfrac{m}{k}$; find k

10. Find sides x and y.　　　　　　　　11. Expand $(x - 3)^3$.

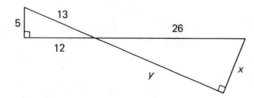

Solve:

12. $4x^2 - 49 = 0$　　　　　　　　　13. $x^3 = 3x^2 + 18x$

Simplify:

14. $\dfrac{10x - 7x^2 + x^3}{x^2 + 4x - 12} \div \dfrac{x^3 - 8x^2 + 15x}{x^2 + 3x - 18}$　　　15. $\dfrac{(-4^0)^2}{4^{-\frac{3}{2}}}$

16. $\dfrac{1}{16^{-\frac{3}{4}}}$

17. $\dfrac{\dfrac{4x^2a}{y^2} + \dfrac{1}{a^2}}{2 - \dfrac{2}{y^2a^2}}$

18. $\dfrac{\dfrac{xy^2}{p} - 4}{a^2y - \dfrac{1}{p}}$

19. $\dfrac{(4000 \times 10^{14})(.007 \times 10^{-23})}{14,000 \times 10^{-20}}$

20. $3\sqrt{50} - 2\sqrt{72} + 3\sqrt{162}$

21. $3\sqrt{\dfrac{3}{7}} + 2\sqrt{\dfrac{7}{3}}$

22. How many 1-yard-square floor tiles would it take to cover the area shown? Dimensions are in yards.

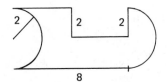

23. Find the perimeter of this figure. Dimensions are in yards.

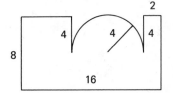

24. Find the equations of lines (a) and (b).

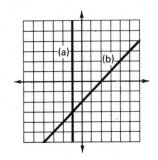

Solve:

25. $\dfrac{3 - 2x}{4} + \dfrac{x}{3} = 5$

26. $.004x - .02 = 2.02$

27. Add: $\dfrac{3}{x} + \dfrac{2}{x + 2} + \dfrac{3x}{x^2 + 3x + 2}$

28. Simplify: $\dfrac{x + 4x}{x}$

29. Expand: $\dfrac{x^{-2}y}{p}\left(\dfrac{x^2p}{y} - \dfrac{3x^2y}{p}\right)$

30. Evaluate: $x^2 - xy - x^3$ if $x = \dfrac{1}{2}$ and $y = \dfrac{1}{3}$.

LESSON 41 *Units • Unit multipliers*

41.A

units   When we attach words to numbers as shown here,

4 ft   6 centimeters   42.78 mph   93 liters

we often call the resulting combinations **denominate numbers.** The Latin prefix for "completely" is *de* and the word for "to name" is *nominare*. Thus *denominate* literally means "completely named." The words are often called the **units** of the denominate numbers. Thus, the units in the denominate numbers above are feet, centimeters, miles per hour and liters.

   We find it convenient to use exponential notation to handle units. If we do this, we can handle units the same way we handle numbers or variables. This is especially useful when we multiply or divide units.

(a) $\text{ft}^2 \cdot \text{ft} = \textbf{ft}^3$ (b) $\dfrac{\text{cm}^2}{\text{cm}^2} = \textbf{1}$

(c) $\dfrac{\text{yd}^3}{\text{yd}} = \textbf{yd}^2$ (d) $\dfrac{\text{in}^3}{\text{in}^2} = \textbf{in}$

41.B

unit multipliers   We remember that any nonzero quantity divided by itself has a value of 1.

(e) $\dfrac{x^2}{x^2} = \textbf{1}$ (f) $\dfrac{24.123}{24.123} = \textbf{1}$ (g) $\dfrac{6\ \text{ft}^2}{6\ \text{ft}^2} = \textbf{1}$ (h) $\dfrac{4\ \text{in}}{4\ \text{in}} = \textbf{1}$

Furthermore, we remember that the product of any quantity and 1 is the quantity itself.

(i) $x^2 ym(1) = \textbf{\textit{x}}^2 \textbf{\textit{ym}}$ (j) $\left(4\,\dfrac{\text{ft}}{\text{sec}}\right)(1) = \textbf{4}\,\dfrac{\textbf{ft}}{\textbf{sec}}$ (k) $(3\ \text{in})(1) = \textbf{3 in}$

We know that 12 inches equals 1 foot, so if we write either

$\dfrac{12\ \text{in}}{1\ \text{ft}}$ or $\dfrac{1\ \text{ft}}{12\ \text{in}}$

we have written an expression whose value is 1, and thus we can multiply any expression by either of these terms without changing the value of the expression. We call these terms **unit multipliers** for two reasons: One is that they contain units, and the second is that they have a value of unity (1). Unit multipliers are very helpful when we want to change one set of units to another set of units.

example 41.B.1   Use a unit multiplier to change 600 inches to feet.

solution   We will use one of the unit multipliers above. We choose the one on the left.

$600\ \text{in} \times \dfrac{12\ \text{in}}{1\ \text{ft}} = \dfrac{7200\ \text{in}^2}{\text{ft}}$

This answer is not incorrect, but it is not what we want. Let's try again and use the other unit multiplier.

$600\ \cancel{\text{in}} \times \dfrac{1\ \text{ft}}{12\ \cancel{\text{in}}} = \textbf{50 ft}$

This time the inches canceled and we got the desired answer.

example 41.B.2   Use unit multipliers to convert 44 square feet to square inches.

solution   We write what was given and use 1 for a denominator.

$$\frac{44 \text{ ft}^2}{1}$$

We note that square feet (ft^2) is in the numerator. Thus, we will use the unit multiplier that has the abbreviation ft in the denominator. We must use two unit multipliers because we are converting from square feet (ft^2) to square inches (in^2).

$$\frac{44 \text{ ft}^2}{1} \times \frac{12 \text{ in}}{1 \text{ ft}} \times \frac{12 \text{ in}}{1 \text{ ft}} = (44)(12)(12) \text{ in}^2$$

This multiplication is relatively easy, but some unit conversion problems will result in very complicated multiplications and divisions. We suggest that these answers not be worked out or that a pocket calculator be used to get the final numerical answer. These problems are designed to teach unit conversion and are not designed for practice in arithmetic.

We have begun our study of unit conversion with problems that can be solved mentally without the use of unit multipliers. It is recommended that the use of unit multipliers not be eschewed for these simple problems because the unit conversion problems that we encounter later will be rather involved. The use of unit multipliers will make these involved conversions straightforward and the experience that we will gain by doing simple problems will prove to be valuable.

example 41.B.3   Use unit multipliers to convert 42 square yards to square inches.

solution   We could use the fact that 1 yard equals 36 inches, but instead, we will go from square yards (yd^2) to square feet (ft^2) to square inches (in^2), a procedure that is recommended because shortcuts can lead to errors.

$$42 \text{ yd}^2 \times \frac{3 \text{ ft}}{1 \text{ yd}} \times \frac{3 \text{ ft}}{1 \text{ yd}} \times \frac{12 \text{ in}}{1 \text{ ft}} \times \frac{12 \text{ in}}{1 \text{ ft}} = (42)(3)(3)(12)(12) \text{ in}^2$$

example 41.B.4   Use unit multipliers to convert 16 cubic miles (mi^3) to cubic inches (in^3).

solution   We will go from cubic miles (mi^3) to cubic feet (ft^3) to cubic inches (in^3).

$$16 \text{ mi}^3 \times \frac{5280 \text{ ft}}{1 \text{ mi}} \times \frac{5280 \text{ ft}}{1 \text{ mi}} \times \frac{5280 \text{ ft}}{1 \text{ mi}} \times \frac{12 \text{ in}}{1 \text{ ft}} \times \frac{12 \text{ in}}{1 \text{ ft}} \times \frac{12 \text{ in}}{1 \text{ ft}}$$

$$= (16)(5280)(5280)(5280)(12)(12)(12) \text{ in}^3$$

It seems that there are quite a few cubic inches in 16 cubic miles.

problem set 41

1. While Doctor Andy operated, he thought of consecutive odd integers. His integers were such that 4 times the sum of the first and fourth was 12 greater than 3 times the sum of the second and third. What were the first four integers on Doctor Andy's list?

2. The formula for chromium chloride is $CrCl_3$. What would be the weight of the chlorine (Cl) in 1256 grams of chromium chloride? (Cr, 52; Cl, 35)

3. The delivery truck unloaded 184 percent more silicon than was required for the experiment. If 1136 tons was unloaded, how many tons was required for the experiment?

4. Bob and Judy found that their horde of nickels and dimes was worth $7. If they had a total of 100 coins, how many coins of each kind did they have?

5. Larry and Shadid rode on their motor scooters at 16 mph until Larry ran out of petrol. Then they walked the rest of the way at 4 mph. If the entire trip was 76 miles and it took a total of 7 hours, how far did they walk and how far did they ride?

Use unit multipliers to convert:

*6. 44 ft^2 to in^2

*7. 42 yd^2 to in^2

*8. 16 mi^3 to in^3

9. $\dfrac{x}{p} - \dfrac{k}{m} = c$; find p

10. $\dfrac{xy}{p} - \dfrac{k}{c} = m$; find p

11. $\dfrac{4p}{x} - \dfrac{xk}{c} = \dfrac{y}{m}$; find c

12. Find sides m and p.

13. Divide $2x^3 - 1$ by $x - 2$.

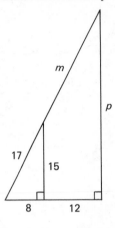

Solve:

14. $16x = -x^3 + 10x^2$

15. $4x^2 - 9x = 0$

16. Find the area of this figure. The dimensions are in centimeters.

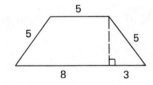

17. This figure is the base of a solid 6 ft high. Find the volume of the solid. Dimensions are in feet.

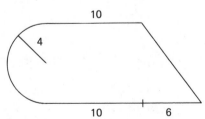

Simplify:

18. $\dfrac{x^3 + 8x^2 + 15x}{x^2 - 4x + 4} \cdot \dfrac{x^2 + 2x - 8}{12x + x^3 + 7x^2}$

19. $32^{-\frac{2}{5}}$

20. $\dfrac{a^2x - \dfrac{a}{x}}{ax - \dfrac{4}{x}}$

21. $\dfrac{(21{,}000 \times 10^{-42})(500{,}000)}{.00015 \times 10^{-7}}$

22. $3\sqrt{\dfrac{5}{7}} - 6\sqrt{\dfrac{7}{5}}$ **23.** $4\sqrt{24}(2\sqrt{6} - 3\sqrt{2})$

24. Find the equation of the line that is perpendicular to $3y + x = -2$ and passes through the point $(-2, -5)$.

Solve:

25. $\dfrac{-3 - x}{2} - \dfrac{x}{2} = 7$ **26.** $2\dfrac{1}{3}x - \dfrac{1}{9} = -\dfrac{1}{18}$

27. Add: $\dfrac{2}{x^2(x - 2)} - \dfrac{2x + 2}{x^2 - 4}$

Simplify:

28. $\dfrac{4x + 8x^2}{4x}$ **29.** $\dfrac{x^2 x^0 x^{-1}(x^{-2})^2 y x^{-3}}{(x^2 y)^{-3} x y x^{-2} x^2}$

30. Evaluate: $x^3 - xy + x^2$ if $x = \dfrac{1}{2}$ and $y = \dfrac{1}{3}$.

LESSON 42 *Estimating with scientific notation*

42.A
estimating with scientific notation

The scientific notation problems that we have encountered thus far have been carefully designed so that the numbers multiply and divide easily and so that the first part of the answer is an integer.

$$\frac{(.0003 \times 10^{-6})(4000)}{(.006 \times 10^{15})(2000 \times 10^4)} \qquad \text{problem}$$

$$= \frac{(3 \times 10^{-10})(4 \times 10^3)}{(6 \times 10^{12})(2 \times 10^7)} \qquad \text{scientific notation}$$

$$= \frac{12 \times 10^{-7}}{12 \times 10^{19}} = \mathbf{1 \times 10^{-26}} \qquad \text{simplified}$$

These problems have been used to help develop the skills required to handle both positive and negative integral exponents in scientific notation. Unfortunately, real-life problems contain numbers that are not so easy to handle. For instance, the answer to the last example in Lesson 41 was

$$(16)(5280)(5280)(5280)(12)(12)(12) \text{ in}^3$$

Multiplying these numbers by hand would be tedious, and we might make a mistake. If we use a calculator that does not have scientific notation, we get an error notation early because the answer is a number that is too large for these calculators to handle. If we use a calculator that has scientific notation for this multiplication, we will get

$$4.0697 \times 10^{15}$$

But we will find to our dismay that we often make mistakes when we use calculators for complicated operations such as this one. Thus, we need to develop a way to see if this answer is reasonable, and we should be able to estimate the answer when a calculator is not available. In this problem, we should be able to estimate an answer between

$$4 \times 10^{14} \quad \text{and} \quad 4 \times 10^{16}$$

This would let us know that our calculator answer of

$$4.0697 \times 10^{15}$$

is a reasonable answer. **There would be no excuse for accepting an answer of**

$$4.0697 \times 10^{21}$$

and blaming the error on the calculator. A mistake that generates an answer

1,000,000 times

the correct answer is totally inexcusable.

example 42.A.1   Estimate the answer to

$$(16)(5280)(5280)(5280)(12)(12)(12)$$

solution   Let's begin by rounding each number to one digit and writing each number in scientific notation.

$$(2 \times 10^1)(5 \times 10^3)(5 \times 10^3)(5 \times 10^3)(1 \times 10^1)(1 \times 10^1)(1 \times 10^1)$$

We used 2×10^1 for 16, 5×10^3 for each 5280, and 1×10^1 for each 12. Multiplying, we get

$$(2)(5)(5)(5) \times 10^{13} = 250 \times 10^{13} \approx \mathbf{3 \times 10^{15}}$$

From this estimate we see that our calculator answer of 4.0697×10^{15} is probably correct.

example 42.A.2   Use scientific notation to help estimate the answer to this problem.

$$\frac{(3728)(470{,}165 \times 10^{-14})}{(278{,}146)(.000713 \times 10^{-5})}$$

solution   We write each entry in scientific notation, rounding to one digit. Then we simplify.

$$\frac{(4 \times 10^3)(5 \times 10^{-9})}{(3 \times 10^5)(7 \times 10^{-9})} \approx \mathbf{1 \times 10^{-2}}$$

example 42.A.3   Use scientific notation to help estimate the answer to this expression.

$$\frac{(.0418765 \times 10^{-14})(41{,}725 \times 10^{43})}{9764 \times 10^{-23}}$$

solution   We use an approximation in scientific notation for each entry. Then we simplify.

$$\frac{(4 \times 10^{-16})(4 \times 10^{47})}{1 \times 10^{-19}} = \frac{16}{1} \times \frac{10^{31}}{10^{-19}} \approx \mathbf{2 \times 10^{51}}$$

problem
set 42

1. Pelagic fish were not usually seen near the reef so the diver was surprised to find that $\frac{3}{17}$ of the fish sighted were pelagic. If the diver saw 2244 fish on the morning dive, how many were pelagic?

2. Only 20 percent of the students were taciturn, as most of them had a penchant for prolixity. If 4800 had a penchant for prolixity, how many students were there in all?

3. It took 600 grams of potassium chlorate to make a batch of 3600 grams of the aggregate. If 43,200 grams of aggregate was required, how much potassium chlorate was needed?

4. The chemical formula for potassium chlorate is $KClO_3$. If 488 grams of this compound was on the scales, how much did the potassium (K) weigh? (K, 39; Cl, 35; O, 16)

5. Bruce and Lynn found that the larger number was 2 greater than 4 times the smaller number. Also, the larger number was 6 smaller than 8 times the smaller number. What were the two numbers?

Estimate the answer to:

*6. $(16)(5280)(5280)(5280)(12)(12)(12)$

*7. $\dfrac{(3728)(470,165 \times 10^{-14})}{(278,146)(.000713 \times 10^{-5})}$

*8. $\dfrac{(.0418765 \times 10^{-14})(41,725 \times 10^{43})}{9764 \times 10^{-23}}$

9. Use unit multipliers to convert 40 cubic yards to cubic inches.

10. $\dfrac{m}{c} - x = \dfrac{p}{m}$; find c

11. $\dfrac{ax}{y} + m = \dfrac{pc}{d}$; find y

12. Divide $4x^3 - 1$ by $x + 2$.

13. Find sides a and b.

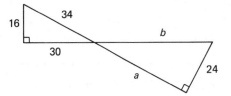

Solve:

14. $-20x = x^3 - 9x^2$

15. $4x^2 - 25 = 0$

Simplify:

16. $\dfrac{x^3 - 3x^2 + 2x}{x^2 + 3x - 4} \div \dfrac{x^3 - 6x + x^2}{15 + 8x + x^2}$

17. $-81^{-\frac{3}{4}}$

18. $\dfrac{\dfrac{m}{x} - 4}{6 - \dfrac{1}{x}}$

19. $2\sqrt{\dfrac{3}{11}} - 5\sqrt{\dfrac{11}{3}}$

20. $3\sqrt{6}(2\sqrt{6} - 4\sqrt{2})$

21. $3\sqrt{20} + 2\sqrt{45} - \sqrt{245}$

22. How many floor tiles 1-inch square would it take to cover this figure? Dimensions are in inches.

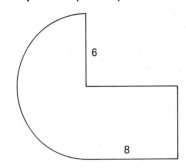

23. This figure is the base of a container whose sides are 100 inches high. How many 1-inch sugar cubes would this container hold? The dimensions are in inches.

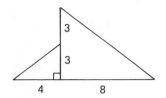

24. Find the equation of the line that passes through the point $(-3, -2)$ and is perpendicular to the line that passes through the points $(-3, -2)$ and $(4, 5)$.

Solve:

25. $\dfrac{-2x - 4}{2} - \dfrac{x}{3} = 5$

26. $3\dfrac{1}{4}x - \dfrac{1}{8} = \dfrac{3}{16}$

27. Add: $\dfrac{3}{x + 2} - \dfrac{4}{x^2 - 4} - \dfrac{3}{x - 2}$

Simplify:

28. $-\dfrac{x^{-2}}{y^2}\left(y^2 x^2 - \dfrac{3x^{-2}}{y^{-2}}\right)$

29. $\dfrac{4x^2 - 4x^4}{4x^2}$

30. Evaluate: $-3x - x^{-2} - x^{-3}$ if $x = \dfrac{1}{2}$.

LESSON 43 *Sine, cosine, and tangent*

43.A

sine, cosine, and tangent

Here we show three triangles. Each one is a right triangle, and each one contains a 30° angle.

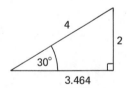
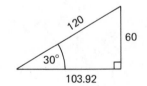
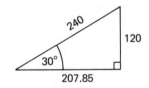

The longest side of a right triangle is always the side opposite the right angle and this side is called the **hypotenuse.** The hypotenuses of these triangles from left to right are 4, 120, and 240.

We remember that right triangles that have one acute angle equal are similar triangles, and that the ratios of corresponding sides of similar triangles are equal. If, for each of the three triangles shown, we write the ratio of the side opposite the 30° angle to the hypotenuse, we get these expressions,

$$\frac{2}{4} \qquad \frac{60}{120} \qquad \frac{120}{240}$$

each of which has a value of .5. **We would find this to be true for this ratio in any right triangle with a 30° angle. In a right triangle we call this ratio, which is the ratio of the side opposite the angle to the hypotenuse, the *sine (sin) of the angle.***

$$\sin 30° = \frac{\text{side opposite the } 30° \text{ angle}}{\text{hypotenuse}}$$

This ratio always equals .5 for a 30° angle. Every angle has a unique value of the ratio of the side opposite to the hypotenuse.

In a right triangle, the ratio of the side adjacent to the angle to the hypotenuse is called the *cosine (cos) of the angle.* For these triangles, the ratios are

$$\frac{3.464}{4} \qquad \frac{103.92}{120} \qquad \frac{207.85}{240}$$

Each of these ratios has a value of approximately .8660, which we round off to .87. **This ratio has the same value in any right triangle with a 30° angle regardless of the size of the triangle.** So we say

$$\cos 30° = \frac{\text{side adjacent to } 30° \text{ angle}}{\text{hypotenuse}} = .87$$

We call the ratio of the side opposite the angle to the side adjacent to the angle the *tangent* (*tan*) of the angle. Thus,

$$\tan 30° = \frac{\text{side opposite the } 30° \text{ angle}}{\text{side adjacent to the } 30° \text{ angle}}$$

This ratio is approximately .5774, which we round off to .58

$$\frac{2}{3.464} = .58 \qquad \frac{60}{103.92} = .58 \qquad \frac{120}{207.85} = .58$$

There is nothing sacrosanct about the words sine, cosine and tangent, and there is no particular reason for them to be defined as they are. Learning which one has which definition is blind memorization, and mnemonics are always helpful for memorizing. On the left, we will write the words sine, cosine, and tangent in that order. On the right, we will use the first letters of the words opposite, hypotenuse and adjacent to form the first letters of a sentence that is easy to remember.

$$\sin A = \frac{\text{opposite}}{\text{hypotenuse}} \qquad \frac{\text{Oscar}}{\text{had}}$$

$$\cos A = \frac{\text{adjacent}}{\text{hypotenuse}} \qquad \frac{\text{a}}{\text{hold}}$$

$$\tan A = \frac{\text{opposite}}{\text{adjacent}} \qquad \frac{\text{on}}{\text{Arthur}}$$

Thus, if we can remember to write down sine, cosine, and tangent in that order and then to write down "Oscar had a hold on Arthur," we have the definitions memorized.

The sines, cosines and tangents of angles are available in tables called **tables of trigonometric functions.** Tables can be constructed that are accurate to any number of digits desired. The value of the cosine of 30° accurate to two places, to four places, and to six places is

$$\cos 30° = .87 \qquad \cos 30° = .8660 \qquad \cos 30° = .866025$$

Which table to use depends on the accuracy necessary and the availability of tables. Most mathematics books, including this one, contain four-digit tables. Later, in Lesson 63, we will discuss how the tables stored in the memory of scientific calculators can be used. After that lesson either the tables or the calculator may be used. Of course, the answers to the problems will differ a little because the tables have been rounded off and are not as accurate as the tables stored in the calculator. When we use the tables we will round off the numbers. This will make the numbers easier to handle and allow us to concentrate on the concepts rather than on the numbers.

example 43.A.1   Find (a) the sine of 39.2°, (b) the cosine of 23.6°, and (c) the tangent of 57.8°.

solution   We use the trigonometric tables in the appendix and round off to find

   (a) sin 39.2° = **.63**   (b) cos 23.6° = **.92**   (c) tan 57.8° = **1.59**

example 43.A.2   Find (a) the sine of A, (b) the cosine of B, (c) the tangent of C.

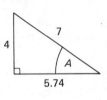 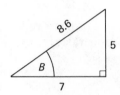 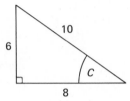

solution

$$\sin A = \frac{\text{opposite}}{\text{hypotenuse}} = \frac{4}{7} = .57$$

$$\cos B = \frac{\text{adjacent}}{\text{hypotenuse}} = \frac{7}{8.6} = .81$$

$$\tan C = \frac{\text{opposite}}{\text{adjacent}} = \frac{6}{8} = .75$$

problem set 43

1. The oligarchs were in control, although they comprised only .0032 of the total population. If there were 1280 oligarchs, what was the total population?

2. The boys believed that temerity was a desirable attribute so the number of boys that were temerarious increased 132 percent in only 1 month. If the temerarious now numbered 9280, how many were temerarious last month?

3. The chemical formula for sodium hydroxide is $NaOH$. What is the weight of the sodium (Na) in 320 grams of sodium hydroxide? (Na, 23; O, 16; H, 1)

4. Twenty-three percent of the mixture was cadmium. If the total weight of the mixture was 3000 grams, what was the weight of the other constituents of the mixture?

5. Connie was 20 miles ahead of Larry when he started after her. If he caught her in 5 hours and traveled twice as fast as she traveled, how far did he have to go to catch her?

6. Use the mnemonic "Oscar had a hold on Arthur" as an aid in writing the definitions of sine A, cosine A, and tangent A.

*7. In the trigonometric tables, find (a) sin 39.2°, (b) cos 23.6°, (c) tan 57.8°.

*8. Use the definitions of sine, cosine, and tangent and the triangles shown to find to two decimal places: (a) sin A, (b) cos B, and (c) tan C.

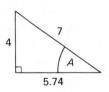 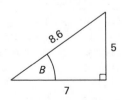 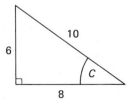

9. Use unit multipliers to convert 4 ft^3 to in^3.

10. $\dfrac{a}{b} - x = \dfrac{p}{z}$; find b

11. $\dfrac{xz}{p} - k = \dfrac{m}{c}$; find p

12. $\dfrac{x}{m} - \dfrac{k}{c} = \dfrac{p}{z}$; find m

13. Use similar triangles to solve for a, b, and c.

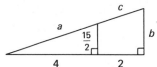

Solve by factoring:

14. $-x^3 - 9x^2 = 20x$

15. $4x^2 - 81 = 0$

16. Expand: $(x - 3)^3$

Simplify:

17. $\dfrac{5x - 6x^2 + x^3}{x^2 + 2x - 3} \cdot \dfrac{x^2 - 9}{x^3 - 8x^2 + 15x}$

18. $16^{-\frac{3}{4}}$

19. $\dfrac{4x + \dfrac{1}{x}}{\dfrac{ay^2}{x} - 4}$

20. $\sqrt{\dfrac{3}{7}} - 4\sqrt{\dfrac{7}{3}}$

21. $3\sqrt{2}(5\sqrt{12} - \sqrt{2})$

22. Find the area of the shaded portion of the figure. Dimensions are in centimeters.

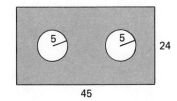

23. Find the equation of the line that passes through the point $(-2, -5)$ that has a slope of $-\frac{1}{7}$.

24. Solve: $\dfrac{5x - 7}{2} - \dfrac{3x - 2}{5} = 4$

25. Estimate: $\dfrac{(47{,}816 \times 10^5)(4923 \times 10^{-14})}{403{,}000}$

26. Solve the system by graphing and then get an exact solution by using either substitution or elimination.

$$\begin{cases} 2x - 3y = -9 \\ x + y = 2 \end{cases}$$

27. Add: $\dfrac{3}{x + 1} - \dfrac{2}{x^2(x + 1)} + \dfrac{3x + 2}{x^2 - 1}$

28. Simplify: $\dfrac{4x^{-2}y}{p^{-2}}\left(\dfrac{2p^{-2}x^2}{y} - \dfrac{4x^{-2}y}{p^2}\right)$

29. Find the distance between $(-2, 5)$ and $(4, 4)$.

30. Evaluate: $x^2 - yx - (x - y)$ if $x = \dfrac{1}{2}$ and $y = \dfrac{1}{4}$.

LESSON *44* *Solving right triangles*

44.A

solving right triangles

Sines, cosines, and tangents are ratios and are always the same for any given angle. For example, the sine, cosine and tangent of 37° given by the author's calculator are as follows:

$$\sin 37° = .601815 \qquad \cos 37° = .7986355 \qquad \tan 37° = .753554$$

Since these ratios depend only on the size of the angle and do not depend on the size of the triangle, we can use them to find missing parts of right triangles. If we use these six- and seven-digit figures in our calculations, we tend to get lost in the arithmetic of the problems. This is true even when four-digit trigonometric tables are used. Thus, for the present we will round off the entries in the four digit tables and concentrate on the process rather than on the exactness of the answers. Later, we will learn how to use a calculator to get more exact answers, but the results obtainable by using rounded off values will be sufficiently accurate for the time being.

example 44.A.1 Find the missing parts of this triangle.

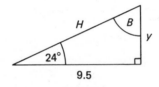

solution The sum of the interior angles of any triangle is 180°. This triangle has one 90° angle and one 24° angle. Thus angle B must be 66° because 90° + 24° + 66° = 180°.

Now we can use either the 24° angle or the 66° angle to find H and y. We decide to use the 24° angle. Beginners often find it difficult to decide whether to use the sine, cosine, or tangent in a particular case. Some find it helpful to use all three and see which ones work out. We will do this. Thus, we first find the sine, cosine, and tangent of 24° from the table in the appendix.

$$\sin 24° = .41 \qquad \cos 24° = .91 \qquad \tan 24° = .45$$

Next we write all three ratios and substitute.

(a) $\sin 24° = \dfrac{\text{opposite}}{\text{hypotenuse}} \longrightarrow .41 = \dfrac{y}{H}$ (two unknowns)

(b) $\cos 24° = \dfrac{\text{adjacent}}{\text{hypotenuse}} \longrightarrow .91 = \dfrac{9.5}{H}$ (one unknown)

(c) $\tan 24° = \dfrac{\text{opposite}}{\text{adjacent}} \longrightarrow .45 = \dfrac{y}{9.5}$ (one unknown)

We see that we can go no further in (a) because we still have the two unknowns y and H but only one equation. However, equations (b) and (c) can be solved because each of these equations contains only one unknown.

(b) $.91 = \dfrac{9.5}{H} \longrightarrow H = \dfrac{9.5}{.91} \longrightarrow \mathbf{H = 10.44}$

(c) $.45 = \dfrac{y}{9.5} \longrightarrow (.45)(9.5) = y \longrightarrow \mathbf{4.275 = y}$

example 44.A.2   Find the missing parts of the triangle.

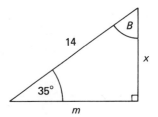

solution   Angle B must be 55° because 90° + 35° + 55° = 180°. This time we decide to use the 35° angle. First, we use the trigonometric tables to find the sine, cosine, and tangent of 35°.

$$\sin 35° = .57 \qquad \cos 35° = .82 \qquad \tan 35° = .70$$

Now we write all three ratios and substitute where possible.

(a) $\sin 35° = \dfrac{\text{opposite}}{\text{hypotenuse}} \quad \longrightarrow \quad .57 = \dfrac{x}{14}$ (one unknown)

(b) $\cos 35° = \dfrac{\text{adjacent}}{\text{hypotenuse}} \quad \longrightarrow \quad .82 = \dfrac{m}{14}$ (one unknown)

(c) $\tan 35° = \dfrac{\text{opposite}}{\text{adjacent}} \quad \longrightarrow \quad .70 = \dfrac{x}{m}$ (two unknowns)

This time we see that we can proceed no further with equation (c) because this equation has two unknowns, x and m. Equations (a) and (b), however, contain only one unknown and can be solved.

(a) $.57 = \dfrac{x}{14} \quad \longrightarrow \quad (.57)(14) = x \quad \longrightarrow \quad \mathbf{7.98 = x}$

(b) $.82 = \dfrac{m}{14} \quad \longrightarrow \quad (.82)(14) = m \quad \longrightarrow \quad \mathbf{11.48 = m}$

example 44.A.3   Find the missing parts of the triangle.

solution   When two sides of a right triangle are given, we can always find the third side by using the Pythagorean theorem.

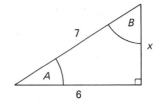

$$7^2 = 6^2 + x^2$$
$$49 = 36 + x^2$$
$$13 = x^2$$
$$\sqrt{13} = x$$

Now to find angle A we will use trigonometric functions. We will try all three functions and see what happens.

(a) $\sin A = \dfrac{\text{opposite}}{\text{hypotenuse}} \quad \longrightarrow \quad \sin A = \dfrac{x}{7}$ (two unknowns)

(b) $\cos A = \dfrac{\text{adjacent}}{\text{hypotenuse}} \quad \longrightarrow \quad \cos A = \dfrac{6}{7}$ (one unknown)

(c) $\tan A = \dfrac{\text{opposite}}{\text{adjacent}} \quad \longrightarrow \quad \tan A = \dfrac{x}{6}$ (two unknowns)

We can't solve equations (a) and (c) because each of these equations has two unknowns. We can solve equation (b), however, and find the cosine of angle A.

$$\text{(b)} \quad \cos A = \frac{6}{7} \quad \longrightarrow \quad \cos A = .8571$$

Now we use the table and find that angle A is approximately $31.0°$.

Angle A = angle whose cosine is $.8571$ = **31.0°**

Now since the sum of angle A and angle B is $90°$, we can find angle B by subtracting $31.0°$ from $90°$.

Angle $B = 90° - 31.0° \quad \longrightarrow \quad$ angle $B =$ **59.0°**

problem set 44

1. A secret number was unearthed in the Mayan ruins. When the number was increased by 5 and this sum multiplied by -2, the result was 18 greater than 6 times the opposite of the number. What was the secret number found in the ruins?

2. Bobby and Garnetta thought of four consecutive odd integers such that 5 times the sum of the first and the third was 22 greater than the product of 8 and the sum of the second and the fourth. Find the numbers.

3. The mixture was composed of antimony and tin, and 430 tons of antimony was required to make 700 tons of the mixture. How many tons of tin was required to make 2800 tons of the mixture?

4. Emelio and Nessie had 550 grams of freon in a container. What was the weight of the carbon (C) in the container given that the formula for freon is CCl_2F_2? (C, 12; Cl, 35; F, 19)

5. The trip was only 200 miles so Bob and Rita rode at 25 mph for awhile. Then they doubled their speed for the last part of the trip so that they could complete the journey in 5 hours. How far did they drive before they increased their speed?

Use trigonometric functions as necessary to find the missing parts of the following triangles.

*6.   *7.   *8.

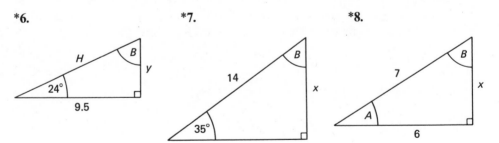

9. Use unit multipliers to convert 4 square miles to square feet.

10. $\dfrac{k}{m} - c = \dfrac{p}{d}$; find m

11. $\dfrac{3k}{m} - \dfrac{d}{p} = \dfrac{L}{c}$; find m

12. $\dfrac{a}{c} + d = \dfrac{x}{m}$; find c

13. Use similar triangles to solve for a and b.

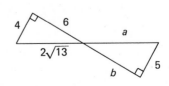

Solving by factoring:

14. $24x = -x^3 + 10x^2$ **15.** $25p^2 - 81 = 0$

16. Divide $x^4 - 1$ by $x + 4$.

Simplify:

17. $\dfrac{14x + 9x^2 + x^3}{x^2 - x - 6} \div \dfrac{x^3 + 12x^2 + 35x}{x^2 + 10x + 25}$ **18.** $\dfrac{1}{-32^{\frac{4}{5}}}$

19. This figure is the base of a solid that is 4 feet high. Find the volume of the solid. The dimensions are in inches.

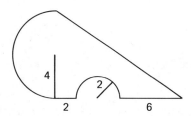

Simplify:

20. $\dfrac{5x^2 + \dfrac{1}{x}}{\dfrac{pm^2}{x} + 5}$ **21.** $3\sqrt{\dfrac{2}{17}} - 5\sqrt{\dfrac{17}{2}}$ **22.** $3\sqrt{6}(2\sqrt{6} - \sqrt{12})$

23. Find the equations of lines a and b.

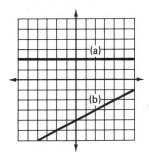

Solve:

24. $\dfrac{5x - 2}{3} - \dfrac{2x - 4}{2} = 6$ **25.** $3\dfrac{1}{5}k + \dfrac{2}{5} = \dfrac{1}{10}$

26. Solve the system by graphing and then get an exact solution by using either substitution or elimination.

$\begin{cases} 2x - y = 5 \\ 4x + 3y = 9 \end{cases}$

27. Add: $\dfrac{x}{x + 2} - \dfrac{2}{x^2 - 4}$

28. Estimate: $\dfrac{(51,463 \times 10^{-14})(748,600 \times 10^{-21})}{7,861,523}$

29. Evaluate: $a^2 - ab^2$ if $a = \dfrac{1}{2}$ and $b = -\dfrac{1}{2}$.

30. Simplify: $-2^0 - 2^2 - (-2)^3 - |-2 - 3^0| - (-2)^3$

LESSON 45 *Difference-of-two-squares theorem*

45.A
difference-of-two-squares theorem

The two equations

$$\text{(a)} \quad x^2 = 4 \qquad \text{and} \qquad \text{(b)} \quad x^2 - 4 = 0$$

are two different forms of the same equation. In (b) the equation is written as the difference of two squares. It can be solved by factoring.

$$x^2 - 4 = 0 \qquad\qquad \text{equation}$$
$$(x + 2)(x - 2) = 0 \qquad\qquad \text{factored}$$
$$x = 2 \quad \text{or} \quad x = -2 \qquad \text{solved}$$

In the same way, any other equation that is in the form

$$p^2 = q^2$$

can be solved by rewriting the equation as the difference of two squares and factoring.

$$p^2 = q^2 \qquad\qquad \text{equation}$$
$$p^2 - q^2 = 0 \qquad\qquad \text{difference-of-two-squares form}$$
$$(p - q)(p + q) = 0 \qquad\qquad \text{factored}$$
$$p = +q \quad \text{or} \quad p = -q \qquad \text{solved}$$

It is important to note that we get two solutions to this equation. One is positive, and the other is negative.

This result is important and useful, and we call it the **difference-of-two-squares theorem.** We state it formally here.

DIFFERENCE-OF-TWO-SQUARES THEOREM

If p and q are real numbers and if $p^2 = q^2$, then

$$p = q \qquad \text{or} \qquad p = -q$$

We find this theorem will permit quick solutions to problems in which a squared variable expression is equal to a constant, as shown here:

EQUATION	SOLUTION
$x^2 = 3$	$x = \pm\sqrt{3}$
$x^2 = 5$	$x = \pm\sqrt{5}$

The same pattern holds if the squared term contains a constant. First, we take the square root of both sides, and then we solve the resulting equation.

EQUATION	SOLUTION
$(x + 2)^2 = 3$	$x + 2 = \pm\sqrt{3} \longrightarrow \mathbf{x = -2 \pm \sqrt{3}}$
$(p - 4)^2 = 5$	$p - 4 = \pm\sqrt{5} \longrightarrow \mathbf{p = 4 \pm \sqrt{5}}$
$\left(m + \dfrac{3}{4}\right)^2 = 7$	$m + \dfrac{3}{4} = \pm\sqrt{7} \longrightarrow \mathbf{m = -\dfrac{3}{4} \pm \sqrt{7}}$

example 45.A.1 Solve $(x + 17)^2 = 2$.

solution We begin our solution by taking the square root of both sides. We remember that there are two answers.

$$(x + 17)^2 = 2 \qquad \text{equation}$$

$$x + 17 = \pm\sqrt{2} \qquad \text{square root of both sides}$$

$$\boldsymbol{x = -17 + \sqrt{2}} \quad \text{or} \quad \boldsymbol{x = -17 - \sqrt{2}} \qquad \text{two solutions}$$

We will check both solutions in the original equation.

Check $-17 + \sqrt{2}$:   Check $-17 - \sqrt{2}$:

$$(-17 + \sqrt{2} + 17)^2 = 2 \qquad\qquad (-17 - \sqrt{2} + 17)^2 = 2$$

$$(\sqrt{2})^2 = 2 \qquad\qquad\qquad (-\sqrt{2})^2 = 2$$

$$2 = 2 \quad \text{Check} \qquad\qquad\qquad 2 = 2 \quad \text{Check}$$

example 45.A.2 Solve $\left(x + \dfrac{2}{5}\right)^2 = 3$.

solution We begin by taking the square root of both sides of the equation.

$$\left(x + \frac{2}{5}\right)^2 = 3 \qquad \text{equation}$$

$$x + \frac{2}{5} = \pm\sqrt{3} \qquad \text{square root of both sides}$$

$$x = -\boldsymbol{\frac{2}{5}} \pm \sqrt{3} \qquad \text{added } -\frac{2}{5} \text{ to both sides}$$

Now we check:

Check $-\dfrac{2}{5} + \sqrt{3}$:   Check $-\dfrac{2}{5} - \sqrt{3}$:

$$\left(-\frac{2}{5} + \sqrt{3} + \frac{2}{5}\right)^2 = 3 \qquad\qquad \left(-\frac{2}{5} - \sqrt{3} + \frac{2}{5}\right)^2 = 3$$

$$(\sqrt{3})^2 = 3 \qquad\qquad\qquad (-\sqrt{3})^2 = 3$$

$$3 = 3 \quad \text{Check} \qquad\qquad\qquad 3 = 3 \quad \text{Check}$$

problem set 45

1. Since knowledge of chemistry is useful even in nonscientific fields of study, a majority of the students elected to take chemistry. If 38 percent did not take chemistry and 248 students did take chemistry, how many students were there in all?

2. The chemical formula for phosphine is PH_3. If the hydrogen (H) in a quantity of phosphine weighed 24 grams, what was the total weight of the phosphine? (P, 31; H, 1)

3. Sergio and Lolita found that bromides took up 36 percent of the storage space in the stockroom. If 5376 bottles did not contain bromides, how many did contain bromides?

4. Roses were $12 a bunch and carnations were $14 a bunch. Amy and Kathy bought a total of 11 bunches and spent $138. How many bunches of each did they buy?

5. The train made the trip in 8 hours, and the bus made all but 80 miles of the trip in 12 hours. What was the rate of each if the rate of the train was 30 mph greater than the rate of the bus?

Use the difference-of-two-squares theorem to find the solution to the following equations.

*6. $x^2 = 3$

*7. $(x + 2)^2 = 3$

*8. $(x + 17)^2 = 2$

*9. $\left(x + \dfrac{2}{5}\right)^2 = 3$

10. Find side x.

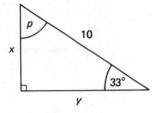

11. Find side p.

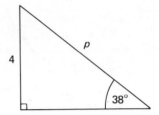

12. Find angles A and B.

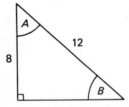

13. Use unit multipliers to convert 100,000 square miles to square inches.

14. Estimate: $\dfrac{(571,652)(40,316)}{214,000 \times 10^6}$

15. Find side d.

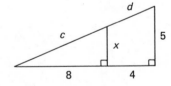

16. $\dfrac{mx}{4} + \dfrac{y}{p} = c$; find p

17. $\dfrac{5x}{p} - k = \dfrac{m}{c}$; find p

Solve by factoring:

18. $x^3 = -5x^2 + 50x$

19. $36x^2 - 25 = 0$

Simplify:

20. $\dfrac{x^3 + 6x^2 - 7x}{x^2 + 4x - 21} \div \dfrac{x^3 - 2x + x^2}{x^2 + 2x - 15}$

21. $-(-27)^{-\frac{5}{3}}$

22. Find the perimeter of this figure. Dimensions are in inches.

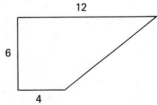

Simplify:

23. $\dfrac{4xp - \dfrac{1}{x}}{6p - \dfrac{p^2}{x}}$

24. $5\sqrt{\dfrac{2}{5}} + 3\sqrt{\dfrac{5}{2}}$

25. $2\sqrt{5}(3\sqrt{15} - 2\sqrt{10})$

26. Find the equation of the line that passes through $(-2, -5)$ whose slope is $-\frac{2}{3}$.

27. Solve: $\dfrac{5x - 2}{7} - \dfrac{x - 3}{5} = 4$

28. Solve the system by graphing and then find an exact solution by using either substitution or elimination.

$$\begin{cases} 2x + 3y = 3 \\ x - 5y = -20 \end{cases}$$

29. Add: $\dfrac{2}{x - 2} - \dfrac{3}{x + 2} - \dfrac{2x}{x^2 - 4}$

30. Evaluate: $a^2 - ab^3$ if $a = -\dfrac{1}{2}$ and $b = \dfrac{1}{2}$.

LESSON 46 *More on radical expressions • Radicals to fractional exponents*

46.A

more on radical expressions

We have been simplifying expressions such as

$$-\sqrt{\dfrac{2}{3}} + 3\sqrt{\dfrac{3}{2}}$$

by first using the quotient-of-square-roots theorem, then rationalizing the denominators by using the denominator-numerator–same-quantity theorem, and finishing by finding a common denominator and adding.

(a) $-\dfrac{\sqrt{2}}{\sqrt{3}} + \dfrac{3\sqrt{3}}{\sqrt{2}}$ quotient-of-square-roots theorem

(b) $-\dfrac{\sqrt{2}}{\sqrt{3}}\dfrac{\sqrt{3}}{\sqrt{3}} + \dfrac{3\sqrt{3}}{\sqrt{2}}\dfrac{\sqrt{2}}{\sqrt{2}} = -\dfrac{\sqrt{6}}{3} + \dfrac{3\sqrt{6}}{2}$ rationalized denominators

(c) $\dfrac{-2\sqrt{6}}{6} + \dfrac{9\sqrt{6}}{6} = \dfrac{\mathbf{7\sqrt{6}}}{\mathbf{6}}$ found common denominator and added

Problems such as this one are good practice problems but will provide even more practice if we add one more term whose simplification requires that we use the product-of-square-roots theorem.

example 46.A.1 Simplify: $3\sqrt{\dfrac{5}{2}} + 3\sqrt{\dfrac{2}{5}} - \sqrt{90}$

solution The last term was contrived so it will contain a $\sqrt{10}$, as will the first two terms. We begin by rationalizing the denominators and simplifying the last term.

$$\frac{3\sqrt{5}}{\sqrt{2}}\frac{\sqrt{2}}{\sqrt{2}} + \frac{3\sqrt{2}}{\sqrt{5}}\frac{\sqrt{5}}{\sqrt{5}} - \sqrt{9}\sqrt{10} \qquad \text{quotient-of-square-roots theorem}$$

$$\frac{3\sqrt{10}}{2} + \frac{3\sqrt{10}}{5} - 3\sqrt{10} \qquad \text{simplified}$$

Now we use 10 as the common denominator and add.

$$\frac{15\sqrt{10}}{10} + \frac{6\sqrt{10}}{10} - \frac{30\sqrt{10}}{10} = \frac{-9\sqrt{10}}{10} \qquad \text{simplified and added}$$

example 46.A.2 Simplify: $3\sqrt{\dfrac{2}{7}} - 5\sqrt{\dfrac{7}{2}} + 3\sqrt{56}$

solution Again we begin by simplifying the radicals

$$\frac{3\sqrt{2}}{\sqrt{7}}\frac{\sqrt{7}}{\sqrt{7}} - \frac{5\sqrt{7}}{\sqrt{2}}\frac{\sqrt{2}}{\sqrt{2}} + 3\sqrt{4}\sqrt{14}$$

$$\frac{3\sqrt{14}}{7} - \frac{5\sqrt{14}}{2} + 6\sqrt{14}$$

As the last step we use 14 as the common denominator and add.

$$\frac{6\sqrt{14}}{14} - \frac{35\sqrt{14}}{14} + \frac{84\sqrt{14}}{14} = \frac{55\sqrt{14}}{14}$$

46.B
radicals to fractional exponents

Many radical expressions are easier to simplify if we first change the radical expressions to expressions that have exponents that are fractions.

example 46.B.1 Simplify: $\sqrt{3\sqrt{3}}$.

solution We begin by replacing both radicals with parentheses, brackets and fractional exponents.

$$[3(3^{\frac{1}{2}})]^{\frac{1}{2}}$$

Next we use the power theorem and get

$$3^{\frac{1}{2}}3^{\frac{1}{4}}$$

We multiply expressions with like bases by adding exponents, so we get as our answer

$$3^{\frac{3}{4}}$$

example 46.B.2 Simplify: $\sqrt[3]{3\sqrt{3}}$.

solution First we replace the radicals with parentheses and fractional exponents.

$$[3(3^{\frac{1}{2}})]^{\frac{1}{3}}$$

next we use the power theorem to simplify.

$$3^{\frac{1}{3}}3^{\frac{1}{6}}$$

Then we multiply the exponentials by adding the exponents.

$$3^{\frac{2}{6}}3^{\frac{1}{6}} = 3^{\frac{3}{6}} = \mathbf{3^{\frac{1}{2}}}$$

example 46.B.3   Simplify: $\sqrt{x^3 y^2}\ \sqrt[3]{xy}.$[†]

solution   We replace the radicals with parentheses and fractional exponents and get

$$(x^3 y^2)^{\frac{1}{2}}(xy)^{\frac{1}{3}}$$

Next we use the power theorem.

$$x^{\frac{3}{2}}y x^{\frac{1}{3}}y^{\frac{1}{3}}$$

We finish by multiplying exponentials with like bases and we get

$$x^{\frac{3}{2}}x^{\frac{1}{3}}y y^{\frac{1}{3}} = x^{\frac{9}{6}}x^{\frac{2}{6}}y^{\frac{3}{3}}y^{\frac{1}{3}} = \mathbf{x^{\frac{11}{6}}y^{\frac{4}{3}}}$$

example 46.B.4   Simplify: $\sqrt[3]{x^5 y^3}\ \sqrt[4]{xy^5}.$

solution   First we replace the radicals with parentheses and fractional exponents,

$$(x^5 y^3)^{\frac{1}{3}}(xy^5)^{\frac{1}{4}}$$

and use the power theorem,

$$x^{\frac{5}{3}}y x^{\frac{1}{4}}y^{\frac{5}{4}}$$

and simplify to get.

$$\mathbf{x^{\frac{23}{12}}y^{\frac{9}{4}}}$$

problem set 46

1. Many did not consider a 140 percent increase to be excessive, as the total was still only 1440. What was the total before the increase?

2. One thousand grams of beryllium and 3000 grams of other elements made up the mixture. If 24,000 kilograms of the mixture was needed, how much beryllium was required?

3. The formula for ammonium chloride is NH_4Cl. If the chlorine (Cl) in a quantity of ammonium chloride weighed 140 grams, what was the weight of the ammonium chloride? (N, 14; H, 1; Cl, 35)

4. The ratio of fairies to elves was 7 to 2 and the number of fairies was 11 greater than 3 times the number of elves. How many of each were there?

5. Bruce and Mark headed north at 11 a.m. at full speed. After 4 hours Bruce was 16 miles ahead. What did Bruce consider to be full speed if Mark's speed was 16 mph?

Simplify:

*6. $3\sqrt{\dfrac{5}{2}} + 3\sqrt{\dfrac{2}{5}} - \sqrt{90}$ ***7.** $3\sqrt{\dfrac{2}{7}} - 5\sqrt{\dfrac{7}{2}} + 3\sqrt{56}$ ***8.** $\sqrt{3\sqrt{3}}$

[†] Even roots of even powers of the variable can be troublesome when the replacements for the variable are negative numbers. Thus, for the present, the domain for problems like this one will be assumed to be the set of positive real numbers.

*9. $\sqrt[3]{3\sqrt{3}}$ **10.** $\sqrt{x^3y^2}\,\sqrt[3]{xy}$ **11.** $\sqrt[3]{x^5y^3}\,\sqrt[4]{xy^5}$

Solve:

12. $(x - \frac{2}{5})^2 = 7$ **13.** $(x - \frac{1}{4})^2 = 5$

14. Find angle B and side y. **15.** Find angle A and side C.

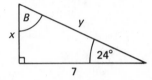 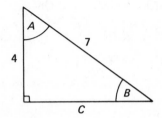

16. Estimate: $\dfrac{(476{,}800)(9{,}016{,}423 \times 10^4)}{408 \times 10^{10}}$

17. $\dfrac{ax}{b} - c = \dfrac{k}{m}$; find m **18.** $\dfrac{x}{m} - \dfrac{yb}{c} = p$; find c

19. Find sides x and y.

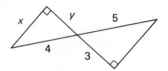

20. Solve $-40x = 13x^2 + x^3$ by factoring.

Simplify:

21. $\dfrac{x^2 + 45 + 14x}{x^3 + 10x^2 + 9x} \cdot \dfrac{x^3 - 2x - x^2}{x^2 - 2x - 35}$ **22.** $8^{-\frac{4}{3}}$

23. This figure is the base of a box that is 3 inches high. How many 1-inch sugar cubes would the box hold? Dimensions are in inches.

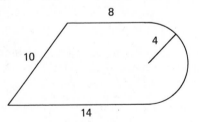

Simplify:

24. $\dfrac{\dfrac{m^2}{x} - x}{\dfrac{p^2}{x} + 2x}$ **25.** $3\sqrt{7}(2\sqrt{14} - \sqrt{7})$

26. Find the equation of the line that goes through $(-2, 4)$ and is perpendicular to the line that goes through $(-5, 7)$ and $(4, 2)$.

Solve:

27. $\dfrac{-x + 3}{2} - \dfrac{2x + 4}{3} = 5$ **28.** $\dfrac{-2x - 5}{4} - \dfrac{x - 2}{3} = 7$

29. Solve the system by graphing and then get an exact solution by using substitution or elimination.
$$\begin{cases} x - y = -3 \\ x + 2y = -2 \end{cases}$$

30. Evaluate: $a^2 - a^3b - ab$ if $a = -\dfrac{1}{2}$ and $b = -\dfrac{1}{4}$

LESSON 47 *Rate unit conversions • More on fractional exponents*

47.A
more on unit conversion

Rate units such as miles per hour or feet per second contain both a distance unit and a time unit. To convert rate units we use distance unit multipliers to convert the distance units, and time unit multipliers to convert the time units.

example 47.A.1   Convert 42 inches per second to miles per hour.

solution   We begin by writing the unit inches per second in fractional form.

$$\frac{42 \text{ in}}{\text{sec}}$$

We will convert the inches to miles first. We will go from inches to feet and from feet to miles.

$$\frac{42 \text{ in}}{\text{sec}} \times \frac{1 \text{ ft}}{12 \text{ in}} \times \frac{1 \text{ mile}}{5280 \text{ ft}}$$

Now our rate is miles per second. We continue by converting seconds to minutes and then minutes to hours.

$$\frac{42 \text{ in}}{\text{sec}} \times \frac{1 \text{ ft}}{12 \text{ in}} \times \frac{1 \text{ mi}}{5280 \text{ ft}} \times \frac{60 \text{ sec}}{\text{min}} \times \frac{60 \text{ min}}{\text{hr}} = \frac{(42)(60)(60)}{(12)(5280)} \frac{\text{mi}}{\text{hr}}$$

If we use a calculator to multiply and divide, we get, if we round off to two decimal places,

$$2.39 \frac{\text{mi}}{\text{hr}}$$

example 47.A.2   Convert 480 miles per hour to feet per second.

solution   Again we begin by writing the given units in fractional form.

$$\frac{480 \text{ mi}}{\text{hr}}$$

Now we will use one unit multiplier to convert miles to feet and two unit multipliers to convert hours to seconds.

$$\frac{480 \text{ mi}}{\text{hr}} \times \frac{5280 \text{ ft}}{\text{mi}} \times \frac{\text{hr}}{60 \text{ min}} \times \frac{\text{min}}{60 \text{ sec}} = \frac{(480)(5280)}{(60)(60)} \frac{\text{ft}}{\text{sec}}$$

Now a calculator can be used to get a decimal answer if desired.

example 47.A.3   Convert 15 inches per second to yards per hour.

solution   We will convert inches to feet to yards, and seconds to minutes to hours.

$$\frac{15 \text{ in}}{\text{sec}} \times \frac{1 \text{ ft}}{12 \text{ in}} \times \frac{1 \text{ yd}}{3 \text{ ft}} \times \frac{60 \text{ sec}}{\text{min}} \times \frac{60 \text{ min}}{\text{hr}} = \frac{(15)(60)(60)}{(12)(3)} \frac{\text{yd}}{\text{hr}}$$

47.B
more on fractional exponents

Some exponent problems are trick problems and cannot be simplified unless one realizes that they are trick problems.

example 47.B.1   Simplify: $\sqrt[3]{27\sqrt{3}}$.

solution   **This is a trick problem that cannot be simplified unless it is recognized. The trick is to write 27 as 3^3 so all the bases will be the same.**

$$\sqrt[3]{3^3\sqrt{3}}$$

Next we replace the radicals with parentheses, brackets and fractional exponents.

$$[3^3(3^{\frac{1}{2}})]^{\frac{1}{3}}$$

Now we use the power theorem and finish by adding the exponents.

$$3^1 3^{\frac{1}{6}} = 3^{\frac{7}{6}}$$

To work these problems, we must always remember simplifications such as

$$4 = 2^2 \qquad 9 = 3^2 \qquad 16 = 2^4 \qquad 27 = 3^3$$

If we do not recognize the problem and make these replacements, the expressions cannot be simplified.

example 47.B.2   Simplify: $\sqrt[6]{4\sqrt{2}}$.

solution   First we write 4 as 2^2.

$$\sqrt[6]{2^2\sqrt{2}}$$

Next we replace the radicals with parentheses, brackets and fractional exponents.

$$[2^2(2^{\frac{1}{2}})]^{\frac{1}{6}}$$

Then we finish by using the power theorem and the product theorem.

$$2^{\frac{2}{6}}2^{\frac{1}{12}} = 2^{\frac{5}{12}}$$

problem set 47

1. Three-sevenths of the assembled throng watched in horror as the tsunami approached. If 2800 did not watch in horror, how many were in the throng?

2. Melissa giggled as she thought about four consecutive odd integers such that 5 times the sum of the first two was 10 less than 7 times the sum of the second and fourth. What were her four integers?

3. Thirty percent of the girls often pondered their muliebrity, and the rest didn't even know what the word meant. If 1400 were ignorant of the meaning, how many girls were there in all?

4. The chemical formula for sulfurous acid is H_2SO_3. If the oxygen in a quantity of sulfurous acid weighed 192 grams, what was the total weight of the sulfurous acid? (H, 1; S, 32; O, 16)

5. Tacitus made the first leg of the trip to Londinium in his chariot at 8 kph and walked the last leg at 4 kph. If the total distance traveled was 48 km and the total time was 7 hours, how far did he walk and how far did he ride?

Convert:

*6. 42 inches per second to miles per hour

*7. 480 miles per hour to feet per second

*8. 15 inches per second to yards per hour

Simplify:

*9. $\sqrt[3]{27\sqrt{3}}$

*10. $\sqrt[6]{4\sqrt{2}}$

11. $\sqrt{x^2 y}\,\sqrt[3]{y^2 x}$

12. $\sqrt{2\sqrt[3]{2}}$

13. $3\sqrt{\dfrac{2}{3}} - 5\sqrt{\dfrac{3}{2}} + 2\sqrt{24}$

14. $3\sqrt{\dfrac{5}{7}} - 2\sqrt{\dfrac{7}{5}} - \sqrt{315}$

Find the missing parts of the following triangles.

15. Find x.

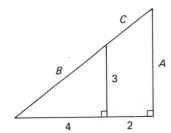

16. Find x.

Solve:

17. $(x - 3)^2 = 5$

18. $\left(x + \dfrac{2}{7}\right)^2 = \dfrac{4}{49}$

19. Estimate: $\dfrac{(36{,}421 \times 10^5)(493{,}025)}{40{,}216 \times 10^7}$

20. $\dfrac{ay}{x} + p = \dfrac{m}{c}$; find x

21. $\dfrac{a}{x} - \dfrac{c}{p} = b$; find p

22. Find side C.

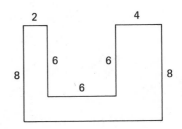

23. Find the perimeter of this figure in inches. The dimensions are in yards.

Simplify:

24. $\dfrac{\dfrac{x^2p}{m} - m}{\dfrac{x}{m} - p}$

25. $\dfrac{-14x + x^3 + 5x^2}{x^2 + 12x + 35} \div \dfrac{x^3 - 2x^2 - 15x}{x^2 + 8x + 15}$

26. $-27^{-\frac{4}{3}}$

27. Solve the system by graphing and then get an exact solution by using substitution or elimination.
$$\begin{cases} x - 4y = 8 \\ 2x - 3y = 9 \end{cases}$$

28. Find the equation of the line that goes through $(-2, 5)$ that has a slope of $\frac{2}{5}$.

29. Solve $20x = -12x^2 - x^3$ by factoring.

30. Evaluate: $-a - a^2 - a^3 - ab$ if $a = -\dfrac{1}{2}$ and $b = \dfrac{1}{5}$.

LESSON 48   *Radical equations*

48.A
radical equations

We have noted that the difference-of-two-squares theorem permits the solution of some equations by taking the square roots of both sides of the equation. We can use this theorem to find the roots to the equation

$$(x + 2)^2 - 5 = 0$$

First, we transform the equation by adding $+5$ to both sides and we get

$$(x + 2)^2 = 5$$

Next we take the square root of both sides, remembering that we get two signs on the right side.

$$x + 2 = \pm\sqrt{5} \qquad \text{square root of both sides}$$
$$x = -2 \pm \sqrt{5} \qquad \text{added } -2 \text{ to both sides}$$

If we wish, we can list the two roots separately by writing

$$x = -2 + \sqrt{5} \qquad \text{and} \qquad x = -2 - \sqrt{5}$$

We use another process to solve equations that contain a square root radical. First we isolate the radical, and then we square both sides of the equation. This process is permitted because it follows directly from the definition of the square root of a positive number.

DEFINITION OF SQUARE ROOT

If x is a positive real number, then $\sqrt{x}$ is the unique positive real number such that $(\sqrt{x})^2 = x$.

But when we do this, we must always check our answers in the original equation because squaring both sides of an equation sometimes generates an equation that has roots that are not roots of the original equation. To see how this is possible, let us begin with the equation

$$x = 2$$

whose only solution is the number 2. Now if we square both sides of this equation, we get the equation

$$x^2 = 4$$

But now we have an equation that can be satisfied by using either $+2$ or -2 as the replacement for x. **The number 2 is a solution to both the original equation and the new equation, but -2 is not a solution to the original equation. By squaring both sides, we have generated an equation that has more solutions than the original equation.**

example 48.A.1   Solve $\sqrt{x - 4} - 2 = 6$.

solution   We begin by **isolating the radical** by adding $+2$ to both sides of the equation, and we get

$$\sqrt{x - 4} = 8$$

Next we square both sides, remembering that $(\sqrt{x - 4})^2 = x - 4$.

$$x - 4 = 64$$

We finish by adding $+4$ to both sides.

$$x = 68$$

Now we must check this solution in the original equation to see if it satisfies the original equation.

$$\sqrt{x - 4} - 2 = 6 \qquad \text{original equation}$$
$$\sqrt{68 - 4} - 2 = 6 \qquad \text{substitute 68 for } x$$
$$\sqrt{64} - 2 = 6 \qquad \text{simplified}$$
$$8 - 2 = 6 \qquad \text{took square root}$$
$$6 = 6 \qquad \text{check}$$

Thus 68 is a solution to the original equation.

example 48.A.2   Solve $\sqrt{x + 4} + 2 = -7$.

solution   We always begin by **isolating the radical.**

$$\sqrt{x + 4} = -9$$

Now we square both sides and solve.

$$x + 4 = 81 \quad \longrightarrow \quad x = 77$$

Now we must check the original equation to see if 77 is a solution.

$$\sqrt{x + 4} + 2 = -7 \qquad \text{original equation}$$
$$\sqrt{77 + 4} + 2 = -7 \qquad \text{substituted}$$
$$\sqrt{81} + 2 = -7 \qquad \text{simplified}$$
$$9 + 2 = -7 \qquad \text{not true}$$

Since 9 + 2 ≠ −7, then 77 is not a solution to the original equation. Thus we find that the original equation has no solution.

example 48.A.3 Solve $\sqrt{x - 3} - 3 = 5$.

solution We begin by **isolating the radical.**

$$\sqrt{x - 3} = 8$$

Now we square both sides and solve.

$$x - 3 = 64 \longrightarrow x = 67$$

We finish by checking 67 in the original equation.

$$\sqrt{67 - 3} - 3 = 5 \longrightarrow \sqrt{64} - 3 = 5 \longrightarrow 8 - 3 = 5 \longrightarrow 5 = 5 \qquad \text{Check}$$

example 48.A.4 Solve $\sqrt{x^2 + 2x + 13} - 3 = x$.

solution We begin by isolating the radical by adding $+3$ to both sides.

$$\sqrt{x^2 + 2x + 13} = x + 3$$

Now we square both sides and solve.

$$x^2 + 2x + 13 = x^2 + 6x + 9 \qquad \text{squared both sides}$$
$$4 = 4x \qquad \text{simplified}$$
$$1 = x \qquad \text{divided by 4}$$

We check by replacing x with 1 in the original equation

$$\sqrt{(1)^2 + 2(1) + 13} - 3 = 1 \qquad \text{replaced } x \text{ with 1}$$
$$\sqrt{16} - 3 = 1 \qquad \text{simplified}$$
$$1 = 1 \qquad \text{check}$$

problem set 48

1. The lieutenant noted that .68 of the frontiersmen at the conclave wore buckskin. If 512 did not wear buckskin, how many frontiersmen were at the conclave?

2. If 140 grams of germanium was required to make 1540 grams of the compound, how many grams of other elements was required to make 6160 grams of the compound?

3. If a crucible contained 972 grams of $FeBr_2$, what was the weight of the iron (Fe) in the crucible? (Fe, 56; Br, 80)

4. In a group of children, 5 times the number of boys was 17 greater than 3 times the number of girls. Also, 6 times the number of girls was 2 greater than the number of boys. How many were boys and how many were girls?

5. The freight train took 20 hours to make the same trip the express train made in 10 hours. Find the speed of each if the express train was 30 mph faster than the freight train.

Solve:

*6. $\sqrt{x - 4} - 2 = 6$ *7. $\sqrt{x + 4} + 2 = -7$

*8. $\sqrt{x^2 + 2x + 13} - 3 = x$

9. Convert 60 miles per hour to feet per second.

Simplify:

10. $\sqrt{2\sqrt[3]{2}}$

11. $\sqrt{m^2 y}\sqrt[3]{m^4 y}$

12. $\sqrt{8\sqrt[3]{2}}$

13. $3\sqrt{\dfrac{2}{5}} - \sqrt{\dfrac{5}{2}} + 2\sqrt{40}$

14. $2\sqrt{\dfrac{7}{11}} - \sqrt{\dfrac{11}{7}} + 2\sqrt{308}$

Solve:

15. $\left(x - \dfrac{2}{7}\right)^2 = 4$

16. $(x - 3)^2 = 16$

17. Find angle C and side b.

18. Find angle A.

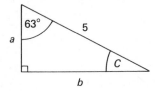

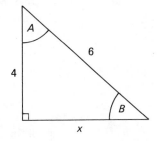

19. Estimate: $\dfrac{(4{,}071{,}623)(51{,}642 \times 10^5)}{200{,}000 \times 10^{-13}}$

20. $\dfrac{b}{x} - \dfrac{c}{p} + k = \dfrac{m}{y}$; find p

21. Find side y.

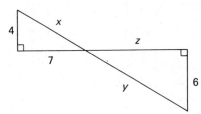

22. Divide $3x^3 - 2$ by $x + 1$.

23. Find the distance between $(-4, -3)$ and $(-8, 2)$.

Simplify:

24. $\dfrac{x^2 - 3x + 2}{x^3 + 4x^2 - 5x} \div \dfrac{x^2 + x - 6}{35x + 12x^2 + x^3}$

25. $-16^{-\frac{3}{4}}$

26. The figure is the base of a solid that is 4 inches high. The dimensions shown are in inches. Find the volume of the solid in cubic inches.

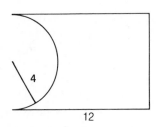

27. Find the distance between $(-2, 5)$ and $(7, -3)$.

28. Simplify: $\dfrac{\dfrac{x-1}{x} - \dfrac{1}{x^2}}{\dfrac{x+2}{x^2} - 4}$

29. Find the equation of the line that passes through the points $(2, 5)$ and $(7, -3)$.

30. Solve: $\dfrac{2x - 3}{2} - \dfrac{x}{2} = \dfrac{5}{3}$

LESSON 49 *Linear intercepts • Transversals*

49.A
linear
intercepts

We remember that a line is completely identified if we know its slope and its intercept. The equation of the line in the figure on the left is

$$y = -\frac{2}{3}x + 2$$

The first number in the equation is $-\frac{2}{3}$. This number is the slope and tells us that the slope is negative and that the ratio of the rise to the run is 2 to 3. The last number in the equation is 2, and this number is the y intercept of the line, which is the y coordinate of the point where the line crosses the y axis.

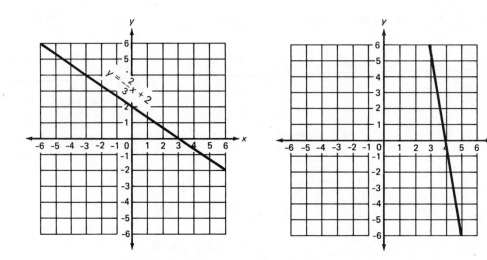

Sometimes it is necessary to find the equation of a line whose intercept is not shown on the graph. The line shown on the right is an example. By drawing the proper triangle, we can determine that the slope of this line is approximately -6. If we replace m with -6 in the equation $y = mx + b$, we get

$$y = -6x + b$$

Now to determine the intercept, we must estimate the coordinates of one point on the line and use these numbers for x and y in the equation. It appears that the line goes through the point $(4, 0)$. We will use these numbers for x and y and solve for b.

$$0 = -6(4) + b \qquad \text{substituted}$$
$$0 = -24 + b \qquad \text{multiplied}$$
$$24 = b \qquad \text{solved}$$

Now that we have the intercept, we can write our estimate of the equation of the line as

$$y = -6x + 24$$

49.B
transversals

If two straight lines in the same plane are intersected by a third straight line at two points, the third line is called a **transversal.** The lines that are intersected do not have to be parallel, as shown on the left. Lines A and B are intersected by line C at two points labeled M and N. Thus, line C is a transversal. On the right, we show the parallel lines D and E that are intersected by line G. Thus line G is also called a transversal. **When parallel lines are cut by a transversal, the corresponding angles are equal.** Thus,

angle H = angle M

angle J = angle P

angle K = angle Q

angle L = angle N

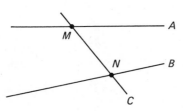

We remember that vertical angles are equal, and thus in this figure we have two groups of four equal angles. We usually use the symbol

$$\angle$$

to mean angle. Thus, we can designate angle H by writing

$$\angle H$$

Using this notation, we can designate the two groups of equal angles by writing

$$\angle H = \angle L = \angle M = \angle N \quad \text{and} \quad \angle K = \angle J = \angle P = \angle Q$$

example 49.B.1   Lines (1) and (2) are parallel. Find sides a and b.

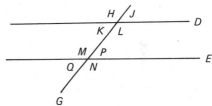

solution   Angle Z equals 27° because it and the 153° angle add to make a straight line, whose angle is 180°. Then angle P is 27° because $\angle P$ and $\angle Z$ are corresponding angles.

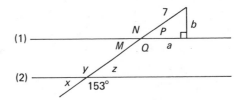

$$\sin 27° = \frac{b}{7} \quad \longrightarrow \quad .45 = \frac{b}{7} \quad \longrightarrow \quad \textbf{3.15} = \textbf{\textit{b}}$$

$$\cos 27° = \frac{a}{7} \quad \longrightarrow \quad .89 = \frac{a}{7} \quad \longrightarrow \quad \textbf{6.23} = \textbf{\textit{a}}$$

$$\tan 27° = \frac{b}{a} \quad \text{(two unknowns)}$$

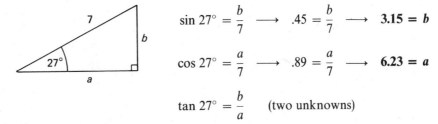

example 49.B.2   Lines (1) and (2) are parallel. Find sides M and N.

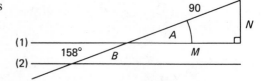

solution   Angle B is $22°$ because $22° + 158° = 180°$. Thus angle A equals $22°$ because $\angle A$ and $\angle B$ are corresponding angles.

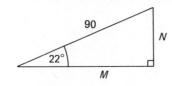

$\sin 22° = \dfrac{N}{90} \longrightarrow .37 = \dfrac{N}{90} \longrightarrow \mathbf{33.3 = N}$

$\cos 22° = \dfrac{M}{90} \longrightarrow .93 = \dfrac{M}{90} \longrightarrow \mathbf{83.7 = M}$

$\tan 22° = \dfrac{N}{M}$ (two unknowns)

problem set 49

1. Five times the sum of a number and -13 is 92 less than 4 times the opposite of the number. What is the number?

2. The number of celebrants increased 160 percent as the saturnalia drew to a close. If there were 1092 celebrants at the end, what was the number before the increase?

3. The formula for beryllium fluoride is BeF_2. If the fluorine (F) in a quantity of this compound weighed 95 grams, what was the total weight of the beryllium fluoride? (Be, 9; F, 19)

4. In Saudi Arabian currency 1 riyal equals 20 qurush. Sheik Ahab had 15 coins whose value was 205 qurush. How many coins of each kind were there?

5. Grant and Bruce went jogging. Grant was 200 yards ahead after 10 minutes. Find Bruce's speed if Grant's speed was 240 yards per minute.

6. Find the equation of the line shown.

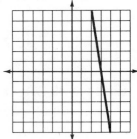

*7. Find side N.

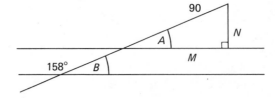

Solve:

8. $\sqrt{x - 4} - 3 = 5$

9. $\sqrt{x^2 - 2x + 5} = x + 1$

10. Covert 200 inches per minute to feet per second.

Simplify:

11. $\sqrt[4]{2}\sqrt[3]{2}$

12. $\sqrt[3]{9}\sqrt{3}$

13. $\sqrt{x^2 y^3}\sqrt{xy^4}$

14. $\sqrt[3]{xy}\sqrt{xy}$

15. $3\sqrt{\dfrac{7}{2}} + 2\sqrt{\dfrac{2}{7}} - 3\sqrt{56}$

16. $5\sqrt{\dfrac{2}{9}} + 3\sqrt{\dfrac{9}{2}} + \sqrt{162}$

Solve:

17. $\left(x - \dfrac{5}{3}\right)^2 = 5$

18. $(x + 2)^2 = 16$

19. Estimate: $\dfrac{(517{,}832 \times 10^{-14})(80{,}123)}{200{,}000 \times 10^{-42}}$

20. $\dfrac{x}{y} + c = \dfrac{m}{x} - d$; find y

21. $\dfrac{2}{x} - \dfrac{3}{y} = \dfrac{m}{p}$; find p

22. Expand: $(x + 1)^3$

23. Simplify: $\dfrac{x^2 - 6x + 5}{20x - 9x^2 + x^3} \div \dfrac{2 - 3x + x^2}{x^3 - 7x^2 + 12x}$

24. Find the distance between $(4, 2)$ and $(-5, 3)$.

25. Find the equation of the line that passes through $(4, 2)$ and $(-5, 3)$.

26. How many 1-inch floor tiles would it take to cover the figure shown. Dimensions are in feet.

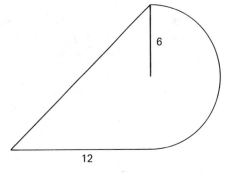

Add:

27. $\dfrac{x}{ay^2} - \dfrac{3x}{a^2y^3} - \dfrac{5}{y^4}$

28. $\dfrac{-3}{x + 3} - \dfrac{3x + 2}{x^2 - 9}$

Solve:

29. $\dfrac{x - 2}{2} - \dfrac{x - 3}{3} = 2$

30. $\dfrac{x}{2} - \dfrac{x + 2}{5} = 1$

LESSON 50   *Quadratic equations · Completing the square · Selected Problems*

50.A
quadratic equations

A quadratic equation in one unknown has the number 2 as the highest power of the variable. The highest power of x in each of the three equations shown here

$$x^2 = -4x + 2 \qquad x - 3 = x^2 \qquad x^2 + 4x - 2 = 0$$

is 2 so these equations are all quadratic equations. We have solved quadratic equations thus far by two methods. To use the first method we factor and then use the zero factor theorem. To review this method, we begin with the equation

$$x^2 + x - 2 = 0$$

First we factor and get

$$(x + 2)(x - 1) = 0$$

and then we set each of the factors equal to zero and solve.

If $x + 2 = 0$ If $x - 1 = 0$

$x = -2$ **$x = 1$**

The other method can be used when the equation is in the form

$$(x - 2)^2 = 3$$

We can solve equations that are in this form by taking the square root of both sides and remembering the $\pm$ sign that appears on the right.

$(x - 2)^2 = 3$ equation

$x - 2 = \pm\sqrt{3}$ square root of both sides

$x = 2 \pm \sqrt{3}$ added $+2$ to both sides

Unfortunately, the factoring method cannot always be used because some equations cannot be factored into binomials whose constants are all integers. For example, the equation

$$x^2 + 4x - 2 = 0$$

cannot be factored. **The inability to factor some quadratic equations is offset by the fact that any quadratic equation can be rearranged into the form**

$$(x + a)^2 = k$$

and then the equation can be solved by taking the square root of both sides, as we have just demonstrated. The method used to accomplish this rearrangement is called **completing the square.** If we rearrange the equation

$$x^2 + 4x - 2 = 0$$

by completing the square, we can change the equation into this form

$$(x + 2)^2 = 6$$

which we can solve by taking the square root of both sides.

$x + 2 = \pm\sqrt{6}$ square root of both sides

$x = -2 \pm \sqrt{6}$ added -2 to both sides

50.B
completing the square

To complete the square, it is necessary to remember the form of a trinomial that is the square of some binomial. On the left, we show several binomials, and on the right we show the trinomials that result when the binomials are squared.

(a) $x + 2$ (a') $(x + 2)^2 = x^2 + 4x + 4$

(b) $x - 5$ (b') $(x - 5)^2 = x^2 - 10x + 25$

(c) $x - 6$ (c') $(x - 6)^2 = x^2 - 12x + 36$

(d) $x + 8$ (d') $(x + 8)^2 = x^2 + 16x + 64$

(e) $x - 3$ (e') $(x - 3)^2 = x^2 - 6x + 9$

In each of these examples, we note that the last term of the trinomial is the square of one-half of the middle term of the trinomial. Thus,

	LAST TERM		$(\frac{1}{2}$ MIDDLE TERM$)^2$
In (a')	4	is	$\left(\dfrac{4}{2}\right)^2$
In (b')	25	is	$\left(-\dfrac{10}{2}\right)^2$
In (c')	36	is	$\left(-\dfrac{12}{2}\right)^2$
In (d')	64	is	$\left(\dfrac{16}{2}\right)^2$
In (e')	9	is	$\left(-\dfrac{6}{2}\right)^2$

This pattern occurs every time we square a binomial. There is nothing to understand. It happens so we will remember it and use it. This pattern is the key to completing the square.
Thus, if we have the expression

$$x^2 + 10x + ?$$

and ask what number should be added if the trinomial is to be the square of some binomial, the answer is 25 because

$$\left(\frac{10}{2}\right)^2 = 25$$

so the completed expression is

$$x^2 + 10x + 25$$

and the binomial that was squared to get this result had to be $x + 5$ because

$$(x + 5)^2 = x^2 + 10x + 25$$

example 50.B.1   Solve $x^2 + 6x - 4 = 0$ by completing the square.

solution   We want to rearrange the equation into the form

$$(x + a)^2 = k$$

We begin by enclosing the first two terms in parentheses and moving the -4 to the right side as $+4$.

$$(x^2 + 6x \quad) = 4$$

Now we want to change the expression inside the parentheses so that it is a perfect square. To do this, we first divide the coefficient of x by 2 and square the result.

$$\left(\frac{6}{2}\right)^2 = 9$$

Then we add this number to both sides of the equation.

$$(x^2 + 6x + 9) = 4 + 9$$

Now the left side can be written as $(x + 3)^2$ and the right side as 13.

$$(x + 3)^2 = 13$$

We finish the solution by taking the square root of both sides and remembering the $\pm$ sign that will appear on the right.

$$x + 3 = \pm\sqrt{13} \qquad \text{square root of both sides}$$

$$\mathbf{x = -3 \pm \sqrt{13}} \qquad \text{added } -3 \text{ to both sides}$$

example 50.B.2   Solve $2x + x^2 - 5 = 0$ by completing the square.

solution   We want to change the form of the equation to

$$(x + a)^2 = k$$

We begin by moving the constant term to the right side and enclosing the other two terms in parentheses.

$$(x^2 + 2x \quad) = 5$$

Now we square $\frac{1}{2}$ of the coefficient of x

$$\left(\frac{2}{2}\right)^2 = 1$$

and add it to both sides

$$(x^2 + 2x + 1) = 5 + 1$$

The left side is a perfect square and the right side is 6.

$$(x + 1)^2 = 6 \qquad \text{simplified}$$

$$x + 1 = \pm\sqrt{6} \qquad \text{square root of both sides}$$

$$\mathbf{x = -1 \pm \sqrt{6}} \qquad \text{added } -1 \text{ to both sides}$$

example 50.B.3   Solve $x^2 = 5x - 5$ by completing the square.

solution   We begin by placing the constant term on the right and enclosing the other two terms in parentheses.

$$(x^2 - 5x \quad) = -5$$

Next we divide -5 by 2 and square the result

$$\left(\frac{-5}{2}\right)^2 = \frac{25}{4}$$

and add $\frac{25}{4}$ to both sides.

$$\left(x^2 - 5x + \frac{25}{4}\right) = -5 + \frac{25}{4}$$

Next we write the left side as a perfect square and simplify the right side.

$$\left(x - \frac{5}{2}\right)^2 = \frac{5}{4} \qquad \text{simplified}$$

$$x - \frac{5}{2} = \pm\sqrt{\frac{5}{4}} \qquad \text{square root of both sides}$$

$$\mathbf{x = \frac{5}{2} \pm \frac{\sqrt{5}}{2}} \qquad \text{added } \frac{5}{2} \text{ to both sides}$$

50.C
selected
problems

The problems in the problem sets were designed to teach the fundamental skills of algebra. Certain types of problems appear again and again. This repetition is provided to permit the development of quickness and sureness in the solution of problems and to insure long-term retention. **All problems in every problem set should be worked;** but, on occasion, exigencies may cause a teacher to wish to shorten some assignments slightly by deleting several problems. Beginning with this problem set, the problems whose omission would be the least harmful are designated by a line drawn under the problem number.

problem
set 50

1. The Two-Steppers were good recruiters as they numbered 10 more than 5 times the number of the Waltzers. Also, there were 10 times as many Two-Steppers as Waltzers. How many of each were there?

2. The first part of the trip was in a surrey at 8 mph and the last part was in a buckboard at 12 mph. If the total trip was 104 miles and took 10 hours, how much of the trip was made in each type of carriage?

3. What is the weight of the sodium (Na) in 348 grams of NaCl? (Na, 23; Cl, 35)

4. Richard and Lynn found three consecutive even integers such that 7 times the sum of the first and third was 48 less than 10 times the second. What were the integers?

5. Hadrian's soldiers increased their wall-building speed by 140 percent. If their new speed was 432 inches per day, what was their old wall-building speed?

Solve by completing the square.

*6. $x^2 + 6x - 4 = 0$ *7. $2x + x^2 - 5 = 0$ *8. $x^2 = 5x - 5$

9. Find the equation of this line.

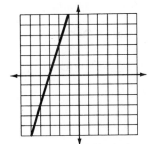

10. Find angle C and side M.

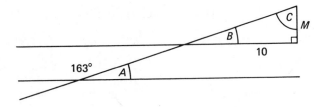

Solve:

11. $\sqrt{x^2 - 4x + 20} = x + 2$ 12. $-5 = -\sqrt{x + 5} + 1$

13. Convert 400 yards per second to miles per hour.

Simplify:

14. $\sqrt[5]{2\sqrt[3]{2}}$ 15. $\sqrt{9\sqrt{3}}$

16. $\sqrt{m^3 y^5} \sqrt[3]{m^2 y^2}$ 17. $5\sqrt{\dfrac{3}{11}} + 2\sqrt{\dfrac{11}{3}} - \sqrt{297}$

18. Estimate: $\dfrac{(746,800 \times 10^{14})(703,916 \times 10^4)}{500,000}$

19. $\dfrac{mxc}{p} - k = \dfrac{2}{r}$; find c

20. $\dfrac{4}{x} - \dfrac{3x}{p} = \dfrac{c}{m}$; find p

21. Find c.

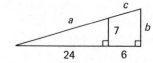

22. Divide $4x^3 + 3x + 5$ by $2x - 3$.

23. Solve $x^3 - 28x = 3x^2$ by factoring.

Simplify:

24. $\dfrac{x^2 + 49 - 14x}{x^3 - 13x^2 + 42x} \cdot \dfrac{x^3 - 4x^2 - 12x}{-35 - 2x + x^2}$

25. $-49^{\frac{3}{2}}$

26. How many 1-inch-square floor tiles would it take to cover this figure? Dimensions are in inches.

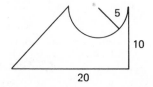

27. Find the equation of the line that passes through $(-2, 5)$ that is perpendicular to the line $3x - 5y = 2$.

28. Find the distance between $(-3, 5)$ and $(-3, 7)$.

29. Solve: $-2(x^0 - x - 3) - |-2^0 - 3| = x - 2(-2 - 4)$

30. Multiply: $\dfrac{x^2 y^{-2}}{z^2 p^0} \left(\dfrac{4p^0 x^{-5} z^2}{y^{-2}} - \dfrac{3x^{-4} y^2}{z^{-2} p} \right)$

LESSON 51 *Imaginary numbers • Product-of-square-roots theorem • Euler's notation • Complex numbers*

51.A
imaginary numbers

We know that $+2$ and -2 are both square roots of 4 because

$$(+2)^2 = 4 \quad \text{and} \quad (-2)^2 = 4$$

We remember that when we use the notation

$$\sqrt{4}$$

we are designating $+2$ because we reserve the notation $\sqrt{4}$ to designate the **principal** or **positive square root** of 4. If we wish to designate -2, which is the negative square root of 4, we can write

$$-\sqrt{4}$$

and if we wish to designate both $+2$ and -2, we use both the plus sign and the minus sign and write

$$\pm\sqrt{4}$$

We remember that if the square root of a number is multiplied by itself, the result is the number. This is the definition of square root.

> The square root of a number is that number which multiplied by itself equals the given number.

Thus it follows that

$$\sqrt{3}\sqrt{3} = 3 \qquad \sqrt{.046}\sqrt{.046} = .046 \qquad \sqrt{3\tfrac{1}{2}}\sqrt{3\tfrac{1}{2}} = 3\tfrac{1}{2}$$

and it also follows that

$$\sqrt{-4}\sqrt{-4} = -4$$

But where on the number line can we locate $\sqrt{-4}$? Any positive or negative number times itself equals a positive number, not a negative number! There is no number on the number line that can be used for x in the following expression.

$$(x)(x) = -4$$

Thus, on the number line, we can locate both $+\sqrt{4}$ and $-\sqrt{4}$ as shown, but we

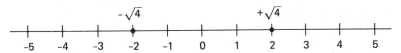

cannot locate $+\sqrt{-4}$ or $-\sqrt{-4}$ on the number line. For this reason, unfortunately, we call $\sqrt{-4}$ an **imaginary number.** We say unfortunately because all numbers are ideas and thus all numbers are really imaginary. The number $\sqrt{4}$ is an idea that we can graph on the number line, and the number $\sqrt{-4}$ is an idea that we cannot graph on the number line. **We call the square roots of negative numbers** *imaginary numbers* **because we cannot find them on the number line. The numbers that we can locate on the number line are called** *real numbers* **to distinguish them from numbers that cannot be found there.**

51.B
product-of-square-roots theorem

When the radicand (the number underneath the radical sign) is a negative number, we cannot always use the product-of-square-roots theorem as we have in the past. There are two versions of the product-of-square-roots theorem. One version applies when at least one radicand is positive.

> PRODUCT-OF-SQUARE-ROOTS THEOREM
> (AT LEAST ONE POSITIVE RADICAND)
>
> If m and n are real numbers and one is a positive number or both are positive numbers,
> $$\sqrt{m}\sqrt{n} = \sqrt{mn}$$

Thus $\sqrt{3}\sqrt{5} = \sqrt{15}$ and $\sqrt{-3}\sqrt{5} = \sqrt{-15}$.

The rule is different when both radicands are negative.

PRODUCT-OF-SQUARE-ROOTS THEOREM
(BOTH RADICANDS NEGATIVE)

If m and n are both negative real numbers

$$\sqrt{m}\sqrt{n} = -\sqrt{mn}$$

---

example 51.B.1   Simplify: (a) $\sqrt{3}\sqrt{2}$ (b) $\sqrt{3}\sqrt{-2}$ (c) $\sqrt{-3}\sqrt{-2}$

solution   In (a) and (b) at least one radicand is positive, so we simplify in both by multiplying the radicands.

$$\text{(a)} \quad \sqrt{3}\sqrt{2} = \sqrt{6} \qquad \text{(b)} \quad \sqrt{3}\sqrt{-2} = \sqrt{-6}$$

In (c) both radicands are negative, and we use a different rule.

$$\text{(c)} \quad \sqrt{-3}\sqrt{-2} = -\sqrt{6}$$

51.C
Euler's notation

The eighteenth-century Swiss mathematician Euler,[†] in addition to other major accomplishments, made four significant contributions to the notations used in mathematics. He introduced the Greek letter Σ to stand for summation. He invented functional notation, and he was the first to use the letter e to represent the base of natural logarithms. And, in addition, he introduced the use of the letter i to represent $\sqrt{-1}$.

$$i = \sqrt{-1}$$

The first three concepts will be used in later courses. The last will be used in this course.

It happens that the solutions of many quadratic equations are numbers that are wholly or in part the square roots of negative numbers. We can use the first part of the theorem on the previous page to write one of these as a real number times $\sqrt{-1}$. For example, if we have

$$\sqrt{-13}$$

we can write this as

$$\sqrt{13(-1)}$$

and use the product-of-square-roots theorem to write

$$\sqrt{13}\sqrt{-1}$$

Instead of writing $\sqrt{-1}$ Euler used the letter i. Thus he would have written the above as

$$\sqrt{13}\,i$$

We surmise that Euler began to use i instead of $\sqrt{-1}$ because i can be made with only one stroke of the pen (if you leave off the dot) while $\sqrt{-1}$ requires 3 strokes of the pen.

[†] Euler is pronounced "oiler."

example 51.C.1   Use Euler's notation to write: (a) $\sqrt{-13}$ (b) $\sqrt{-4}$ (c) $\sqrt{3}\sqrt{-2}$

solution

$$\text{(a)} \quad \sqrt{-13} = \sqrt{13(-1)} = \sqrt{13}i$$

$$\text{(b)} \quad \sqrt{-4} = \sqrt{4(-1)} = 2i$$

$$\text{(c)} \quad \sqrt{3}\sqrt{-2} = \sqrt{3}\sqrt{2}i = \sqrt{6}i$$

51.D
complex numbers

A complex number is a number that has a real part and an imaginary part. Thus

$$4 + 5i \quad \text{and} \quad 5i + 4$$

are complex numbers. **We say that when we write the real part first, as we did on the left, we have written the complex number in *standard form*.** We add complex numbers by adding the real parts to find the real part of the sum and the imaginary parts to find the imaginary part of the sum.

example 51.D.1   Add $4 + 6i + 3i - 2$.

solution   We add real parts to real parts and imaginary parts to imaginary parts and get

$$2 + 9i$$

example 51.D.2   Add $(4 - 3i) + (6 - 4i) - (2 - 5i)$.

solution   First we remove the parentheses and get

$$4 - 3i + 6 - 4i - 2 + 5i$$

Now we add the like parts and get

$$8 - 2i$$

Since $\sqrt{-1}$ times $\sqrt{-1}$ equals -1 then ii equals -1 and i^2 equals -1. We must keep this in mind when simplifying expressions that have i as a multiple factor.

example 51.D.3   Simplify: $2ii - 3iii + 2i - 4 - \sqrt{-4}$

solution   We pair the i's as follows

$$2(ii) - 3i(ii) + 2i - 4 - \sqrt{-4}$$

Now we remember that each $(ii) = i^2 = -1$, so we have

$$2(-1) - 3i(-1) + 2i - 4 - \sqrt{4}i$$

Now we simplify

$$-2 + 3i + 2i - 4 - 2i$$

and finish by adding like parts

$$-6 + 3i$$

example 51.D.4   Simplify: $3i^3 + 2i^2 + 7i + 4 + 2i^5$

solution   We will begin by expanding and pairing the i's.

$$3i(ii) + 2(ii) + 7i + 4 + 2(ii)(ii)i$$

Next we remember that each (ii) equals -1

$$3i(-1) + 2(-1) + 7i + 4 + 2(-1)(-1)i$$

Now we multiply and get

$$-3i - 2 + 7i + 4 + 2i$$

and lastly we simplify and get

$$\mathbf{2 + 6i}$$

problem set 51

1. Times were so hard that three-seventeenths of the knights did not even have a gauntlet to fling down. If 72 knights were in this category, how many knights were there in all?

2. Five hundred twenty-eight students had limited vocabularies and thus found that lucubration was almost impossible. If these students comprised 22 percent of the total, how many students were there in all?

3. The weight of the sodium (Na) in a quantity of Na_2SO_4 was 115 grams. What was the total weight of the Na_2SO_4? (Na, 23; S, 32; O, 16)

4. Horatio gleefully fingered his horde of quarters and dimes. If their total value was \$6.50 and he had 35 coins, how many coins of each type did he have?

5. Buggs made the trip at 40 miles per hour during the first hiatus. Jake made the same trip at 50 miles per hour during the second hiatus. How long was the trip if it took Buggs 2 hours longer than it took Jake?

Simplify:

*6. $4 + 6i + 3i - 2$

*7. $2ii - 3iii + 2i - 4 - \sqrt{-4}$

*8. $3i^3 + 2i^2 + 7i + 4 + 2i^5$

9. $-3i^2 - 2i + i^3 - 3$

Solve by completing the square.

10. $x^2 = -x + 1$

11. $-4 = -x^2 - 3x$

12. Find the equation of the line.

13. Find side m.

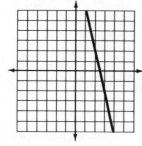

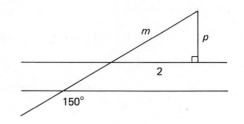

14. Solve: $\sqrt{x - 3} + 5 = 2$

15. Convert 20 inches per hour to miles per minute.

Simplify:

16. $\sqrt[5]{3\sqrt{3}}$

17. $\sqrt{9\sqrt[3]{3}}$

18. $\sqrt{x^2y^2m}\sqrt[4]{xym^2}$

19. $3\sqrt{\dfrac{2}{13}} + 3\sqrt{\dfrac{13}{2}} + 3\sqrt{104}$

20. $(-27)^{-\frac{5}{3}}$

21. Estimate: $\dfrac{(4,941,625)(7,041,683)}{.00007142 \times 10^{-5}}$

22. $\dfrac{x}{p} - c = \dfrac{k}{m}$; find p **23.** $\dfrac{3}{p} - \dfrac{x}{R_1} = \dfrac{1}{R_2}$; find R_2

24. How many 1-inch-square floor tiles would it take to cover this figure? Dimensions are in feet.

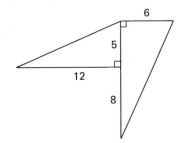

25. Find the equation of the line through $(-4, -2)$ that is perpendicular to the line $3x + 2y = 5$.

26. Find the distance between $(-2, -5)$ and $(-2, -8)$.

27. Multiply: $\dfrac{x^{-2}y}{p^0}\left(\dfrac{4p^2}{x^{-2}y} - \dfrac{2px^2y}{y^2}\right)$ **28.** Divide $2x^3 - 2x + 4$ by $x + 2$.

Solve:

29. $\dfrac{-x - 2}{4} - \dfrac{x + 2}{3} = 3$

30. $-2(x^0 - x - 2) - |-2| + 30(-x - x^0) = -x$

In the appendix there are four enrichment lessons on the use of the scientific calculator. Each lesson contains a homework problem set. If these lessons are to be included, the first one should be taught at this point so that the repetitive problems on this topic at the end of future lessons will be effective. The other three calculator lessons should be taught after Lessons 55, 67, and 74.

LESSON 52 *Chemical mixtures*

52.A
chemical mixture problems

In chemical mixture problems two different constituents are mixed to get a desired result. Often one of the constituents is water, and the other is alcohol, or iodine, or antifreeze, or whatever else is mixed with the water. The key equation in one of these problems comes from a simple statement about one of the constituents. The statement can be made about either constituent. Thus, if the constituents are water and iodine, the statement can be made about water

> water poured in + water dumped in = water total

or the statement can be made about iodine.

> iodine poured in + iodine dumped in = iodine total

We will restrict our investigation of these problems to two basic types. The first type is discussed in this lesson and the second type will be discussed in Lesson 61.

example 52.A.1   A druggist has one solution that is 10% iodine and another that is 50% iodine. How much of each should the druggist use to get 100 milliliters (ml) of a mixture that is 20% iodine?

solution   A picture is helpful.

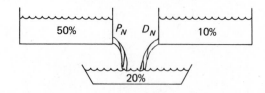

As shown in the figure we will pour in some of the 50% solution and dump in some of the 10% solution. We can work the problem by considering either iodine or water. First we will work it by considering only the iodine. Our equation in words is

$$\text{iodine poured in} + \text{iodine dumped in} = \text{iodine total}$$

Next we write three parentheses that we will use as **"mixture containers."**

$$(\quad) + (\quad) = (\quad)$$

Many people find that using "mixture containers" is helpful, and the use of these containers is recommended. Now we put the requisite **mixture** in each container. We use P_N in the first one, D_N in the second one, and 100 in the third one.

$$(P_N) + (D_N) = (100)$$

Next we multiply each parentheses by the proper decimal number so that each term represents only iodine.

$$\text{(a)} \quad .5(P_N) + .1(D_N) = .2(100)$$

This equation has two unknowns, so we need one more equation. The equation is

$$\text{(b)} \quad P_N + D_N = 100$$

We solve equation (b) to find that $P_N = 100 - D_N$. Then we substitute $100 - D_N$ for P_N in equation (a) and solve.

$$
\begin{aligned}
.5(P_N) + .1(D_N) &= .2(100) && \text{equation (a)} \\
.5(100 - D_N) + .1D_N &= .2(100) && \text{substituted} \\
50 - .5D_N + .1D_N &= 20 && \text{multiplied} \\
.4D_N &= 30 && \text{simplified} \\
D_N &= \textbf{75 ml} && \text{divided by .4}
\end{aligned}
$$

And since $P_N + D_N = 100$,

$$P_N = \textbf{25 ml}$$

Thus the druggist should use 25 ml of the 50 percent solution and 75 ml of the 10 percent solution. If we wish to work the problem by considering only water, the word statement would be

$$\text{water poured in} + \text{water dumped in} = \text{water final}$$

The entries in the mixture containers are exactly the same,

$$(P_N) + (D_N) = (100)$$

but the decimal multipliers are different because this time each term represents water.

$$\text{(c)}\quad .5(P_N) + .9(D_N) = .8(100)$$

The second equation is the same as before

$$\text{(b)}\quad P_N + D_N = 100$$

Either substitution or elimination can be used to solve equations (c) and (b), and the same answers will result.

$$P_N = 25 \text{ ml} \qquad D_N = 75 \text{ ml}$$

Thus the druggist should use 25 ml of the 50 percent solution and 75 ml of the 10 percent solution.

example 52.A.2   A chemist has one solution that is 10% salt and 90% water and another solution that is only 2% salt. How many milliliters of each should the chemist use to make 1400 ml of a solution that is 6% salt?

solution   We can use the same diagram as in the last problem. Only the percents are different.

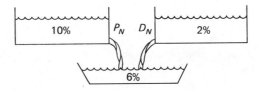

This time we decide to work the problem by considering water. Thus our equation in words is

$$\text{water poured in} + \text{water dumped in} = \text{water total}$$

Next we write the parentheses that we call mixture containers. Their use in these problems is always helpful.

$$(\quad) + (\quad) = (\quad)$$

Mixture containers always contain mixtures. We will use P_N and D_N to represent the unknown mixtures.

$$(P_N) + (D_N) = (1400)$$

This time we use decimal multipliers so that the decimal times the mixture equals water.

$$.9(P_N) + .98(D_N) = .94(1400)$$

and we get

$$\text{(a)}\quad .9P_N + .98D_N = 1316$$

We have two unknowns, so we need another equation, which is

$$\text{(b)}\quad P_N + D_N = 1400$$

We solve (b) for P_N and substitute into (a) and solve.

$$
\begin{aligned}
.9(1400 - D_N) + .98D_N &= 1316 && \text{substitution} \\
1260 - .9D_N + .98D_N &= 1316 && \text{multiplied} \\
.08D_N &= 56 && \text{added} -1260
\end{aligned}
$$

$$D_N = 700 \text{ ml}$$

Thus $P_N = 1400 - 700 = $ **700 ml.** Thus the chemist should use 700 ml of the 10 percent solution and 700 ml of the 2 percent solution.

problem set 52

*1. A druggist has one solution that is 10% iodine and another that is 50% iodine. How much of each should the druggist use to get 100 ml of a mixture that is 20% iodine?

*2. A chemist has one solution that is 10% salt and 90% water and another solution that is only 2% salt. How many milliliters of each should the chemist use to make 1400 ml of a solution that is 6% salt?

3. What is the weight of the sodium (Na) in 1580 grams of $Na_2S_2O_3$? (Na, 23; S, 32; O, 16)

4. Four times the number of yellows equaled 76 reduced by 6 times the number of reds. If there were 4 more yellows than reds, how many of each were there?

5. Queen Hatshepsut rode a litter at 2 kilometers per hour for the first part of the journey to the necropolis. She was going to be late so she changed to an 8 kilometer per hour chariot for the last part of the trip. If it was 28 kilometers to the necropolis and the trip took 8 hours, how far did she ride in the litter and how far did she ride in the chariot?

Simplify:

6. $4i^2 - 3i + 2$

7. $3i^5 - i + 5 - \sqrt{-9}$

8. $\sqrt{-16} - 2i^2 - 2i$

9. $2i^4 + \sqrt{-9} - 3i^3$

Solve by completing the square:

10. $2x + 5 = x^2$

11. $x^2 - 5x = 2$

12. Find the equation of the line.

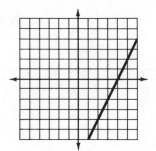

13. Find P.

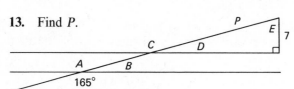

Solve:

14. $\sqrt{x^2 + 2x + 10} = x + 2$

15. $3 = -5 + \sqrt{x - 3}$

16. Convert 200 inches per hour to miles per minute.

Simplify:

17. $2\sqrt{2\sqrt[4]{2}}$

18. $3\sqrt{9\sqrt[4]{3}}$

19. $\sqrt{4x^3y^5}\sqrt[3]{8xy^2}$

20. $3\sqrt{\dfrac{2}{13}} - 5\sqrt{\dfrac{13}{2}} + \sqrt{104}$

21. Estimate: $\dfrac{(987,612 \times 10^5)(413,280)}{(74,630)(400)}$

22. $\dfrac{x}{p} - c + \dfrac{a}{b} = m$; find x

Simplify:

23. $\dfrac{x^2 + 3x - 28}{21x + 10x^2 + x^3}$

24. $-16^{\frac{5}{4}}$

25. $\dfrac{\dfrac{x^2y}{p^2} - p}{\dfrac{m}{p} - \dfrac{1}{p^2}}$

26. The figure shown is the base of a solid 6 ft high. Find the volume in cubic feet. The dimensions are in yards.

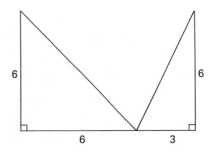

27. Add: $-\dfrac{3}{x - 3} + \dfrac{2x + 4}{x^2 - 9}$

28. Find the equation of the line that passes through the point $(-2, -7)$ that is parallel to the line $2x + 3y = 4$.

29. Divide $3x^3 - x$ by $x - 4$.

30. Solve: $\dfrac{3x + 4}{2} - \dfrac{2x - 5}{3} = 4$

ER 52-1, 52-2

LESSON 53 *Metric unit conversions • English to metric • Weight combination by percent*

53.A
metric units

The meter is the basic unit of length in the metric system. A meter is just a little longer than a yard. There are 100 centimeters in a meter, and there are 1000 meters in a kilometer, which is a little more than half a mile. From these relationships, we get two metric unit multipliers that can be used for distance and area conversions.

(a) $\dfrac{100 \text{ cm}}{1 \text{ m}}$ or $\dfrac{1 \text{ m}}{100 \text{ cm}}$ and   (b) $\dfrac{1000 \text{ m}}{1 \text{ km}}$ or $\dfrac{1 \text{ km}}{1000 \text{ m}}$

example 53.A.1 Convert 7560 centimeters to kilometers.

solution We will go from centimeters (cm) to meters (m) to kilometers (km).

$$7560 \text{ cm} \times \frac{1 \text{ m}}{100 \text{ cm}} \times \frac{1 \text{ km}}{1000 \text{ m}} = \frac{7560}{(100)(1000)} \text{ km} = .0756 \text{ km}$$

This problem illustrates the beauty of the metric system. All we have to do to change units in the metric system is to move the decimal point.

example 53.A.2 Convert 15,740,000 square centimeters to square kilometers.

solution We will use two multipliers to go from square centimeters (cm^2) to square meters (m^2) and two more to go to square kilometers (km^2).

$$15{,}740{,}000 \text{ cm}^2 \times \frac{1 \text{ m}}{100 \text{ cm}} \times \frac{1 \text{ m}}{100 \text{ cm}} \times \frac{1 \text{ km}}{1000 \text{ m}} \times \frac{1 \text{ km}}{1000 \text{ m}} = \frac{15{,}740{,}000}{1 \times 10^{10}} \text{ km}^2$$

$$= .001574 \text{ km}^2$$

Again we note that all we had to do was to move the decimal point.

53.B
English to metric

We need to know only one equivalence to change from English units of length to metric units of length.

$$2.54 \text{ cm} = 1 \text{ in}$$

This equivalence is exact because 1 inch is exactly 2.54 centimeters long. This definition of the inch is official and has been official for almost 100 years. Thus to go from English units to metric units, we first go to inches in the English system. Then we convert to centimeters.

example 53.B.1 Convert 32 yards to meters.

solution **Some tables give a direct conversion from yards to meters. We disdain these conversions because the tables are not always available. We need only remember that 2.54 centimeters equals 1 inch to make any English to metric conversion involving length, area, or volume.** Thus, in this problem we will use unit multipliers as required to go from yards to feet to inches to centimeters to meters.

$$32 \text{ yd} \times \frac{3 \text{ ft}}{1 \text{ yd}} \times \frac{12 \text{ in}}{1 \text{ ft}} \times \frac{2.54 \text{ cm}}{1 \text{ in}} \times \frac{1 \text{ m}}{100 \text{ cm}} = \frac{(32)(3)(12)(2.54)}{100} \text{ m}$$

example 53.B.2 Convert .042 square kilometers to square miles.

solution We need to use two unit multipliers in each step as we go from square kilometers to square meters to square centimeters to square inches to square feet to square miles.

$$.042 \text{ km}^2 \times \frac{1000 \text{ m}}{1 \text{ km}} \times \frac{1000 \text{ m}}{1 \text{ km}} \times \frac{100 \text{ cm}}{1 \text{ m}} \times \frac{100 \text{ cm}}{1 \text{ m}} \times \frac{1 \text{ in}}{2.54 \text{ cm}} \times \frac{1 \text{ in}}{2.54 \text{ cm}}$$

$$\times \frac{1 \text{ ft}}{12 \text{ in}} \times \frac{1 \text{ ft}}{12 \text{ in}} \times \frac{1 \text{ mi}}{5280 \text{ ft}} \times \frac{1 \text{ mi}}{5280 \text{ ft}}$$

$$= \frac{(.042)(1000)(1000)(100)(100)}{(2.54)(2.54)(12)(12)(5280)(5280)} \text{ mi}^2$$

53.C
weight combination by percent

Thus far, we have investigated the relative weights of the elements in chemical compounds by using ratios. For example, in a molecule of the compound

$$Na_2S_2O_3$$

there are two atoms of sodium (Na), two atoms of sulfur (S), and three atoms of oxygen (O). If we use the gram atomic weights of these elements, we can find the gram molecular weight of the molecule. (Na, 23; S, 32; O, 16)

Two atoms of sodium:	2×23 =	46 grams
Two atoms of sulfur:	2×32 =	64 grams
Three atoms of oxygen:	3×16 =	48 grams
Gram atomic weight of a molecule =		158 grams

The decimal part of the total made up by each element can be found by dividing the atomic weight of the element by the molecular weight of the molecule.

$$\text{Sodium} = \frac{46}{158} = .29 \qquad \text{Sulfur} = \frac{64}{158} = .41 \qquad \text{Oxygen} = \frac{48}{158} = .30$$

The percent of each element by weight is found by moving each of these decimal points two places to the right. If we do this, we get

Sodium = 29%   Sulfur = 41%   Oxygen = 30%

The sum of these percents is 100 percent. Sometimes, because we have rounded off, the sum will not be exactly 100 percent.

example 53.C.1   What percent by weight of $Na_2S_2O_3$ is sodium (Na)?

solution   From the above, the compound is **29%** sodium.

problem set 53

*1.   What percent by weight of $Na_2S_2O_3$ is sodium (Na)? (Na, 23; S, 32; O, 16)

2.   A chemist has one solution that is 20% alcohol and another that is 60% alcohol. How much of each should the chemist use to get 100 ml of a solution that is 52% alcohol?

3.   The hospital pharmacist wanted 250 ml of a solution that was 72% iodine. How many milliliters of a 40% solution should be mixed with how many milliliters of an 80% solution to get the desired result?

4.   The results totaled 80, and the number of good results totaled 8 more than 5 times the number of bad results. How many results were good and how many were bad?

5.   The automobile was traveling at 50 mph and had already gone 200 miles when the airplane set out in pursuit. If the airplane overtook the automobile in 4 hours, what was the speed of the airplane?

Convert:

*6.   7560 centimeters to kilometers   *7.   32 yards to meters

*8.   15,740,000 square centimeters to square kilometers

*9.   .042 square kilometers to square miles

Simplify:

10. $-\sqrt{-4} + 2 + 2i^5$ 11. $2i^2 + 5i + 4 + \sqrt{-9}$

12. $-4i^5 + 2\sqrt{-16}$ 13. $2i^3 - i^4 + 3i^2$

Solve by completing the square:

14. $x^2 - 5 = 5x$ 15. $-x^2 = -6x - 6$

16.   Find side m.

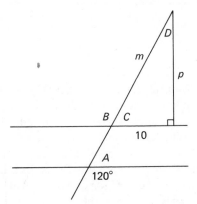

Solve:

17. $\sqrt{x - 11} - 1 = 6$ **18.** $\sqrt{x^2 + 2x + 5} - 3 = x$

Simplify:

19. $\sqrt[5]{2\sqrt[3]{2}}$ **20.** $\sqrt{81\sqrt[4]{3}}$

21. $\sqrt[5]{x^2 y} \sqrt[3]{xy^2}$ **22.** $-4^{-\frac{5}{2}}$

23. $3\sqrt{\dfrac{2}{9}} - 2\sqrt{\dfrac{9}{2}} - 2\sqrt{50}$ **24.** Estimate: $\dfrac{(2,135,820)(4,913,562)}{801,394,026}$

25. $\dfrac{x}{y} - \dfrac{m}{p} + \dfrac{k}{c} = 0$; find p **26.** $\dfrac{p}{x} + c = d$; find x

27. Find B.

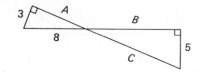

28. Divide $x^3 - 2x + 2$ by $x + 1$.

29. Find the area of the figure shown in square inches. Dimensions are in feet.

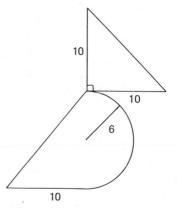

30. Find the solution to this system by graphing. Then find the exact solution by using either substitution or elimination.
$$\begin{cases} 2x - 3y = -9 \\ 5x + 3y = 3 \end{cases}$$

ER 53-3, 53-4

LESSON 54 *Rectangular coordinates • Polar coordinates*

54.A
rectangular and polar coordinates

We can use two different methods to describe the location of a point on the coordinate plane with respect to the origin. One method is to use one number to give the location of the point to the right or the left of the origin and another number to give the location of the point above or below the origin. When we do this, we say we are using **rectangular coordinates.** To associate the numbers with the proper directions we can use parentheses and ordered pairs (x, y), or we can omit the parentheses and use letters (often i, j and k) to designate directions. In this book we will use $+R$ for right, $-R$ for left, $+U$ for up, and $-U$ for down. We will demonstrate this notation by using rectangular coordinates to locate the four points on the graph below. For point (a) we write

$4R + 3U$; for point (b) we write $-3R + 5U$; for point (c) we write $-3R - 3U$; and for point (d) we write $5R - 2U$.

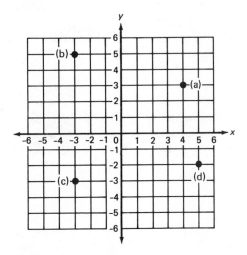

The other way to describe the location of a point is to use a distance at a given angle. The angle is measured counterclockwise from the line that normally designates the $+x$ axis. We say that we are using **polar coordinates** when we designate the location of a point by using an angle and a distance.

It is interesting to note that seafarers and air navigators have always measured their angles clockwise from due north, while mathematicians, for some reason, measure their angles counterclockwise from due East!

We will use notations such as

$$4\underline{/57°} \qquad\qquad 7\underline{/230°}$$

to designate the magnitudes and angles. The first number gives the magnitude and the second number gives the angle. Thus the above would be read from left to right as 4 at 57° and 7 at 230°. In many books the authors use ordered pairs to designate the magnitudes and the angles. In these books the magnitudes and angles shown above would be written as the ordered pairs

$$(4, 57°) \qquad\qquad\qquad (7, 230°)$$

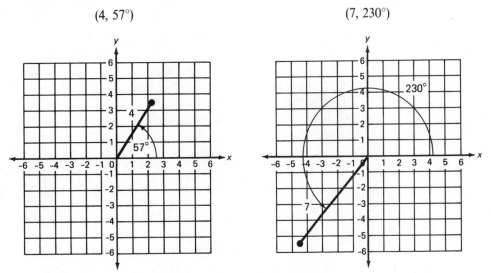

In mathematics and science, a **vector** is defined as a quantity that has both a magnitude and a direction. Since each of the lines shown in the figures has a magnitude and a direction, we can call these lines vectors. In the left figure, we designate the point by using the vector 4/57°; and in the right figure, we designate the point by using the vector 7/230°.

We can use sines, cosines, and tangents to help us convert from rectangular coordinates to polar coordinates or from polar coordinates to rectangular coordinates. For the present we will concentrate on learning how to convert from polar coordinates to rectangular coordinates.

example 54.A.1   Change 4/57° to rectangular coordinates.

solution   We draw the vector that designates the point and **then complete the triangle by drawing a vertical line from the point to the x axis.**

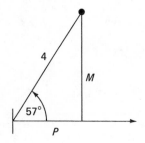

We can find M and P by using the sine and cosine.

$$\text{(a)} \quad \sin 57° = \frac{M}{4} \qquad \text{(b)} \quad \cos 57° = \frac{P}{4}$$

We solve these equations for M and P and get

$$\text{(a)} \quad 4 \sin 57° = M \qquad \text{(b)} \quad 4 \cos 57° = P$$

We use the trigonometric tables in the appendix to get the sine and cosine and then multiply.

$$4(.84) = M \qquad 4(.54) = P$$
$$3.36 = M \qquad 2.16 = P$$

Thus the point is 2.16 to the right of the origin and 3.36 above the origin, so its location can be described in rectangular coordinates by writing

$$\mathbf{2.16R + 3.36U}$$

example 54.A.2   Change 7/230° to rectangular coordinates.

solution   We draw the vector and then complete the triangle **by drawing a vertical line from the end of the vector to the x axis.** Since 230° is 50° more than 180°, the angle in our triangle is 50°.

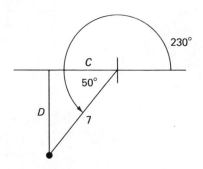

We can find D and C by using the sine and cosine.

$$\text{(a)} \quad \sin 50° = \frac{D}{7} \quad \text{(b)} \quad \cos 50° = \frac{C}{7}$$

We solve these equations for D and C and get

$$7 \sin 50° = D \qquad 7 \cos 50° = C$$

We get the sine and cosine from the trigonometric tables and then multiply.

$$7(.77) = D \qquad 7(.64) = C$$
$$5.39 = D \qquad 4.48 = C$$

Thus, our point is 4.48 to the left of the origin and 5.39 below the origin. We indicate this in rectangular coordinates by writing

$$-4.48R - 5.39U$$

It is helpful if we note that only the sine and the cosine of the angle are used when we break a vector into its components. If H is the hypotenuse and θ is the angle, then one component is

$$H \cos \theta \qquad \text{and the other component is} \qquad H \sin \theta$$

example 54.A.3   Change $42\underline{/340°}$ to rectangular coordinates.

solution   We draw the vector and then complete the triangle **by drawing a vertical line from the end of the vector to the x axis.** Since 340° is 20° less than 360°, the angle in the triangle is 20°.

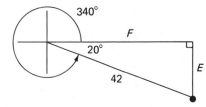

Again we need to use only the sine and the cosine.

(a) $\sin 20° = \dfrac{E}{42}$ (b) $\cos 20° = \dfrac{F}{42}$

$\qquad 42 \sin 20° = E \qquad\qquad\qquad 42 \cos 20° = F$

$\qquad\quad 42(.34) = E \qquad\qquad\qquad\quad 42(.94) = F$

$\qquad\qquad 14.28 = E \qquad\qquad\qquad\qquad 39.48 = F$

Thus our point is 39.48 to the right and 14.28 down. We indicate this by writing

$$39.48R - 14.28U$$

problem set 54

1.   Sarah wants to mix 400 liters of a solution that is 57.5% iodine. She has two solutions available. One is 20% iodine and the other is 70% iodine. How many liters of each should Sarah use?

2.   Selby was in the next stall and she needed 150 ml of a solution that was 30% glycerine. The two solutions available were 10% glycerine and 40% glycerine. How many milliliters of each should Selby use?

3.   What percent by weight of potassium chlorate is potassium (K)? The chemical formula for potassium chlorate is $KClO_3$. (K, 39; Cl, 35; O, 16)

4.   The ratio of ducks to geese was 5 to 4. Four times the number of ducks was 40 greater than 3 times the number of geese. How many of each kind of fowl was present?

5.   The sports car was twice as fast as the truck and took 3 hours less to make the trip. If the truck traveled at 50 miles per hour, how long was the trip?

*6.   Change $4\underline{/57°}$ to rectangular coordinates.

*7.   Change $7\underline{/230°}$ to rectangular coordinates.

*8.   Change $42\underline{/340°}$ to rectangular coordinates.

9.   Change 100 square kilometers to square centimeters.

10. Change 100 square feet to square centimeters.

11. Change 60 miles per hour to kilometers per second.

Simplify:

12. $-\sqrt{-4} - 2i^3 + 4i^4$

13. $3i^3 + 2i - 4i^2 + \sqrt{-9}$

14. $-3i + 2i^2 - 2 + i$

15. $2i^2 + 2i - 2\sqrt{-25}$

Solve by completing the square:

16. $x^2 = 7 + 3x$

17. $-7x = -x^2 + 3$

18.   Find the equation of line 18.

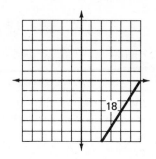

Solve:

19. $\dfrac{4x - 2}{5} - \dfrac{x - 3}{2} = 7$

20. $\sqrt{x^2 - x - 5} + 1 = x$

21. $\sqrt{x - 2} - 11 = 1$

22. $\dfrac{x}{k} - cm = \dfrac{p}{c}$; find k

23. $\dfrac{a}{p} - x + \dfrac{c}{m} = y$; find m

Simplify:

24. $\sqrt[3]{4\sqrt{2}}$

25. $\sqrt[5]{9\sqrt[3]{3}}$

26. $\sqrt[6]{xy}\sqrt[3]{xy^2}$

27. $-81^{\frac{1}{4}}$

28. $2\sqrt{\dfrac{3}{7}} - 5\sqrt{\dfrac{7}{3}} + 2\sqrt{84}$

29. $\dfrac{x^2 p - \dfrac{x}{p^2}}{\dfrac{x^2 y}{p^2} - x}$

30.   Multiply: $\dfrac{4x^{-2}yp}{m^2 y^{-1}}\left(\dfrac{p^{-1}m^2 y}{16x^{-2}y} - \dfrac{2x^2 y^0 p}{m^{-2}}\right)$

ER 54-5, 54-6

LESSON 55   *Advanced abstract equations • Word problems and quadratic equations*

55.A
abstract equations

Equations that contain many variables and that have no numerical answer are often called **abstract equations.** We remember that when an equation has rational terms (fractions), the solution is facilitated if we begin by eliminating the denominators. Thus to solve the equation

$$\frac{a}{x} + \frac{y}{m} = c$$

for x, we begin by multiplying every numerator by xm and then canceling the denominators.

$$\frac{a}{x} \cdot xm + \frac{y}{m} \cdot xm = c \cdot xm \qquad \text{multiplied}$$

$$am + yx = cxm \qquad \text{canceled denominators}$$

$$am = cxm - yx \qquad \text{added } -xy$$

$$am = x(cm - y) \qquad \text{factored out } x$$

$$\frac{am}{cm - y} = x \qquad \text{divided}$$

The same procedure is used for more complicated equations. Parentheses sometimes help us to avoid mistakes.

example 55.A.1 Solve $\dfrac{a + b}{x} + \dfrac{y}{m} = k$ for m.

solution We can prevent some mistakes when we cancel if we enclose sums in the numerators in parentheses. **Then, after we cancel, it is important to eliminate the parentheses by multiplying.** In this problem, we begin by enclosing $a + b$ in parentheses. Then we multiply each term by mx, cancel, and then eliminate the parentheses by multiplying.

$$\frac{(a + b)}{x} mx + \frac{y}{m} mx = kmx \qquad \text{multiplied}$$

$$(a + b)m + y(x) = kmx \qquad \text{cancelled}$$

$$am + bm + yx = kmx \qquad \text{multiplied}$$

$$yx = kmx - am - bm \qquad \text{added } -am - bm$$

$$yx = m(kx - a - b) \qquad \text{factored out } m$$

$$\frac{yx}{kx - a - b} = m \qquad \text{divided}$$

example 55.A.2 Solve $\dfrac{mp}{c} + \dfrac{d + e}{x} = d$ for x.

solution We use parentheses around $d + e$ and then multiply by cx.

$$\frac{mp}{c} cx + \frac{(d + e)}{x} cx = dcx \qquad \text{multiplied}$$

$$mpx + dc + ec = dcx \qquad \text{canceled and multiplied}$$

$$dc + ec = dcx - mpx \qquad \text{added } -mpx$$

$$dc + ec = x(dc - mp) \qquad \text{factored out } x$$

$$\frac{dc + ec}{dc - mp} = x \qquad \text{divided}$$

example 55.A.3   Solve $\dfrac{mx}{y} + \dfrac{d}{a+b} = p$ for a.

solution   We begin by multiplying every numerator by $y(a+b)$ and canceling.

$$\frac{mxy(a+b)}{y} + \frac{dy(a+b)}{a+b} = py(a+b) \qquad \text{multiplied}$$

$$mxa + mxb + dy = pya + pyb \qquad \text{canceled and multiplied}$$

$$mxa - pya = pyb - mxb - dy \qquad \text{moved three terms}$$

$$a(mx - py) = pyb - mxb - dy \qquad \text{factored out } a$$

$$a = \frac{pyb - mxb - dy}{mx - py} \qquad \text{divided}$$

55.B
word problems and quadratic equations

Algebra books often contain problems whose solutions require the solutions of quadratic equations. Consecutive integer problems in which two of the integers are multiplied often lead to quadratic equations.

example 55.B.1   Find three consecutive integers such that the product of the first and the third is 4 greater than 4 times the second.

solution   As always we begin by writing the integers as

$$N \qquad N+1 \qquad N+2$$

Now we write the equation.

$$N(N+2) - 4 = 4(N+1) \qquad \text{equation}$$

$$N^2 + 2N - 4 = 4N + 4 \qquad \text{multiplied}$$

$$N^2 - 2N - 8 = 0 \qquad \text{simplified}$$

$$(N-4)(N+2) = 0 \qquad \text{factored}$$

$$N = 4, -2 \qquad \text{solved}$$

Thus, we get two solutions: **4, 5, 6** and **−2, −1, 0**.

problem set 55

*1. Find three consecutive integers such that the product of the first and the third is 4 greater than 4 times the second.

2. Find three consecutive even integers such that the product of the first and the third is 24 less than 9 times the second.

3. One solution was 5% bromide and the other was 40% bromide. How much of each should be used to get 60 ml of a solution that is 12% bromide?

4. The formula for carbon tetrachloride is CCl_4. What is the weight of the carbon in 1368 grams of carbon tetrachloride? (C, 12; Cl, 35)

5. Find three consecutive multiples of 3 such that 6 times the first is 48 greater than 4 times the third.

*6. $\dfrac{a+b}{x} + \dfrac{y}{m} = k$; find m \qquad\qquad *7. $\dfrac{mp}{c} + \dfrac{d+e}{x} = d$; find x

*8. $\dfrac{mx}{y} + \dfrac{d}{a+b} = p$; find a

9. Change $40\underline{/325°}$ to rectangular form.

10. Change $10\underline{/200°}$ to rectangular form.

11. Change 4 cubic yards to cubic meters.

12. Change 1000 feet per second to kilometers per minute.

Simplify:

13. $4i^5 - 2\sqrt{-9} - 2i^4$

14. $4 + 2i^2 + 3i - \sqrt{-4}$

15. $2i^3 + 2i^4 + 2 - 2i$

16. $-3i^6 - 2i - 2 - 2i^2$

Solve by completing the square.

17. $x^2 = 5x + 5$

18. $x^2 - 6 = 6x$

Solve:

19. $\sqrt{x-2} + 4 = 2$

20. $\sqrt{x^2 - 2x + 14} - 12 = x$

Simplify:

21. $\sqrt{16\sqrt{2}}$

22. $\sqrt[4]{27\sqrt[3]{3}}$

23. $\sqrt[4]{x^2 y}\sqrt{x^5 y^2}$

24. $-81^{\frac{5}{4}}$

25. $4\sqrt{\dfrac{2}{11}} + 2\sqrt{\dfrac{11}{2}} - 4\sqrt{198}$

26. Estimate: $\dfrac{(4{,}183{,}256)(704{,}185 \times 10^{-42})}{802{,}164 \times 10^{30}}$

27. Find B.

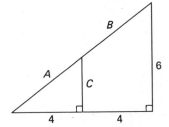

28. Find the volume of the prism shown. The dimensions are in feet.

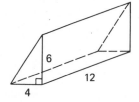

29. Simplify: $\dfrac{apx^2 - 10pa + 3xpa}{x^2 + 5x - 14} \cdot \dfrac{x^2 - 21 + 4x}{-20ap + xpa + ax^2p}$

30. Find the distance between $(-2, 4)$ and $(-10, -4)$.

ER 55-7, 55-8

If the enrichment lessons on the scientific calculator are to be included, Enrichment Lesson 2 on multiplication and division should be included at this point.

LESSON 56 *Surface area*

56.A
surface area

The surface area of a solid is the sum of the areas of the surfaces of the solid.

example 56.A.1 Find the surface area of the rectangular box shown. Dimensions are in inches.

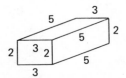

solution The box has six faces—two ends, two sides, and a top and a bottom.

One side = 5 in × 2 in = 10 in²   Both sides   = 20 in²
One end = 3 in × 2 in = 6 in²   Both ends   = 12 in²
Top = 3 in × 5 in   = 15 in²   Top and bottom = 30 in²
                                        Surface area   = **62 in²**

example 56.A.2 Find the surface area of the prism shown. Dimensions are in feet.

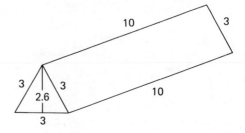

solution The prism has two ends and three rectangular faces.

$$\text{One end} = \frac{3 \text{ ft} \times 2.6 \text{ ft}}{2} = 3.9 \text{ ft}^2 \qquad \text{Both ends} = 7.8 \text{ ft}^2$$

One face = 10 ft × 3 ft = 30 ft²   Three faces = 90.0 ft²
                                         Surface area = **97.8 ft²**

example 56.A.3 Find the surface area of the right circular cylinder shown. Dimensions are in meters.

solution The cylinder has two ends that are circles. The area of one end is πR^2, so the area of both ends is

$$\pi R^2 + \pi R^2 \approx (3.14)(4 \text{ m})^2 + (3.14)(4 \text{ m})^2$$
$$= 100.48 \text{ m}^2$$

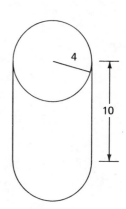

We can easily calculate the lateral surface area if we think of the cylinder as a tin can which we can cut down the dotted line and then press flat.

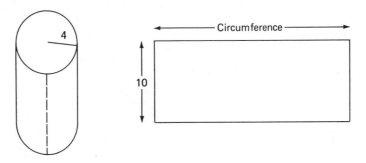

We note that the height of the rectangle is 10 meters, and that the length is the circumference of the circle which is π times the diameter. The radius of this cylinder is 4 meters so the diameter is 8 meters.

$$\text{Circumference} = \pi D = (3.14)(8) = 25.12 \text{ m}$$

Thus the area is $10 \text{ m} \times 25.12 \text{ m} = 251.2 \text{ m}^2$.
Thus the total surface area is $100.48 \text{ m}^2 + 251.2 \text{ m}^2 = \textbf{351.68 m}^2$.

problem set 56

1. Find three consecutive even integers such that the product of the first and the second is 8 greater than the product of -10 and the third.

2. The chemical formula for methylene bromide is CH_2Br_2. If the bromine (Br) in a container of methylene bromide weighs 320 grams, what is the total weight of the methylene bromide? What percent by weight of the compound is bromine? (C, 12; H, 1; Br, 80)

3. The walk into the country at 4 mph was leisurely, but the ride back at 20 mph on a scooter was a little scary. If the total time of the trip was 12 hours, how far was the walk?

4. It was necessary to mix 1000 gallons that was 56% fluorine. If one solution was 20% fluorine and another was 80% fluorine, how much of each one should be used?

5. They just kept coming. Their approach was inexorable. Finally, there were $3\frac{3}{5}$ as many as were desired. If 1440 came, how many were desired?

Find the surface areas of the following solids.

6.

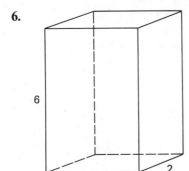

6
4
2
Dimensions in inches

7.

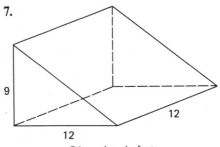

9
12
12
Dimensions in feet

8. Find the surface area of this right circular cylinder.

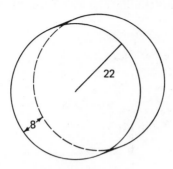

Dimensions in meters

9. $\dfrac{a + x}{b} - \dfrac{c}{m} = \dfrac{p}{k}$; find x

10. $\dfrac{m}{a + c} - \dfrac{x}{m} = p$; find c

11. Convert $10\underline{/210°}$ to rectangular form.

12. Convert $20\underline{/60°}$ to rectangular form.

13. Convert 60 kilometers per hour to inches per second.

14. Convert 400 cubic yards to cubic centimeters.

Simplify:

15. $3i^5 + 2\sqrt{-25} - 3i^2$

16. $2i^4 - 3i^3 + 2i + 4$

17. $\sqrt[6]{4\sqrt[5]{2}}$

18. $\sqrt{y^4}\sqrt{xy^2}$

19. $\sqrt{25\sqrt[3]{5}}$

20. $4\sqrt{\dfrac{3}{4}} - 2\sqrt{\dfrac{4}{3}} - 2\sqrt{27}$

Solve by completing the square:

21. $x^2 = 7x + 7$

22. $-8x - 8 = -x^2$

Solve:

23. $\sqrt{x + 1} + 1 = 1$

24. $\sqrt{x^2 - 2x + 21} - 1 = x$

25. Divide $4x^4 - 1$ by $x - 3$.

26. Simplify: $\dfrac{\dfrac{ax}{y^2} - \dfrac{yp}{x}}{\dfrac{yp}{xy} - \dfrac{1}{y^2}}$

27. Find the equation of the line that passes through $(-5, -5)$ that is perpendicular to the line that passes through $(5, -2)$ and $(-3, -3)$.

Solve:

28. $\dfrac{x - 3}{4} - \dfrac{2 - x}{8} = 6$

29. $\dfrac{a - 5}{2} - \dfrac{3 - a}{4} = 1$

30. Add: $\dfrac{x}{a^2y} - \dfrac{3x + 2}{a^2y(x - 1)} - \dfrac{4}{x^2 - 1}$

ER 56-9, 56-10

LESSON 57 *Ideal gas laws*

57.A
ideal
gas laws

Ideal gas law problems are simple algebra problems that require the use of ratio equations for their solution. The equation that can solve any of these problems is called the **general ideal gas law equation.** In this equation,

$$\frac{P_1 V_1}{T_1} = \frac{P_2 V_2}{T_2}$$

P stands for pressure, V stands for volume, and T stands for temperature. If we omit one of these variables, the resulting equation is called either Charles' law or Boyle's law or Gay-Lussac's law. These names are not important, and we won't worry about which is which. The equations for these laws are

$$\text{(a)} \quad \frac{P_1}{T_1} = \frac{P_2}{T_2} \qquad\qquad \text{(b)} \quad \frac{V_1}{T_1} = \frac{V_2}{T_2} \qquad\qquad \text{(c)} \quad P_1 V_1 = P_2 V_2$$

In (a) we have omitted the symbol V. In (b) we have omitted the symbol P, and in (c) we have omitted the symbol T. The units that we will use in these problems are.

For pressure: torr or pounds per square inch or atmospheres

For volume: liters or milliliters or cubic centimeters or quarts

For temperature: kelvins[†]

These units will be studied in chemistry and physics, and we won't discuss what they mean because we are studying algebra—not chemistry or physics.

However, we must be always careful to use the same units throughout a particular problem. The problems will be worded so that the units need not be considered. We will just use the numbers.

example 57.A.1   Four liters of an ideal gas at a temperature of 800 kelvins had a pressure of 100 pounds per square inch. If the volume were increased to 10 liters and the temperature reduced to 600 kelvins, what would the pressure be?

solution   We begin by writing the general gas law equation.

$$\frac{P_1 V_1}{T_1} = \frac{P_2 V_2}{T_2}$$

The symbols P_1, V_1, and T_1 represent the original pressure, volume and temperature; and the symbols P_2, V_2, and T_2 represent the final pressure, volume, and temperature. If we replace these letters with the proper numbers, we get

$$\frac{(100)(4)}{800} = \frac{P_2(10)}{600}$$

we can solve this equation easily by multiplying both sides by $\frac{600}{10}$.

$$\frac{600}{10} \cdot \frac{(100)(4)}{800} = \frac{P_2(10)}{600} \cdot \frac{600}{10} \longrightarrow 30 = P_2$$

[†] A *kelvin* (abbreviated K) is a unit of temperature. To designate 400 of these units we can write either 400 kelvins or 400 K. The word degree is not used with kelvin as it is when we write 680 degrees Celsius (680°C). Absolute zero is zero kelvin which is approximately equal to $-273°C$.

If P_1 was given in pounds per square inch, then P_2 will be in pounds per square inch, for the same units must be used for the same variable everywhere in a problem. Thus our final pressure is

$$P_2 = 30 \text{ pounds per square inch}$$

example 57.A.2   The initial pressure of a quantity of an ideal gas was 400 torr, and the temperature was 1200 kelvins. The volume was held constant. What was the pressure if the temperature was decreased to 900 kelvins?

solution   First we write the gas law as

$$\frac{P_1 V_1}{T_1} = \frac{P_2 V_2}{T_2}$$

If the volume is held constant, we can mentally divide both sides of the equation by this constant and eliminate volume. Now we have

$$\frac{P_1}{T_1} = \frac{P_2}{T_2}$$

Next we insert the given values for the variables.

$$\frac{400}{1200} = \frac{P_2}{900}$$

We can solve for P_2 by multiplying both sides of the equation by 900.

$$900 \cdot \frac{400}{1200} = \frac{P_2}{900} \cdot 900 \quad \longrightarrow \quad 300 = P_2$$

Since P_1 was given in torr, then P_2 must also be in torr because the same units are always used for a variable everywhere in a problem. Thus our answer is

$$P_2 = 300 \text{ torr}$$

example 57.A.3   The temperature of a quantity of ideal gas was held constant in an experiment. The original pressure was 7 atmospheres and the original volume was 42 liters. If the volume was reduced to 10 liters, what was the final pressure?

solution   We write the ideal gas law as

$$\frac{P_1 V_1}{T_1} = \frac{P_2 V_2}{T_2}$$

Since the temperature is constant, we omit the symbol T from both sides of the equation and write

$$P_1 V_1 = P_2 V_2$$

Now we replace the symbols with the given numbers.

$$(7)(42) = (P_2)(10)$$

We solve by dividing by 10.

$$\frac{(7)(42)}{10} = \frac{P_2(10)}{10} \quad \longrightarrow \quad P_2 = 29.4 \text{ atmospheres}$$

The final pressure is in atmospheres because we used atmospheres as the units of pressure for P_1.

*1. Four liters of an ideal gas at a temperature of 800 kelvins had a pressure of 100 pounds per square inch. If the volume were increased to 10 liters and the temperature reduced to 600 kelvins, what would the pressure be?

*2. The temperature of a quantity of ideal gas was held constant in an experiment. The original pressure was 7 atmospheres and the original volume was 42 liters. If the volume was reduced to 10 liters, what was the final pressure?

*3. The initial pressure of a quantity of an ideal gas was 400 torr and the temperature was 1200 kelvins. The volume was held constant. What was the pressure if the temperature was decreased to 900 kelvins?

4. To get 1000 gallons of a mixture that was 35.2% alcohol, it was necessary to mix some 20% alcohol solution with some 40% alcohol solution. How much of each was required?

5. Find four consecutive odd integers such that the product of the third and fourth is 49 greater than the product of the first and the number 10.

6. Find the surface area of this right circular cylinder. The dimensions are in meters.

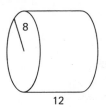

7. $\dfrac{a}{x-y} - \dfrac{c}{p} = m$; find y

8. $\dfrac{x-a}{p} - c = \dfrac{k}{d}$; find a

9. Convert $4\underline{/40°}$ to rectangular form.

10. Convert $40\underline{/330°}$ to rectangular form.

11. Convert 40 centimeters per second to miles per hour.

12. Convert 1000 cubic centimeters to cubic feet.

Simplify:

13. $3i^3 - 2i^2 + i^4 - 5$

14. $-2\sqrt{-9} - 3i^2 + 2i - 2$

Solve by completing the square:

15. $-5x - 6 = -x^2$

16. $-6x + x^2 = -5$

17. Find the equation of the line.

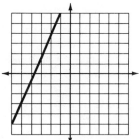

Solve:

18. $\sqrt{x-2} + 2 = 3$

19. $\sqrt{x^2 - x + 13} - 1 = x$

20. $\dfrac{x-2}{4} - \dfrac{x}{3} = 5$

21. $\dfrac{x}{4} - \dfrac{3x+1}{2} = 3$

Simplify:

22. $\sqrt{9\sqrt[3]{3}}$

23. $\sqrt{x^4 \sqrt[3]{x^2 y}}$

24. $\sqrt[3]{4}\sqrt[5]{2}$

25. $3\sqrt{\dfrac{2}{5}} + 7\sqrt{\dfrac{5}{2}} - 2\sqrt{40}$

26. $16^{-\frac{5}{4}}$

27. $\dfrac{\dfrac{x^2 y}{p^5 z} - 1}{\dfrac{x}{p^5} - \dfrac{4}{z}}$

28. Solve: $-2(-x^0 - 4^0) - 3x(2 - 6^0) = (x)(-2 - 3^2 - 2) - x(-2 - 2^0)$

29. Solve $-28x + x^3 = 3x^2$ by factoring.

30. Add: $\dfrac{x}{a^2} - \dfrac{x + 2}{a(a + 2)}$

ER 57-11, 57-12

LESSON 58   *Lead coefficients • More on completing the square*

58.A
lead coefficients

The number in front of the x^2 term in a quadratic equation is called the **coefficient** of the x^2 term. When a quadratic equation is written in descending powers of the variable, the x^2 term comes first and is the lead term. Thus, the coefficient of this term is often called the **lead coefficient of the equation.**

$$\text{(a)} \quad 4x^2 + 3x + 5 = 0 \qquad \text{(b)} \quad x^2 - 3x + 5 = 0$$

In equation (a), we say that the lead coefficient is 4. The coefficient of the x^2 term in equation (b) is 1 and is not written because x^2 has an understood coefficient of 1. We say that equation (b) has a **unity lead coefficient** because mathematicians often say unity instead of saying 1.

58.B
more on completing the square

The coefficient of the x^2 term in the following equation is understood to be 1, and we say that this polynomial equation has a unity lead coefficient.

$$x^2 - 3x + 5 = 0$$

We have discussed the method of completing the square to solve quadratic equations whose lead coefficient is 1. **If the lead coefficient is not 1, then the first step is to divide every term on both sides of the equation by the lead coefficient. The result will be an equation with a unity lead coefficient which we can solve by completing the square.**

example 58.B.1   Solve $4x^2 + 3x - 3 = 0$ by completing the square.

solution   The coefficient of the lead term is not 1, but is 4. Thus, we begin by dividing every term by 4. Then we place parentheses around the first two terms and move the constant to the right side of the equation.

$$x^2 + \frac{3}{4}x - \frac{3}{4} = 0 \qquad \text{divided by 4}$$

$$\left(x^2 + \frac{3}{4}x \qquad \right) = \frac{3}{4} \qquad \text{rearranged}$$

Next we multiply the coefficient of x by $\frac{1}{2}$ and square the result.

$$\left(\frac{3}{4} \cdot \frac{1}{2}\right)^2 = \left(\frac{3}{8}\right)^2 = \frac{9}{64}$$

Then we add $+\frac{9}{64}$ to both sides of the equation and get

$$\left(x^2 + \frac{3}{4}x + \frac{9}{64}\right) = \frac{3}{4} + \frac{9}{64}$$

Now we write the left side as $(x + \frac{3}{8})^2$ and simplify the sum and solve.

$$\left(x + \frac{3}{8}\right)^2 = \frac{57}{64} \qquad \text{simplified}$$

$$x + \frac{3}{8} = \pm\sqrt{\frac{57}{64}} \qquad \text{square root of both sides}$$

$$x = -\frac{3}{8} \pm \frac{\sqrt{57}}{8} \qquad \text{added } -\frac{3}{8} \text{ to both sides}$$

example 58.B.2 Solve $5x^2 - x - 2 = 0$ by completing the square.

solution The first step is to divide every term by 5.

$$x^2 - \frac{1}{5}x - \frac{2}{5} = 0$$

Next we move the constant to the right side and use parentheses.

$$\left(x^2 - \frac{1}{5}x \quad\right) = \frac{2}{5}$$

Then we multiply $-\frac{1}{5}$ by $\frac{1}{2}$ and square the result.

$$\left(-\frac{1}{5} \cdot \frac{1}{2}\right)^2 = \left(-\frac{1}{10}\right)^2 = \frac{1}{100}$$

Next we add $\frac{1}{100}$ to both sides of the equation

$$\left(x^2 - \frac{1}{5}x + \frac{1}{100}\right) = \frac{2}{5} + \frac{1}{100}$$

Now we simplify and solve.

$$\left(x - \frac{1}{10}\right)^2 = \frac{41}{100} \qquad \text{simplified}$$

$$x - \frac{1}{10} = \pm\sqrt{\frac{41}{100}} \qquad \text{square root of both sides}$$

$$x = \frac{1}{10} \pm \frac{\sqrt{41}}{10} \qquad \text{solved}$$

problem set 58

1. Six liters of an ideal gas at a temperature of 600 kelvins had a pressure of 4 atmospheres. If the volume was increased to 8 liters and the pressure decreased to 3 atmospheres, what was the final temperature?

2. The carbon in a container of C_6H_8NCl weighed 360 grams. What did the entire container of the compound weigh? (C, 12; H, 1; N, 14; Cl, 35)

3. In the compound of Problem 2, what percent by weight of the compound was chlorine (Cl)?

4. Two solutions are to be mixed to make 50 ml of a solution that is 16% bromine. One solution is 10% bromine and the other solution is 40% bromine. How much of each solution should be used?

5. After 3 hours the racer was 15 miles ahead of the trotter. How far did the trotter trot if the speed of the racer was 20 mph?

Solve by completing the square:

*6. $4x^2 + 3x - 3 = 0$ ***7.** $5x^2 - x - 2 = 0$ **8.** $3x^2 - 4 = -2x$

9. Find the surface area of the box in square inches. Dimensions are in feet.

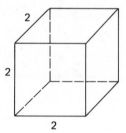

<u>10.</u> $\dfrac{x - a}{p} - c = \dfrac{y}{k}$; find p **11.** $\dfrac{a}{x - p} - c = \dfrac{y}{k}$; find p

12. Convert $4/\underline{220°}$ to rectangular form.

13. Convert $10/\underline{315°}$ to rectangular form.

14. Convert 70 meters per second to miles per hour.

15. Convert 40 cubic feet to cubic centimeters.

Simplify:

<u>16.</u> $\sqrt[4]{9}\sqrt[5]{3}$ **17.** $\sqrt[7]{4\sqrt{2}}$

18. $\sqrt{x^2 y^3}\sqrt[3]{xy^5}$ **19.** $-16^{-\frac{3}{4}}$

20. $5\sqrt{\dfrac{5}{11}} - 2\sqrt{\dfrac{11}{5}} + 3\sqrt{220}$ **21.** $5i^3 - 6i^8 + 2\sqrt{-25} - 4i^2$

22. Estimate: $\dfrac{(40{,}621{,}857)(6{,}031{,}824)}{19{,}610 \times 10^{-24}}$

23. Simplify: $\dfrac{x^2 + 16 + 10x}{x^2 + 11x + 24} \cdot \dfrac{x^2 + 8x + 15}{x^2 - 25}$

24. This shape is the base of a container that is 4 feet high. How many 1-inch sugar cubes will it hold? Dimensions are in inches.

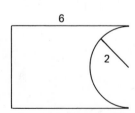

25. Simplify: $\dfrac{\dfrac{x^2 a^2}{m} - \dfrac{4}{p^3}}{\dfrac{xa}{mp^3} - 5}$

26. Solve by graphing and then find an exact solution by using either substitution or elimination.
$$\begin{cases} 3x + 4y = -4 \\ x - 5y = 10 \end{cases}$$

27. Divide $4x^3 - 5$ by $x + 2$.

Solve:

28. $\dfrac{3 - x}{2} - \dfrac{4x}{3} = 7$

<u>29.</u> $x^0 - 2x - 5(x - 3^0) = -2x^0 - 7$

30. Add: $\dfrac{b}{a(y + 1)} + \dfrac{cx}{a^2(y + 1)}$

ER 58-13, 58-14

LESSON 59   *Experimental data • Simultaneous equations with fractions and decimals • Rectangular form to polar form*

59.A
experimental
data

Linear equations in two unknowns are important because they have applications in real-life problems. In chemistry and other advanced courses that use mathematics, we will find that the relationship between two variables can often be expressed by using equations. Happily, many of these equations are linear equations, and thus their graphs are straight lines. We can look at the graphs and get a pictorial representation of the relationship between the two variables being considered.

Laboratory experiments are used to confirm theories. The data points obtained in these experiments are graphed and circled to indicate that they are experimental data points. If the theory indicates that the relationship is linear, we estimate the line indicated by the data points. Then we write the equation of the line, using the variables of the data rather than x and y.

example 59.A.1   The data graphed come from an experiment about bronze and copper. The line represented by the data points has been estimated. Write the equation of this line that gives bronze as a function of copper. Note that the two scales are different.

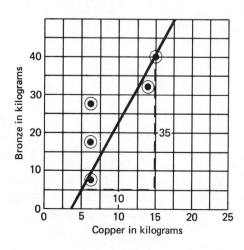

solution   Bronze is graphed vertically and copper is graphed horizontally, so in the equation we will replace y with B and x with C. Thus our equation is

$$B = mC + b$$

From the triangle the slope is 3.5 so we have

$$\textbf{\textit{B} = 3.5\textit{C} + \textit{b}}$$

Now we will use the coordinates of one point on the line so we can solve for b. We will use the point (15, 40). Thus we will replace C with 15 and will replace B with 40.

$$(40) = 3.5(15) + b \quad\longrightarrow\quad 40 = 52.5 + b \quad\longrightarrow\quad b = -12.5$$

Thus our final equation is

$$\textbf{\textit{B} = 3.5\textit{C} − 12.5} \qquad \textbf{kilograms}$$

This equation approximates the line indicated by the data points. The values of m and b that we have found are not exact.

example 59.A.2   The data graphed come from an experiment that relates nitrogen and sulfur. Write the equation that expresses nitrogen as a function of sulfur. Note the extreme difference in the horizontal and vertical scales.

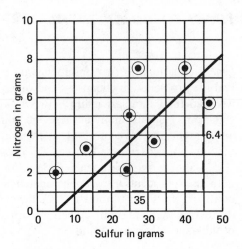

solution   Nitrogen is plotted vertically, so our equation is

$$N = mS + b$$

The slope is about $\frac{6.4}{35}$, which approximately equals .18, so we can write

$$N = .18S + b$$

We choose the point (10, 1) and replace S with 10 and N with 1 to get

$$(1) = (.18)(10) + b \quad\longrightarrow\quad b = -.8$$

so the complete equation is

$$\textbf{\textit{N} = .18\textit{S} − .8} \qquad \textbf{grams}$$

This equation approximates the line indicated by the data points. It is not exact. It is possible to get only approximate equations from experimental data. Another try would probably yield slightly different values for m and b.

59.B
simultaneous equations with fractions and decimals

Some systems of simultaneous equations that contain fractions and decimal numbers look very difficult. If we remember, as the first step, to change these equations to equations in which all numbers are integers, the solution will be much easier.

example 59.B.1 Solve: (a) $\begin{cases} \dfrac{x}{2} + \dfrac{3y}{5} = -\dfrac{2}{5} \\[2mm] \text{(b)} \quad .06x - .2y = 1.04 \end{cases}$

solution We can eliminate the denominators in the fractional equation by multiplying every term by 10.

$$\text{(a)} \quad \frac{x}{2}(10) + \frac{3y}{5}(10) = -\frac{2}{5}(10) \qquad \longrightarrow \qquad \text{(a')} \quad 5x + 6y = -4$$

And in the other equation we can change the decimal coefficients to integers if we multiply every term by 100.

$$\text{(b)} \quad .06x(100) - .2y(100) = 1.04(100) \quad \longrightarrow \quad \text{(b')} \quad 6x - 20y = 104$$

Now we can solve these equations by using the elimination method.

$$\begin{array}{lllll}
\text{(a')} & 5x + 6y = -4 & \longrightarrow & (-6) & \longrightarrow & -30x - 36y = 24 \\
\text{(b')} & 6x - 20y = 104 & \longrightarrow & (5) & \longrightarrow & \underline{30x - 100y = 520} \\
& & & & & -136y = 544 \\
& & & & & \mathbf{y = -4}
\end{array}$$

We find x by replacing y with -4 in equation (a').

$$
\begin{array}{ll}
5x + 6y = -4 & \text{equation (a')} \\
5x + 6(-4) = -4 & \text{replaced } y \text{ with } -4 \\
5x - 24 = -4 & \text{multiplied} \\
5x = 20 & \text{added 24} \\
x = 4 & \text{divided}
\end{array}
$$

Thus, our solution is the ordered pair **(4, −4)**.

59.C
rectangular form to polar form

We can convert a number from rectangular form to polar form by using the rectangular coordinates to find the tangent of the angle. Then we use the trigonometric tables to find the angle whose tangent we have found. The hypotenuse of the triangle can be found by using trigonometric functions or the theorem of Pythagoras. The theorem of Pythagoras produces exact answers but is difficult to use when the lengths of the sides are not whole numbers. The use of the sine and cosine to find the hypotenuse is not ideal because this method propagates errors made when the angle was determined. However, the sine and cosine can be used easily even when the sides are not whole numbers, and many prefer to use the sine and the cosine for this reason.

example 59.C.1 Convert $-5R - 3U$ to polar form.

solution   We begin by locating the point on the coordinate plane and then drawing the triangle. We remember that $-5R$ means 5 to the left and $-3U$ means 3 down.

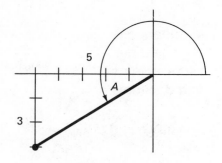

First we find the tangent of angle A.

$$\tan A = \frac{\text{opposite}}{\text{adjacent}} \longrightarrow \tan A = \frac{3}{5} \longrightarrow \tan A = .6$$

Then we use the table to find that angle A is approximately $31.0°$.

$$\text{Angle whose tangent is } .6 = 31.0°$$

Now we can use either the sine of $31°$ or the cosine of $31°$ to find H. We decide to use the sine.

$$\sin 31° = \frac{3}{H} \longrightarrow H = \frac{3}{\sin 31°} \longrightarrow H = \frac{3}{.5150} \longrightarrow H = 5.83$$

Finally we write the coordinates of the point in polar form and remember to add $180°$ so the point will be in the third quadrant.

$$-5R - 3U = \mathbf{5.83\underline{/211°}}$$

If we use the Pythagorean theorem, we can find the exact length of H.

$$H = \sqrt{3^2 + 5^2} = \sqrt{34}$$

Thus the location of the point can also be designated by writing

$$\mathbf{\sqrt{34}\underline{/211°}}$$

problem set 59

1. Find four consecutive integers such that the product of the first and the fourth is 22 less than the product of 10 and the opposite of the third.

2. What is the weight of the oxygen (O) in 460 grams of the compound whose formula is C_2H_6O? (C, 12; H, 1; O, 16)

3. Rasputin ran part of the way at 8 mph and walked the rest of the way at 3 mph. If the total trip was 41 miles and the total time was 7 hours, how far did he run and how far did he walk?

4. One solution was 90% iodine and the other was 70% iodine. How much of each should be used to get 100 liters of a solution that is 78% iodine?

5. In an experiment with a quantity of an ideal gas, the temperature was held constant. The initial pressure and volume were 14 atmospheres and 10 liters. If the pressure was increased to 20 atmospheres, what was the final volume?

The figures show data points from two experiments for which the theory predicted a linear relationship between the variables. Write the equation for the line in the figure on the left that gives bronze as a function of copper and for the figure on the right that gives nitrogen as a function of sulfur.

*6.

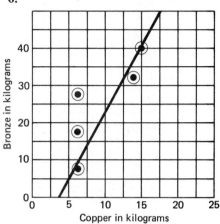

*7.

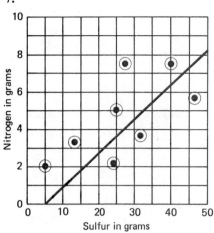

*8. Solve: $\begin{cases} \dfrac{x}{2} + \dfrac{3y}{5} = -\dfrac{2}{5} \\ .06x - .2y = 1.04 \end{cases}$

*9. Convert $-5R - 3U$ to polar form.

10. Convert $30\ \underline{/330^\circ}$ to rectangular form.

Solve by completing the square.

11. $2x^2 = x + 5$ **12.** $3x^2 - 5 = 2x$ **13.** $3x^2 - 4 = x$

14. Find the surface area of this box in square inches. Dimensions are in feet.

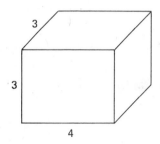

15. $\dfrac{p - s}{m} - \dfrac{c}{4} + x = 0$; find m **16.** $\dfrac{m}{p + s} - \dfrac{c}{x} + 4 = 0$; find p

17. Convert 40 inches per second to meters per hour.

18. Convert 18,000 cubic centimeters to cubic feet.

Simplify:

19. $i^4 + 5 + 3\sqrt{-9} - 2\sqrt{-4}$ **20.** $3i^5 - 2i^2 - 4 - i$

Solve:

21. $\sqrt{x^2 - 4x + 39} + 1 = x + 4$ **22.** $\sqrt{x - 3} + 2 = -4$

Simplify:

23. $\sqrt[5]{3\sqrt{3}}$ **24.** $\sqrt[4]{4\sqrt{2}}$ **25.** $-16^{-\frac{3}{4}}$

26. $\sqrt{x^2 y}\ \sqrt[3]{y^5 x^4}$ **27.** $\sqrt{\dfrac{2}{13}} - 4\sqrt{\dfrac{13}{2}} + 3\sqrt{234}$

28. Solve: $\dfrac{x-4}{2} - \dfrac{x}{3} - 2 = \dfrac{5}{2}$ **29.** Add: $\dfrac{x}{y} + \dfrac{x^2+2}{y^2a} + \dfrac{x^3}{y(a+y)}$

30. Simplify: $\dfrac{(x^0yp)^{-3}x^3yp^0p^{-2}}{x^2p^2y^0(y^{-3}p)^2x^{-3}y^0p^2}$

ER59-15, 59-16

LESSON 60 *Direct and inverse variation*

60.A
direct and inverse variation

When the statement of a problem says that A **varies directly as** B or that A **is directly proportional to** B, the equation

$$A = kB$$

is implied. When the statement says that A **varies inversely** as B or that A **is inversely proportional** to B, the equation

$$A = \frac{k}{B}$$

is implied. The constant k is called the **constant of proportionality. Note that k is always in the numerator. In direct variation, both variables are in the numerators; and in inverse variation, one variable is in the numerator and the other is in the denominator.** The key to working variation problems is recognizing the equation implied by the statement. The following examples should be helpful.

STATEMENT	IMPLIED EQUATION
The number of boys varied directly as the number of girls.	$B = kG$
The price varied inversely as the number.	$P = \dfrac{k}{N}$
The resistance is directly proportional to the length.	$R = kL$
RPM is inversely proportional to the number of teeth.	$RPM = \dfrac{k}{N_t}$
The water produced varied directly as the amount of hydrogen burned	$W = kH_B$

Direct and inverse variation problems are four-step problems. **The first step is recognizing that the words varies directly (is directly proportional to) and varies inversely (is inversely proportional to) imply equations of the forms**

$$A = kB \qquad \text{and} \qquad A = \frac{k}{B}$$

The next step is to find k. In order to find k, the problem must give sample values of A and B. The third step is to replace k in the equation with the proper number, and the last step is to reread the problem, make the final substitution, and then solve.

example 60.A.1   The number of boys in every classroom of a school varies directly as the number of girls. In one room, there are 8 boys and 2 girls. If, in another room, there are 5 girls, how many boys are in this room?

solution   (1)   We write the equation implied by the words **varies directly.**

$$B = kG$$

(2)   We use 8 for boys and 2 for girls and solve for k.

$$(8) = k(2) \longrightarrow k = 4$$

(3)   We replace k in the equation with 4.

$$B = 4G$$

(4)   Now we use 5 for G and solve for B.

$$B = 4(5) \longrightarrow \mathbf{B = 20\ boys}$$

example 60.A.2   Revolutions per minute (RPM) vary inversely as the number of teeth in the gear. If 40 teeth results in 100 RPM, what would be the RPM if the gear had 30 teeth?

solution   We will use the same four steps as in the last problem.

(1)   First we write the equation implied by the words **varies inversely.**

$$RPM = \frac{k}{N_t}$$

(2)   Next we find k.

$$100 = \frac{k}{40} \longrightarrow k = 4000$$

(3)   Then we replace k with 4000.

$$RPM = \frac{4000}{N_t}$$

(4)   Now we solve for RPM if the number of teeth is 30.

$$RPM = \frac{4000}{30} \longrightarrow \mathbf{RPM = 133\tfrac{1}{3}}$$

example 60.A.3   The number of clowns was directly proportional to the number of performers. If there were 40 clowns when there were 20,000 performers, how many clowns would 12,000 performers produce?

solution   We will again use four steps.

(1) $C = kP$ wrote the implied equation

(2) $40 = k(20,000) \longrightarrow k = \dfrac{1}{500}$ solved for k

(3) $C = \dfrac{1}{500}P$ replaced k with $\dfrac{1}{500}$

(4) $C = \dfrac{1}{500} \times 12,000 \longrightarrow \mathbf{C = 24\ clowns}$ found C when P equals 12,000

problem set 60   *1.   The number of boys in every classroom of a school varies directly as the number of girls. In one room there are 8 boys and 2 girls. If in another room there are 5 girls, how many boys are in this room?

*2. Revolutions per minute (RPM) vary inversely as the number of teeth in the gear. If 40 teeth results in 100 RPM, what would be the RPM if the gear had 30 teeth?

*3. The number of clowns was directly proportional to the number of performers. If there were 40 clowns when there were 20,000 performers, how many clowns would 12,000 performers produce?

4. The volume of a quantity of an ideal gas was held constant. The initial pressure and temperature were 400 torr and 500 kelvins. What would the pressure be if the temperature was increased to 1000 kelvins?

5. There were 50 more pugnacious students than trepid students. In fact, twice the number of pugnacious exceeded 3 times the number of trepid by 60. How many students were in each category?

6. Convert $4\underline{/135°}$ to rectangular form.   **7.** Convert $-2R + 4U$ to polar form.

8. Solve: $\begin{cases} \dfrac{2}{5}x + \dfrac{3}{2}y = 34 \\ .02x + .3y = 6.2 \end{cases}$

Solve by completing the square:

9. $4x^2 - 3 = x$ **10.** $3x^2 = 2x + 1$

11. The data points shown came from an experiment involving sodium (Na) and carbon (C). Write the equation that expresses sodium as a function of carbon: $Na = mC + b$.

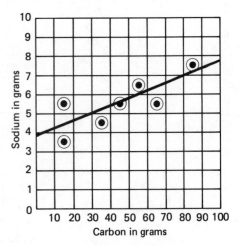

12. Find the surface area of this pyramid which has 4 equal triangular faces and a square bottom. Units are in inches.

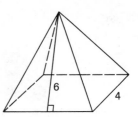

13. $\dfrac{a}{x + y} = c + \dfrac{m}{d}$; find x **14.** $\dfrac{xy + c}{m} + d = \dfrac{k}{z}$; find c

Simplify:

15. $3i^3 - \sqrt{-4} + 3\sqrt{-9} - 2 + i^2$ **16.** $-3i^4 - 2i^2 + 2 - 3\sqrt{-25}$

17. Convert 60 centimeters per second to miles per hour.

18. Convert 1,400,000 cubic centimeters to cubic yards.

Simplify:

19. $\sqrt[3]{27\sqrt{3}}$

20. $\sqrt[4]{81}\sqrt[3]{3}$

21. $81^{-\frac{3}{4}}$

22. $\sqrt{m^2 p}\sqrt[3]{m^5 p^4}$

23. $5\sqrt{\dfrac{2}{9}} + 3\sqrt{\dfrac{9}{2}} - 5\sqrt{8}$

24. Estimate: $\dfrac{(5,162,348)(.0000165)}{.003217642}$

25. Find K.

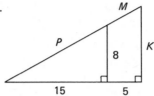

26. Find the number of 1-inch-square tiles necessary to cover this figure. Dimensions are in feet.

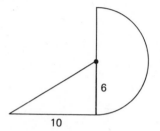

27. Find an exact solution by using substitution or elimination:
$$\begin{cases} 2x - 3y = -3 \\ 2x + y = 8 \end{cases}$$

28. Solve $x^3 + 50x = 15x^2$ by factoring.

29. Find the equation of the line that has a slope of $\frac{2}{5}$ and passes through the point $(-40, 2)$.

30. Multiply: $\dfrac{3x^{-4}y^0 y^2}{x^2 x}\left(\dfrac{2x^2 y^2}{y^4} - \dfrac{y^0 x^0 p^{-2}}{x^4 y^4}\right)$

ER 60-17

LESSON 61 *Chemical mixture type B*

61.A
chemical mixture type B

In the chemical mixture problems worked thus far we have mixed two solutions of different percentage concentrations to get a mixture that has a different percentage concentration than either of the original solutions. It is interesting to note that the percentage concentration of the final mixture must fall between the percentage concentrations of the two solutions used. **For example, if we pour in some 15% iodine solution**

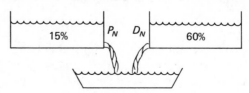

and dump in some 60% iodine solution we can get a final mixture whose concentration is somewhere between 15% iodine and 60% iodine. It is impossible to mix these solutions and get a final concentration greater than 60% iodine or less than 15% iodine.

For want of a better name, we will call problems like the one above type A chemical mixture problems. Now we will look at another kind of chemical mixture problem. We will call these type B problems. In a type B problem, we will begin with a given mixture and add to this mixture or remove something from this mixture. **As in type A problems, the equation will make a statement about one of the components of the mixture.**

example 61.A.1   How much water must be evaporated from 100 gallons of a 10% brine solution to get a 40% brine solution?

solution   Usually, we would begin by deciding whether to work the problem in water or in salt. For this example problem, we will work the problem both ways to demonstrate that both ways give the same answer. First, we write the true statements that consider either water or salt.

CONSIDERING WATER   CONSIDERING SALT

Water one − water out = water final   Salt one − salt out = salt final

$$W_1 - W_o = W_f \qquad\qquad S_1 - S_o = S_f$$

Now we write the mixture containers. Their use is very important in mixture problems.

$$(\quad) - (\quad) = (\quad) \qquad (\quad) - (\quad) = (\quad)$$

Next we make the entries in the containers and we are careful to use neither W or S as a variable. Our original mixture was 100 gallons so this goes in the first container. We evaporated some, so E_o goes into the second container. The final mixture is the original mixture minus what was evaporated, so $100 - E_o$ goes into the last container.

CONSIDERING WATER   CONSIDERING SALT

$$(100) - (E_o) = (100 - E_o) \qquad (100) - (E_o) = (100 - E_o)$$

Now we multiply each container by the proper decimal so that the containers times the decimals in the left example represent water and those in the right example represent salt. *No salt was evaporated,* so the multipliers for the center containers are 1 and 0.

CONSIDERING WATER   CONSIDERING SALT

$$.9(100) - 1(E_o) = .6(100 - E_o) \qquad .1(100) - 0(E_o) = .4(100 - E_o)$$

Note that if a mixture is .9 water, then it is .1 salt; and if it is .6 water, then it is .4 salt. Now we solve.

$$90 - E_o = 60 - .6E_o \qquad 10 = 40 - .4E_o$$

$$30 = .4E_o \qquad\qquad -30 = -.4E_o$$

$$75 = E_o \qquad\qquad 75 = E_o$$

Thus both approaches give the same result. We must evaporate 75 gallons of water to get a mixture that is 40% salt.

example 61.A.2   When Frank and Mark finished milking, they found that they had 900 pounds of milk that was 2 percent butterfat. How much butterfat did they have to add to raise the butterfat content to 8 percent? (Whole milk is a mixture of skim milk and butterfat.)

solution   This time we are adding something, so the final mixture will weigh more than the original mixture. We remember that the decimals for 2 percent and 8 percent are .02 and .08. We decide to work this problem in butterfat.

$$\text{Butterfat one} + \text{butterfat added} = \text{butterfat final}$$

$$B_1 + B_A = B_f$$

Now we write down the mixture containers and use P_N for the amount added.

$$(900) + (P_N) = (900 + P_N)$$

Now since we decided to work the problem in butterfat, our decimals are .02, 1, and .08.

$$.02(900) + 1(P_N) = .08(900 + P_N) \qquad \text{equation}$$
$$18 + P_N = 72 + .08P_N \qquad \text{multiplied}$$
$$.92P_N = 54 \qquad \text{added} -.08P_N \text{ and } -18$$
$$P_N = \textbf{58.7 lb} \qquad \text{rounded off}$$

Thus 58.7 pounds of butterfat should be added to get a mixture that is 8% butterfat.

example 61.A.3   Virginia and Campbell had 100 kilograms of a 20% glycol solution. How much of a 40% glycol solution should they add to get a solution that is 35% glycol?

solution   We decide to make the statement about glycol.

$$\text{glycol}_1 + \text{glycol added} = \text{glycol final}$$

Next we write the mixture containers. In the first, we place 100; in the second P_N for "poured in," and in the last we write $100 + P_N$:

$$(100) + (P_N) = (100 + P_N)$$

Now we multiply each container by the proper percentage.

$$.2(100) + .4(P_N) = .35(100 + P_N)$$

Then we multiply and solve.

$$20 + .4P_N = 35 + .35P_N \qquad \text{multiplied}$$
$$.05P_N = 15 \qquad \text{simplified}$$
$$P_N = \textbf{300 kg} \qquad \text{divided}$$

Thus if they pour in 300 kg of a 40% glycol mixture, the result will be a mixture that is 35% glycol.

problem set 61

1. The rate of decomposition varied directly as the amount of substance present. When the amount was 5 kilograms, the rate was .005 kilogram per second. What was the rate of decomposition when the amount was .3 kilogram?

2. The volume of a quantity of ideal gas was kept constant in an experiment. The final temperature was 600 kelvins (K) and the final pressure was 300 torr. What was the original pressure if the original temperature was 1000 K?

*3. How much water must be evaporated from 100 gallons of a 10% brine solution to get a 40% brine solution?

*4. When Frank and Mark finished milking, they found that they had 900 pounds of milk that was 2 percent butterfat. How much butterfat did they have to add to raise the butterfat content to 8 percent? (Whole milk is a mixture of skim milk and butterfat.)

*5. Virginia and Campbell had 100 kilograms of a 20% glycol solution. How much of a 40% glycol solution should they add to get a solution that is 35% glycol?

6. Convert $20 \underline{/165°}$ to rectangular form.

7. Convert $6R - 2U$ to polar form.

8. Solve: $\begin{cases} \dfrac{2}{3}x - \dfrac{2}{5}y = -4 \\ .2x + .9y = 19.2 \end{cases}$

Solve by completing the square:

9. $-4x - 4 = -5x^2$

10. $-x = 7 - 2x^2$

11. The data points shown came from an experiment that involved lead (Pb) and antimony (Sb). Write the equation that expresses lead as a function of antimony: $Pb = mSb + b$.

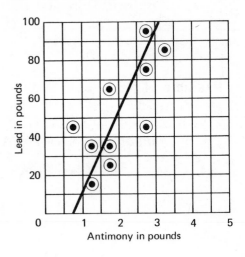

12. Find the surface area of this right circular cylinder. Dimensions are in centimeters.

13. $\dfrac{(m + c + x)b}{k} + \dfrac{a}{d} = p$; find c

14. $\dfrac{4y}{2a + x} + \dfrac{m}{c} = d$; find a

Simplify:

15. $-2i^2 - 3i - 4 - 2\sqrt{-4}$

16. $8i^4 - 2i^3 - 2i - 6 - 4\sqrt{-16}$

17. Convert 40 cubic centimeters per second to cubic inches per hour.

18. Convert 4 cubic feet to cubic centimeters.

Simplify:

19. $\sqrt{32\sqrt{2}}$

20. $\sqrt[3]{8\sqrt[3]{2}}$

21. $-8^{-\frac{5}{3}}$

22. $\sqrt{x^5 y}\,\sqrt[4]{y^2 x}$

23. $3\sqrt{\dfrac{2}{3}} - 4\sqrt{\dfrac{3}{2}} + 8\sqrt{24}$

24. Estimate: $\dfrac{(41,685,231)(.0012846 \times 10^{-14})}{.001998 \times 10^{-10}}$

25. Find C.

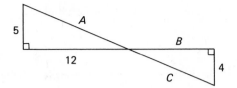

26. The figure shown is the base of a container that is 6 ft high. How many one inch sugar cubes will this container hold? Dimensions are in feet.

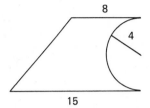

27. Solve by graphing and then find an exact solution by using substitution or elimination:
$$\begin{cases} x - 2y = -6 \\ x + y = -1 \end{cases}$$

28. Solve $-15x = -x^3 + 2x^2$ by factoring.

29. Find the equation of the line that passes through $(-2, 4)$ that is perpendicular to $2x + 3y = 5$.

30. Solve: $-2^2 - 3^2 - (2^0)^2 - (-2)^0 = -x(-2x^0 - 5^0)2^2$

ER 61-18

LESSON 62 *Complex roots of quadratic equations*

62.A

complex roots of quadratic equations

When we solve quadratic equations by completing the square, we will find that we get some solutions of the form

$$x = -3 \pm \sqrt{-5}$$

Now since the square root of a negative number can be written by using Euler's notation, we can write this answer as

$$x = -3 \pm \sqrt{5}\,i$$

In this lesson we will complete the square to solve quadratic equations whose solutions are complex numbers.

example 62.A.1   Solve $-x + 3x^2 + 5 = 0$.

solution   As the first step we write the equation in standard form:

$$3x^2 - x + 5 = 0$$

Then we divide every term by 3 so that the coefficient of x^2 will be 1.

$$x^2 - \frac{1}{3}x + \frac{5}{3} = 0$$

Next we write the parentheses and move the constant term to the right side.

$$\left(x^2 - \frac{1}{3}x \qquad\right) = -\frac{5}{3}$$

Now we multiply the coefficient of x by $\frac{1}{2}$ and square the product.

$$\left(-\frac{1}{3}\cdot\frac{1}{2}\right)^2 = \frac{1}{36}$$

Then we add $\frac{1}{36}$ to both sides of the equation.

$$\left(x^2 - \frac{1}{3}x + \frac{1}{36}\right) = -\frac{5}{3} + \frac{1}{36}$$

Now we simplify and solve for x.

$$\left(x - \frac{1}{6}\right)^2 = -\frac{59}{36} \qquad \text{simplified}$$

$$x - \frac{1}{6} = \pm\sqrt{-\frac{59}{36}} \qquad \text{square root of both sides}$$

$$x = \frac{1}{6} \pm \frac{\sqrt{59}}{6}i \qquad \text{solved}$$

example 62.A.2   Solve $-2x + 5x^2 = -3$.

solution   First we write the equation in standard form

$$5x^2 - 2x + 3 = 0$$

Next we get a unity lead coefficient by dividing by 5.

$$\left(x^2 - \frac{2}{5}x \qquad\right) = -\frac{3}{5}$$

Now we multiply $-\frac{2}{5}$ by $\frac{1}{2}$ and square this product.

$$\left(-\frac{2}{5}\cdot\frac{1}{2}\right)^2 = \frac{1}{25}$$

Now we add $\frac{1}{25}$ to both sides of the equation.

$$\left(x^2 - \frac{2}{5}x + \frac{1}{25}\right) = -\frac{3}{5} + \frac{1}{25}$$

Then we simplify and finish the solution.

$$\left(x - \frac{1}{5}\right)^2 = -\frac{14}{25} \qquad \text{simplified}$$

$$x - \frac{1}{5} = \pm\sqrt{-\frac{14}{25}} \qquad \text{square root}$$

$$x = \frac{1}{5} \pm \frac{\sqrt{14}}{5}i \qquad \text{solved}$$

1. The number of victories varied inversely as the skill of the opponents. The team won 8 games when the opponents had a skill factor of 2. How many victories could be expected when the opponents' average skill factor was 8?

2. The initial state for a quantity of an ideal gas was a pressure of 600 torr, a temperature of 300 kelvins and a volume of 2 liters. If the volume was increased to 4 liters and the pressure decreased to 400 torr, what was the final temperature?

3. The vat contained 40 liters of a 5% salt solution. How much of a 20% salt solution should be added to get a 10% salt solution?

4. Part of the journey was by sleigh at 8 mph and the rest was by truck at 20 mph. If the total journey was 152 miles and the total time was 10 hours, what part of the trip was by sleigh?

5. In the chemical compound CH_4ON_2, what percent of the total weight is nitrogen (N)? (C, 12; H, 1; O, 16; N, 14)

6. Convert $20\underline{/340°}$ to rectangular form.

7. Convert $-2R + 5U$ to polar form.

8. Solve: $\begin{cases} \dfrac{1}{3}x - \dfrac{2}{3}y = -1 \\ .02x + .4y = 2.58 \end{cases}$

Solve by completing the square:

*9. $-x + 3x^2 + 5 = 0$

*10. $-2x + 5x^2 = -3$

11. The data points shown come from an experiment that involved bismuth (Bi) and mercury (Hg). Write the equation that expresses bismuth as a function of mercury: $Bi = mHg + b$.

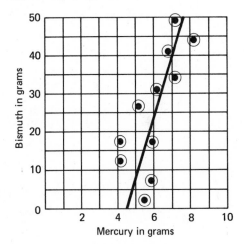

12. Find the surface area of the right circular cylinder shown. Dimensions are in inches.

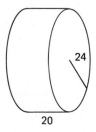

13. $\dfrac{(a + b)m}{c} - k = \dfrac{p}{r}$; find b

14. $\dfrac{6x}{2y + 4a} - c = \dfrac{p}{r}$; find y

Simplify:

15. $3i^3 - \sqrt{-4} - 2 - \sqrt{9} + 3$

16. $-3i^3 + 2i - 4 - 3i^2 - 2\sqrt{9}$

17. Convert 600 cubic centimeters per minute to cubic feet per second.

18. Convert 20 cubic yards to cubic centimeters.

Simplify:

19. $\sqrt{16\sqrt{2}}$

20. $\sqrt[4]{4\sqrt[5]{2}}$

21. $-4^{\frac{5}{2}}$

22. $\sqrt{xy^7}\sqrt[3]{x^5y}$

23. $3\sqrt{\dfrac{2}{5}} - 5\sqrt{\dfrac{5}{2}} - 3\sqrt{40}$

24. Estimate: $\dfrac{(.000618427 \times 10^{14})(7{,}891{,}642)}{3{,}728{,}196{,}842}$

25. Find B.

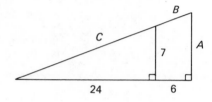

26. This is the floor of a container 10 feet tall. How many 1-inch cubes will this container hold? Dimensions are in feet.

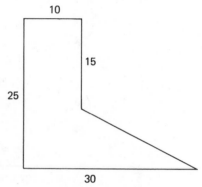

27. Solve by graphing and then find an exact solution by using substitution or elimination:
$$\begin{cases} 2x - 3y = -9 \\ 2x + 3y = -3 \end{cases}$$

28. Solve $50x + x^3 = 15x^2$ by factoring.

29. Find the equation of the line whose slope is $-\frac{2}{7}$ and that passes through the point $(-5, -7)$.

30. Solve: $\dfrac{3+x}{2} - \dfrac{2}{7} = 4$

ER 62-19

LESSON 63 *Addition of vectors*

63.A
addition
of vectors

To add two vectors, we first write the vectors in rectangular form. Then we add the horizontal components to find the horizontal component of the sum and add the vertical components to find the vertical component of the sum. The process is easier to understand if we think of each vector as describing a journey, as in the following example.

example 63.A.1 Flying Arrow left the village and traveled 20 miles on a heading of 20°. From this point he went 40 miles on a heading of 210°. How far did he end up from the village?

solution We need to add the vectors $20\underline{/20°}$ and $40\underline{/210°}$. **We remember that when we break up a vector k into components, one component is $k \sin \theta$ and the other component is $k \cos \theta$.**

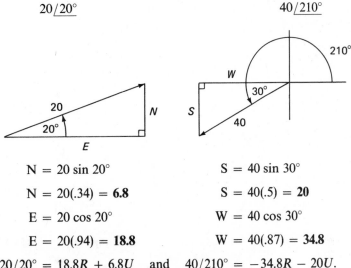

$$20\underline{/20°} \qquad\qquad 40\underline{/210°}$$

N = 20 sin 20°	S = 40 sin 30°
N = 20(.34) = **6.8**	S = 40(.5) = **20**
E = 20 cos 20°	W = 40 cos 30°
E = 20(.94) = **18.8**	W = 40(.87) = **34.8**

Thus $\qquad 20\underline{/20°} = 18.8R + 6.8U \quad$ and $\quad 40\underline{/210°} = -34.8R - 20U$.
Now we add the vectors by adding like components algebraically.

$$\begin{array}{r} 18.8R + 6.8U \\ -34.8R - 20.0U \\ \hline -16R - 13.2U \end{array}$$

Thus, Flying Arrow ended up 16 miles west and 13.2 miles south of the village.

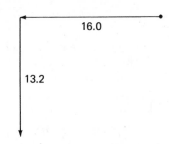

In this problem we had to find the value of

$$\text{(a)} \quad 20 \sin 20° \qquad \text{and} \qquad \text{(b)} \quad 40 \cos 30°$$

Both of these calculations can be performed on a pocket calculator without having to write down the value of sin 20° or cos 30°. To find (b)[†]:

	DEPRESS	DISPLAY
(1)	40	40
(2)	$\times$	40
(3)	30	30
(4)	COS	.8660254
(5)	$=$	34.641016

[†] The procedure is slightly different for a calculator that uses reverse Polish notation, such as the Hewlett-Packard.

The same procedure is used to solve (a) except that in step 4, we depress SIN instead of COS. Of course, the answers obtained by using the calculator will differ a little from those obtained by using the trigonometric tables. This should be remembered if the trigonometric tables are used, for all answers to future problems sets will be obtained by using a calculator and rounding off.

example 63.A.2   Add $30\underline{/55°}$ and $10\underline{/170°}$.

solution   We can think of these as trips of 30 miles and 10 miles in the directions given. First we find the horizontal and vertical components of each vector.

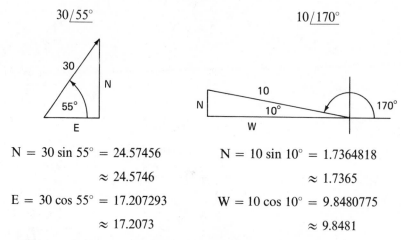

$$30\underline{/55°} \qquad\qquad\qquad\qquad 10\underline{/170°}$$

$$N = 30 \sin 55° = 24.57456 \qquad N = 10 \sin 10° = 1.7364818$$
$$\approx 24.5746 \qquad\qquad\qquad \approx 1.7365$$
$$E = 30 \cos 55° = 17.207293 \qquad W = 10 \cos 10° = 9.8480775$$
$$\approx 17.2073 \qquad\qquad\qquad \approx 9.8481$$

Next we write the vectors in rectangular form and add algebraically.

$$\begin{array}{r} 17.2073R + 24.5746U \\ -9.8481R + 1.7365U \\ \hline 7.3592R + 26.3111U \end{array}$$

If we round the sum of the vectors to two decimal places we get **7.36R + 26.31U.**

63.B
inverse trigonometric functions

The inverse trigonometric function keys on the calculator can be used to find the angle if we know the sine, the cosine or the tangent of the angle. The magnitude of the vector found in the problem above is

$$\sqrt{(7.36)^2 + (26.31)^2} = 27.32$$

The tangent of the angle is

$$\frac{26.31}{7.36} = 3.575$$

If we use the inverse tangent key on the calculator, we can find that the angle whose tangent is 3.575 is approximately 74.37°. Thus, we can write the answer to the last problem in polar form as

$$27.32\underline{/74.37°}$$

problem set 63

1. Odysseus found that his troubles varied directly as his distance from his home island of Ithaca. If he had 20 troubles when he was 400 miles from home, how many troubles did he have when only 60 miles from home?

2. The pressure of a container of ideal gas was held constant for an experiment. The initial temperature was 800 kelvins and the initial volume was 20 liters. If the final volume was reduced to 12 liters, what was the final temperature?

3. Somehow salt got in the rainbarrel, for the 50 gallons it contained was found to be 4% salt. How much pure water must be added to reduce the salt content to 1%?

4. Oedipus beat Rex to the goal by 4000 feet. If Rex ran at 20 feet per second and Oedipus ran at 40 feet per second, what was the length of the race course?

5. The weight of the carbon in a container of C_3H_7Cl was 48 grams. What was the total weight of the compound? (C, 12; H, 1; Cl, 35)

*6. Flying Arrow left the village and traveled 20 miles on a heading of 20°. From this point he went 40 miles on a heading of 210°. How far did he end up from the village?

*7. Add $30\underline{/55°}$ and $10\underline{/170°}$.

8. Write $7.3R + 26.34U$ in polar form.

9. Solve: $\begin{cases} \dfrac{1}{5}x - \dfrac{5}{2}y = -48 \\ .4x + .05y = 5 \end{cases}$

Solve by completing the square:

<u>10.</u> $-x = -2x^2 - 5$

11. $3x^2 = -4 + 2x$

12. The data points shown come from an experiment that involved molybdenum (Mo) and zirconium (Zr). Write the equation that expresses molybdenum as a function of zirconium: $Mo = mZr + b$.

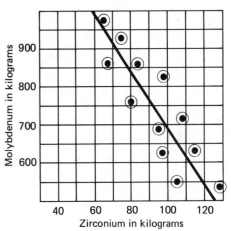

13. Find the surface area of the prism shown in square centimeters. The dimensions are in meters.

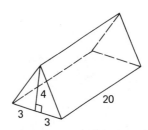

14. $\dfrac{p + zy}{m} - c = \dfrac{a}{b}$; find z

<u>15.</u> $\dfrac{p + zy}{m} - c = \dfrac{a}{b}$; find m

Simplify:

16. $4i^3 - i^5 + 2i^2 - \sqrt{-16}$

17. $3 - 2i^5 - 3i^4 + \sqrt{-4} - i$

18. Convert 400 centimeters per minute to yards per second.

19. Convert 4 cubic miles to cubic kilometers.

Simplify:

20. $\sqrt[5]{3\sqrt[4]{3}}$ **21.** $\sqrt[5]{4\sqrt[4]{2}}$ **22.** $\dfrac{4}{(-27)^{-\frac{2}{3}}}$

23. $\sqrt[4]{xy^2}\sqrt{x^3y}$ **24.** $3\sqrt{\dfrac{2}{7}} + 5\sqrt{\dfrac{7}{2}} - 3\sqrt{126}$

25. Estimate: $\dfrac{(476{,}158 \times 10^{22})(79{,}318{,}642)}{(983{,}704)(514.0 \times 10^{-14})}$

26. Find C.

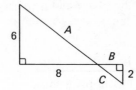

27. How many 1-centimeter squares would it take to cover the solid triangle? Dimensions are in meters.

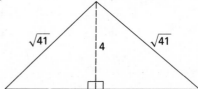

28. Find the equation of the line that passes through $(2, -7)$ and is parallel to $5x + 4y = 7$.

Solve:

29. $-2^0 - 2^2 - 2^2(-2 - 1^0)x - 3x - 7x^0y^0 - 4 = 2$

30. $\dfrac{4x + 5}{3} - \dfrac{x}{7} = 2$

ER 63-20

LESSON 64 *Complex fractions · Complex numbers*

64.A

complex fractions

Any fraction that contains more than one fraction line is called a complex fraction. Thus, this expression

$$\dfrac{\dfrac{1}{b} + x}{\dfrac{a}{b}}$$

is a complex fraction. We simplify by adding in the numerator and then by multiplying the denominator and the numerator by $\frac{b}{a}$.

$$\cfrac{\dfrac{1}{b} + x}{\dfrac{a}{b}} \quad \longrightarrow \quad \cfrac{\dfrac{1 + bx}{b}}{\dfrac{a}{b}} \quad \longrightarrow \quad \cfrac{\dfrac{1 + bx}{b} \cdot \left(\dfrac{b}{a}\right)}{\dfrac{a}{b} \cdot \left(\dfrac{b}{a}\right)} \quad \longrightarrow \quad \dfrac{1 + bx}{a}$$

The next example is just a little more involved.[†]

example 64.A.1   Write the following as a simple fraction: $a + \cfrac{1}{\dfrac{1}{b} + x}$

solution   We begin by simplifying the second term.

$$\cfrac{1}{\dfrac{1}{b} + x} \quad \longrightarrow \quad \cfrac{1}{\dfrac{1 + bx}{b}} \quad \longrightarrow \quad \cfrac{1 \cdot \dfrac{b}{1 + bx}}{\dfrac{1 + bx}{b} \cdot \dfrac{b}{1 + bx}} \quad \longrightarrow \quad \dfrac{b}{1 + bx}$$

Now we add the first term and the second term.

$$a + \dfrac{b}{1 + bx} \qquad\qquad \text{two terms}$$

$$\dfrac{}{1 + bx} + \dfrac{}{1 + bx} \qquad\qquad \text{new denominators}$$

$$\dfrac{a(1 + bx)}{1 + bx} + \dfrac{b}{1 + bx} = \dfrac{a + abx + b}{1 + bx} \qquad\qquad \text{added}$$

example 64.A.2   Write $\dfrac{a}{x} + \cfrac{4}{1 + \dfrac{b}{x}}$ as a simple fraction.

solution   First we simplify the second term.

$$\dfrac{a}{x} + \cfrac{4}{\dfrac{x + b}{x}} \quad \longrightarrow \quad \dfrac{a}{x} + \dfrac{4x}{x + b}$$

Now we use $x(x + b)$ as a common denominator and add.

$$\dfrac{}{x(x + b)} + \dfrac{}{x(x + b)} \qquad\qquad \text{new denominators}$$

$$\dfrac{a(x + b)}{x(x + b)} + \dfrac{4x^2}{x(x + b)} \qquad\qquad \text{new numerators}$$

$$\dfrac{ax + ab + 4x^2}{x(x + b)} \qquad\qquad \text{added}$$

[†] The arrows indicate successive steps in the simplification process. Equal signs could have been used instead of the arrows.

64.B
**complex
numbers**

We remember that we say that any number that can be graphed on the number line is a real number.

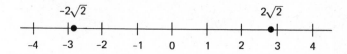

On this line we have graphed $2\sqrt{2}$ and $-2\sqrt{2}$, so both of these numbers are real numbers. Square roots of negative numbers or numbers that have i as a factor such as

$$2i \qquad \sqrt{-3} \qquad \sqrt{-1} \qquad \sqrt{-142} \qquad \sqrt{7}\,i$$

cannot be graphed on the number line. We say that these numbers are imaginary numbers to distinguish them from the real numbers. If a number has a real part and an imaginary part such as

$$-4 + 2i \qquad 5 - 3i \qquad 7\sqrt{2} + 6i$$

we call the numbers complex numbers. **If the real part is written first and the imaginary part is written second, we say that the complex number is written in _standard form_. We** define standard form to be the form

$$a + bi$$

where both a and b are real numbers. Since zero is a real number, then

$$0 + 2i \qquad \text{which is} \qquad 2i$$

and
$$7 - 0i \qquad \text{which is} \qquad 7$$

are complex numbers written in standard form. Thus, every real number is a complex number whose imaginary part is zero, and every imaginary number is a complex number whose real part is zero. For this reason, we can say that all of the following real numbers and imaginary numbers are also complex numbers.

$$-7 \qquad 2i \qquad -3\sqrt{2} \qquad -\sqrt{2}\,i \qquad 5 \qquad -\frac{3}{4}i$$

example 64.B.1 Simplify $\sqrt{-2}\sqrt{-3}$.

solution **We remember that when both radicands are negative, the radicands cannot be multiplied to change the form of the expression. First we must use Euler's notation.**

$$\sqrt{2}\,i \cdot \sqrt{3}\,i \qquad \text{which is} \qquad \sqrt{6}\,i^2$$

But i^2 equals -1, so our answer is

$$-\sqrt{6}$$

example 64.B.2 Simplify $4i^3 - 2i^4 + 2\sqrt{-9} + \sqrt{-3}\sqrt{-3}$.

solution Let's use two steps.

$$4(ii)i - 2(ii)(ii) + 2\sqrt{9}\,i + \sqrt{3}\,i\sqrt{3}\,i$$

Now we remember that $i^2 = -1$ and we write

$$-4i - 2 + 6i - 3$$

Now we add like parts and write the result in standard form.

$$-5 + 2i$$

example 64.B.3 Multiply $(4 + 2i)(3 - 2i)$.

solution We will use the vertical format.

$$
\begin{array}{r}
4 + 2i \\
3 - 2i \\
\hline
12 + 6i \\
-8i - 4i^2 \\
\hline
12 - 2i - 4i^2
\end{array}
$$

Now since i^2 equals -1, then $-4i^2$ equals $+4$, so our answer is

$$16 - 2i$$

example 64.B.4 Multiply $(5i - 2)(4 + 3i)$.

solution This time we use the horizontal format and get

$$(5i - 2)(4 + 3i) = 20i + 15i^2 - 8 - 6i$$

Now we simplify and remember that $15i^2$ equals -15, so we get

$$-23 + 14i$$

problem set 64

1. The total weight of the carbon varied inversely as the total weight of the fluorine. The carbon weighed 300 grams when the fluorine weighed only 2 grams. What was the weight of the carbon when the fluorine weighed .5 gram?

2. An amount of an ideal gas was placed in a container whose volume was constant. The pressure was found to be 700 torr and the temperature was 400 kelvins. If the pressure was increased to 2800 torr, what was the final temperature?

3. Two containers are on the shelf. The first one contains a 30% iodine solution and the other contains an 80% iodine solution. How much of each should be used to get 50 liters of a solution that is 40% iodine?

4. The ride in was quick, as the speed was 400 kilometers per hour. The ride out at 100 kilometers per hour was much more relaxing. If the total time in and out was 40 hours, what was the distance in?

5. The chemical formula for methyl iodide is CH_3I. What percent of the total weight of this compound is iodine (I)? (C, 12; H, 1; I, 127)

Simplify:

*6. $a + \dfrac{1}{\dfrac{1}{b} + x}$

*7. $\dfrac{a}{x} + \dfrac{4}{1 + \dfrac{b}{x}}$

8. $\dfrac{m}{c} + \dfrac{8}{2 + \dfrac{m}{c}}$

*9. $4i^3 - 2i^4 + 2\sqrt{-9} + \sqrt{-3}\sqrt{-3}$

10. $\sqrt{-4} + \sqrt{-2}\sqrt{-2} - 4i^3$

11. $i^4 - 3i^2 - 2\sqrt{-2}\sqrt{-3}$

12. $2\sqrt{-9} - 3i^4 + 2\sqrt{3}\sqrt{-3} + i$

*13. $(4 + 2i)(3 - 2i)$

*14. $(5i - 2)(4 + 3i)$

15. $(2i - 4)(i + 2)$

16. Add: $10\underline{/10°} + 30\underline{/150°}$

17. Write $-4R - 6U$ in polar form.

18. Solve: $\begin{cases} \frac{3}{2}x - \frac{1}{5}y = 28 \\ .02x + .4y = 4.4 \end{cases}$

Solve by completing the square:

19. $3x^2 = -2x - 5$

20. $-3x + 2x^2 = -7$

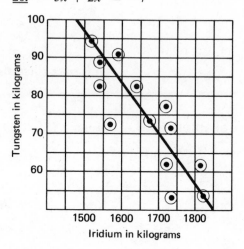

21. The data points shown came from an experiment that involved tungsten (W) and iridium (Ir). Write the equation that expresses tungsten as a function of iridium: $W = mIr + b$.

22. Find the surface area of the right circular cylinder. Dimensions are in centimeters.

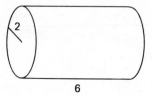

23. Convert 400 cubic centimeters per second to cubic inches per hour.

Simplify:

24. $\sqrt[3]{9\sqrt[3]{3}}$

25. $2\sqrt{\frac{7}{5}} - 3\sqrt{\frac{5}{7}} + 2\sqrt{140}$

26. $\frac{a(b + c)}{x} - m = \frac{d}{f}$; find x

27. $\frac{a(b + c)}{x} - m = \frac{d}{f}$; find c

28. Solve by graphing and then find an exact solution by using substitution or elimination.
$\begin{cases} x - 3y = -6 \\ 2x + 5y = 15 \end{cases}$

29. Solve $56x = -15x^2 - x^3$ by factoring.

30. Estimate: $\dfrac{(146{,}842 \times 10^2)\,(.0007892)}{(96{,}478 \times 10^{14})(.000712 \times 10^{42})}$

ER 64-21

LESSON 65   *Advanced substitution*

65.A
advanced substitution

We have used substitution to solve systems of equations that are derived from uniform motion word problems. In the systems studied thus far, such as

$$R_W T_W + 210 = R_R T_R \qquad R_W = 4 \qquad R_R = 6 \qquad T_W + T_R = 5$$

the values of two of the variables have always been given. In this system of equations we have been told that R_W equals 4 and that R_R equals 6. Now we will investigate the solution of a system of four equations in which none of the values of the variables is given. In a later lesson, we will find that these equations will permit the solution of a new type of uniform motion word problem.

      In the problem sets the problems of this type will be exactly like the two problems that we will work here. The numbers and the subscripts will change, but otherwise the problems will be the same.

example 65.A.1   Find the values of all four variables in this system of equations.

      (a) $R_W T_W = 6$ (b) $R_B T_B = 6$ (c) $R_B = 3R_W$ (d) $T_B = 2 - T_W$

solution   We will use either equation (a) or equation (b) as our base equation and will substitute the other three equations into the base equation. Equation (c) determines which of the first two equations will be the base equation. Equation (c) has two forms:

      (c) $R_B = 3R_W$ (c′) $\dfrac{R_B}{3} = R_W$

If equation (b) is used as the base equation, we would use form (c) and replace R_B with $3R_W$. If equation (a) is used as the base equation, we would use form (c′) and replace R_W with $R_B/3$. Either way will work, but the second approach introduces a fraction, which complicates the solution. Thus, we will use equation (b) as the base equation and replace R_B with $3R_W$ and T_B with $2 - T_W$.

      (b) $R_B T_B = 6$ base equation

      (d) $(3R_W)(2 - T_W) = 6$ substituted

      (e) $6R_W - 3R_W T_W = 6$ multiplied

Now we have used three of the four equations.

$$R_W T_W = 6 \qquad \cancel{R_B T_B = 6} \qquad \cancel{R_B = 3R_W} \qquad \cancel{T_B = 2 - T_W}$$

Next we use the remaining equation by replacing $R_W T_W$ in equation (e) with 6.

$$6R_W - 3 \cdot 6 = 6 \qquad \text{substituted}$$

$$6R_W - 18 = 6 \qquad \text{multiplied}$$

$$6R_W = 24 \qquad \text{added} + 18$$

$$\mathbf{R_W = 4} \qquad \text{divided}$$

We did not decide to solve for R_W when we began the solution. We just began substituting and simplifying, and it turned out that this approach produced the value of R_W. Now we will use this value of R_W to help us find the other three variables.

$$R_B = 3R_W \qquad \text{equation (c)}$$

$$\mathbf{R_B = 3(4) = 12} \qquad \text{substituted}$$

Now we will use $R_W = 4$ and $R_B = 12$ in equations (a) and (b) to find T_W and T_B.

$R_W T_W = 6$	equation (a)		$R_B T_B = 6$	equation (b)
$4T_W = 6$	substituted		$12T_B = 6$	substituted
$T_W = \dfrac{3}{2}$	divided		$T_B = \dfrac{1}{2}$	divided

example 65.A.2   Find the values of all four variables in this system of equations.

$$\text{(a)} \quad R_P T_P = 693 \qquad \text{(b)} \quad R_C T_C = 165 \qquad \text{(c)} \quad R_P = 3R_C \qquad \text{(d)} \quad T_P + T_C = 12$$

solution   This problem is just like the last one except that equation (d) must be rearranged before it can be used. We look at equation (c) and decide to use equation (a) as the base equation.

$$R_P T_P = 693 \qquad \text{base equation}$$
$$(3R_C)(12 - T_C) = 693 \qquad \text{substituted}$$
$$36R_C - 3R_C T_C = 693 \qquad \text{multiplied}$$

Each time we get to this point in one of these problems, we will have a double variable. This time it is $R_C T_C$, and equation (b) tells us that $R_C T_C = 165$. We substitute 165 for $R_C T_C$ and solve.

$$36R_C - 3(165) = 693 \qquad \text{substituted 165}$$
$$36R_C - 495 = 693 \qquad \text{multiplied}$$
$$36R_C = 1188 \qquad \text{added 495}$$
$$R_C = 33 \qquad \text{divided}$$

Again, we note that we did not select the variable R_C in the beginning. We just substituted until we got a solution for one of the four variables. It just happened to be R_C. Now we find R_P.

$$R_P = 3R_C \qquad \text{equation (c)}$$
$$R_P = 3(33) = 99 \qquad \text{substituted}$$

Now we will use $R_P = 99$ and $R_C = 33$ to solve for T_P and T_C.

$R_P T_P = 693$	equation (a)		$R_C T_C = 165$	equation (b)
$99T_P = 693$	substituted		$33T_C = 165$	substituted
$T_P = 7$	divided		$T_C = 5$	divided

problem set 65

1. The weight of the silicon varied directly as the weight of the phosphorus. When the silicon weighed 400 kilograms, the phosphorus weighed only 100 kilograms. What was the weight of the silicon when the phosphorus weighed only 12 kilograms?

2. The temperature of a fixed amount of an ideal gas was held constant at 600 K.[†] The initial pressure and volume were 800 torr and 2 liters. What was the final pressure if the volume was reduced to .1 liter?

[†] If the value of the temperature is constant, it can be eliminated from the equation. The fact that the temperature is 600 K is not relevant.

3. The solution came up to the 500-ml mark on the beaker. If the solution was 84% alcohol, it was necessary to evaporate how much alcohol so that what is left would be only 80% alcohol?

4. The ignorant exceeded the erudite by 400. In fact, the ignorant numbered 100 more than 4 times the number of erudite. How many fell into each category?

5. Four percent of the phosgene combined with other chemicals. If 1920 kilograms did not combine, how much did combine?

Solve:

*6. $R_W T_W = 6,\ R_B T_B = 6,\ R_B = 3R_W,\ T_B = 2 - T_W$

*7. $R_P T_P = 693,\ R_C T_C = 165,\ R_P = 3R_C,\ T_P + T_C = 12$

8. $R_1 T_1 = 120,\ R_2 T_2 = 120,\ R_1 = 2R_2,\ T_1 + T_2 = 6$

Simplify:

9. $x + \dfrac{1}{a + \dfrac{b}{c}}$

10. $\dfrac{4}{c} + \dfrac{1}{a + \dfrac{1}{b}}$

11. $x + \dfrac{a}{1 + \dfrac{1}{a}}$

12. $\sqrt{-4} - \sqrt{-3}\,\sqrt{-3} + 2i^5 - 4$

13. $(5i - 2)(2i - 3)$

14. $(-i - 3)(-2i + 4)$

15. Add: $20\underline{/45^\circ} + 10\underline{/210^\circ}$

16. Write $3R - 5U$ in polar form.

17. Solve: $\begin{cases} \dfrac{2}{7}x - \dfrac{1}{6}y = 1 \\ .3x + .07y = .84 \end{cases}$

Solve by completing the square:

18. $-2 = -3x^2 - 7x$

19. $2x^2 - 4 = -5x$

20. The data points shown came from an experiment that involved potassium (K) and radium (Ra). Write the equation that expresses potassium as a function of radium: $K = mRa + b$.

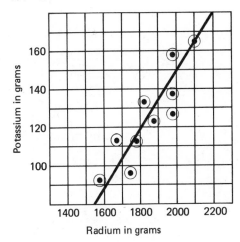

21. Find the surface area of the pyramid. Include the area of the bottom. Dimensions are in feet.

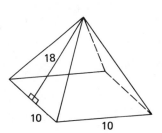

22. Convert 600 cubic feet per hour to cubic inches per minute.

Simplify:

23. $\sqrt{x^2 \sqrt{y^3}}$ **24.** $\sqrt[6]{4\sqrt[5]{2}}$

25. $\sqrt{\dfrac{2}{9}} - 3\sqrt{\dfrac{9}{2}} - 2\sqrt{50}$ **26.** $\dfrac{x}{a(b+c)} - m = \dfrac{d}{f}$; find x

27. $\dfrac{x}{a(b+c)} - m = \dfrac{d}{f}$; find b

28. Simplify: $\dfrac{32 - 12x + x^2}{x^3 + x^2 - 20x} \div \dfrac{x^2 - x - 56}{45x + 14x^2 + x^3}$

29. Estimate: $\dfrac{(47{,}123 \times 10^5)(980)(476)}{(.00134)(576 \times 10^5)}$

30. Add: $\dfrac{4}{x+2} - \dfrac{3}{x^2 - 4}$

ER 65-22

LESSON 66 *Signs of fractions • Cones and spheres*

66.A
signs
of fractions

Every fraction has three signs. If one of the signs is not written it is understood to be a plus sign. One of the signs is in front of the fraction, and the other two are above and below, as shown here.

$$-\frac{+3}{+4}$$

In this case, it is unnecessary to record the plus signs, for the fraction can be written with just one sign as

$$-\frac{3}{4}$$

Any two of the three signs of a fraction may be changed without changing the value of the fraction.

$$\text{(a)}\quad -\frac{+3}{+4} \qquad \text{(b)}\quad +\frac{-3}{+4} \qquad \text{(c)}\quad +\frac{+3}{-4} \qquad \text{(d)}\quad -\frac{-3}{-4}$$

Each of these four notations designates the same number which is

$$-\frac{3}{4}$$

We find that the ability to change signs is often helpful when we add fractions.

example 66.A.1   Add: $\dfrac{1}{x-3} - \dfrac{7a}{-x+3}$

solution   The fractions can be added if we can make the denominator of the second fraction $x - 3$. To do this we must change both signs below and then we must either change the sign on top or the sign in front. We choose to change the sign in front.

$$\frac{1}{x-3} + \frac{7a}{x-3}$$

Now the denominators are the same, so we add and get

$$\frac{1 + 7a}{x - 3}$$

example 66.A.2   Add: $\dfrac{4x + 5}{x - 3} + \dfrac{2x - 3}{3 - x}$

solution   We will change the second fraction by changing all signs below and all signs above. Thus, the + in front of the second fraction remains unchanged.

$$\frac{4x + 5}{x - 3} + \frac{-2x + 3}{x - 3}$$

Now the denominators are the same, so we add the numerators and get

$$\frac{2x + 8}{x - 3}$$

66.B
cones and spheres

The number that tells how many times a diameter can be laid out on the perimeter of a circle is the number pi for which we use the symbol

$$\pi$$

Thus, we can understand the formula for the circumference of a circle because it comes directly from the definition of π.

$$C = \pi D$$

Unfortunately, not all formulas can be understood. The formula for the area of a circle is one of these. We just memorize the formula and use it when we need to calculate the area of a circle. This formula is usually developed in the second semester of college calculus and is

$$\text{Area} = \pi r^2$$

where r is the radius of the circle.

In our investigations of volume and surface area, we have tried to use developments that can be understood and have avoided formulas when possible. Unfortunately, this is not always possible, and sometimes we must use formulas blindly. Necessary formulas can usually be looked up in a book of mathematical tables or an engineer's handbook. We will discuss two shapes whose area and volume require the use of formulas that must be looked up. These are the sphere and the right circular cone. The formulas are given here.

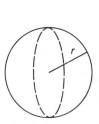

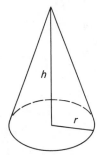

SPHERE   RIGHT CIRCULAR CONE

Surface area $= 4\pi r^2$ Lateral surface area $= \pi r \sqrt{r^2 + h^2}$

Volume $= \dfrac{4}{3}\pi r^3$ Volume $= \dfrac{1}{3}\pi r^2 h$

example 66.B.1 Find (a) the surface area and (b) the volume of a sphere whose radius is 4 ft.

solution We use the formulas given above

$$\text{(a)} \quad A = 4\pi r^2 = (4)(3.14)(4 \text{ ft})^2 = \textbf{200.96 ft}^2$$

$$\text{(b)} \quad V = \frac{4}{3}\pi r^3 = \frac{4}{3}(3.14)(4 \text{ ft})^3 = \textbf{267.95 ft}^3$$

example 66.B.2 Find (a) the lateral surface area and (b) the volume of a right circular cone, where $r = 4$ in and $h = 10$ in.

solution We use the formulas for a cone.

$$\text{(a)} \quad A = \pi r\sqrt{r^2 + h^2} = (3.14)(4)\sqrt{(4)^2 + (10)^2} = \textbf{135.28 in}^2$$

$$\text{(b)} \quad V = \frac{1}{3}\pi r^2 h = \frac{(3.14)(4 \text{ in})^2(10 \text{ in})}{3} = \textbf{167.47 in}^3$$

problem set 66

1. Cobalt varied inversely as the amount of uranium. When there was 5 grams of cobalt, the mass of the uranium was 20 grams. How much cobalt was there when only 2 grams of uranium was present?

2. The initial pressure, volume, and temperature of a quantity of ideal gas was 400 torr, 6 liters, and 200 kelvins. What was the temperature if the pressure was increased to 800 torr and the volume increased to 60 liters?

3. The solution had to be exactly 36% arsenic. Two solutions were available. One was 60% arsenic, and the other was only 20% arsenic. How much of each should be used to get 200 liters of solution that is 36% arsenic?

4. Cleon had a 10-mile head start when Deborah set out in pursuit in a chariot. If Cleon was walking at 4 miles per hour and the speed of the chariot was 6 miles per hour, how long did it take Deborah to catch Cleon?

5. The laboratory assistant stumbled upon a container of methyl bromide, CH_3Br. If the methyl bromide weighed 950 grams, what did the bromine (Br) weigh? (C, 12; H, 1; Br, 80)

Add:

*6. $\dfrac{1}{x-3} - \dfrac{7a}{-x+3}$

*7. $\dfrac{4x+5}{x-3} + \dfrac{2x-3}{3-x}$

8. $\dfrac{4}{x^2-4} - \dfrac{2x}{x-2}$

*9. Find the surface area of a sphere whose radius is 4 ft.

*10. Find the lateral surface area of a cone whose radius is 4 inches and whose height is 10 inches.

Solve:

11. $R_T T_T = 300$, $R_P T_P = 1200$, $R_P = 8R_T$, $T_T = T_P + 3$

12. $R_P T_P = 624$, $R_T T_T = 364$, $T_P = T_T - 4$, $R_P = 4R_T$

Simplify:

13. $ax + \dfrac{1}{a + \dfrac{1}{x}}$

14. $\dfrac{m}{x} + \dfrac{1}{x + \dfrac{1}{x}}$

15. $-\sqrt{-3}\sqrt{-2} + \sqrt{-4} - \sqrt{-3}\sqrt{-3} - 2i^3$

16. $(4i - 2)(3 + 5i)$ **17.** Add: $4\underline{/28°} + 10\underline{/35°}$

18. Write $8R + 4U$ in polar form.   **19.** Solve: $\begin{cases} \dfrac{3}{8}x - \dfrac{1}{4}y = -2 \\ .012x + .02y = .496 \end{cases}$

Solve by completing the square:

20. $-3 = -2x^2 - 6x$ **21.** $5x^2 - 4 = -5x$

22. Find the surface area of the right circular cylinder in square centimeters. Dimensions are in meters.

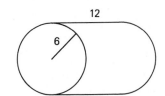

23. The data points shown came from an experiment that involved calcium (Ca) and magnesium (Mg). Write the equation that expresses magnesium as a function of calcium: $Mg = mCa + b$.

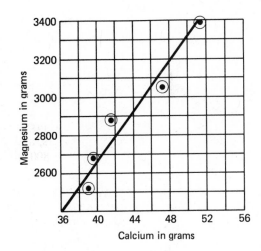

24. Convert 10 cubic inches per hour to cubic centimeters per minute.

Simplify:

25. $\sqrt[3]{x^{\frac{1}{2}}\sqrt{x^2}}$ **26.** $\sqrt[3]{x^{\frac{1}{5}}\sqrt[4]{x}}$

27. $3\sqrt{\dfrac{2}{5}} + 7\sqrt{\dfrac{5}{2}} - 6\sqrt{40}$ **28.** $\dfrac{px - y}{m} - c = \dfrac{k}{d}$; find y

29. $\dfrac{m}{px - y} + \dfrac{k}{d} = -c$; find x

30. Estimate: $\dfrac{(40{,}213 \times 10^5)(748{,}609 \times 10^{-30})}{(.164289)(506{,}217 \times 10^2)}$

ER 66-23

LESSON 67   *Radical denominators*

67.A
radical denominators

When a pair of two-part expressions are the same in all respects except that the signs in the middle are different, we say that each of the expressions is the **conjugate** of the other expression. Thus,

$$-4 + \sqrt{3} \quad \text{is the conjugate of} \quad -4 - \sqrt{3}$$
$$-4 - \sqrt{3} \quad \text{is the conjugate of} \quad -4 + \sqrt{3}$$
$$-2 + bx \quad \text{is the conjugate of} \quad -2 - bx$$
$$-2 - bx \quad \text{is the conjugate of} \quad -2 + bx$$
$$a + b \quad \text{is the conjugate of} \quad a - b$$
$$a - b \quad \text{is the conjugate of} \quad a + b$$

We have studied the products of conjugates and noted that the results are always the difference of two squares.

$$
\begin{array}{ccc}
(a) & (b) & (c) \\
\end{array}
$$

$$
\begin{array}{ccc}
-4 + \sqrt{3} & a + b & -2 + bx \\
\underline{-4 - \sqrt{3}} & \underline{a - b} & \underline{-2 - bx} \\
16 - 4\sqrt{3} & a^2 + ab & 4 - 2bx \\
\underline{\quad + 4\sqrt{3} - 3} & \underline{\quad - ab - b^2} & \underline{\quad + 2bx - b^2x^2} \\
16 \qquad - 3 & a^2 \qquad - b^2 & 4 \qquad - b^2x^2 \\
\end{array}
$$

In this lesson we are interested in expressions that contain square root radicals such as (a) above. We note that the product of the conjugates is $16 - 3$, which equals 13, a whole number. We can use this observation to help us simplify expressions such as the following:

$$\frac{1}{-4 + \sqrt{3}}$$

We can eliminate the radical in the denominator by multiplying above and below by the conjugate of the denominator. Again, we find another use for the most important rule in algebra, the denominator-numerator–same-quantity theorem, as we can use it to simplify this expression. **We say that an expression containing square root radicals is in simplified form when no radicand has a perfect square factor and when no radicals are in the denominator.**

example 67.A.1   Simplify: $\dfrac{1}{-4 + \sqrt{3}}$

solution   We can eliminate the radical in the denominator if we multiply above and below by $-4 - \sqrt{3}$, which is the conjugate of $-4 + \sqrt{3}$.

$$\frac{1}{-4 + \sqrt{3}} \cdot \frac{(-4 - \sqrt{3})}{(-4 - \sqrt{3})} = \frac{-4 - \sqrt{3}}{16 + 4\sqrt{3} - 4\sqrt{3} - 3} = \frac{-4 - \sqrt{3}}{13}$$

Note that we were unable to get rid of the radical expression. All that we could do was to eliminate radical expressions from the denominator.

example 67.A.2   Simplify: $\dfrac{3}{2\sqrt{3} + \sqrt{2}}$

solution   We simplify by multiplying above and below by the conjugate of the denominator.

$$\frac{3}{2\sqrt{3} + \sqrt{2}} \cdot \frac{(2\sqrt{3} - \sqrt{2})}{(2\sqrt{3} - \sqrt{2})} = \frac{6\sqrt{3} - 3\sqrt{2}}{12 - 2} = \frac{6\sqrt{3} - 3\sqrt{2}}{10}$$

problem set 67

1. The number of Danish invaders repulsed varied directly as the number of Danes who attacked. If 12,400 attacked and 2000 were repulsed, how many attacked if 3000 were repulsed?

2. The temperature of a quantity of ideal gas was held constant at 700 K. If the pressure of 800 torr was increased to 1200 torr, what was the final volume if the initial volume was 300 cubic centimeters?

3. The original mixture weighed 400 pounds and was 20% fertilizer. How much of another mixture of 80% fertilizer should be added so that the result would be 32% fertilizer?

4. There were 75 more nurses than doctors. In fact, 8 times the number of nurses exceeded 10 times the number of doctors by 140. How many of each were there?

5. When the reaction was complete, the researchers found that 40 percent of the iridium had not reacted. If 240 grams had reacted, how much had not reacted?

Simplify:

*6. $\dfrac{1}{-4 + \sqrt{3}}$

*7. $\dfrac{3}{2\sqrt{3} + \sqrt{2}}$

8. $\dfrac{2}{3\sqrt{5} - 3}$

Add:

9. $\dfrac{4x + 2}{x - 2} - \dfrac{3}{2 - x}$

10. $\dfrac{4}{x^2 - 9} + \dfrac{2}{x - 3}$

11. Solve: $R_P T_P = 1062$, $R_T T_T = 295$, $T_P = T_T - 2$, $R_P = 6R_T$

Simplify:

12. $4x + \dfrac{a}{x + \dfrac{a}{b}}$

13. $a + \dfrac{x}{m + \dfrac{1}{x}}$

14. $(5 - i)(6 + 2i)$

15. $\sqrt{-4} - \sqrt{-9} + \sqrt{-2}\sqrt{2} - 4i^2$

16. Add: $10\underline{/217°} + 8\underline{/227°}$

17. Write $-4R - 3U$ in polar form.

18. Solve: $\begin{cases} \dfrac{2}{7}x - \dfrac{2}{5}y = 0 \\ .2x - .04y = 2.4 \end{cases}$

Solve by completing the square:

19. $-x = -1 - 4x^2$

20. $3x^2 + 5 = -2x$

21. The figure is the base of a container 5 feet high. How many 1-inch sugar cubes will the container hold? Dimensions are in feet.

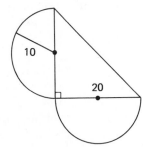

22. Convert 400 cubic centimeters per hour to cubic inches per minute.

23. The data points shown come from an experiment that involved silver (Ag) and gold (Au). Write the equation that expresses silver as a function of gold: $Ag = mAu + b$.

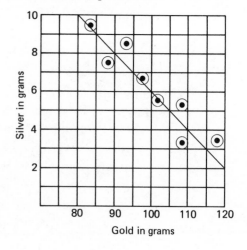

Simplify:

24. $\sqrt{2\sqrt[6]{4}}$

25. $\sqrt[5]{x^2yp}\,\sqrt[3]{xy^2}$

26. $4\sqrt{\dfrac{3}{11}} + 2\sqrt{\dfrac{11}{3}} - 2\sqrt{297}$

27. $\dfrac{a(b+c)}{x} - \dfrac{m}{y} = p$; find x

28. $\dfrac{x}{a(b+c)} - \dfrac{y}{m} = p$; find c

29. Find the equation of the line that passes through $(-2, 4)$ and $(5, 7)$.

30. Find the distance between $(-2, 4)$ and $(5, 7)$.

ER 67-24

If the enrichment lessons on the scientific calculator are to be included, Enrichment Lesson 3 on roots and exponentials should be included at this point.

LESSON 68 *Gas law problems*

68.A

gas law problems In Lesson 57 we learned that the gas law for a fixed amount of an ideal gas can be written as

$$\frac{P_1V_1}{T_1} = \frac{P_2V_2}{T_2}$$

We have been substituting numbers in these equations and then solving. Many people believe it is a better procedure to solve the equation for the desired variable and then substitute. We will do that in the next two example problems. Also we will use values for the variables such as $.0004 \times 10^{-16}$ so that we can get practice in scientific notation as well as in rearranging equations.

example 68.A.1   A quantity of an ideal gas had initial values of pressure, volume and temperature of 40×10^4 atmospheres, $.0003 \times 10^{-6}$ cm^3, and 700×10^4 K. Find the final temperature if the final pressure was $.0004 \times 10^{15}$ atmospheres and the final volume was $.015 \times 10^{-14}$ cm^3. Begin by solving the equation for T_2.

solution   We need to solve the ideal gas law for T_2 in terms of the other variables. We begin by writing the ideal gas law.

$$\frac{P_1 V_1}{T_1} = \frac{P_2 V_2}{T_2}$$

Next we solve for T_2. **As always in a fractional equation, our first step is to eliminate the denominators.** We do this by multiplying both sides by $T_1 T_2$.

$$\cancel{T_1} T_2 \frac{P_1 V_1}{\cancel{T_1}} = \frac{P_2 V_2}{\cancel{T_2}} T_1 \cancel{T_2} \quad \longrightarrow \quad T_2 P_1 V_1 = P_2 V_2 T_1$$

Now we complete our solution for T_2 by dividing both sides by $P_1 V_1$.

$$\frac{T_2 \cancel{P_1 V_1}}{\cancel{P_1 V_1}} = \frac{P_2 V_2 T_1}{P_1 V_1} \quad \longrightarrow \quad T_2 = \frac{P_2 V_2 T_1}{P_1 V_1}$$

We finish by inserting the given values of P_2, V_2, T_1, P_1, and V_1 and simplifying.

$$T_2 = \frac{(.0004 \times 10^{15})(.015 \times 10^{-14})(700 \times 10^4)}{(40 \times 10^4)(.0003 \times 10^{-6})}$$

$$= \frac{(4 \times 10^{11})(15 \times 10^{-17})(7 \times 10^6)}{(4 \times 10^5)(3 \times 10^{-10})}$$

$$= 35 \times 10^5 = 3.5 \times 10^6$$

Thus the final temperature is 3.5×10^6. **The temperature we used for T_1 was in kelvins so the temperature for T_2 will also be in kelvins. The same units are always used throughout a problem.**

$$T_2 = 3.5 \times 10^6 \text{ K}$$

example 68.A.2   A quantity of an ideal gas had initial values of $.003 \times 10^{14}$ lb/in^2, $.007$ cm^3 and 7000 K. Find P_2 if the final volume was 3×10^2 cm^3, and the final temperature was $.003 \times 10^7$ K. Solve the equation for P_2 as the first step.

solution   We write the ideal gas law equation.

$$\frac{P_1 V_1}{T_1} = \frac{P_2 V_2}{T_2}$$

Now, as always with fractional equations, we begin by eliminating the denominators.

$$\cancel{T_1} T_2 \frac{P_1 V_1}{\cancel{T_1}} = \frac{P_2 V_2}{\cancel{T_2}} T_1 \cancel{T_2} \quad \longrightarrow \quad T_2 P_1 V_1 = P_2 V_2 T_1$$

Now we solve for P_2 by dividing by $V_2 T_1$.

$$\frac{T_2 P_1 V_1}{V_2 T_1} = \frac{P_2 \cancel{V_2 T_1}}{\cancel{V_2 T_1}} \quad \longrightarrow \quad \frac{T_2 P_1 V_1}{V_2 T_1} = P_2$$

We finish by inserting the numbers and simplifying.

$$\frac{(.003 \times 10^7)(.003 \times 10^{14})(.007)}{(3 \times 10^2)(7000)} = \frac{(3 \times 10^4)(3 \times 10^{11})(7 \times 10^{-3})}{(3 \times 10^2)(7 \times 10^3)}$$

$$= 3 \times 10^7$$

The units of P_2 are pounds per square inch because P_1 was in pounds per square inch. Thus,

$$P_2 = 3 \times 10^7 \text{ lb/in}^2$$

problem set 68

*1. A quantity of an ideal gas had initial values of pressure, volume and temperature of 40×10^4 atmospheres, $.0003 \times 10^{-6}$ cm³, and 700×10^4 K. Find the final value of temperature if the final pressure was $.0004 \times 10^{15}$ atmospheres and the final volume was $.015 \times 10^{-14}$ cm³. Begin by solving the equation for T_2.

*2. A quantity of an ideal gas had initial values of $.003 \times 10^{14}$ lb/in², $.007$ cm³, and 7000 K. Find P_2 if the final volume was 3×10^2 cm³ and the final temperature was $.003 \times 10^7$ kelvins. Solve the equation for P_2 as the first step.

3. The borax varied inversely as the tungsten. If 400 tons of borax went with 5 tons of tungsten, how much borax went with 25 tons of tungsten?

4. Two solutions were available. One was 80% alcohol and the other was 40% alcohol. How much of each should be used to get 2000 ml of a solution that is 64% alcohol?

5. The bus headed north at noon at 50 miles per hour. At 2 p.m. the train headed north from the same station at 70 miles per hour. What time was it when the train got within 40 miles of the bus?

Add:

6. $\dfrac{1}{x - 5} - \dfrac{2x - 3}{5 - x}$

7. $\dfrac{2x + 3}{x - 2} + \dfrac{2x}{2 - x} - \dfrac{3}{-x + 2}$

8. $\dfrac{3x + 5}{x - 5} + \dfrac{2}{x^2 - 25}$

Simplify:

9. $\dfrac{4}{3 - \sqrt{2}}$

10. $\dfrac{2}{5 - 3\sqrt{2}}$

11. $\dfrac{2}{3 - 2\sqrt{8}}$

12. Solve: $R_g T_g = 171$, $R_R T_R = 171$, $R_R = 3R_g$, $T_R = T_g - 6$

Simplify:

13. $m + \dfrac{1}{m + \dfrac{1}{m}}$

14. $x + \dfrac{a}{x + \dfrac{1}{a}}$

15. $(4i - 2)(2i - 4)$

16. $-\sqrt{-2}\sqrt{-2} + 2i^3 - i^2$

17. Add: $20\underline{/30°} + 60\underline{/210°}$

18. Write $2R - 3U$ in polar form.

19. Solve: $\begin{cases} \dfrac{2}{3}x + \dfrac{2}{5}y = 28 \\ -.05x - .2y = -5.5 \end{cases}$

Solve by completing the square:

<u>20.</u> $3x^2 + 8 = 5x$

21. $3x^2 + 8 = -5x$

22. Find the surface area of this box in square feet. Dimensions are in feet.

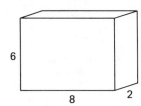

6

8   2

23. Convert 4 cubic feet per minute to cubic inches per hour.

24. The data points shown come from an experiment that involved chromium (Cr) and vanadium (V). Write the equation that expresses chromium as a function of vanadium: $Cr = mV + b$.

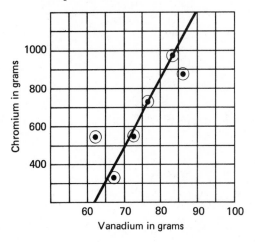

Simplify:

25. $\sqrt{27\sqrt[4]{3}}$

26. $\sqrt[3]{xm^5}\ \sqrt[4]{xm^2}$

27. $2\sqrt{\dfrac{9}{2}} + 5\sqrt{\dfrac{2}{9}} - 5\sqrt{50}$

28. $\dfrac{a(x + y)}{m} - \dfrac{c}{z} = k$; find x

29. Solve: $\dfrac{x + 2}{5} - \dfrac{3x - 3}{4} = 2$

ER 68-25, 68-26, 68-27

LESSON 69 *Advanced abstract equations*

69.A
advanced abstract equations

Sometimes it is necessary to use the distributive property as the first step in solving a fractional equation.

example 69.A.1   Solve $x = p\left(\dfrac{1}{a} + \dfrac{m}{b}\right)$ for b.

solution   We begin by multiplying to eliminate the parentheses. If we do this, we get

$$x = \frac{p}{a} + \frac{mp}{b}$$

Next we want to eliminate the denominators. We can do this if we multiply every numerator by the least common multiple of the denominators.

$$x \cdot ab = \frac{p}{a} \cdot ab + \frac{mp}{b} \cdot ab \quad \longrightarrow \quad xab = pb + mpa$$

Now we place all terms that have a factor of b on one side, then factor. Then we divide.

$$xab - pb = mpa \qquad \text{added } -pb$$

$$b(xa - p) = mpa \qquad \text{factored}$$

$$b = \frac{mpa}{xa - p} \qquad \text{divided}$$

example 69.A.2 $xp = m\left(\dfrac{b}{1 + c} + \dfrac{a}{p}\right)$; find c

solution   We begin by multiplying so we can eliminate the parentheses.

$$xp = \frac{mb}{1 + c} + \frac{ma}{p}$$

Now to eliminate the denominators, we multiply every numerator by the LCM of the denominators,

$$p(1 + c)xp = \frac{mbp(1 + c)}{1 + c} + \frac{ma}{p}p(1 + c)$$

and now we cancel the denominators and multiply to get

$$xp^2 + xp^2c = mbp + ma + mac$$

Next we place all terms with a c factor on one side, factor out the c, and divide.

$$xp^2c - mac = mbp + ma - xp^2 \qquad \text{added } -xp^2 - mac$$

$$c(xp^2 - ma) = mbp + ma - xp^2 \qquad \text{factored out } c$$

$$c = \frac{mbp + ma - xp^2}{xp^2 - ma} \qquad \text{divided}$$

problem set 69

1. The initial values of pressure, volume and temperature of a quantity of an ideal gas were $.001 \times 10^{-13}$ torr, $.04 \times 10^{14}$ liters, and 4×10^3 K. What was the final temperature if the final pressure and volume were $.04 \times 10^5$ torr and 500 liters? Solve for T_2 as the first step.

2. The beaker contained 400 ml of a solution that was 20% alcohol. How many milliliters of a 50% alcohol solution must be added so that the result will be 26% alcohol?

3. Tourist tickets for the flight from Rhinelander to Hodag were $50 each while first class tickets were $100 each. If 60 people paid $5000, how many flew as tourists and how many flew first class?

4. The container contained 680 grams of the compound $CaSO_4$. What was the weight of the sulfur (S) in the compound? (Ca, 40; S, 32; O, 16)

5. Find three consecutive even integers such that 4 times the product of the first and the third is 28 greater than the product of -10 and the sum of the second and the third.

*6. $x = p\left(\dfrac{1}{a} + \dfrac{m}{b}\right)$; find b *7. $xp = m\left(\dfrac{b}{1 + c} + \dfrac{a}{p}\right)$; find c

Add:

8. $\dfrac{3x - 2}{x - 2} - \dfrac{3x}{2 - x}$ 9. $\dfrac{4}{x^2 + 8x + 12} + \dfrac{4x - 5}{x + 2}$

Simplify:

10. $\dfrac{2}{\sqrt{2} - 4}$

11. $\dfrac{2}{3\sqrt{12} - 2}$

12. $\dfrac{2}{2\sqrt{3} - 2}$

13. Solve: $R_B T_B = 65$, $R_X T_X = 104$, $R_X = 2R_B$, $T_X = T_B - 1$

Simplify:

<u>**14.**</u> $xy + \dfrac{a}{1 + \dfrac{a}{b}}$

15. $\dfrac{m}{y} + \dfrac{x}{a + \dfrac{1}{y}}$

16. $(2 - 3i)(5 - 6i)$

17. $-\sqrt{-4} - \sqrt{-2}\sqrt{-3} + \sqrt{-9}$

18. Add: $4\underline{/340°} + 6\underline{/320°}$

19. Write $-4R + 5U$ in polar form.

20. Solve: $\begin{cases} \dfrac{1}{5}x - \dfrac{1}{4}y = -2 \\ .07x + .3y = 5.5 \end{cases}$

Solve by completing the square:

21. $3x^2 + 1 = 4x$

<u>**22.**</u> $-4x + 7 = -3x^2$

23. Find the surface area of this right circular cylinder. Dimensions are in inches.

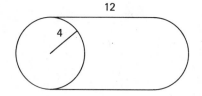

24. The data points shown come from an experiment that involved cobalt (Co) and nickel (Ni). Write the equation that expresses cobalt as a function of nickel: $Co = mNi + b$.

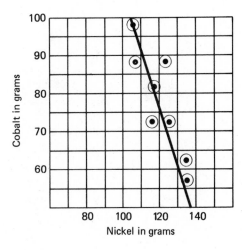

Simplify:

<u>**25.**</u> $\sqrt{8\sqrt[4]{2}}$

26. $\sqrt[5]{x^4 y^3} \, \sqrt[4]{x^2 y}$

27. $\sqrt{\dfrac{7}{2}} + 2\sqrt{\dfrac{2}{7}} - 2\sqrt{126}$

28. Divide $4x^3 - 2$ by $x - 1$.

29. Find the equation of the line that passes through $(-2, 5)$ and is perpendicular to the line $4x + 3y = 5$.

30. Solve: $\dfrac{4x - 2}{5} - \dfrac{3x - 2}{4} = 10$

ER 69-28, 69-29, 69-30

LESSON 70 *Quadratic formula*

70.A

quadratic
formula

We have found that some quadratic equations can be solved by factoring and then using the zero factor theorem. We will use this procedure to solve $x^2 + 2x - 15 = 0$. First we factor and get

$$(x - 3)(x + 5) = 0$$

Now, from the zero factor theorem we know that if the product of two factors equals zero, one of the factors must be zero. So

$$\text{If } x - 3 = 0 \quad \text{If } x + 5 = 0$$
$$x = 3 \qquad\qquad x = -5$$

We have also found that quadratic equations that cannot be factored, such as $x^2 + 3x - 3 = 0$, can be rearranged into the form

$$\left(x + \frac{3}{2}\right)^2 = \frac{21}{4}$$

This form of the equation can be solved by taking the square root of both sides of the equation.

$$\left(x + \frac{3}{2}\right)^2 = \frac{21}{4} \qquad\qquad \text{equation}$$

$$x + \frac{3}{2} = \pm\sqrt{\frac{21}{4}} \qquad\qquad \text{square root of both sides}$$

$$x = -\frac{3}{2} \pm \frac{\sqrt{21}}{2} \qquad\qquad \text{simplified}$$

This method is called *completing the square* and can be used to solve any quadratic equation. There is a quicker method, however, that we can use. We can complete the square on the general form of the quadratic equation and derive a formula whose use will give us the same answer. We begin this derivation by writing a general quadratic equation that uses the letters a, b, and c as the constants.

$$ax^2 + bx + c = 0$$

Next we give x^2 a unity coefficient by dividing every term by a and we get

$$x^2 + \frac{b}{a}x + \frac{c}{a} = 0$$

Now we move $\dfrac{c}{a}$ to the right side and use parentheses.

$$\left(x^2 + \frac{b}{a}x + \qquad\right) = \qquad -\frac{c}{a}$$

Note that we placed the $-\dfrac{c}{a}$ well to the right of the equals sign. Now we multiply $\dfrac{b}{a}$ by $\frac{1}{2}$ and square the result.

$$\left(\frac{b}{a} \cdot \frac{1}{2}\right)^2 = \frac{b^2}{4a^2}$$

Next add $\frac{b^2}{4a^2}$ inside the parentheses and also to the other side of the equation. On the right we are careful to place $\frac{b^2}{4a^2}$ in front of $-\frac{c}{a}$.

$$\left(x^2 + \frac{b}{a}x + \frac{b^2}{4a^2}\right) = \frac{b^2}{4a^2} - \frac{c}{a}$$

Next we write the parentheses term as a squared term and combine $\frac{b^2}{4a^2}$ and $-\frac{c}{a}$.

$$\left(x + \frac{b}{2a}\right)^2 = \frac{b^2 - 4ac}{4a^2}$$

Finally, we take the square root of both sides and then solve for x.

$$x + \frac{b}{2a} = \pm\sqrt{\frac{b^2 - 4ac}{4a^2}} \qquad \text{took square roots}$$

$$x = -\frac{b}{2a} \pm \frac{\sqrt{b^2 - 4ac}}{2a} \qquad \text{solved for } x$$

$$x = \frac{-b \pm \sqrt{b^2 - 4ac}}{2a} \qquad \text{added}$$

This result is called the quadratic formula and should be memorized. It will be used in many higher mathematics courses.

The derivation of the quadratic formula will be required in future problem sets. This derivation requires only simple algebraic manipulations, and the requirement that a student be able to perform this derivation is not unreasonable.

example 70.A.1   Use the quadratic formula to find the roots of the equation $3x^2 - 2x + 5 = 0$.

solution   The formula is $x = \dfrac{-b \pm \sqrt{b^2 - 4ac}}{2a}$

If we write the given equation just below the general quadratic equation

$$ax^2 + bx + c = 0 \qquad \text{general equation}$$
$$3x^2 - 2x + 5 = 0 \qquad \text{given equation}$$

We note the following correspondences between the equations

$$a = 3 \qquad b = -2 \qquad c = 5$$

If we use these numbers for a, b, and c in the quadratic formula, we get

$$x = \frac{-(-2) \pm \sqrt{(-2)^2 - 4(3)(5)}}{2(3)}$$

$$= \frac{2 \pm \sqrt{-56}}{6} = \frac{1}{3} \pm \frac{\sqrt{14}}{3}i$$

example 70.A.2   Solve $x^2 = 3x + 28$ by using the quadratic formula.

solution   The formula is $x = \dfrac{-b \pm \sqrt{b^2 - 4ac}}{2a}$

We rearrange the given equation so that it is in standard form, and we write it just below the general quadratic equation.

$$ax^2 + bx + c = 0 \qquad \text{general equation}$$
$$x^2 - 3x - 28 = 0 \qquad \text{given equation}$$

We note the following correspondences between the coefficients.

$$a = 1 \qquad b = -3 \qquad c = -28$$

We use these values in the quadratic formula and simplify.

$$x = \frac{3 \pm \sqrt{9 - 4(1)(-28)}}{2(1)} \longrightarrow x = \frac{3 \pm \sqrt{9 + 112}}{2} \longrightarrow x = \frac{3 \pm \sqrt{121}}{2}$$

So

$$x = \frac{3 + 11}{2} \qquad \text{or} \qquad x = \frac{3 - 11}{2}$$

$$= \frac{14}{2} \qquad\qquad\qquad = -\frac{8}{2}$$

$$= 7 \qquad\qquad\qquad\qquad = -4$$

This means that the factors of the original equation are $(x - 7)(x + 4)$ and that the equation could have been solved by factoring. **But this shows that the quadratic formula can be used to solve any quadratic equation—even those that can be solved by factoring.**

problem set 70

1. The temperature of a quantity of an ideal gas was held constant at 740 kelvins. Find the final pressure if the initial pressure and volume were $40{,}000 \times 10^{-3}$ torr and 4000×10^{-2} cm^3 and the final volume was 8000×10^{-2} cm^3. Begin by solving the equation for P_2.

2. The number of folk dancers varied directly as the number of people who attended the festival. If 4800 attended the festival and 240 were folk dancers, how many attended when the folk dancers totaled 600?

3. One solution was 10% fluorine and the other was 30% fluorine. How much of each should be used to get 200 ml of a solution that is 17% fluorine?

4. Sandy ran out at 8 mph and rode back to Ski Island in a jitney at 24 mph. If the total trip took 8 hours, how far did she run and how far did she ride?

5. In the compound H_2CO_3, what percent of the weight of the compound is hydrogen (H)? (H, 1; C, 12; O, 16)

*6. Begin with $ax^2 + bx + c = 0$ and derive the quadratic formula.

Use the quadratic formula to solve:

*7. $3x^2 - 2x + 5 = 0$

*8. $x^2 = 3x + 28$

9. $2x^2 = -x - 4$

10. $r = m\left(\dfrac{1}{x + c} + \dfrac{3}{y}\right)$; find x

11. Add: $\dfrac{4x + 5}{x - 2} - \dfrac{4}{2 - x}$

12. Simplify: $\dfrac{4}{2 - 3\sqrt{12}}$

13. Solve: $R_M T_M = 160$, $R_P T_P = 400$, $R_P = 2R_M$, $T_P = T_M + 1$

Simplify:

14. $x + \dfrac{x}{1 + \dfrac{1}{x}}$

15. $a + \dfrac{b}{a + \dfrac{a}{b}}$

16. $-3i^3 + 2\sqrt{-2}\sqrt{2} - \sqrt{-9}$

17. $(-i - 1)(-3i + 2)$

18. Add: $20\underline{/70°} + 10\underline{/40°}$

19. Write $4R - 4U$ in polar form.

20. Solve: $\begin{cases} \dfrac{3}{8}x - \dfrac{1}{2}y = 2 \\ .06x - .2y = -.64 \end{cases}$

21. Solve $3x^2 - 2x + 5 = 0$ by completing the square.

22. Find the surface area of this right circular cylinder. Dimensions are in meters.

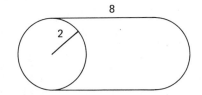

23. The data shown came from an experiment that involved lead (Pb) and boron (B). Write the equation that expresses lead as a function of boron: $Pb = mB + b$.

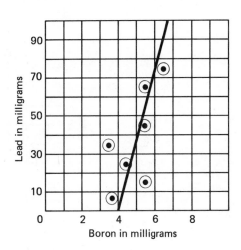

Simplify:

24. $\sqrt{81\sqrt{3}}$

25. $\sqrt[3]{x^5 y^6}\sqrt{xy^3}$

26. $3\sqrt{\dfrac{2}{5}} + 3\sqrt{\dfrac{5}{2}} - 6\sqrt{40}$

27. Expand $(x - 2)^3$

28. Find B.

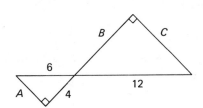

29. Simplify: $\dfrac{x^2 y - \dfrac{1}{y}}{\dfrac{x^2}{y} - 6}$

30. Estimate: $\dfrac{(-471,635 \times 10^5)(.0071893)}{-125,000,000}$

ER 70-31, 70-32, 70-33

LESSON 71 Lines from experimental data • Negative angles

71.A
lines from experimental data

In the problems thus far that deal with experimental data points, the line indicated by these points has already been estimated and drawn. In science courses, it is necessary to do one's own estimate of the line indicated by the data points. In future problems of this type, the data points will be graphed, but the line will not be drawn. It will be necessary to estimate the location of the line indicated by the data points and draw the line.

example 71.A.1 Draw an estimate of the line indicated by the data points shown on the left. Then write the equation that expresses salt as a function of carbon.

$$S = mC + b$$

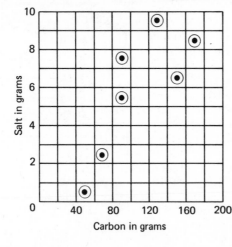   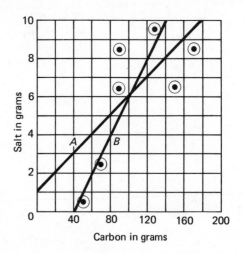

solution There is never an exact answer to these problems. The experimental data points are often scattered, and we can only estimate the position of the line. On the right we have drawn two lines, either of which could be said to represent the line indicated by the data. We have labeled the lines A and B. We exercise care in determining the slopes of these lines because each horizontal square has a value of 20, while the vertical squares each have a value of 1. We used triangles (not shown in the figure) to get the slopes.

Slope of $A = \dfrac{3}{60}$ or .05   Slope of $B = \dfrac{6}{60}$ or .1

Equation: $S = .05C + b$ $S = .1C + b$

The intercept of line A appears to be 1, and we will use the coordinates of the point (100, 6) to calculate the intercept of line B.

$$S = .05C + b \qquad\qquad S = .1C + b$$
$$\qquad\qquad\qquad\qquad\qquad 6 = .1(100) + b$$
$$b = 1 \text{ (by inspection)} \qquad\quad 6 = 10 + b$$
$$\qquad\qquad\qquad\qquad\qquad -4 = b$$

Equation A: So equation B is:

$$S = .05C + 1 \qquad\qquad\qquad S = .1C - 4$$

These equations appear to be very different, but the data points were very scattered. These equations are much closer to each other than they are to equations in which the numbers are greatly different such as

$$S = 475C - 23 \qquad \text{or} \qquad S = -578C - 460$$

Your answers to problems like these in the problem sets should be approximately the same as the answers given in the appendix. They will almost never be exactly the same as the answers given in the appendix.

71.B
negative angles

Rectangular coordinates designate the location of a point by giving the distance of the point to the left or right of the origin and the distance of the point above or below the origin. Thus the coordinates

$$4R + 3U$$

tell us that the location of the point is 4 units to the right of the origin and 3 units above the origin. We can locate the same point by saying that it is 5 units from the origin at an angle of 36.87°.

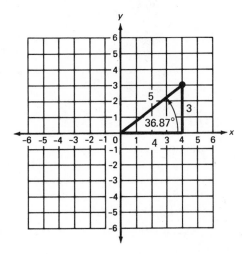

We remember to measure the angle counterclockwise from the positive x axis because this is the way mathematicians measure positive angles. **There is only one way to designate a point in rectangular coordinates, but we can use either positive or negative angles when we use polar coordinates. Negative angles are also measured from the positive x axis, but they are measured in the clockwise direction.**

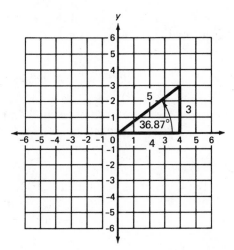
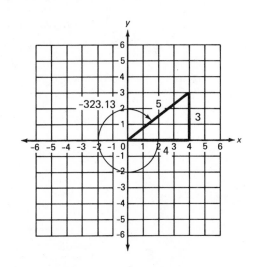

Since a full revolution is 360°, the negative angle is

$$-(360° - 36.87°) = -323.13°$$

Thus we can say that these two forms designate the same point.

$$5\underline{/36.87°} \quad \text{equals} \quad 5\underline{/-323.13°}$$

example 71.B.1   Write $4\underline{/-210°}$ in rectangular coordinates.

solution   **We always measure the angle first, and then the length, and then we draw the triangle.**

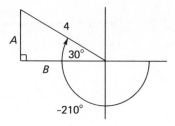

Now we solve for A and B.

$$\sin 30° = \frac{A}{4} \quad \longrightarrow \quad A = 4 \sin 30° \quad \longrightarrow \quad A = 4(.5) \quad \longrightarrow \quad A = 2$$

$$\cos 30° = \frac{B}{4} \quad \longrightarrow \quad B = 4 \cos 30° \quad \longrightarrow \quad B = 4(.8660) \quad \longrightarrow \quad B = 3.46$$

Thus we can write

$$4\underline{/-210°} = -3.46R + 2U$$

problem
set 71

1. A container contained 500 ml of a solution that was 52% water. How much water should be removed from the solution so that the remainder would be only 40% water?

2. Pollyana felt that 31,314 things were felicific. If this was 3.07 times the number of things considered felicific by the average person, how many things did the average person think were felicific?

3. Only 34 percent of the people in the mob carried a flambeau. If 5412 did not carry a flambeau, how many were in the mob?

4. Flotsam and jetsam littered the beach. The pieces of flotsam numbered 160 more than the pieces of jetsam, and 6 times the number of pieces of jetsam outnumbered the number of pieces of flotsam by 40. How many pieces of each were there?

5. A quantity of an ideal gas was confined in a container of fixed volume. If the initial pressure and temperature were 500×10^5 torr and $.0004 \times 10^7$ K, what was the pressure when the temperature was changed to $.0002 \times 10^5$ K? Begin by solving the equation for P_2.

*6. Draw an estimate of the line indicated by the data points shown. Then write the equation that expresses salt as a function of carbon.

$$S = mC + b$$

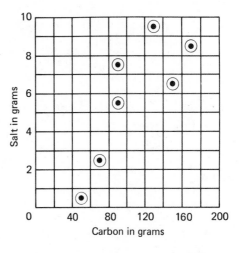

*7. Write $4\underline{/-210°}$ in rectangular coordinates.

8. Write $-60R - 20U$ in polar form.

9. Begin with $ax^2 + bx + c = 0$ and complete the square to derive the quadratic formula.

Use the quadratic formula to solve:

10. $5x^2 = -7 - 2x$

11. $3x^2 + 7x = -3$

12. $a = x\left(\dfrac{1}{m_1} + \dfrac{y}{m_2}\right)$; find m_1

13. Add: $\dfrac{2x + 3}{x^2 - 2x - 8} + \dfrac{3x - 2}{x - 4}$

14. Simplify: $\dfrac{3}{2 + 3\sqrt{20}}$

15. Solve: $R_P T_P = 693$, $R_C T_C = 165$, $R_P = 3R_C$, $T_P = T_C + 2$

Simplify:

16. $ab + \dfrac{b}{b + \dfrac{1}{b}}$

17. $x^2 + \dfrac{y}{y + \dfrac{1}{xy}}$

18. $(-3i - 5)(i + 5)$

19. $-4i^3 - 3i^2 + \sqrt{-9} - \sqrt{-3}\sqrt{-3}$

20. Solve: $\begin{cases} \dfrac{1}{4}x - \dfrac{1}{5}y = 2 \\ .03x - .4y = -1.64 \end{cases}$

21. Solve $2x^2 - x + 4 = 0$ by completing the square.

22. How many tiles 1 centimeter square would it take to cover the figure at the right? The dimensions are in meters.

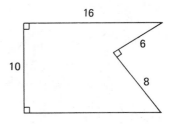

Simplify:

23. $\sqrt[5]{4\sqrt{2}}$

24. $\dfrac{-2^0}{-8^{-\frac{4}{3}}}$

25. $\sqrt[4]{a^5 y}\sqrt{ay^4}$

26. $2\sqrt{\dfrac{3}{5}} + 4\sqrt{\dfrac{5}{3}} - 2\sqrt{135}$

27. Find B.

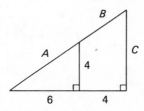

28. Solve: $(-2)^0 - 2^2 - 2 - 2^0 - |-2 - 2| - 2^3 = -2(-2x - 2)$

29. Find the equation of the line that passes through $(-7, 0)$ that is perpendicular to the line $4y - 3x = 1$.

30. Estimate: $\dfrac{(-35,123 \times 10^4)(-2,184,536)}{-798 \times 10^{-15}}$

ER 71-34, 71-35, 71-36

LESSON 72 *More on radical denominators*

72.A

more on radical denominators

In Lesson 67 we learned to eliminate the radical in the denominator of expressions such as

$$\frac{1}{2 - 3\sqrt{2}}$$

by multiplying above and below by the conjugate of the denominator.

$$\frac{1}{2 - 3\sqrt{2}} \cdot \frac{(2 + 3\sqrt{2})}{(2 + 3\sqrt{2})} = \frac{2 + 3\sqrt{2}}{4 - 6\sqrt{2} + 6\sqrt{2} - 18}$$

$$= \frac{2 + 3\sqrt{2}}{-14} = \frac{-2 - 3\sqrt{2}}{14}$$

If the original expression contains radicals above and below such as

$$\frac{4 + \sqrt{3}}{2 - 3\sqrt{3}}$$

we still multiply above and below by the conjugate of the denominator. This will eliminate the radicals in the denominator but not in the numerator.

example 72.A.1 Simplify: $\dfrac{4 + \sqrt{3}}{2 - 3\sqrt{3}}$

solution We remember that an expression that contains square root radicals is in simplified form when no radicand has a perfect square as a factor and no radicals are in the denominator. We can rationalize the denominator if we multiply above and below by $2 + 3\sqrt{3}$.

$$\frac{4 + \sqrt{3}}{2 - 3\sqrt{3}} \cdot \frac{2 + 3\sqrt{3}}{2 + 3\sqrt{3}}$$

We have two multiplications to perform, one above and one below. Many people find it easier to do these multiplications separately and then to write the answer. We will do this.

ABOVE	BELOW
$4 + \sqrt{3}$	$2 - 3\sqrt{3}$
$2 + 3\sqrt{3}$	$2 + 3\sqrt{3}$
$8 + 2\sqrt{3}$	$4 - 6\sqrt{3}$
$12\sqrt{3} + 9$	$+ 6\sqrt{3} - 27$
$17 + 14\sqrt{3}$	$4 \qquad - 27 = -23$

Thus, our simplification is

$$\frac{17 + 14\sqrt{3}}{-23} \quad \text{or} \quad \frac{-17 - 14\sqrt{3}}{23}$$

example 72.A.2   Simplify: $\dfrac{4 - \sqrt{2}}{4 + 3\sqrt{2}}$

solution   We will multiply above and below by $4 - 3\sqrt{2}$.

$$\frac{4 - \sqrt{2}}{4 + 3\sqrt{2}} \cdot \frac{4 - 3\sqrt{2}}{4 - 3\sqrt{2}}$$

We have two multiplications to perform.

ABOVE	BELOW
$4 - \sqrt{2}$	$4 - 3\sqrt{2}$
$4 - 3\sqrt{2}$	$4 + 3\sqrt{2}$
$16 - 4\sqrt{2}$	$16 - 12\sqrt{2}$
$- 12\sqrt{2} + 6$	$+ 12\sqrt{2} - 18$
$22 - 16\sqrt{2}$	$16 \qquad - 18 = -2$

Thus the simplification can be written as

$$\frac{22 - 16\sqrt{2}}{-2} \quad \text{which simplifies to} \quad \mathbf{-11 + 8\sqrt{2}}$$

example 72.A.3   Simplify: $\dfrac{3\sqrt{12} - 2\sqrt{3}}{3\sqrt{3} - 2\sqrt{2}}$

solution   We will multiply above and below by $3\sqrt{3} + 2\sqrt{2}$.

$$\frac{3\sqrt{12} - 2\sqrt{3}}{3\sqrt{3} - 2\sqrt{2}} \cdot \frac{3\sqrt{3} + 2\sqrt{2}}{3\sqrt{3} + 2\sqrt{2}}$$

Again we have two multiplications to perform.

ABOVE   BELOW

$$
\begin{array}{ll}
3\sqrt{12} - 2\sqrt{3} & \qquad\qquad 3\sqrt{3} - 2\sqrt{2} \\
3\sqrt{3} \;\; + 2\sqrt{2} & \qquad\qquad 3\sqrt{3} + 2\sqrt{2} \\
\hline
9\sqrt{36} - 18 & \qquad\qquad 27 \;\;\; - 6\sqrt{6} \\
\qquad\qquad\quad + 6\sqrt{24} - 4\sqrt{6} & \qquad\qquad\quad + 6\sqrt{6} - 8 \\
\hline
54 \;\;\; - 18 \;\; + 12\sqrt{6} - 4\sqrt{6} = 36 + 8\sqrt{6} & \quad 27 \qquad\qquad - 8 = 19
\end{array}
$$

Thus, our result is

$$\frac{36 + 8\sqrt{6}}{19}$$

problem set 72

1. The chemist calculated that the carbon (C) in the compound C_2H_5Br weighed 48 grams. What was the total weight of the compound? (C, 12; H, 1; Br, 80)

2. Twice the number of pansies exceeded 4 times the number of daisies by 8. Also, 7 times the number of daisies was 4 less than 3 times the number of pansies. How many of each were there?

3. The wolf loped for a while at 16 mph and finished the journey by trotting at 12 mph. If the total trip was 256 miles, and he loped for 2 more hours than he trotted, how far did he trot?

4. The number of frangibles varied inversely with the strength of the clay. If 50 were frangible when the clay strength measured 50, how many were frangible when the clay strength dropped to 25?

5. The pressure of a quantity of ideal gas was held constant at 1100 torr. The initial temperature and volume were 700×10^5 kelvins and .0004 liters. What was the final temperature if the final volume was .08 liter? Begin by solving the equation for T_2.

Simplify:

*6. $\dfrac{4 + \sqrt{3}}{2 - 3\sqrt{3}}$ *7. $\dfrac{4 - \sqrt{2}}{4 + 3\sqrt{2}}$ *8. $\dfrac{3\sqrt{12} - 2\sqrt{3}}{3\sqrt{3} - 2\sqrt{2}}$

9. Draw an estimate of the line indicated by the data points. Write the equation that gives hydrogen (H) as a function of carbon (C): $H = mC + b$.

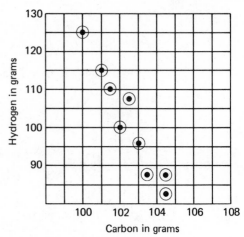

10. Add: $8\underline{/-20°} + 6\underline{/320°}$

11. Write $4R + 9U$ in polar form.

12. Begin with $ax^2 + bx + c = 0$ and complete the square to derive the quadratic formula.

Use the quadratic formula to solve:

13. $2x^2 + 5 = -5x$

14. $2x^2 - 4 = -5x$

15. $xc = px\left(\dfrac{1}{km_1} - \dfrac{1}{m_2}\right)$; find m_1

16. Add: $\dfrac{3x - 2}{x - 2} - \dfrac{4x - 3}{2 - x}$

17. Solve: $R_A T_A = 160$, $R_B T_B = 240$, $R_B = 2R_A$, $T_A + T_B = 7$

Simplify:

18. $ax^2 - \dfrac{a}{a - \dfrac{1}{ax}}$

19. $(3i + 2)(i - 4) - \sqrt{-9}$

20. Solve the system by graphing and then find an exact solution by using either substitution or elimination.
$$\begin{cases} 2x + 3y = 6 \\ x - 2y = 4 \end{cases}$$

21. Solve $x^2 + 6 = 3x$ by completing the square.

22. Find the surface area of the box in square centimeters. The dimensions are in meters

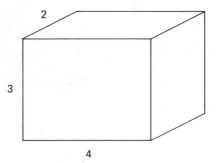

Simplify:

23. $3\sqrt{9\sqrt{3}}$

24. $\dfrac{-2^2}{-16^{-\frac{3}{4}}}$

25. $\sqrt{a^2 x^0 y x^{\frac{1}{4}} y^2}$

26. $2\sqrt{\dfrac{5}{8}} + 3\sqrt{\dfrac{8}{5}} - 2\sqrt{40}$

27. $\dfrac{x^3 - 16x - 6x^2}{x^2 - 8x - 20} \cdot \dfrac{-50 - 5x + x^2}{x^3 - 5x^2 - 24x}$

28. Find the equation of the line that passes through $(-5, 2)$ that has a slope of $-\frac{2}{5}$.

29. Estimate: $\dfrac{(4{,}168{,}214)(572)(891)(142)}{(74{,}612{,}000)(42)(.005)}$

30. Solve: $\dfrac{3x - 5}{2} - \dfrac{x}{3} = 6$

ER 72-37, 72-38, 72-39

LESSON 73 *Uniform motion with both distances given*

73.A
uniform motion problems

The distance diagrams and equations for the uniform motion problems that we have worked thus far have been similar to one of the following.

(a) D_A / D_B (b) D_A D_B / 120   (c) D_A 40 / D_B

$$R_A T_A = R_B T_B \qquad R_A T_A + R_B T_B = 120 \qquad R_A T_A + 40 = R_B T_B$$

In (a), the distances traveled were equal. In (b) the sum of the distances traveled equaled 120; and in (c) the distance traveled by B was 40 greater than the distance traveled by A. **In some problems the distance traveled by each object is given. In these problems, we get two distance equations.** Each of the equations will contain two different unknowns so two more equations will be needed. One of these equations will be an equation about rates, and the other equation will be an equation about times.

example 73.A.1   Friar Tuck rode the 24 miles to the fair in Nottingham at a leisurely pace. He stayed too long at the fair and had to double his speed on the way back in order to get home in time. If his total traveling time was 9 hours, how fast did he travel in each direction? What were the times?

solution   **When both distances are given, we can write two distance equations.**

D_G / 24 D_B / 24

(a) $R_G T_G = 24$ \qquad (b) $R_B T_B = 24$

We have four unknowns but only two equations. We reread the problem to get the other two equations. One equation is a rate equation and one equation is a time equation.

(c) $R_B = 2R_G$ \qquad (d) $T_G + T_B = 9$

We will use equation (b) as our base equation.

$$R_B T_B = 24 \qquad \text{equation (b)}$$
$$(2R_G)(9 - T_G) = 24 \qquad \text{substituted}$$
$$18R_G - 2R_G T_G = 24 \qquad \text{multiplied}$$
$$18R_G - 2(24) = 24 \qquad \text{substituted}$$
$$18R_G = 72 \qquad \text{simplified}$$
$$R_G = 4 \text{ mph} \qquad \text{divided}$$

Since $R_B = 2R_G$, then $R_B = 8$ **mph.**
Now we will use these values in equations (a) and (b) to find T_G and T_B.

$$R_G T_G = 24 \quad \text{equation (a)} \qquad R_B T_B = 24 \quad \text{equation (b)}$$
$$(4)T_G = 24 \quad \text{substituted} \qquad (8)T_B = 24 \quad \text{substituted}$$
$$T_G = 6 \text{ hours} \quad \text{divided} \qquad T_B = 3 \text{ hours} \quad \text{divided}$$

example 73.A.2   Atalanta could run four times as fast as her challenger could run. For this reason, she could run 80 miles in 2 hours less than it took her challenger to run 28 miles. How fast could each of them run? How long did they run?

solution   **Both distances were given, so we will have two distance diagrams and two distance equations.**

$$(a) \quad \overset{D_A}{\underset{80}{\longmapsto}} \quad R_A T_A = 80 \qquad\qquad (b) \quad \overset{D_C}{\underset{28}{\longmapsto}} \quad R_C T_C = 28$$

We reread the problem and write the rate equation and the time equation.

$$(c) \quad R_A = 4R_C \qquad\qquad (d) \quad T_A = T_C - 2$$

We will use equation (a) as our base equation.

$$\begin{array}{ll}
R_A T_A = 80 & \text{equation (a)} \\
(4R_C)(T_C - 2) = 80 & \text{substituted} \\
4R_C T_C - 8R_C = 80 & \text{multiplied} \\
4(28) - 8R_C = 80 & \text{substituted} \\
32 = 8R_C & \text{simplified} \\
\mathbf{4\ mph} = \mathbf{R_C} & \text{divided}
\end{array}$$

Since Atalanta could run 4 times as fast as her competitor, her rate was **16 mph.** Now we use these rates in equations (a) and (b) to find the times.

$$\begin{array}{llll}
R_A T_A = 80 & \text{equation (a)} & R_C T_C = 28 & \text{equation (b)} \\
16T_A = 80 & \text{substituted} & 4T_C = 28 & \text{substituted} \\
\mathbf{T_A = 5\ hours} & \text{divided} & \mathbf{T_C = 7\ hours} & \text{divided}
\end{array}$$

problem set 73

*1. Friar Tuck rode the 24 miles to the fair in Nottingham at a leisurely pace. He stayed too long at the fair and had to double his speed on the way back in order to get home in time. If his total traveling time was 9 hours, how fast did he travel in each direction? What were his times?

*2. Atalanta could run four times as fast as her challenger could run. For this reason, she could run 80 miles in 2 hours less than it took her challenger to run 28 miles. How fast could each of them run? How long did they run?

3. The initial pressure, volume, and temperature of a quantity of an ideal gas were $.004 \times 10^5$ torr, $.02 \times 10^4$ liters, and $.06 \times 10^6$ kelvins. Find the final temperature if the final pressure and volume were 400×10^{-5} torr and 500×10^4 liters. Begin by solving for T_2.

4. Two thousand liters of a solution was 92% alcohol. How much alcohol should be extracted to reduce the concentration to 80% alcohol?

5. Find three consecutive odd integers such that 4 times the product of the second and third is 12 greater than 20 times the sum of the first and second.

Simplify:

6. $\dfrac{2 - \sqrt{3}}{-\sqrt{3} - 2}$

7. $\dfrac{3\sqrt{2} - 4}{\sqrt{2} - 3}$

8. $\dfrac{4\sqrt{2} - 5}{2 - 3\sqrt{8}}$

9. Draw an estimate of the line indicated by the data points and write the equation that gives nitrogen (N) as a function of fluorine (F): $N = mF + b$.

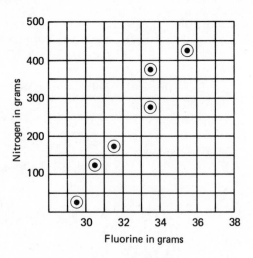

10. Add: $4\underline{/-135°} + 6\underline{/200°}$

11. Write $-5R - 5U$ in polar form.

12. Begin with $ax^2 + bx + c = 0$ and complete the square to derive the quadratic formula.

Use the quadratic formula to solve:

13. $-3x^2 - x = 4$

14. $-x - 3x^2 = -4$

15. $a = m\left(\dfrac{1}{pc} - \dfrac{k}{x}\right)$; find x

16. $\dfrac{a}{m} = c\left(\dfrac{1}{x_1} + \dfrac{b}{x_2}\right)$; find x_1

Simplify:

17. $ax - \dfrac{a}{x - \dfrac{x}{a}}$

18. $(2i - 3)(i - 3) + \sqrt{-9} + 3i^3$

19. Add: $\dfrac{3x - 2}{x^2 + 7x + 10} - \dfrac{1}{x + 5}$

20. Solve: $\begin{cases} \dfrac{3}{2}x - \dfrac{2}{5}y = 2 \\ 3x + .5y = 17 \end{cases}$

21. Solve $3x^2 - x = -7$ by completing the square.

22. Find the surface area of the right circular cylinder in square centimeters. Dimensions are in meters.

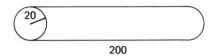

Simplify:

23. $2\sqrt{4\sqrt{2}}$

24. $\dfrac{-3^0(-3^0)^2}{-9^{-\frac{3}{2}}}$

25. $\sqrt{4x^2y^5}\sqrt[3]{8y^5x}$

26. $2\sqrt{\dfrac{6}{7}} + 3\sqrt{\dfrac{7}{6}} - 3\sqrt{42}$

27. Find the equation of the line that passes through $(-2, 0)$ and is parallel to the line $4x - y = 7$.

Solve:

28. $\sqrt{x - 5} - 2 = 7$

29. $\dfrac{4x - 3}{7} - \dfrac{x - 2}{3} = 5$

ER 73-40, 73-41, 73-42

LESSON 74 *Factorable denominators and sign changes*

74.A
factorable denominators

We have discussed how the addition of rational expressions can sometimes be facilitated by making sign changes in one or more of the expressions. For example, if we wish to add the expressions

$$\frac{3x + 2}{x - 5} + \frac{x - 3}{5 - x}$$

we can do so easily if we change the sign in front of the second expression and all the signs in the denominator of the second expression.

$$\frac{3x + 2}{x - 5} - \frac{x - 3}{x - 5}$$

Now the denominators are the same, and we can add the numerators and get

$$\frac{3x + 2 - x + 3}{x - 5} = \frac{2x + 5}{x - 5}$$

We have also learned to add algebraic expressions in which it is helpful to factor one or more denominators as the first step. In this lesson we will add expressions that require factoring in one term and sign changes in another term.

example 74.A.1   Add: $\dfrac{x + 3}{x^2 - x - 6} - \dfrac{3}{3 - x}$

solution   We need to factor the first denominator and change signs in the second denominator. We do this and get

$$\frac{x + 3}{(x - 3)(x + 2)} + \frac{3}{x - 3}$$

We see that the second denominator needs an $x + 2$ factor to make the denominators the same, so we multiply above and below by $x + 2$.

$$\frac{x + 3}{(x - 3)(x + 2)} + \frac{3}{(x - 3)} \cdot \frac{(x + 2)}{(x + 2)}$$

Now both denominators are the same and we can finish by adding the numerators.

$$\frac{x + 3 + 3x + 6}{(x - 3)(x + 2)} = \frac{4x + 9}{x^2 - x - 6}$$

We could have left the denominator in factored form, but we chose to multiply it out.

example 74.A.2   Add: $\dfrac{x + 7}{x^2 + 2x + 1} - \dfrac{3}{-1 - x}$

solution   This problem requires that we factor the denominator of the first expression and change the signs in the second expression. We factor first and get

$$\frac{x + 7}{(x + 1)(x + 1)} - \frac{3}{-1 - x}$$

To permit changing the signs in the second denominator, we must change the sign in front or the sign above. We choose to change the sign in front, and now we have.

$$\frac{x + 7}{(x + 1)(x + 1)} + \frac{3}{x + 1}$$

Now we need another $x + 1$ factor in the second denominator so we multiply above and below by $x + 1$. Then we add the numerators.

$$\frac{x + 7}{(x + 1)(x + 1)} + \frac{3}{(x + 1)} \cdot \frac{(x + 1)}{(x + 1)} = \frac{x + 7 + 3x + 3}{(x + 1)(x + 1)} = \frac{4x + 10}{(x + 1)(x + 1)}$$

This time we left the denominator in factored form.

example 74.A.3   Add: $\dfrac{x - 3}{x^2 - 2x - 15} + \dfrac{x + 2}{5 - x}$

solution   Again we find that one denominator must be factored and the signs must be changed in the other denominator. We do this and get

$$\frac{x - 3}{(x - 5)(x + 3)} - \frac{x + 2}{x - 5}$$

Now we need another $x + 3$ factor in the second denominator.

$$\frac{x - 3}{(x - 5)(x + 3)} - \frac{(x + 2)}{(x - 5)} \cdot \frac{(x + 3)}{(x + 3)}$$

We finish by adding and simplifying, and we get

$$\frac{x - 3 - x^2 - 5x - 6}{(x - 5)(x + 3)} = \frac{-x^2 - 4x - 9}{(x - 5)(x + 3)}$$

problem set 74

1. David traveled 120 miles in 1 hour less than it took Emily to travel 360 miles. Emily could do this because she drove twice as fast as David drove. Find the rates of both and the times of both.

2. A quantity of an ideal gas was confined in a container whose volume was fixed at .05 liters. The initial pressure and temperature were $.0036 \times 10^{-2}$ torr and 50×10^7 K. If the temperature was changed to 40×10^4 K, find the final pressure. Begin by solving for P_2.

3. The time to complete the job varied inversely as the number of men working. When 500 men worked, the job could be completed in 10 days. How long would 200 men take to complete the job?

4. The bus headed north 2 hours before the train headed north from the same town. The rate of the train was 60 mph and the rate of the bus was 40 mph. How long did it take the train to get 20 miles ahead of the bus?

5. Sandra bought lillies for $4 each and poinsettias for $6 each. She bought 2 fewer lillies than poinsettias and spent a total of $192. How many of each did she buy?

Add:

*6. $\dfrac{x + 7}{x^2 + 2x + 1} - \dfrac{3}{-1 - x}$

*7. $\dfrac{x - 3}{x^2 - 2x - 15} + \dfrac{x + 2}{5 - x}$

Simplify:

8. $\dfrac{2 - \sqrt{2}}{3 + \sqrt{2}}$

9. $\dfrac{3 - \sqrt{3}}{4 + \sqrt{3}}$

10. $\dfrac{2 - 3\sqrt{3}}{3 - 2\sqrt{12}}$

11. Draw an estimate of the line indicated by the data points and write the equation that gives vanadium (V) as a function of potassium (K): $V = mK + b$.

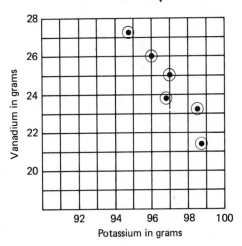

12. Begin with $ax^2 + bx + c = 0$ and complete the square to derive the quadratic formula.

13. Add: $10\underline{/30°} + 10\underline{/-380°}$

14. Write $-2R - 8U$ in polar form.

15. Use the quadratic formula to find the roots of $-2x + 4 = -5x^2$.

16. $\dfrac{m}{x} = cm\left(\dfrac{a}{x} + \dfrac{b}{y}\right)$; find y

17. $\dfrac{p}{c} = -b\left(\dfrac{1}{d} + \dfrac{a}{f}\right)$; find d

Simplify:

18. $a^2 y + \dfrac{a^2}{a + \dfrac{a}{y}}$

19. $(2 + i)(i - 4) - \sqrt{-16}$

20. Solve: $\begin{cases} \dfrac{1}{4}x + \dfrac{1}{3}y = 15 \\ .02x + .2y = 6.4 \end{cases}$

21. Solve $3x^2 + 2 = -x$ by completing the square.

22. A semicircle is cut from a right triangle. Find the area of the figure shown. Dimensions are in inches.

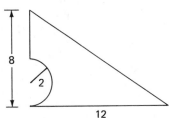

Simplify:

23. $4\sqrt{2\sqrt[3]{2}}$

24. $\dfrac{-2^0(-2^0)}{-4^{-3/2}}$

25. $\sqrt[4]{mp^5}\,\sqrt[3]{m^2 p^4}$

26. $3\sqrt{\dfrac{7}{8}} + 2\sqrt{\dfrac{8}{7}} - 2\sqrt{56}$

27. Estimate: $\dfrac{(4,813,000)(59,132)(12)}{(1000)(1000)(500)}$

28. Find the equation of the line through $(5, 7)$ that is perpendicular to the line that passes through $(-2, 4)$ and $(-3, 5)$.

Solve:

29. $\sqrt{3x - 5} - 2 = 7$ 30. $\dfrac{4x + 7}{2} - \dfrac{5x}{3} = 4$

ER 74-43, 74-44, 74-45

If the enrichment lessons on the scientific calculator are to be included, Enrichment Lesson 4 on scientific notation should be taught at this point.

LESSON 75 *Using both substitution and elimination • Negative vectors*

75.A
using both substitution and elimination

We have found that we can use either the substitution method or the elimination method to solve a system of two equations in two unknowns such as

$$\begin{cases} 3x + 2y = 8 \\ x + 3y = -2 \end{cases}$$

When we have to solve a system of three equations in three unknowns, it is sometimes helpful if we begin by using substitution and then finish by using elimination.

example 75.A.1   Use substitution and elimination as necessary to solve this system of equations.

$$\begin{cases} x = 2y & \text{(a)} \\ x + y + z = 9 & \text{(b)} \\ x - 3y - 2z = -8 & \text{(c)} \end{cases}$$

solution   Equation (a) tells us that x is equal to $2y$. Thus we will substitute $2y$ for x in equations (b) and (c) and then simplify.

(b) $(2y) + y + z = 9 \quad\longrightarrow\quad 3y + z = 9$ (d)

(c) $(2y) - 3y - 2z = -8 \quad\longrightarrow\quad -y - 2z = -8$ (e)

Now we have two equations in y and z that can be solved by using either substitution or elimination. We will use elimination and will multiply the top equation by 2.

(d) $3y + z = 9 \quad\longrightarrow\quad (2) \quad\longrightarrow\quad 6y + 2z = 18$

(e) $-y - 2z = -8 \quad\longrightarrow\quad (1) \quad\longrightarrow\quad \dfrac{-y - 2z = -8}{5y \qquad = 10}$

$$y = 2$$

We can use 2 for y in either equation (d) or (e) to find z. This time we will use both equations to show that either one will give the desired result

EQUATION (d)		EQUATION (e)	
$3y + z = 9$	equation	$-y - 2z = -8$	equation
$3(2) + z = 9$	substituted	$-(2) - 2z = -8$	substituted
$6 + z = 9$	multiplied	$-2z = -6$	simplified
$z = 3$	solved	$z = 3$	solved

Finally, we can use 2 for y and 3 for z in any of the first three equations to find x. This time we will use all three.

EQUATION (a)	EQUATION (b)	EQUATION (c)	
$x = 2y$	$x + y + z = 9$	$x - 3y - 2z = -8$	equation
$x = 2(2)$	$x + 2 + 3 = 9$	$x - 3(2) - 2(3) = -8$	substituted
$x = 4$	$x = 4$	$x = 4$	solved

Thus we find that the solution to this system of three equations in three unknowns is the ordered triple **(4, 2, 3)**.

example 75.A.2   Use substitution and elimination as necessary to solve this system of equations.

$$\begin{cases} 2x + 2y - z = 12 & \text{(a)} \\ 3x - y + 2z = 21 & \text{(b)} \\ x - 3z = 0 & \text{(c)} \end{cases}$$

solution   If we solve equation (c) for x, we get

$$x = 3z$$

Next, in equations (a) and (b), we will replace x with $3z$ and then simplify.

(a) $2(3z) + 2y - z = 12 \longrightarrow 5z + 2y = 12$

(b) $3(3z) - y + 2z = 21 \longrightarrow 11z - y = 21$

We can eliminate y if we multiply equation (e) by 2 and add.

(d) $5z + 2y = 12 \longrightarrow 5z + 2y = 12$
(e) $11z - y = 21 \longrightarrow \underline{22z - 2y = 42}$
$$27z \qquad = 54$$
$$z = 2$$

Now we can use 2 for z in either equation (d) or equation (e) to find y. We choose to use equation (d).

$5z + 2y = 12$	equation (d)
$5(2) + 2y = 12$	substituted
$2y = 2$	simplified
$y = 1$	divided

Now we can use 2 for z and 1 for y in any of the original equations to find x. Equation (c) is the simplest so we will use it.

$$x - 3z = 0 \qquad \text{equation (c)}$$

$$x - 3(2) = 0 \qquad \text{substituted}$$

$$x = 6 \qquad \text{solved}$$

Thus, we find that the solution to this system of three equations in three unknowns is the ordered triple **(6, 1, 2)**.

75.B
negative vectors

In Lesson 71, we noted that there is only one way to use rectangular coordinates to designate the location of a point, but more than one form of polar coordinates is possible because either positive angles or negative angles may be used.

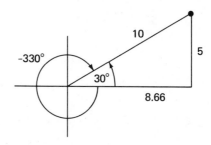

Since the point is 8.66 units to the right of the origin and 5 units above the origin, we can designate its location with rectangular coordinates by writing

$$8.66R + 5U$$

If we wish to use polar coordinates to name the same point, we can use either a positive angle or a negative angle, so we can write either

$$10/30° \qquad \text{or} \qquad 10/-330°$$

To make matters even more confusing, we note that it is also possible to use negative magnitudes to locate a point.

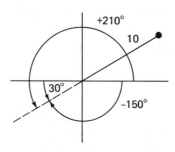

We see if we begin by turning through an angle of $+210°$ or $-150°$, we are pointing away from the point. Now if we back up 10 units we are on the point. Thus, the point can also be designated by using negative magnitudes and writing

$$-10/210° \qquad \text{or} \qquad -10/-150°$$

example 75.B.1   Add: $-4/-20° + 5/135°$

solution   We begin by drawing the vectors. We lay out the angles first and then the magnitudes.

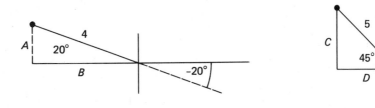

$$A = 4 \sin 20° = 4(.3420) \longrightarrow A = 1.368$$
$$B = 4 \cos 20° = 4(.9397) \longrightarrow B = 3.7588$$
$$C = 5 \sin 45° = 5(.7071) \longrightarrow C = 3.5355$$
$$D = 5 \cos 45° = 5(.7071) \longrightarrow D = 3.5355$$

Thus we go $3.7588 + 3.5355 = 7.2943$ to the left and $1.368 + 3.5355 = 4.9035$ up and our answer is

$$-7.29R + 4.90U$$

problem set 75

1. Claudius could walk the 32 miles to Pompeii in 2 hours more than it took Tiberius to drive the 72 miles to the sea. Find the rate of each and the time of each if the rate of Tiberius was 3 times that of Claudius.

2. The initial pressure, volume, and temperature of a quantity of an ideal gas were 700×10^5 torr, 700×10^{-7} liter, and 56×10^4 K. What would be the final volume if the pressure were changed to 3500×10^4 torr and the temperature changed to 8000×10^5 K? Begin by solving the equation for V_2.

3. The elixir was much too powerful as 500 ml of it tested at 64% alcohol. How much alcohol had to be evaporated to reduce the alcohol concentration to 40%?

4. The chemical formula for sodium hypochlorite is NaClO. Benjamin had 280 grams of chlorine (Cl) available. What would be the total weight of the oxygen (O) and sodium (Na) that he would need to make a batch of sodium hypochlorite? (Na, 23; O,16; Cl, 35)

5. When the chips were down, Chester found that 4 times the number of blue chips exceeded 3 times the number of red chips by 14. Also, 6 times the number of red chips exceeded the number of blue chips by 7. How many chips were down?

Solve:

*6. $\begin{cases} x = 2y \\ x + y + z = 9 \\ x - 3y - 2z = -8 \end{cases}$

*7. $\begin{cases} 2x + 2y - z = 12 \\ 3x - y + 2z = 21 \\ x - 3z = 0 \end{cases}$

*8. Add: $-4\underline{/-20°} + 5\underline{/135°}$

9. Write $-4R - 1U$ in polar form.

10. Add: $\dfrac{5x + 2}{x^2 + 3x - 10} - \dfrac{2x}{2 - x}$

Simplify:

11. $\dfrac{2\sqrt{2} - 1}{1 - \sqrt{2}}$

12. $\dfrac{3 - \sqrt{2}}{4 + 2\sqrt{8}}$

13. Draw an estimate of the line indicated by the data points and write the equation that expresses aluminum (Al) as a function of boron (B): $Al = mB + b$.

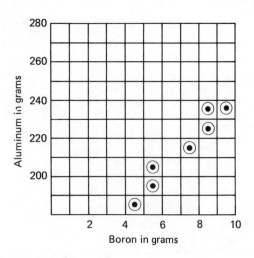

14. Begin with $ax^2 + bx + c = 0$ and derive the quadratic formula.

15. $\dfrac{a}{c + x} = m\left(\dfrac{1}{r} + \dfrac{1}{t}\right)$; find x

16. $\dfrac{a}{c + x} = m\left(\dfrac{1}{r} + \dfrac{1}{t}\right)$; find t

17. Solve $7x^2 - x - 1 = 0$ by completing the square.

Simplify:

18. $x + \dfrac{a}{a + \dfrac{a}{x}}$

19. $(3 - i)(2 - i) - 2i^2 - \sqrt{-9}$

20. Solve: $\begin{cases} \dfrac{1}{3}x + \dfrac{2}{3}y = 31 \\ .02x + .7y = 27.6 \end{cases}$

21. The figure is the base of a container 5 feet high. How many 1-inch sugar cubes will the container hold? Dimensions are in inches.

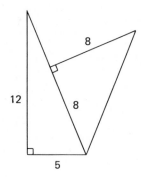

Simplify:

22. $5\sqrt{25\sqrt{5}}$

23. $\dfrac{-1^0(-1^0)}{-4^{-\frac{5}{2}}}$

24. $\sqrt[6]{my^3}\,\sqrt[4]{m^3y}$

25. $2\sqrt{\dfrac{3}{8}} + 3\sqrt{\dfrac{8}{3}} - 2\sqrt{216}$

26. Estimate: $\dfrac{(120)(37,842)(9000)}{(12)(12)(400,000)}$

27. Find the equation of the line that passes through $(-5, -7)$ and $(2, 4)$.

Solve:

28. $\sqrt{2x - 2} + 7 = 9$

29. $\dfrac{15x - 2}{3} - \dfrac{2x - 5}{4} = 3$

ER 75-46, 75-47

LESSON 76 *Advanced radical equations*

76.A
higher-order roots

Thus far, we have restricted our investigation of radical equations to equations that contain one square root radical such as

$$\sqrt{x - 1} = 4$$

We can solve this equation for x if we begin by squaring both sides of the equation.

$$(\sqrt{x - 1})^2 = (4)^2$$

Then we simplify and complete the solution.

$$x - 1 = 16$$

$$x = 17$$

As the final step, we must check the solution in the original equation because squaring both sides of an equation sometimes generates an equation that has more solutions than the original equation.

$$\sqrt{(17) - 1} = 4$$

$$\sqrt{16} = 4$$

$$4 = 4 \qquad \text{Check}$$

It can be shown that if both sides of an equation are raised to the nth power ($n = 2, 3, 4, \ldots$), the solutions of the original equation (if any exist) are also solutions of the resulting equation. This permits us to solve many radical equations by isolating the radical and then raising both sides to the integral power that will eliminate the radical.

example 76.A.1 Solve: $\sqrt[3]{x^3 + 6x^2 - 4} - x - 2 = 0$

solution We begin by isolating the radical on one side of the equation.

$$\sqrt[3]{x^3 + 6x^2 - 4} = x + 2$$

Now we raise both sides of the equation to the third power to eliminate the radical.

$$(\sqrt[3]{x^3 + 6x^2 - 4})^3 = (x + 2)^3$$

$$x^3 + 6x^2 - 4 = x^3 + 6x^2 + 12x + 8 \qquad \text{multiplied}$$

$$-4 = 12x + 8 \qquad \text{simplified}$$

$$-12 = 12x \qquad \text{added } -8$$

$$-1 = x \qquad \text{divided}$$

Now we must check the answer in the original equation.

$$\sqrt[3]{(-1)^3 + 6(-1)^2 - 4} - (-1) - 2 = 0$$

$$\sqrt[3]{1} + 1 - 2 = 0$$

$$1 + 1 - 2 = 0 \qquad \text{Check}$$

76.B
multiple radicals

When an equation contains more than one square root radical, it is sometimes necessary to square both sides of the equation more than once to eliminate all the radicals.

example 76.B.1   Solve: $\sqrt{k - 5} - \sqrt{k} + 1 = 0$

solution   We begin by rearranging the equation so that $\sqrt{k - 5}$ is isolated on the left side.

$$\sqrt{k - 5} = \sqrt{k} - 1$$

Now we square both sides to eliminate the radical on the left.

$$(\sqrt{k - 5})^2 = (\sqrt{k} - 1)^2$$

On the left we get $k - 5$ and on the right we must multiply $\sqrt{k} - 1$ by $\sqrt{k} - 1$ and get $k - 2\sqrt{k} + 1$.

$$k - 5 = k - 2\sqrt{k} + 1$$

Now we rearrange so that $\sqrt{k}$ is isolated.

$$-6 = -2\sqrt{k}$$

$$3 = \sqrt{k}$$

Next we square both sides again and get

$$\mathbf{9 = k}$$

Now we must check this value of k in the original equation.

$$\sqrt{(9) - 5} - \sqrt{(9)} + 1 = 0 \qquad \text{substituted}$$

$$\sqrt{4} - 3 + 1 = 0 \qquad \text{simplified}$$

$$0 = 0 \qquad \text{check}$$

example 76.B.2   Solve: $\sqrt{s - 8} + \sqrt{s} = 2$

solution   We begin by changing the equation so that $\sqrt{s - 8}$ is isolated on the left side.

$$\sqrt{s - 8} = 2 - \sqrt{s}$$

We will square both sides to eliminate the square root radical on the left side.

$$(\sqrt{s - 8})^2 = (2 - \sqrt{s})^2 \qquad \text{square both sides}$$

$$s - 8 = 4 - 4\sqrt{s} + s \qquad \text{multiplied}$$

$$-12 = -4\sqrt{s} \qquad \text{simplified}$$

$$3 = \sqrt{s} \qquad \text{divided by } -4$$

Now to finish, we square both sides again and get

$$9 = s$$

as a solution. Now we must check this value of s in the original equation.

$$\sqrt{(9) - 8} + \sqrt{9} = 2 \qquad \text{substituted}$$
$$\sqrt{1} + 3 = 2 \qquad \text{simplified}$$
$$4 = 2 \qquad \text{not true}$$

Our answer does not check so the original equation has no solution.

problem set 76

1. Hahira could drive 320 miles in twice the time it took Sylvester to drive 240 miles. The speed of Sylvester was 20 mph greater than that of Hahira. What were the speeds and times of both?

2. Enjoyment for some varies inversely as the cost. If enjoyment measured 500 when the cost was $10, what did enjoyment measure when the cost was reduced to $1?

3. The drink cloyed the appetite quickly because it was 30% sugar. If 50 liters was on hand, how much water should be added to reduce the sugar concentration to 3% sugar?

4. Rasputin searched for three consecutive multiples of 7 such that the sum of the first and 4 times the third exceeded 3 times the second by 133. What were the multiples?

5. Camilla ran and Quitman walked. Thus, Camilla could complete the trip in 6 hours while Quitman took 12 hours. What was Camilla's rate if she ran 4 miles per hour faster than Quitman walked?

Solve:

*6. $\sqrt[3]{x^3 + 6x^2 - 4} - x - 2 = 0$

*7. $\sqrt{k - 5} - \sqrt{k + 1} = 0$

*8. $\sqrt{s - 8} + \sqrt{s} = 2$

9. $\begin{cases} x + y + z = 7 \\ 2x - y - z = -4 \\ z = 2y \end{cases}$

10. $\begin{cases} 2x + y - z = 7 \\ x - 2y + z = -2 \\ 2y + z = 0 \end{cases}$

11. Add: $-4\underline{/-30°} + 6\underline{/-200°}$

12. Write $-2R - 7U$ in polar form.

13. Add: $\dfrac{7x + 2}{x^2 - 2x - 15} - \dfrac{2}{5 - x}$

Simplify:

14. $\dfrac{2 - \sqrt{2}}{2\sqrt{2} - 1}$

15. $\dfrac{3 + 2\sqrt{5}}{1 - \sqrt{5}}$

16. Use the quadratic formula to solve: $-8x - 1 = 2x^2$

17. $\dfrac{a}{x} = m\left(\dfrac{a}{R_1} + \dfrac{b}{R_2}\right)$; find R_1

18. $\dfrac{a}{x} = m\left(\dfrac{a}{R_1} + \dfrac{b}{R_2}\right)$; find R_2

Simplify:

19. $a + \dfrac{a}{a + \dfrac{a}{x}}$

20. $-\sqrt{-9} - 3i^3 - 2i^4 + 2$

21. Solve: $\begin{cases} \dfrac{3}{7}x + \dfrac{2}{5}y = 10 \\ .03x - .2y = -1.58 \end{cases}$

22. Solve $-7x - 1 = 2x^2$ by completing the square.

23. Find the surface area of this right circular cylinder in square centimeters. Dimensions are given in meters.

24. Convert 4000 cubic centimeters per second to cubic feet per minute.

Simplify:

25. $\sqrt[4]{4\sqrt[3]{2}}$

26. $\dfrac{(-3)^0(-3^0)}{-9^{-\frac{3}{2}}}$

27. $(2\sqrt{5} + 5)(5\sqrt{20} - 1)$

28. Find the equation of the line that passes through $(-2, 4)$ that is perpendicular to $5x + 4y = 3$.

29. Find the distance between $(-3, 8)$ and $(5, -2)$.

30. Solve: $\dfrac{3x - 1}{4} - \dfrac{x - 5}{7} = 1$

ER 76-48, 76-49

LESSON 77 *Force vectors*

77.A

force vectors at a point

In the study of physical science much effort is devoted to understanding the relationships that force, mass, weight, velocity, and acceleration have to one another. In this study, it is helpful to use vectors to represent forces; and, if we do this, we can add the force vectors that act on a point to find the resultant force.

example 77.A.1   Two ropes are attached to a point and pulled on with the directions and magnitudes shown. What is the resultant force on the point?

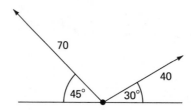

solution If we complete the triangles, we get a picture of the problem. The left vector pulls to the left and up, and the right vector pulls to the right and up.

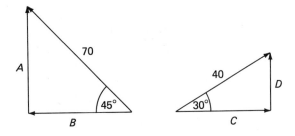

Next we solve for *A*, *B*, *C*, and *D*.

$$A = 70 \sin 45° \longrightarrow A = 70(.7071) = 49.497 \qquad C = 40 \cos 30° = (40)(.8660)$$
$$= 34.64$$

$$B = 70 \cos 45° \longrightarrow B = 70(.7071) = 49.497 \qquad D = 40 \sin 30° = 40(.5) = 20$$

Thus the total left-right force is $\qquad -49.497 + 34.64 = -14.857$

and the total up-down force is $\qquad +49.497 + 20 = 69.497$

Thus the resultant force is

$$-14.86R + 69.50U$$

To express the resultant force in polar coordinates we need to find the angle and the hypotenuse of the triangle.

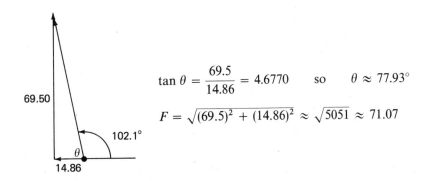

$$\tan \theta = \frac{69.5}{14.86} = 4.6770 \qquad \text{so} \qquad \theta \approx 77.93°$$

$$F = \sqrt{(69.5)^2 + (14.86)^2} \approx \sqrt{5051} \approx 71.07$$

Thus the resultant force is $71.07 \underline{/102.07°}$. This is the angle and pull that must be used with a single rope to get the same result as obtained by using the two ropes.

example 77.A.2 Here a point is acted on by a push force and a pull force as shown. Find the resultant force on the point.

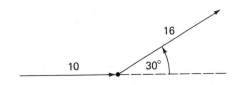

solution   The point can't tell the difference between a push from the left and a pull from the right. So we can redraw the problem, showing both vectors as pull vectors.

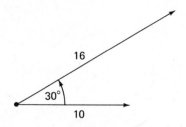

The horizontal vector is $10R + 0U$ and we find the horizontal and vertical components of the other vector to be $13.856R$ and $8U$.

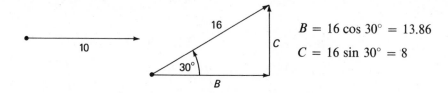

$$B = 16 \cos 30° = 13.86$$
$$C = 16 \sin 30° = 8$$

The resultant vector is the sum of the two vectors:

$$\begin{array}{r} 10 \quad R + 0U \\ 13.86R + 8U \\ \hline 23.86R + 8U \end{array}$$

$$\tan \theta = \frac{8}{23.86} \approx .335 \quad \longrightarrow \quad \theta \approx 18.54°$$

$$F = \sqrt{(23.86)^2 + 8^2} \quad \longrightarrow \quad F \approx \sqrt{633.3} \quad \longrightarrow \quad F \approx 25.17$$

So the resultant force in polar form is

$$\underline{25.17\angle 18.54°}$$

problem set 77

1. Rebecca covered the 135 miles in 4 fewer hours than it took Jean to cover 945 miles. If Jean's speed was 3 times that of Rebecca, what was the speed of each and for how long did each one travel?

2. Cathy and Ernie had four consecutive even integers. They found that the product of the second and fourth was 16 less than the product of -3 and the sum of the first and third. What were their integers?

3. Fourteen percent of the compound was pure chlorine. If 3440 grams was not chlorine, what was the total weight of the compound?

4. The assonance was uncanny because 42 percent of the chords sounded alike. If 232 chords had different sounds, how many chords were there in all?

5. The quotient of two numbers was 2 and their product was 200. What were the numbers?

6. The value of the fraction was 5. Twice the numerator was 6 greater than 7 times the denominator. What was the fraction?

Solve:

7. $\sqrt{x-9} + \sqrt{x} = 3$ **8.** $\sqrt{x-8} + \sqrt{x} = 4$

9. $\sqrt{k-24} = 6 - \sqrt{k}$

10. Draw an estimate of the line indicated by the data points and then write the equation that expresses steel as a function of iron: $S = mI + b$.

11. Solve: $\begin{cases} x - 2y - z = -9 \\ 2x - y + 2z = 7 \\ 3x - y = 0 \end{cases}$

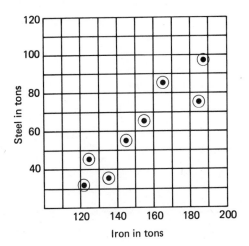

12. The two forces act on a point as shown. Find the resultant force.

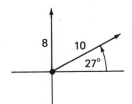

13. Add: $\dfrac{x-3}{x^2 + 5x - 14} + \dfrac{3x}{2-x}$

Simplify:

14. $\dfrac{3 - \sqrt{5}}{\sqrt{5} + 2}$ **15.** $\dfrac{2 + 2\sqrt{2}}{3 - 3\sqrt{2}}$ **16.** $\dfrac{4 - 3\sqrt{2}}{1 - \sqrt{2}}$

17. Use the quadratic formula to solve: $x^2 = -x - 1$

18. $mc = a\left(\dfrac{1}{x} + \dfrac{1}{R_1}\right)$; find R_1 **19.** $mc = a\left(\dfrac{1}{x} + \dfrac{1}{R_1}\right)$; find c

Simplify:

20. $a + \dfrac{a}{a + \dfrac{x}{a}}$ **21.** $-5i^3 - \sqrt{-9} + \sqrt{-3}\sqrt{-3}$

22. Find the surface area of the box in square centimeters. Dimensions are in meters.

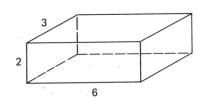

23. Convert 400 cubic inches per second to cubic centimeters per minute.

Simplify:

24. $2\sqrt[5]{4\sqrt[6]{2}}$

25. $\dfrac{-2^0(2^{-2})}{4^{-\frac{3}{2}}}$

26. $(2\sqrt{4} - 2)(3\sqrt{9} - 2)$

27. $4\sqrt{\dfrac{5}{8}} - 3\sqrt{\dfrac{8}{5}} + 2\sqrt{40}$

28. Estimate: $(.0471632 \times 10^{-5})(1200)(1200)$

29. Find v if $\dfrac{\sqrt{mv}}{e} = p$ and if $m = 4 \times 10^7$, $e = 500$, and $p = 100 \times 10^{-14}$.

ER 77-50, 77-51

LESSON 78   *Metric volume*

78.A

metric volume

When we convert units of volume, it is necessary to use three unit multipliers for each step in the process. For example, to convert 7 cubic meters (m^3) to cubic centimeters (cm^3), we must use three unit multipliers, as shown here.

$$7\,m^3 \times \frac{100 \text{ cm}}{m} \times \frac{100 \text{ cm}}{m} \times \frac{100 \text{ cm}}{m} = 7{,}000{,}000 \text{ cm}^3$$

The cubic meter is a rather large unit of volume, and the cubic centimeter is a rather small unit of volume. The **liter** is used to fill the requirement for an intermediate unit of volume and is about the size of a quart. **A liter is defined to equal the volume of 1000 cubic centimeters.**

$$1 \text{ liter} = 1000 \text{ cubic centimeters (cm}^3)$$

Because a cubic centimeter is one one-thousandth of a liter, a cubic centimeter is often called a milliliter and is abbreviated ml.

$$1 \text{ cm}^3 = 1 \text{ ml}$$

We will use the terms cubic centimeter and milliliter interchangeably in this book because they are used interchangeably in chemistry and other science courses. **Because of the way liters are defined, only one unit multiplier is required to make volume unit conversions between cubic centimeters (milliliters) and liters.**

example 78.A.1   Convert 14,000 ml to liters.

solution   **We need only one unit multiplier.** We can use either

$$\text{(a)} \quad \frac{1000 \text{ ml}}{1 \text{ liter}} \qquad \text{or} \qquad \text{(b)} \quad \frac{1 \text{ liter}}{1000 \text{ ml}}$$

We will use form (b) so that the ml units will cancel.

$$14{,}000 \text{ ml} \times \frac{1 \text{ liter}}{1000 \text{ ml}} = 14 \text{ liters}$$

example 78.A.2   Convert 4 cubic feet to liters.

solution We will go from feet to inches to cubic centimeters to liters. **Note that to go from cubic centimeters to liters, we need only one unit multiplier.**

$$4 \text{ ft}^3 \times \frac{12 \text{ in}}{1 \text{ ft}} \times \frac{12 \text{ in}}{1 \text{ ft}} \times \frac{12 \text{ in}}{1 \text{ ft}} \times \frac{2.54 \text{ cm}}{1 \text{ in}} \times \frac{2.54 \text{ cm}}{1 \text{ in}} \times \frac{2.54 \text{ cm}}{1 \text{ in}} \times \frac{1 \text{ liter}}{1000 \text{ cm}^3}$$

$$= \frac{(4)(12)(12)(12)(2.54)(2.54)(2.54)}{1000} \text{ liters}$$

example 78.A.3 Convert 4.7 liters to cubic inches.

solution We will go from liters to cubic centimeters to cubic inches.

$$4.7 \text{ liters} \times \frac{1000 \text{ cm}^3}{1 \text{ liter}} \times \frac{1 \text{ in}}{2.54 \text{ cm}} \times \frac{1 \text{ in}}{2.54 \text{ cm}} \times \frac{1 \text{ in}}{2.54 \text{ cm}} = \frac{(4.7)(1000)}{(2.54)^3} \text{ in}^3$$

problem set 78

1. Profligate Pauline found that the total cost varied directly as the number purchased. The cost was $2142 when she purchased 3. What would it cost her to purchase 10?

2. Darnley could travel 900 kilometers in 5 times the time it took Essex to travel 120 kilometers. Find the rates and times of both if Darnley's speed exceeded that of Essex by 10 kph.

3. What is the percent weight of the carbon (C) in a quantity of carbon dioxide, CO_2? (C, 12; O, 16)

4. The two jugs stood side by side on the shelf. One contained a solution that was 60% antiseptic and the solution in the other one was 90% antiseptic. How much of each should be used to get 50 milliliters of a solution that is 78% antiseptic?

5. When the hoplites collected their weapons, they found that the ratio of swords to spears was 2 to 7. Further, they found that 5 times the number of swords exceeded the number of spears by 120. How many of each did they have?

*6. Convert 14,000 ml to liters.   *7. Convert 4 cubic feet to liters.

Solve:

8. $\sqrt{x^2 - 4x + 4} = x + 2$ 9. $\sqrt{s} = 4 - \sqrt{s + 8}$

10. $\begin{cases} 2x - y + 2z = 3 \\ x - y - 2z = -6 \\ 3x - y = 0 \end{cases}$

11. The two forces act on a point as shown. Find the resultant force.

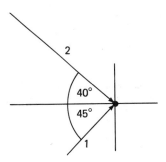

12. Add: $\dfrac{4x + 3}{x^2 - 9} - \dfrac{2x}{3 - x}$

Simplify:

13. $\dfrac{2 - \sqrt{2}}{4 + \sqrt{2}}$ **14.** $\dfrac{-2 - \sqrt{3}}{2\sqrt{3} + 2}$ **15.** $\dfrac{-1 - \sqrt{2}}{-5 - \sqrt{2}}$

16. Begin with $ax^2 + bx + c = 0$ and derive the quadratic formula by completing the square.

17. $\dfrac{a}{x} = m\left(\dfrac{a}{R_1} + \dfrac{b}{R_2}\right)$; find a **18.** $\dfrac{a}{x} = m\left(\dfrac{a}{R_1} + \dfrac{b}{R_2}\right)$; find b

Simplify:

19. $ax - \dfrac{ax}{a - \dfrac{a}{x}}$

20. $\sqrt{-4} - 3i^2 - 2i^4 + 2 - \sqrt{-2}\sqrt{-2}$

21. Solve: $\begin{cases} \dfrac{1}{5}x - \dfrac{1}{4}y = -6 \\ .2x + .2y = 12 \end{cases}$

22. Use the quadratic formula to solve $3x^2 - 1 = 2x$.

23. Find the surface area of the pyramid. Include the square bottom. Dimensions are in feet.

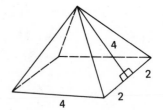

24. Convert 40 ml per second to cubic inches per hour.

Simplify:

25. $3\sqrt{9^2\sqrt{3}}$ **26.** $2\sqrt{\dfrac{7}{2}} + 3\sqrt{\dfrac{2}{7}} - 2\sqrt{126}$

27. $(4\sqrt{2} + 3)(5\sqrt{2} - 4)$ **28.** Estimate: $\dfrac{(41,852)(5062)(7184)}{492,000,000}$

29. Find the equation of the line that passes through $(5, 7)$ whose slope is $\frac{1}{5}$.

30. Find the distance between $(4, -2)$ and $(4, -6)$.

ER 78-52, 78-53

LESSON 79 *Direct and inverse variation as ratios*

79.A

direct variation as a ratio

When a problem states that variable A varies directly as variable B, we know that the relationship implied is

$$A = kB$$

and that the first step in solving the problem is to find the value of the constant of proportionality k.

example 79.A.1 A varies directly as B. If A is 50 when B is 5, what is the value of A when B is 7?

solution   The equation implied is

$$A = kB$$

and if we use 50 for A and 5 for B, we find that the constant k equals 10.

$$50 = k5 \longrightarrow k = 10$$

Now we can completely define the relationship by writing

$$A = 10B$$

We are asked to find the value of A when B is 7, so we replace B with 7 and get

$$A = 10(7) \longrightarrow \boldsymbol{A = 70}$$

There is another way to work this problem because the statement that A varies directly as B also implies the equation

$$\frac{A_1}{A_2} = \frac{B_1}{B_2}$$

Note that both A's are on the same side and that A_1 and B_1 are both on top. We have been given initial values of A_1 and B_1 of 50 and 5 and a value of 7 for B_2. If we insert these values, we get

$$\frac{50}{A_2} = \frac{5}{7}$$

We solve this equation by first cross multiplying and then dividing both sides by 5.

$$7 \cdot 50 = 5A_2 \longrightarrow \frac{7 \cdot 50}{5} = \frac{5A_2}{5} \longrightarrow 70 = A_2$$

It is important to understand that there are two ways to work these problems; some upper-level science authors use the first approach while others use the second approach.

example 79.A.2   Cost varies directly as the number purchased. If 12 can be purchased for \$78, how much would 42 cost?

solution   The two equations that we can use are

$$\text{(a)} \quad C = kN \qquad \text{and} \qquad \text{(b)} \quad \frac{C_1}{C_2} = \frac{N_1}{N_2}$$

We will first use equation (a) and three steps.

     (a)   Step 1 $C = kN \longrightarrow 78 = k12 \longrightarrow k = 6.5$

              Step 2 $C = 6.5N$

              Step 3 $C = 6.5(42) \longrightarrow \boldsymbol{C = \$273}$

Now we will use equation (b).

$$\text{(b)} \quad \frac{C_1}{C_2} = \frac{N_1}{N_2} \longrightarrow \frac{78}{C_2} = \frac{12}{42} \longrightarrow 78 \cdot 42 = 12C_2$$

$$\frac{78 \cdot 42}{12} = \frac{12C_2}{12} \longrightarrow \boldsymbol{\$273 = C_2}$$

79.B

inverse variation as a ratio

If A varies inversely as B, the following equation is implied.

$$A = \frac{k}{B}$$

This statement also implies the inverted ratio

$$\frac{A_1}{A_2} = \frac{B_2}{B_1}$$

Note that both A's are on the same side and both B's are on the other side but the B's are in the inverted form. We will work the next problem using both formats.

example 79.B.1   Blues vary inversely as yellows squared. If 100 blues go with 2 yellows, how many blues go with 10 yellows?

solution   The two relationships implied are

$$\text{(a)} \quad B = \frac{k}{Y^2} \quad \text{and} \quad \text{(b)} \quad \frac{B_1}{B_2} = \frac{Y_2^2}{Y_1^2}$$

We will use equation (a) first. We begin by finding k.

Step 1 $B = \dfrac{k}{Y^2} \quad \longrightarrow \quad 100 = \dfrac{k}{(2)^2} \quad \longrightarrow \quad k = 400$

Step 2 $B = \dfrac{400}{Y^2}$

Step 3 $B = \dfrac{400}{(10)^2} \quad \longrightarrow \quad B = 4$

Thus, 4 blues go with 10 yellows. Now we will rework the problem, using equation (b).

$$\frac{B_1}{B_2} = \frac{Y_2^2}{Y_1^2} \quad \longrightarrow \quad \frac{100}{B_2} = \frac{(10)^2}{(2)^2} \quad \longrightarrow \quad 400 = 100 B_2 \quad \longrightarrow \quad 4 = B_2$$

We get the same answer with either approach!

problem set 79

*1.   Cost varies directly as the number purchased. If 12 can be purchased for $78, how much would 42 cost? Work twice—once using the direct variation form and again using ratio form.

*2.   Blues vary inversely as yellows squared. If 100 blues go with 2 yellows, how many blues go with 10 yellows? Work twice—once using the direct variation form and again using the ratio form.

3.   The ratio of acrobats in blue to acrobats in pink was 4 to 5. Those that wore blue numbered 1200 fewer than twice the number that wore pink. How many wore blue and how many wore pink?

4.   The place was crowded because $3\frac{1}{4}$ times as many people had come as the fire marshal would permit. If 650 had come, how many would the fire marshal permit?

5.   Dogs were $40 each and cats cost only $2 each. If Doerun bought 30 animals and paid a total of $820, how many animals of each kind did he buy?

Solve:

6. $\sqrt{x^2 - x - 2} - x + 2 = 0$

7. $\sqrt{p + 20} + \sqrt{p} = 10$

8. $\sqrt{s - 18} + \sqrt{s - 36} = 0$

9. $\begin{cases} x + y + z = 8 \\ 2x - 3y - z = -6 \\ 2x - z = 0 \end{cases}$

10. Add: $4\underline{/60°} - 6\underline{/-200°}$

11. Write $-2R + 6U$ in polar form.

12. Add: $\dfrac{4x + 2}{x^2 - 6x - 16} - \dfrac{3}{x - 8}$

Simplify:

13. $\dfrac{3\sqrt{2} - 1}{1 + \sqrt{2}}$

14. $\dfrac{2 - 3\sqrt{2}}{3 - 2\sqrt{2}}$

15. Find B.

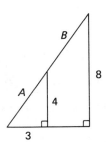

16. Solve $-3x^2 + 2 = -3x$ by completing the square.

17. $a = xm\left(\dfrac{p}{y} + \dfrac{q}{c}\right)$; find y

18. $a = xm\left(\dfrac{p}{y} + \dfrac{q}{c}\right)$; find m

Simplify:

19. $3a - \dfrac{3}{a - \dfrac{3}{a}}$

20. $-\sqrt{-4} + \sqrt{-9} - i^3 + \sqrt{-2}\sqrt{-2} - 4i^4$

21. Solve: $\begin{cases} \dfrac{2}{3}x - \dfrac{1}{4}y = 6 \\ .07x + .06y = 1.32 \end{cases}$

22. Use the quadratic formula to solve $-x^2 = -x - 5$.

23. Convert 600 cubic centimeters per minute to cubic feet per hour.

24. The figure is the base of a container 6 ft high. Find the number of 1-inch sugar cubes this container will hold. Dimensions are in inches.

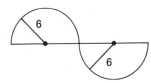

Simplify:

25. $\sqrt[4]{x^5 y}\sqrt{xy^3}$

26. $\dfrac{-2^0(-2)^0}{-(4)^{-\frac{3}{2}}}$

27. Solve by graphing and then use substitution or elimination to get an exact solution:
$\begin{cases} x - 3y = 6 \\ 2x + y = -1 \end{cases}$

28. Add: $\dfrac{3}{ax} + \dfrac{3x}{a^2x} + \dfrac{7x}{x+a}$ **29.** Solve: $x^3 = 4x^2 + 32x$

30. Find the equation of the line that passes through $(-2, -3)$ which is perpendicular to $-x - y - 1 = 0$.

ER 79-54, 79-55

LESSON 80 Division of complex numbers

80.A
complex numbers and real numbers

A complex number is a number that has a real part and an imaginary part. When the real part is written first, we say that we have written the **complex number in standard form.** Thus, the general expression for a complex number in standard form is

$$a + bi$$

The letter a can be any real number, and the letter b can be any real number. Thus, all of these numbers

$$\text{(a)} \quad -\sqrt{2} + 3i \qquad \text{(b)} \quad -\frac{4\sqrt{3}}{5} + 2\sqrt{3}i \qquad \text{(c)} \quad 3 - \frac{23}{\sqrt{2}}i$$

are complex numbers in standard form because all the replacements for a and b are real numbers. If a equals zero, we can say that these numbers

$$\text{(d)} \quad +3i \qquad \text{(e)} \quad +2\sqrt{3}i \qquad \text{(f)} \quad -\frac{23}{\sqrt{2}}i$$

are all complex numbers whose real part equals zero. If the coefficient of the imaginary part of a complex number is zero, we get a complex number that has only a real part, such as the following:

$$\text{(g)} \quad -\sqrt{2} \qquad \text{(h)} \quad -\frac{4\sqrt{3}}{2} \qquad \text{(i)} \quad 3$$

Thus, we see that every real number is a complex number whose imaginary part is zero, and every imaginary number is a complex number whose real part is zero. **Thus, the set of real numbers is a subset of the set of complex numbers, and the set of imaginary numbers is also a subset of the set of complex numbers.**

The complex number

$$\frac{4 - 3i}{5}$$

is not in standard form because it is not in the form $a + bi$. However, it takes only a slight change to write it in standard form as

$$\frac{4}{5} - \frac{3}{5}i$$

80.B
products of complex conjugates

We have noted that the product of a number and its conjugate has the form $a^2 - b^2$.

$$
\begin{array}{ll}
a + b & 3 + 5\sqrt{2} \\
a - b & 3 - 5\sqrt{2} \\
\hline
a^2 + ab & 9 + 15\sqrt{2} \\
\quad\; - ab - b^2 & \quad\; - 15\sqrt{2} - 50 \\
\hline
a^2 \qquad - b^2 & 9 \qquad\quad - 50
\end{array}
$$

If we multiply a complex number in standard form by its conjugate, we get an answer in the form $a^2 + b^2$. The sign change between the numbers is caused by the presence of an i^2 factor in the second part of the product.

$$
\begin{array}{ll}
a + bi & 3 + 4i \\
a - bi & 3 - 4i \\
\hline
a^2 + abi & 9 + 12i \\
\quad\; - abi - b^2 i^2 & \quad\; - 12i - 16i^2 \\
\hline
a^2 \qquad\; + b^2 & 9 \qquad\quad + 16
\end{array}
$$

We note that neither product has an imaginary part. We will find that we can use this fact to eliminate the i factor in the denominator of a fraction of complex numbers.

80.C
division of complex numbers

The notation

$$
\frac{x^2 - 2x + 7}{x + 3}
$$

indicates that $x^2 - 2x + 7$ is to be divided by $x + 3$. There is a format and a procedure we can use to perform this division.

$$
\begin{array}{r}
x - 5 \\
x + 3 \overline{\smash{\big)}\, x^2 - 2x + 7} \\
\underline{x^2 + 3x} \\
-5x + 7 \\
\underline{-5x - 15} \\
22
\end{array}
$$

By using this procedure, we find that $\dfrac{x^2 - 2x + 7}{x + 3}$ equals $x - 5 + \dfrac{22}{x + 3}$.

If we encounter the expression

$$
\frac{2 + 3i}{4 - 2i}
$$

we see that this notation indicates that $2 + 3i$ is to be divided by $4 - 2i$. **Unfortunately, there is no format or procedure that we can use to perform this division. However, we can change the form of the expression by multiplying above and below by the conjugate of the denominator. The resulting expression will have a denominator that is a real number.**

example 80.C.1   Simplify: $\dfrac{2 + 3i}{4 - 2i}$

solution We can change the denominator to a rational number if we multiply above and below by $4 + 2i$, which is the **conjugate of the denominator.**

$$\frac{2 + 3i}{4 - 2i} \cdot \frac{4 + 2i}{4 + 2i}$$

We have two multiplications indicated, one above and one below. We will use the vertical format for each multiplication.

<table>
<tr><td align="center">ABOVE</td><td align="center">BELOW</td></tr>
</table>

ABOVE

$$\begin{array}{r} 2 + 3i \\ 4 + 2i \\ \hline 8 + 12i \\ 4i + 6i^2 \\ \hline 8 + 16i + 6i^2 = 2 + 16i \end{array}$$

BELOW

$$\begin{array}{r} 4 - 2i \\ 4 + 2i \\ \hline 16 - 8i \\ + 8i - 4i^2 \\ \hline 16 + 4 = 20 \end{array}$$

Thus, we can write our answer as

$$\frac{2 + 16i}{20} = \frac{1 + 8i}{10}$$

This answer is not in the preferred form of $a + bi$. We can write this complex number in standard form if we write

$$\frac{1}{10} + \frac{4}{5}i$$

example 80.C.2 Simplify: $\dfrac{4 - 2i}{2i - 3}$

solution Although it is not necessary, we will begin by writing the denominator in standard form as

$$\frac{4 - 2i}{-3 + 2i}$$

We can change the denominator to a real number if we multiply above and below by $-3 - 2i$.

$$\frac{4 - 2i}{-3 + 2i} \cdot \frac{-3 - 2i}{-3 - 2i}$$

We have two multiplications to perform.

ABOVE

$$\begin{array}{r} 4 - 2i \\ -3 - 2i \\ \hline -12 + 6i \\ - 8i + 4i^2 \\ \hline -12 - 2i + 4i^2 = -16 - 2i \end{array}$$

BELOW

$$\begin{array}{r} -3 + 2i \\ -3 - 2i \\ \hline 9 - 6i \\ 6i - 4i^2 \\ \hline 9 + 4 = 13 \end{array}$$

Thus, the new form of the expression is

$$\frac{-16 - 2i}{13}$$

which can be written in standard form as follows:

$$-\frac{16}{13} - \frac{2}{13}i$$

problem set 80

1. Monkeys varied directly as the turtles squared. When there were 2 turtles, there were 100 monkeys. How many monkeys were there when there were 5 turtles? Work the problem once using the variation format and again using the ratio format.

2. The number of macaws varied inversely as the number of apes squared. When there were 4 macaws, there were 10 apes. How many macaws were there when there were only 2 apes?

3. Roger made the 375-mile trip in 10 hours less than it took Judy. This was because he traveled 3 times as fast as Judy traveled. How fast did each travel and for how long did each travel?

4. Wittlocoodee had one solution that was 10% glycol and another that was 40% glycol. How much of each should she use to get 200 liters of solution that was 19% glycol?

5. The curmudgeon chortled with glee when the results were announced because only 60% had made it. If 1120 had not made it, how many had tried?

Simplify

*6. $\dfrac{2 + 3i}{4 - 2i}$

*7. $\dfrac{4 - 2i}{2i - 3}$

8. $\dfrac{3 - i}{2 + 5i}$

Solve:

9. $\sqrt{x^2 - x + 30} - 3 = x$

10. $\sqrt{p - 48} = 12 - \sqrt{p}$

11. $\begin{cases} x + 2y - 3z = 5 \\ 2x - y - z = 0 \\ y - 3z = 0 \end{cases}$

12. The two forces act on a point as shown. Find the resultant force.

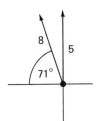

13. Add: $\dfrac{7x - 2}{x^2 - 9} + \dfrac{3x}{3 - x}$

Simplify:

14. $\dfrac{\sqrt{2} - 5}{\sqrt{2} - 2}$

15. $\dfrac{2\sqrt{3} - 1}{1 - 3\sqrt{3}}$

16. $\dfrac{1 + \sqrt{2}}{3 - \sqrt{2}}$

17. Find B.

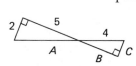

18. Begin with $ax^2 + bx + c = 0$ and derive the quadratic formula.

Simplify:

19. $2a^2 - \dfrac{3a}{a + \dfrac{1}{a}}$

20. $\sqrt{-9} - \sqrt{-2}\sqrt{-2} + \sqrt{-2}\sqrt{2} - 3i^3 - 2i^2$

21. Solve: $\begin{cases} \dfrac{2}{3}x - \dfrac{1}{3}y = 6 \\ .15x + .01y = .84 \end{cases}$

22. Solve $2 = -2x^2 - 3x$ by using the quadratic formula.

23. Divide $4x^3 - x + 2$ by $x - 4$.

24. Find the surface area of the cylinder in square centimeters. Dimensions are in meters.

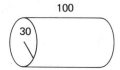

Simplify:

25. $3\sqrt{9\sqrt[4]{3}}$

26. $2\sqrt{\dfrac{1}{5}} - 3\sqrt{5} + 3\sqrt{20}$

27. Add: $\dfrac{x}{x + y} + \dfrac{3}{x^2 y} + \dfrac{2}{xy}$

28. Solve: $-4^2 - 3^0 - 2^0(x - y^0) - 3^0(-2x - 5) = 7$

29. Find the distance between $(-2, 8)$ and $(5, -3)$.

ER 80-56, 80-57

LESSON 81 *Algebraic simplifications*

algebraic
simplifications

There are three rules of algebra that stand above all the rest. Two of the rules cannot be used unless an equals sign is present, for these rules apply only to equations. We state them informally by saying:

  1. The same quantity can be added to both sides of an equation without changing the answer(s) to the equation.
  2. Every term on both sides of an equation can be multiplied by the same quantity (except zero) without changing the answer(s) to the equation.

The third rule can be used on the individual terms of an equation and can also be used on expressions that are not terms of an equation. With few exceptions, if no equals sign is present, this rule is the only rule that can be used.

  3. The denominator and the numerator of an algebraic expression can be multiplied by the same quantity (except zero) without changing the value of the expression. We call this rule the **denominator-numerator–same-quantity theorem.**

In Lesson 64 we found that an expression such as

$$a + \cfrac{1}{1 + \cfrac{1}{x}}.$$

can be written as a simple fraction by using the denominator–numerator–same-quantity theorem several times. We review this procedure by first simplifying $1 + \frac{1}{x}$ and we get

$$a + \cfrac{1}{\cfrac{x + 1}{x}}$$

Next we use the same rule again to simplify the second term and get

$$a + \cfrac{x}{x + 1}$$

We finish by using the rule again to change the form of the first term and then we add the two terms.

$$\frac{a(x + 1)}{x + 1} + \frac{x}{x + 1} = \frac{ax + a + x}{x + 1}$$

In this lesson we will discuss the repeated use of the denominator-numerator–same-quantity theorem to simplify expressions that are just a little more complicated.

example 81.A.1   Simplify: $\cfrac{a}{1 + \cfrac{a}{1 + \cfrac{1}{x}}}$

solution   We begin by adding $1 + \frac{1}{x}$, and we get

$$\cfrac{a}{1 + \cfrac{a}{\cfrac{x + 1}{x}}}$$

Next, we simplify the second term of the denominator and get

$$\cfrac{a}{1 + \cfrac{xa}{x + 1}}$$

Next we add the two terms in the denominator and get

$$\cfrac{a}{\cfrac{x + 1 + xa}{x + 1}}$$

and we finish by multiplying above and below by the reciprocal of the denominator.

$$\frac{a(x + 1)}{x + 1 + xa}$$

example 81.A.2   Simplify: $\cfrac{b}{a + \cfrac{c}{x + \cfrac{1}{y}}}$

solution   We will use the same procedure we used in the last problem. We begin by adding $x + \frac{1}{y}$ and getting

$$\cfrac{b}{a + \cfrac{c}{\cfrac{xy + 1}{y}}}$$

Next, we simplify the second term in the denominator and get

$$\cfrac{b}{a + \cfrac{cy}{xy + 1}}$$

Now we add the two terms in the denominator and get

$$\cfrac{b}{\cfrac{axy + a + cy}{xy + 1}}$$

and we finish by multiplying above and below by the reciprocal of the denominator.

$$\frac{b(xy + 1)}{axy + a + cy}$$

problem set 81

1. The discipline quotient varied inversely as the square of the number of unruly students. If the quotient was 300 when the unruly students totaled 5, what was the quotient when the unruly students totaled 10? Use the ratio format to solve.

2. The plane could fly 1920 miles in 4 more hours than it took the racer to drive 320 miles. The speed of the plane was 3 times the speed of the racer. Find the times of both and the speeds of both.

3. The initial pressure, temperature, and volume of a quantity of an ideal gas were 5 atmospheres, 540 K, and 250 cubic feet. What would the temperature be if the pressure were increased to 50 atmospheres and the volume reduced to 200 cubic feet?

4. After running for awhile at 10 mph, Mingo slowed to a walk at 5 mph. If he traveled 60 miles in 8 hours, how far did he run and how far did he walk?

5. The total weight of the sodium monohydrogen phosphate, Na_2HPO_4, was 852 grams. What was the weight of the sodium (Na) in this amount of the compound? (Na, 23; H, 1; P, 31; O, 16) What percent of the compound was sodium?

Simplify:

*6. $\cfrac{a}{1 + \cfrac{a}{1 + \cfrac{1}{x}}}$

*7. $\cfrac{b}{a + \cfrac{c}{x + \cfrac{1}{y}}}$

8. $\cfrac{m}{x + \cfrac{1}{y + \cfrac{a}{x}}}$

9. $\dfrac{2 - 3i}{4 + i}$

10. $\dfrac{5 - i}{2 - 3i}$

Solve:

11. $\sqrt{x^2 - x + 47} - 5 = x$

12. $\sqrt{x + 24} + \sqrt{x} = 12$

13. $\begin{cases} 2x - 2y - z = 16 \\ 3x - y + 2z = 5 \\ -y + 3z = 0 \end{cases}$

14. Add: $5\underline{/70°} - 30\underline{/-20°}$

15. Write $-4R + 7U$ in polar form.

16. Add: $\dfrac{3x - 2}{x^2 - 9} + \dfrac{x}{3 - x}$

Simplify:

17. $\dfrac{3\sqrt{2} - 1}{1 + \sqrt{2}}$

18. $\dfrac{-2 - \sqrt{5}}{2 + 2\sqrt{5}}$

19. Find B.

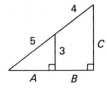

20. $\dfrac{z}{m^2} = \dfrac{p}{m}\left(\dfrac{x}{a} + y\right)$; find a

21. $\dfrac{z}{m^2} = \dfrac{p}{m}\left(\dfrac{x}{a} + y\right)$; find p

22. Simplify: $3i^3 - 2i^2 - \sqrt{-9} + \sqrt{-2}\sqrt{2}$

23. Solve: $\begin{cases} \dfrac{2}{5}x - \dfrac{1}{3}y = -1 \\ .07x + .2y = 2.15 \end{cases}$

24. Begin with $ax^2 + bx + c = 0$ and complete the square to derive the quadratic formula.

25. Solve $-3x^2 - 2 = 5x$ by using the quadratic formula.

Simplify:

26. $\sqrt{x^5 y}\ \sqrt[3]{x^2 y^5}$

27. $3\sqrt{\dfrac{2}{9}} + 3\sqrt{\dfrac{9}{2}} - 4\sqrt{50}$

28. $\dfrac{-2^0}{-4^{-5/2}}$

29. Solve by graphing and then get an exact solution by using substitution or elimination:
$\begin{cases} 4x - 3y = -3 \\ 4x + 3y = 6 \end{cases}$

30. Convert 400 cubic centimeters per second to cubic feet per minute.

ER 81-58, 81-59

LESSON 82 **Variable exponents**

82.A
product theorem with variables

The two major rules for exponents are the product theorem and the power theorem. The product theorem tells us that to multiply exponentials whose bases are equal, we add the exponents.

$$x^2 \cdot x^5 = x^7$$

We can see the reason for this rule if we break the factors down and get

$$xx \cdot xxxxx$$

for there are seven x factors, and we can indicate seven x factors by writing

$$x^7$$

Since variables represent unspecified numbers, we use the same rule when the exponents are variables. If the bases are the same, the exponentials are multiplied by adding the exponents. Thus

$$x^a \cdot x^b = x^{a+b}$$

Until now we have avoided using exponents that contain slash fractions as slash fractions are sometimes difficult to read and beginners often find them troublesome to interpret. They are used in advanced mathematics books because they are much easier to typeset with modern typesetting equipment and also because they take up less space. Thus we will begin to use slash fractions in this lesson to prepare for their use in later courses in science and mathematics.

example 82.A.1 Simplify: $x^a y^{2x} x^{a/2} y^{3x/4}$

solution We simplify by adding the exponents of like bases.

$$x^{a+a/2} y^{2x+3x/4}$$

Now since

$$a + \frac{a}{2} = \frac{3a}{2} \quad \text{and} \quad 2x + \frac{3x}{4} = \frac{11x}{4}$$

we finish by writing

$$x^{3a/2} y^{11x/4}$$

example 82.A.2 Simplify: $\dfrac{x^{a-2} y^{a+4}}{x^{a/2} y^{2a}}$

solution We begin by writing all exponentials in the numerator.

$$x^{a-2} x^{-a/2} y^{a+4} y^{-2a}$$

Now we finish by adding the exponents of x and the exponents of y and get

$$x^{a/2-2} y^{-a+4}$$

82.B
power theorem with variables

The power theorem tells us that when we have a notation such as

$$(x^3)^2$$

we simplify by multiplying the exponents and get

$$x^6$$

We can see the reason for this rule if we remember that a quantity squared means to multiply the quantity by itself. Thus,

$$(x^3)^2 \quad \text{means} \quad (x^3)(x^3) \quad \text{which is } x^6$$

The rule is the same when the exponents are variables.

example 82.B.1   Simplify: $(x^{a+2})^2$

solution   We multiply $a + 2$ by 2 and get

$$x^{2a+4}$$

example 82.B.2   Simplify: $\dfrac{(x^a)^b(y^a)^{b+2}}{x^{-a}}$

solution   First we use the power theorem in the numerator twice and get

$$\frac{x^{ab}y^{ab+2a}}{x^{-a}}$$

We finish by bringing up the x^{-a} from below, and we get

$$x^{ab+a}y^{ab+2a}$$

problem set 82

1. The number who were resentful varied directly as the number of invidious comparisons that were made. When 1200 were resentful, there had been 300 invidious comparisons made. When 100 invidious comparisons were made, how many were resentful? Work once using ratio and again using variation.

2. The 420 kilometers to Merida took twice as long as the 270 kilometers to Tulum. The disparity occurred because Raul drove 20 kilometers per hour faster when he went to Tulum. What were his times and rates to both destinations?

3. Americus won the race with Ashburn by 400 yards. If Americus' speed was 5 yards per second and her time was 400 seconds, how fast did Ashburn run?

4. Millsap observed that 4 times the number of igneous rocks exceeded 8 times the number of sedimentary rocks by 80. He also noted that 10 times the number of sedimentary rocks exceeded the number of igneous rocks by 140. How many rocks were igneous and how many were sedimentary?

5. Lynn and Laws knew that the confection should be exactly 8% sugar. It was to be made using one component that was 5% sugar and another that was 20% sugar. How much of each should they use to get 800 ml of the confection?

Simplify:

*6. $x^a y^{2x} x^{a/2} y^{3x/4}$

*7. $\dfrac{x^{a-2}y^{a+4}}{x^{a/2}y^{2a}}$

*8. $(x^{a+2})^2$

*9. $\dfrac{(x^a)^b(y^a)^{b+2}}{x^{-a}}$

10. $\dfrac{a}{x + \dfrac{1}{a + \dfrac{1}{x}}}$

11. $\dfrac{b}{a + \dfrac{b}{a + \dfrac{a}{b}}}$

12. $\dfrac{2 - 4i}{1 + i}$

13. $\dfrac{3 + 5i}{2 - 2i}$

Solve:

14. $\sqrt{k} + \sqrt{k + 32} = 8$

15. $\begin{cases} 2x + 3y - z = -3 \\ x + 2y = 0 \\ x - 2y + z = -2 \end{cases}$

16. Two forces act on a point as shown. Find the resultant force.

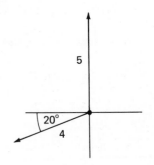

17. Add: $\dfrac{5}{x} + \dfrac{6}{x^2 - 4} - \dfrac{3x}{2 - x}$

Simplify:

18. $\dfrac{-2\sqrt{2} - 2}{4 + \sqrt{2}}$

<u>**19.**</u> $\dfrac{-\sqrt{3} - 3}{2 - \sqrt{3}}$

20. $p = \dfrac{a}{x} - c\left(\dfrac{a}{m} - y\right)$; find m

<u>**21.**</u> $p = \dfrac{a}{x} - c\left(\dfrac{a}{m} - y\right)$; find c

22. Simplify: $\sqrt{-2}\,\sqrt{2} - \sqrt{3}\,\sqrt{3} - \sqrt{-9} + 2i^2 - 3i^4$

23. Solve: $\begin{cases} x - \dfrac{2}{5}y = 11 \\ \\ -.05x - .2y = 1.65 \end{cases}$

24. Solve $2x^2 - 5x = 5$ by completing the square.

Simplify:

<u>**25.**</u> $\sqrt{\dfrac{7}{4}} + 2\sqrt{\dfrac{4}{7}} - 5\sqrt{63}$

26. $\dfrac{-2^0(-3^0)}{-27^{-2/3}}$

27. $(\sqrt[3]{x^2 y})^4$

28. Convert 4 liters per second to cubic centimeters per hour.

29. Find the distance between $(-3, 2)$ and $(5, -7)$.

30. Find the equation of the line that goes through $(-3, 2)$ and $(5, -7)$.

ER 82-60, 82-61

LESSON 83   *Solutions of equations*

83.A

**linear
equations**

The degree of a term of a polynomial is the sum of the exponents of the variables of the term. Thus

$$x^3 \qquad \text{is a third-degree term}$$
$$x^2ymp \qquad \text{is a fifth-degree term}$$
$$47x \qquad \text{is a first-degree term}$$

The degree of a polynomial equation is the same as the degree of the highest-degree term of the equation. Thus,

$$3x^2 + 2x + 5 = 0 \qquad \text{is a second-degree equation}$$
$$3x^3 + 2x^2 + 3 = 0 \qquad \text{is a third-degree equation}$$
$$xy + 5 = 0 \qquad \text{is a second-degree equation}$$
$$y = 3x + 2 \qquad \text{is a first-degree equation}$$

Thus far, we have restricted our equation solving efforts to first- and second-degree polynomial equations. First-degree equations in two or more unknowns are called linear equations because the graph of a first-degree equation in two unknowns is a straight line. The equation

$$3y + 2x = 4$$

is a first-degree polynomial equation in two unknowns, and thus the graph of this equation is a straight line. There is an infinite number of ordered pairs of x and y that will satisfy this equation, and the graph of these points is the line. When we have a system of two linear equations in two unknowns, there are three possibilities. The first is that the equations are equations of two lines that intersect, as shown in (a). Each of these lines has an infinite number of ordered pairs of x and y that satisfy its equation, but only one ordered pair, $(-1, 1)$, satisfies both equations and thus lies on both lines. A system of linear equations whose graphs intersect in one point is called a **consistent system** of equations. In (b) we have graphed both lines and find they have the same graph.

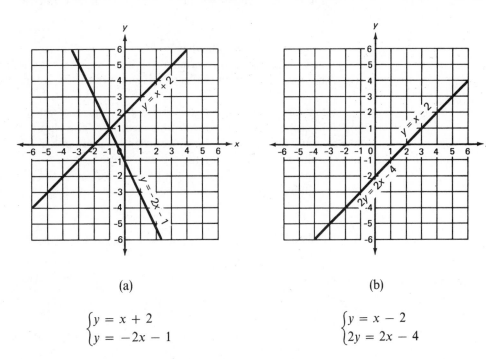

(a)   (b)

$$\begin{cases} y = x + 2 \\ y = -2x - 1 \end{cases} \qquad\qquad \begin{cases} y = x - 2 \\ 2y = 2x - 4 \end{cases}$$

Thus, the coordinates of any point that satisfy one of these equations satisfy the other equation. As we see, there is an infinite number of points whose coordinates satisfy both equations. We say that this system of equations is a **dependent system** of equations.

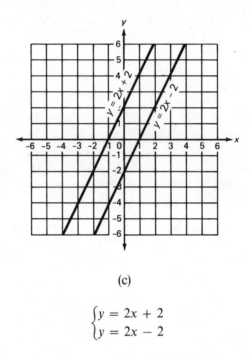

(c)

$$\begin{cases} y = 2x + 2 \\ y = 2x - 2 \end{cases}$$

The last possibility is that the lines have the same slopes but different intercepts and thus are parallel lines. Both of the lines in (c) have a slope of 2 and thus are parallel. These lines never cross and have no point in common. Systems of linear equations that have no solutions are called **inconsistent systems** of equations.

The words consistent, inconsistent, and dependent are not important. The important point is that given a system of two linear equations in two variables there are three possibilities.

1. The lines cross and thus have one point in common.
2. The equations are two forms of the equation of the same line.
3. The lines are parallel and thus have no point(s) in common.

If a system has one solution the lines will cross and the system can be solved by graphing or by using either the substitution method or the elimination method. If the lines do not cross, an attempt to find a solution by using either substitution or elimination must result in failure because there is no solution.

An attempt to solve a dependent system will cause the elimination of all variables, and we will end up with a true numerical statement such as

$$0 = 0 \quad \text{or} \quad 5 = 5$$

We will demonstrate by trying to solve system (b) by using both substitution and elimination.

SUBSTITUTION   ELIMINATION

$$2(x - 2) = 2x - 4 \qquad\qquad y = x - 2 \;\longrightarrow\; (2) \;\longrightarrow\; 2y = 2x - 4$$
$$2x - 4 = 2x - 4 \qquad\qquad 2y = 2x - 4 \;\longrightarrow\; (-1) \;\longrightarrow\; \underline{-2y = -2x + 4}$$
$$0 = 0 \quad \text{True} \qquad\qquad\qquad\qquad\qquad\qquad\qquad\qquad 0 = 0 \quad \text{True}$$

A system of two equations whose graphs are parallel has no solution, so an attempt to solve one of these systems will also result in failure. Again, the variables will be eliminated, but this time the final result will be a false numerical statement such as

$$0 = 5 \qquad \text{or} \qquad -3 = 7$$

We will demonstrate by trying to find a solution for system (c).

SUBSTITUTION   ELIMINATION

$$
\begin{array}{ll}
2x - 2 = \quad 2x + 2 \\
\underline{-2x \qquad\quad -2x} \\
\qquad -2 = +2 \quad \text{False}
\end{array}
$$

$$
\begin{array}{lll}
y = 2x + 2 & \longrightarrow \ (-1) \ \longrightarrow & -y = -2x - 2 \\
y = 2x - 2 & \longrightarrow \ (1) \ \longrightarrow & \underline{\ y = \quad 2x - 2} \\
& & \ 0 = -4 \quad \text{False}
\end{array}
$$

The graph of a system of three linear equations in three unknowns is a graph of three planes. If the three planes intersect in a common point as in the following figure, an ordered triple (x, y, z) exists that will satisfy all three equations. We say that the equations are **consistent equations.** We reserve the use of the word **inconsistent** to describe equations that do not have any ordered triples that will satisfy all three equations.

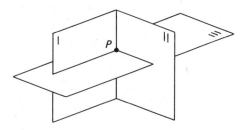

The equations of the three parallel planes of the figure on the left cannot be satisfied by one ordered triple (x, y, z), for there is no point that is common to all three planes. Thus, these equations are called inconsistent equations. Some equations are neither consistent nor inconsistent. The equations of the three planes of the figure on the right are neither consistent nor inconsistent equations. These planes intersect to form a line, and thus there is an infinite number of ordered triples (x, y, z) that satisfies all three equations.

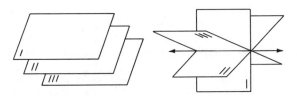

problem set 83

1. The number who were admired varied inversely with the number who had a proclivity for bragging. When 80 bragged, only 20 were admired. How many were admired when 10 bragged? Work using ratio and again using variation.

2. **The temperature in the gas law equations must be in kelvins, and a temperature in kelvins is a number 273 greater than the number that designates the same temperature in degrees Celsius.** Thus, 50 degrees Celsius (50°C) equals $273 + 50 = 323$ kelvins. If in an experiment the volume was held constant and the initial pressure and temperature were 4 atmospheres and 50°C, what would be the final temperature in degrees Celsius if the pressure were increased to 8 atmospheres?

3. When the compound was analyzed, its chemical formula was determined to be $Li_2Ca_2O_7$. If the weight of the lithium (Li) was 56 grams, what was the total

weight of the compound? (Li, 7; Ca, 40; O, 16) What percent of the compound was lithium?

4. The chemist pondered. She had 160 ml of a solution that was 10% glycol. The other solution available was 30% glycol. How much of the other solution should she use to get a solution that was 22% glycol?

5. It really took 150% more borax than had been agreed on. If the agreement was for 140,000 tons of borax, how many tons did it really take?

Simplify:

6. $y^b x^{c+2} y^{b/3} x^{-2-p}$

7. $\dfrac{y^b x^{c+3}}{y^{b/3} x^{2-p}}$

8. $(x^{a+3})^b x^{-2b+4}$

9. $\dfrac{(y^a)^{b+2} y^{-ab}}{y^{-2+a}}$

10. $\dfrac{p}{a + \dfrac{m}{1 + \dfrac{1}{am}}}$

11. $\dfrac{1}{p - \dfrac{b}{b - \dfrac{1}{x}}}$

12. $\dfrac{2 - 3i}{1 + i}$

13. $\dfrac{3 + 4i}{3 - 3i}$

Solve:

14. $\sqrt{s - 48} + \sqrt{s} = 8$

15. $\begin{cases} 3x - y - 2z = -6 \\ 2x - y + z = 2 \\ -y + z = 0 \end{cases}$

16. Add: $-6\underline{/-150°} + 4\underline{/20°}$

17. Write $3R + 8U$ in polar form.

18. Add: $\dfrac{4}{x^2} - \dfrac{2}{-x(x - 3)} + \dfrac{5x}{3 - x}$

19. Simplify: $\dfrac{4\sqrt{2} - 5}{3\sqrt{2} + 2}$

20. Multiply: $(2 + 3\sqrt{20})(4 - 5\sqrt{45})$

21. $\dfrac{m}{c} + a = p\left(\dfrac{1}{x} + \dfrac{b}{y}\right)$; find x

22. $\dfrac{m}{c} + x = p\left(\dfrac{1}{x} + \dfrac{b}{y}\right)$; find c

23. Simplify: $\sqrt{-3}\sqrt{3} - \sqrt{2}\sqrt{2} - \sqrt{-4} + 3i^2 - 2i^5$

24. Solve $-2x^2 - x = 3$ by using the quadratic formula.

25. Draw an estimate of the line indicated by the data points and then write the equation that expresses yttrium as a function of boron: $Y = mB + b$.

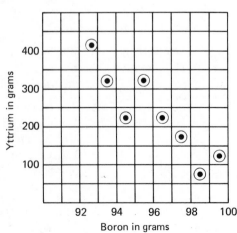

26. Find the surface area of the right circular cylinder in square feet. Dimensions are in feet.

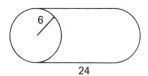

27. Simplify: $3\sqrt{\dfrac{5}{3}} + 2\sqrt{\dfrac{3}{5}} - 4\sqrt{60}$

28. Convert 40 centimeters per second to miles per hour.

29. Solve: $-2^0[-2 - 3(x - 2^2)] - 3(2x - 5^0) = 7 - 2(2x + 2)$

30. Simplify: $-3\{[(-2 - 3) - 2] - 2[-4(-3 - 2^0)]\} + 2^0(-3)$

ER 83-62, 83-63

LESSON 84 *Systems of nonlinear equations*

84.A
systems of nonlinear equations

We know that the graph of a first-degree polynomial equation in two unknowns is a straight line. Each of the variables in the equation

$$y = 2x - 4$$

has an understood exponent of 1, so this equation is a first-degree equation. Also, the equation has two unknowns and thus the equation has a straight line as its graph. The equation

$$3x - 2y + z = 17$$

is also a first-degree polynomial equation, but it has three unknowns. The graph of this equation is not a line but a plane because the presence of three unknowns requires that we use three dimensions to picture it. Nonetheless, we still call this equation a linear equation, for we give this name to any first degree polynomial equation that has two or more unknowns. This is the reason that the equations

$$4x + 2y - 3z - 6k = 14$$

$$3x + 7y - 2z - 6p + 3m - n = 21$$

can be called linear equations even though each has more than three unknowns.
Equations that have variables at least one of whose exponents is not 1, such as

$$x^2 + y = 5 \qquad \text{or} \qquad x^2 + y^2 = 7$$

and equations in which some terms are products of variables, such as

$$xy = 4 \qquad \text{and} \qquad 4x - xy = 7$$

have graphs that are not straight lines, and thus we call these equations **nonlinear equations.** The most important nonlinear figures are called **conic sections** because they can be formed by a plane cutting a single cone or a double cone. Four important conic sections are the **circle,** the **ellipse,** the **parabola** and the **hyperbola.** The figures on the next page show how a plane can cut cones to form conic sections, and two typical equations for each section are given. It is interesting to note that if the plane is tangent to

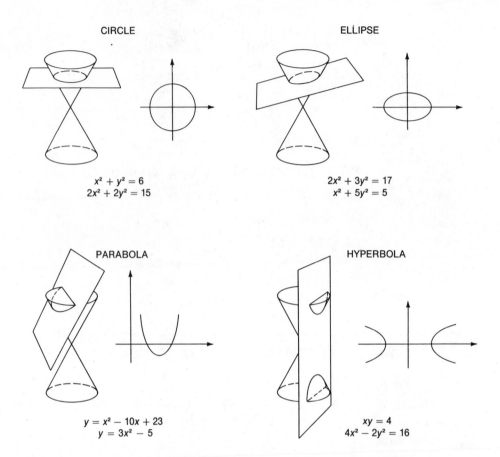

CIRCLE

$$x^2 + y^2 = 6$$
$$2x^2 + 2y^2 = 15$$

ELLIPSE

$$2x^2 + 3y^2 = 17$$
$$x^2 + 5y^2 = 5$$

PARABOLA

$$y = x^2 - 10x + 23$$
$$y = 3x^2 - 5$$

HYPERBOLA

$$xy = 4$$
$$4x^2 - 2y^2 = 16$$

the cone, the set of points where the plane and the cone touch form a straight line; and for this reason, the straight line can also be considered to be a conic section.

A nonlinear system of equations is a system in which at least one of the equations is a nonlinear equation. These systems often have more than one solution because the graphs intersect in more than one point.

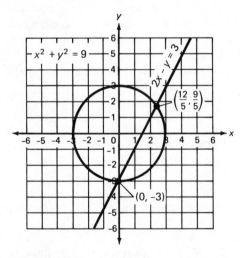

In this figure, we show the two points of intersection of the graphs of a circle and a straight line. The coordinates of either point will satisfy both the equation of the line and the equation of the circle. The study of the graphs of conic sections is a topic for a more advanced course, but the solutions of systems of nonlinear equations is a topic suitable for study at this level.

To solve systems of nonlinear equations, we use the substitution method and the elimination method just as we have done to solve systems of linear equations. Some nonlinear systems lend themselves more readily to the use of one of these methods than to the other, as we will see. Also we will find that it is important to remember that these systems often have more than one solution. We will begin by solving for the points of intersection of the circle and the line shown on the last page.

example 84.A.1 Solve the system: $\begin{cases} x^2 + y^2 = 9 & \text{(a)} \\ 2x - y = 3 & \text{(b)} \end{cases}$

solution We cannot use elimination because the variables in (a) are squared, and the variables in (b) are not. Thus, we will solve equation (b) for y and then square both sides.

$$\text{(b)} \qquad 2x - 3 = y \qquad \text{solved for } y$$

$$4x^2 - 12x + 9 = y^2 \qquad \text{squared both sides}$$

Now we can use $4x^2 - 12x + 9$ instead of y^2 in equation (a).

$$x^2 + y^2 = 9 \qquad \text{equation (a)}$$

$$x^2 + (4x^2 - 12x + 9) = 9 \qquad \text{substituted}$$

$$5x^2 - 12x = 0 \qquad \text{simplified}$$

$$x(5x - 12) = 0 \qquad \text{factored}$$

If $x = 0$ \qquad\qquad If $5x - 12 = 0$

$\qquad x = 0$ \qquad\qquad\qquad $5x = 12$

$$x = \frac{12}{5}$$

We have found two values of x. We will use equation (b) to find a value for y for each value of x.

$$2x - 3 = y \quad \text{equation (b)} \qquad\qquad 2x - 3 = y \quad \text{equation (b)}$$

$$2(0) - 3 = y \quad \text{substituted} \qquad\qquad 2\left(\frac{12}{5}\right) - 3 = y \quad \text{substituted}$$

$$-3 = y \quad \text{solved} \qquad\qquad\qquad \frac{9}{5} = y \quad \text{solved}$$

Thus, the ordered pair $(0, -3)$ is a solution, and $(\frac{12}{5}, \frac{9}{5})$ is also a solution.

example 84.A.2 Solve the system: $\begin{cases} BT_D + 5T_D = 25 & \text{(a)} \\ BT_D - 5T_D = 15 & \text{(b)} \end{cases}$

solution These same equations will be used in a later lesson to solve a problem about a boat in a river. B stands for the speed of the boat, and T_D stands for the time downstream. We will solve this system twice. The first time we add the equations just as they are and eliminate the variable T_D.

$$\begin{array}{ll} \text{(a)} & BT_D + 5T_D = 25 \\ \text{(b)} & \underline{BT_D - 5T_D = 15} \\ & 2BT_D \qquad\quad = 40 \qquad \text{added the equations} \end{array}$$

$$\text{(c)} \qquad\qquad BT_D = 20 \qquad \text{divided by 2}$$

We do not yet have a solution, for BT_D has two variables. The next step is to use 20 as a replacement for BT_D in either equation (a) or equation (b). We decide to use equation (a), so we get

$$20 + 5T_D = 25 \qquad \text{equation (a)}$$
$$5T_D = 5 \qquad \text{added} -20$$
$$T_D = 1 \qquad \text{divided}$$

Now we can use 1 for T_D in either equation (a), (b), or (c) to solve for B. We decide to use equation (c).

$$BT_D = 20 \qquad \text{equation (c)}$$
$$B(1) = 20 \qquad \text{substituted}$$
$$B = 20 \qquad \text{solved}$$

We could have used a less involved procedure had we noticed that we could eliminate the double variable BT_D in the beginning if we multiply either equation (a) or equation (b) by -1. We choose to multiply (b) by -1.

$$\begin{array}{llll}
\text{(a)} & BT_D + 5T_D = 25 & \longrightarrow & (1) \longrightarrow \quad BT_D + 5T_D = 25 \\
\text{(b)} & BT_D - 5T_D = 15 & \longrightarrow & (-1) \longrightarrow -BT_D + 5T_D = -15 \\
& & & \overline{\ 10T_D = 10} \\
& & & \qquad\qquad T_D = 1
\end{array}$$

Now we can use 1 for T_D in either (a) or (b). We will use (a).

$$BT_D + 5T_D = 25 \qquad \text{equation (a)}$$
$$B(1) + 5(1) = 25 \qquad \text{substituted}$$
$$B + 5 = 25 \qquad \text{simplified}$$
$$B = 20 \qquad \text{solved}$$

problem set 84

1. The number that were improvident varied directly as the number that were thoughtless. If 800 were improvident when 2400 were thoughtless, how many were improvident when only 9 were thoughtless? Work once using ratio and again using variation.

2. Pippin could cover 640 miles in twice the time it took le Bref to cover 280 miles. If Pippin's rate exceeded that of le Bref by 20 miles per hour, find the rates and times of both men.

3. More were hoydens than were demure. In fact, 5 times the number that were hoydens exceeded the number that were demure by 90. Also, 3 times the number of demure was only 10 greater than the number of hoydens. How many of each were there?

4. The final mixture had to be exactly 34% gravel. Two piles were available. One was 10% gravel and the other was 50% gravel. How much of each should be used to get 400 cubic feet of the desired composition?

5. Lothario looked for consecutive odd integers. He wanted three such that the product of the first and last was 25 less than the product of 10 and the opposite of the second. What integers did he want?

Solve:

*6. $\begin{cases} BT_D + 5T_D = 25 \\ BT_D - 5T_D = 15 \end{cases}$ ***7.** $\begin{cases} x^2 + y^2 = 9 \\ 2x - y = 3 \end{cases}$

Simplify:

8. $\dfrac{a^x(b^{y-2})^x}{a^{2x}(b^{-2})^x}$ **9.** $\dfrac{a^x b^{x/3} b^{-2}}{a^{x/2}}$ **10.** $\dfrac{x}{a + \dfrac{b}{c + \dfrac{x}{m}}}$

11. $\dfrac{a}{2 + \dfrac{c}{c + \dfrac{b}{c}}}$ **12.** $\dfrac{2 - 2i}{3 - 5i}$ **13.** $\dfrac{4 - \sqrt{2}\,i}{3 + \sqrt{2}\,i}$

Solve:

14. $\sqrt{k - 32} + \sqrt{k} = 8$ **15.** $\begin{cases} x + 2y + 2z = 6 \\ 2x - y + 3z = 6 \\ y - z = 0 \end{cases}$

16. The two forces act on the point as shown. Find the resultant force.

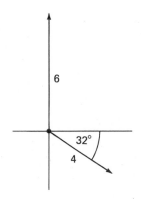

17. Add: $\dfrac{2x + 3}{x - a} - \dfrac{4}{a - x}$

Simplify:

18. $\dfrac{3 - 2\sqrt{2}}{5 - \sqrt{2}}$ **19.** $\dfrac{4 + \sqrt{3}}{2 - 2\sqrt{3}}$

20. $c = m\left(\dfrac{d}{c} - p\right)$; find p **21.** $c = m\left(\dfrac{d}{c} - p\right)$; find m

22. Simplify: $-\sqrt{-2}\sqrt{2} - 3i^3 + 2i + \sqrt{-2}\sqrt{-2} - \sqrt{-9}$

23. Solve: $\begin{cases} \dfrac{1}{9}x + \dfrac{1}{3}y = 3 \\ .3x - .04y = 2.46 \end{cases}$

24. Begin with $ax^2 + bx + c = 0$ and derive the quadratic formula by completing the square.

25. Draw an estimate of the line indicated by the data points and then write the equation that expresses sodium (Na) as a function of magnesium (Mg): $Na = mMg + b$.

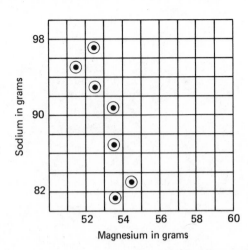

26. Solve by graphing and then find an exact solution by using substitution or elimination:
$$\begin{cases} x - 4y = -8 \\ 3x + y = 6 \end{cases}$$

27. Find the surface area of the pyramid. Dimensions are in meters. Include the area of the square bottom.

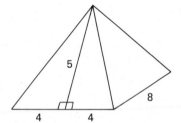

28. Convert 100 cubic feet per second to cubic centimeters per minute.

Simplify:

29. $3\sqrt{\dfrac{5}{12}} + 3\sqrt{\dfrac{12}{5}} + 3\sqrt{240}$

30. $\dfrac{-2^0}{-4^{-5/2}}$

ER 84-64, 84-65

LESSON 85 *Greater than • Trichotomy and transitive axioms • Irrational roots*

85.A
greater than

One number is said to be greater than another number if its graph on the number line lies to the right of the graph of the other number. Thus, we see that -1 is greater than -4 because the graph of -1 lies to the right of the graph of -4.

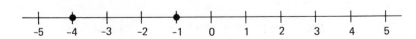

The so-called new mathematics of the 1960's introduced the number line at the elementary algebra level and used it extensively to give students a visual representation of the relationships between real numbers. Before this time, the concept of **greater than** was explained using only words and symbols. These explanations were abstract, and many students had difficulty in understanding why -1 is greater than -4. When we use the number line definition of greater than, we can look at the number line and **see** that -1 is greater than -4. The number line is very useful.

Graphing inequalities on a number line affords practice with the concept of greater than and also allows us to remember the definitions of domain and of the subsets of the set of real numbers. Recall that the domain of an equation or inequality is the set of permissible replacement values of the variable.

example 85.A.1　　Graph $x < 3$, $D = \{$Positive integers$\}$.

solution　　We are asked to graph the numbers that are less than 3, but are allowed to consider only positive integers as replacements for the variable. The solution is

because there are only two positive integers that are less than 3.

example 85.A.2　　Graph $x > -3$, $D = \{$Reals$\}$.

solution　　We are asked to designate all real numbers that are greater than -3.

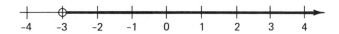

The open circle at -3 indicates that -3 is not a member of the solution set of this inequality.

85.B
trichotomy and transitive axioms

Some fundamental concepts of mathematics are difficult to remember because they are so self-evident. The trichotomy axiom and the transitive axiom are good examples. The trichotomy axiom can be demonstrated by having someone write a number on a piece of paper. Then let that person turn the paper over and write a number on the other side. There are then exactly three possibilities:

1. The second number is the same number as the first number.
2. The second number is greater than the first number.
3. The second number is less than the first number.

These statements are self-evident but are not trivial. They tell us that the real numbers are arranged in order, and thus we can say that the real numbers constitute an **ordered set.** We say that these three statements form the trichotomy axiom because trichotomy comes from the Greek word *trikha*, which means "in three parts."

The transitive axiom also has three parts. It is also self-evident but not trivial. If Arthur is larger than Billy and Billy is larger than Susan, then Arthur is larger than Susan. The same statement can be made using smaller than or the same size as instead of larger than. Real numbers behave just like people in this respect because this thought also applies to real numbers. We state both of these axioms formally in the box below.

AXIOMS	
For any real numbers a, b, and c:	
TRICHOTOMY AXIOM	Exactly one of the following is true: $a < b$, $\quad a = b$, $\quad$ or $\quad a > b$
TRANSITIVE AXIOM	If $a > b$ and $b > c$, then $a > c$. If $a < b$ and $b < c$, then $a < c$. If $a = b$ and $b = c$, then $a = c$.

example 85.B.1   Graph $x \not\leq -4$, $D = \{$Negative integers$\}$.

solution   We are asked to graph the negative integers that are not less than or equal to -4. **By the trichotomy axiom, if the numbers are not less than or equal to -4, they must be greater than -4.** Thus we can say the same thing by writing

$$x > -4$$

In the graph we show the negative integers that are greater than -4.

Note that we do not use an open circle at -4.

example 85.B.2   Graph $-x + 4 \not> 2$, $D = \{$Reals$\}$.

solution   Not greater than means equal to or less than, so we replace the given symbol with $\leq$.

$$-x + 4 \leq 2$$

Now we isolate x by adding -4 to both sides. We do not reverse the inequality symbol when we add a negative number to both sides.

$$\begin{array}{r} -x + 4 \leq 2 \\ \underline{-4 -4} \\ -x \leq -2 \end{array}$$

We need to solve for x, not for $-x$. Thus, we multiply both sides by -1 and reverse the inequality symbol.

$$x \geq 2$$

We finish by graphing this inequality, and we indicate all numbers that equal 2 or are greater than 2.

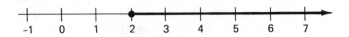

example 85.B.3   Graph $-x - 4 \not\geq -2$, $D = \{$Negative integers$\}$.

solution We begin by replacing the symbol for not greater than or equal to with the symbol for less than.

$$-x - 4 < -2$$

Now we add $+4$ to both sides and get

$$-x < +2$$

Next we multiply both sides by -1 and reverse the inequality symbol and graph the result.

$$x > -2$$

The solution is -1 because this is the only negative integer that is greater than -2.

85.C
irrational roots

The solution for the points of intersection of the circle and the line in the last lesson depended on the solution to the quadratic equation

$$5x^2 - 12x = 0$$

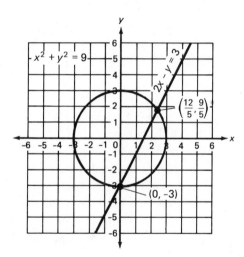

This equation can be factored, and thus the solutions to this equation are rational numbers. Some equations that describe the points of intersection of a line and a circle cannot be factored because the solutions to these equations are irrational numbers. These solutions can be found quickly and easily by using the quadratic formula. Many people always use this formula when solving real-life quadratic equations because so many of these equations cannot be solved by factoring.

example 85.C.1 Solve: $\begin{cases} x^2 + y^2 = 9 & \text{(a)} \\ y - x = 1 & \text{(b)} \end{cases}$

solution We will begin by solving equation (b) for y and then we will square both sides.

$$y - x = 1 \qquad \text{equation (b)}$$

$$y = x + 1 \qquad \text{solved for } y$$

$$y^2 = x^2 + 2x + 1 \qquad \text{squared both sides}$$

Now we will replace y^2 in equation (a) with $x^2 + 2x + 1$.

$$x^2 + (x^2 + 2x + 1) = 9 \qquad \text{substituted}$$
$$2x^2 + 2x - 8 = 0 \qquad \text{simplified}$$
$$x^2 + x - 4 = 0 \qquad \text{divided by 2}$$

This equation cannot be solved by factoring so we will use the quadratic formula.

$$x = \frac{-b \pm \sqrt{b^2 - 4ac}}{2a} \longrightarrow x = \frac{-1 \pm \sqrt{1 - 4(1)(-4)}}{2} \longrightarrow x = -\frac{1}{2} \pm \frac{\sqrt{17}}{2}$$

$$\longrightarrow x = -\frac{1}{2} + \frac{\sqrt{17}}{2} \quad \text{and} \quad x = -\frac{1}{2} - \frac{\sqrt{17}}{2}$$

Now we could use either equation (a) or equation (b) to find the values of y. We will use the equation of the line to find y because it has no squared terms and is easier to use.

$$y = x + 1 \qquad\qquad y = x + 1 \qquad\qquad \text{equation (b)}$$

$$y = \left(-\frac{1}{2} + \frac{\sqrt{17}}{2}\right) + 1 \quad y = \left(-\frac{1}{2} - \frac{\sqrt{17}}{2}\right) + 1 \quad \text{substituted}$$

$$y = \frac{1}{2} + \frac{\sqrt{17}}{2} \qquad\qquad y = \frac{1}{2} - \frac{\sqrt{17}}{2} \qquad\qquad \text{simplified}$$

Thus, the ordered pairs of x and y that satisfy the given system are

$$\left(-\frac{1}{2} + \frac{\sqrt{17}}{2}, \frac{1}{2} + \frac{\sqrt{17}}{2}\right) \quad \text{and} \quad \left(-\frac{1}{2} - \frac{\sqrt{17}}{2}, \frac{1}{2} - \frac{\sqrt{17}}{2}\right)$$

**problem
set 85**

1. The pressure of a perfect gas was held constant at 450 millimeters of mercury. The volume was 400 liters and the temperature was 1000 kelvins. What was the volume when the temperature was increased to 2000 kelvins?

2. Cuthbert rode jauntily out of town on his scooter at 20 miles per hour. Halfway to the swamp his scooter broke down, and he had to push it at 2 miles per hour all the way home. If he was gone for 11 hours, how far was it to the swamp?

3. The large jug contains 100 liters of a solution that was only 50% alcohol. How many liters of a 20% solution should be added to get a solution that is 23% alcohol?

4. The total weight of the carbonic acid H_2CO_3 was 372 grams. What was the weight of the carbon (C) in this amount of acid? (H, 1; C, 12; O, 16)

5. After the count was completed and the tally made, the observers were amazed to find that only .14 of the ducks were ring-necked. If 120,000 ducks were tallied, how many were ring-necked?

Graph:

*6. $x < 3, D = \{\text{Positive integers}\}$ *7. $x \not\le -4, D = \{\text{Negative integers}\}$

*8. $-x + 4 \not> 2, D = \{\text{Reals}\}$

*9. $-x - 4 \not\ge -2, D = \{\text{Negative integers}\}$

Solve:

10. $\begin{cases} BT_D + 3T_D = 60 \\ BT_D - 3T_D = 36 \end{cases}$ *11. $\begin{cases} x^2 + y^2 = 9 \\ y - x = 1 \end{cases}$

Simplify:

12. $\dfrac{x^{2a}(y^b)^{2a}x^{a/3}}{y^{ba/3}}$

13. $\dfrac{(x^{a+2})^2}{x^{2-a}}$

14. $\dfrac{1}{x + \dfrac{a}{x + \dfrac{1}{a}}}$

15. $\dfrac{m}{a - \dfrac{x}{a - \dfrac{x}{a}}}$

16. $\dfrac{3 + 2i}{5 - i}$

17. $\dfrac{2 - 3i}{-5 + i}$

Solve:

18. $\sqrt{p + 48} = 8 - \sqrt{p}$

19. $\begin{cases} x + 2y - z = 0 \\ 3x + y - 2z = 3 \\ 2x - z = 0 \end{cases}$

20. Add: $-10\underline{/-40°} + 10\underline{/-220°}$

21. Write $-3R - 10U$ in polar form.

22. Simplify: $\dfrac{-3 - 2\sqrt{3}}{1 - 3\sqrt{3}}$

23. $\dfrac{x + 2}{y} - c = m\left(\dfrac{a}{b} + x\right)$; find b

Simplify:

24. $-\sqrt{-16} - \sqrt{3}\sqrt{-3} + \sqrt{-3}\sqrt{-3}$

25. $5\sqrt{\dfrac{2}{7}} + 3\sqrt{\dfrac{7}{2}} - 2\sqrt{56}$

26. Solve $-3x^2 - 1 = 6x$ by completing the square.

27. Convert 40 square inches per minute to square yards per hour.

28. Solve $-2x^2 - 1 = 6x$ by using the quadratic formula.

29. Solve by graphing and then find an exact solution by using substitution or elimination:
$$\begin{cases} 2x - 5y = -15 \\ 3x + 4y = -4 \end{cases}$$

30. Find the equation of the line that passes through $(7, 2)$ that is perpendicular to the line $4x - 6y = 25$.

ER 85-66, 85-67

LESSON 86   *Slope formula*

86.A
slope
formula

The slope of a line is defined to be the change in the y coordinate divided by the change in the x coordinate as we move from one point on the line to another point on the line. We will use the points $(-3, 4)$ and $(5, -1)$ to investigate. When we graph the points and draw the triangle, we see that the slope is negative and that the rise over the run is 5 over 8, so the slope of this line is

$$-\frac{5}{8}$$

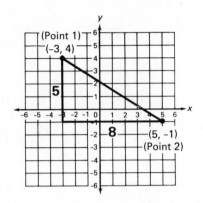

We can get the same answer if we use the definition given in bold print on the last page. We will do it twice to demonstrate that we get the same answer when we move from point 1 to point 2 as we do if we move from point 2 to point 1. The definition defines the slope as

$$m = \frac{\text{change in the } y \text{ coordinate}}{\text{change in the } x \text{ coordinate}}$$

First, we will move from point 1 to point 2. When we do, the y coordinate changes from 4 to -1, a change of -5.

$$\text{Change in } y = -5$$

The x coordinate changes from -3 to $+5$, a change of $+8$.

$$\text{Change in } x = +8$$

Thus, when we move from point 1 to point 2, we find

$$m = \frac{\text{change in } y \text{ coordinate}}{\text{change in } x \text{ coordinate}} = \frac{-5}{+8} = -\frac{5}{8}$$

If we move from point 2 to point 1, the y coordinate changes from -1 to $+4$, a change of $+5$; and the x coordinate changes from 5 to -3, a change of -8.

$$m = \frac{\text{change in } y \text{ coordinate}}{\text{change in } x \text{ coordinate}} = \frac{+5}{-8} = -\frac{5}{8}$$

We get the same answer both ways because $+5$ divided by -8 is the same number as -5 divided by $+8$.

If we call the points point 1 and point 2 with coordinates (x_1, y_1) and (x_2, y_2),

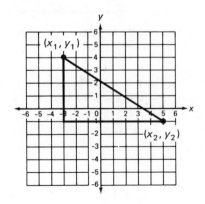

we see that we can find the slope by using either of these formulas:

$$m = \frac{y_2 - y_1}{x_2 - x_1} \quad \text{or} \quad m = \frac{y_1 - y_2}{x_1 - x_2}$$

It makes no difference which subscript comes first as long as the subscripts are in the same order above and below. It also makes no difference which point is called point 1 and which point is called point 2. Of course, the y's must always be in the numerator.

It is very easy to make a mistake in signs when using the slope formula, in which case the slope found will be incorrect.

example 86.A.1   Use the slope formula to find the slope of the line that passes through $(2, -5)$ and $(-3, 7)$.

solution   We will work the problem twice. The first time $(2, -5)$ will be point 1. We can prevent some mistakes if we first identify the coordinates.

$$x_1 = 2 \quad y_1 = -5 \quad x_2 = -3 \quad y_2 = 7$$

Now we use the formula.

$$m = \frac{y_2 - y_1}{x_2 - x_1} \quad \longrightarrow \quad m = \frac{7 - (-5)}{-3 - (2)} = \frac{7 + 5}{-5} = -\frac{12}{5}$$

Now we will find the slope again but this time we will let $(-3, 7)$ be point 1.

$$x_1 = -3 \quad y_1 = 7 \quad x_2 = 2 \quad y_2 = -5$$

We will get the same answer as we did previously.

$$m = \frac{y_2 - y_1}{x_2 - x_1} \quad \longrightarrow \quad m = \frac{-5 - (7)}{2 - (-3)} = \frac{-5 - 7}{2 + 3} = -\frac{12}{5}$$

example 86.A.2   Use the slope formula to find the slope of the line that passes through $(-5, 100)$ and $(-33, 57)$.

solution   We decide to let $(-5, 100)$ be point 1. Thus we get

$$x_1 = -5 \quad y_1 = 100 \quad x_2 = -33 \quad y_2 = 57$$

Now we use the formula to find the slope.

$$m = \frac{y_1 - y_2}{x_1 - x_2} \quad \longrightarrow \quad m = \frac{100 - (57)}{-5 - (-33)} = \frac{100 - 57}{-5 + 33} = \frac{43}{28}$$

We see that the slope formula will yield the slope, but we see that it will yield the incorrect slope unless care is used in handling the signs. Thus many people believe that the graphical method of finding the slope is the most reliable.

problem
set 86

1. Juarez's speed was 4 times as great as that of Benito. Thus, Juarez could travel 1440 miles in only 4 hours more than it took Benito to travel 120. Find the speed of both and the times of both.

2. The hours slept varied inversely as the intensity of the hypnophobia. When hypnophobia measured 400 on the H scale, Trishnutt could sleep 10 hours. How long could she sleep when the reading was 200? Work once using ratio and again using variation.

3. The mixture in the brown barrel was 80% sand, and the mixture in the blue barrel was 30% sand. How many cubic inches of each should have been used to get 600 cubic inches of a mixture that was 40% sand?

4. A few were holographs but most had been printed on a press. Ten times the number of holographs was 102 less than the number of printed ones. Also 100 times the number of holographs exceeded the number of printed ones by 168. How many of each kind were there?

5. To bottle oxygen the manufacturer separated the oxygen from $KClO_3$. What percent by weight of this compound is oxygen (O)? If 576 grams of oxygen were needed, how many grams of $KClO_3$ were required? (K, 39; Cl, 35; O, 16)

6. Write a short explanation of why the slope formula works, and include an explanation of why either point can be point 1 or point 2.

*7. Use the slope formula to find the slope of the line that passes through $(-5, 100)$ and $(-33, 57)$.

8. Graph $x \not< -4$, $D = \{\text{Integers}\}$

9. Graph $-x + 2 \leq -3$, $D = \{\text{positive integers}\}$

Solve:

10. $\begin{cases} BT_D + 4T_D = 36 \\ BT_D - 4T_D = 12 \end{cases}$

11. $\begin{cases} x^2 + y^2 = 4 \\ x - 2y = 1 \end{cases}$

Simplify:

12. $\dfrac{m^a m^{2a+2} y^{-b}}{y^{2-b}}$

13. $\dfrac{(m^{a+3})^2 y}{m^6 y^{b+1}}$

14. $\dfrac{m}{x + \dfrac{x}{m + \dfrac{1}{x}}}$

15. $\dfrac{a}{a + \dfrac{a}{a + \dfrac{b}{a}}}$

16. $\dfrac{5i - 2}{-1 - i}$

17. $\dfrac{-3 + 2i}{-2 - i}$

Solve:

18. $\sqrt{z} - \sqrt{z - 45} = 5$

19. $\begin{cases} x + y - 2z = 7 \\ 3x - y - z = 3 \\ 2x + z = 0 \end{cases}$

20. The two forces act on the point as shown. Find the resultant force.

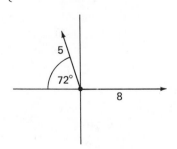

Simplify:

21. $\dfrac{2 - \sqrt{3}}{\sqrt{3} + 2}$ **22.** $\dfrac{3 - 2\sqrt{3}}{2 - 2\sqrt{3}}$ **23.** $-3i^3 - \sqrt{-2}\sqrt{-3}$

24. $3\sqrt{\dfrac{7}{3}} + 3\sqrt{\dfrac{3}{7}} - 4\sqrt{189}$ **25.** $\dfrac{y + 4}{m} = p\left(\dfrac{a}{b} + \dfrac{1}{c}\right)$; find c

26. Solve $-3x^2 - 4 = 2x$ by using the quadratic formula.

27. Convert 10 kilometers per hour to inches per second.

28. Begin with $ax^2 + bx + c = 0$ and derive the quadratic formula by completing the square.

Simplify:

29. $3\sqrt{9\sqrt[4]{3}}$ **30.** $\dfrac{-9^{-3/2}}{-(-27)^{-2/3}}$

ER 86-68, ER 86-69

LESSON 87   *The distance formula • PV = nRT*

87.A
the distance formula

In Lesson 10, we discussed the fact that the square drawn on the hypotenuse of a right triangle has the same area as the sum of the areas of the squares drawn on the other two sides.

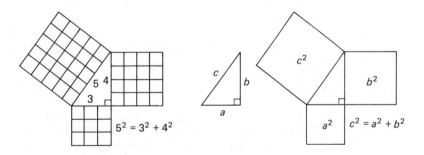

From this geometric approach, we can deduce the algebraic formula

$$c^2 = a^2 + b^2$$

where c is the length of the hypotenuse and a and b are the lengths of the other two sides. We have been using this formula to find the length of the missing side in a right triangle when the other two sides are given. If the distance between two points is required, we know that we can find the answer by graphing the points, drawing the triangle, and then using the algebraic form of this relationship, which is called the Pythagorean theorem. To find the distance between $(4, -4)$ and $(-2, 3)$, we first graph the points.

Then we draw the triangle and find the
lengths of the vertical and horizontal sides.
Now we use 6 for a and 7 for b and solve
for c.

$$c^2 \doteq 6^2 + 7^2 \longrightarrow$$

$$c = \sqrt{6^2 + 7^2} \longrightarrow c = \sqrt{85}$$

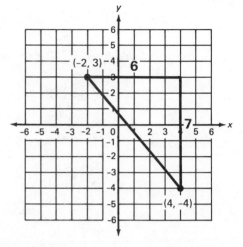

We note that the length of the vertical side of the triangle is 7, which is the difference in the y coordinates of the points,

$$+3 \text{ to } -4 \text{ is a distance of } 7$$

and that the length of the horizontal side of the triangle is 6, which is the difference in
the x coordinates of the points.

$$-2 \text{ to } +4 \text{ is a distance of } 6$$

We will call the points point 1 with coordinates (x_1, y_1) and point 2 with coordinates
(x_2, y_2). Either point can be point 1.

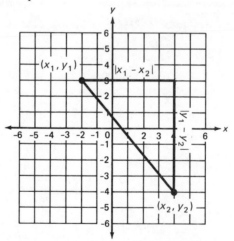

We see that the lengths of the sides are the absolute values of the differences in the
coordinates. Since any number squared is positive, the absolute value notation is not
required. Thus, the distance between the points can be represented by

$$D = \sqrt{(x_1 - x_2)^2 + (y_1 - y_2)^2}$$

This is called the **distance formula.** It is simply an algebraic expression that says that
the distance is the square root of the sum of the squares of the sides of the triangle. In
this formula either point can be used as point 1, and thus the following formula will
produce the same result as the formula above.

$$D = \sqrt{(x_2 - x_1)^2 + (y_2 - y_1)^2}$$

example 87.A.1   Use the distance formula to find the distance between $(4, -2)$ and $(-5, 3)$.

solution   It is easy to make mistakes in sign. Thus, we will be careful and begin by writing down the values of x_1, y_1, x_2, and y_2.

$$x_1 = 4 \qquad y_1 = -2 \qquad x_2 = -5 \qquad y_2 = 3$$

Now we place these values in one of the two forms of the distance formula and simplify

$$D = \sqrt{(x_1 - x_2)^2 + (y_1 - y_2)^2} \quad \longrightarrow \quad D = \sqrt{[4 - (-5)]^2 + [-2 - (3)]^2}$$
$$D = \sqrt{81 + 25} \quad \longrightarrow \quad D = \sqrt{106}$$

87.B
the relationship PV = nRT

Some people say that unfamiliar concepts are difficult concepts and that familiar concepts are easy concepts. These people have confused the words **difficult** and **different**. We have found that algebraic concepts that seem difficult at first are really just different, and they become familiar concepts after we have worked with them for awhile. **Thus we can correctly say that algebra is not difficult—it is just different.**

The same is true for chemistry, physics, and other sciences. This book was written to permit the student to develop a firm understanding of the foundations of algebra and to prepare for the concepts of advanced courses in mathematics and science. We have introduced some problems that will be encountered in chemistry. These problems are part mathematics and part chemistry. Their introduction at this level permits the mathematical part to be taught and allows some of the newness of these problems to rub off. Another kind of problem that will be encountered in both physics and chemistry uses the relationship

$$PV = nRT$$

To use this formula, values for four of the five variables must be given; the formula is then used to find the value of the variable that is unknown. The letter R stands for a constant whose value depends on the units used for the other variables. We will always use the units listed below for the other variables, and thus R will always be .0821. The symbols represent the following:

$$P = \text{pressure in atmospheres}$$
$$V = \text{volume in liters}$$
$$n = \text{number of moles of gas}$$
$$R = \text{a constant (.0821)}$$
$$T = \text{temperature in kelvins}$$

We will not worry about the units. We will insert the numbers for the known values and then simplify.

example 87.B.1   Use the relationship $PV = nRT$ to find the number of moles in an amount of gas when the temperature is 273 K, the pressure is 1 atmosphere, and the volume is 8 liters ($R = .0821$).

solution   Since we are finding n, we will begin by solving the abstract equation for n. We do this by dividing both sides by RT.

$$\frac{PV}{RT} = \frac{nRT}{RT} \quad \longrightarrow \quad \frac{PV}{RT} = n$$

Now we make the indicated substitutions and simplify.

$$\frac{(1)(8)}{(.0821)(273)} = n$$

If a calculator is handy, this reduces to 0.357 moles. A mole is a measure of how many atoms of gas are present. This concept will be discussed in great detail in the first course in chemistry.

example 87.B.2 Use the formula $PV = nRT$ to find the volume of .832 moles of a gas at a pressure of 3 atmospheres and a temperature of 400 K ($R = .0821$).

solution We begin by solving the equation for V. We do this by dividing both sides by P.

$$\frac{PV}{P} = \frac{nRT}{P} \longrightarrow V = \frac{nRT}{P}$$

Now we make the indicated substitutions and simplify.

$$V = \frac{(.832)(.0821)(400)}{3}$$

And if a calculator is handy, we find the volume is

$$V = \textbf{9.11 liters}$$

problem set 87

1. Find three consecutive even integers such that the product of the first and third is 12 greater than the product of 6 and the opposite of the second.

2. The tin (Sn) in the tin II chromate $SnCrO_4$ weighed 595 grams. What was the total weight of the zinc chromate? (Sn, 119; Cr, 52; O, 16)

3. David's speed was 20 miles per hour less than Gretchen's speed. Thus, David could travel 400 miles in one-half the time it took Gretchen to travel 1120 miles. Find the speeds of both and the times of both.

4. The initial pressure, volume, and temperature of a quantity of an ideal gas were recorded as 740 millimeters of mercury, 10 liters, and 300 kelvins. Find the final volume if the final pressure was 1480 millimeters of mercury and the temperature was 1200 kelvins. Begin by solving for V_2.

*5. Use the relationship $PV = nRT$ to find the number of moles in a quantity of gas when the temperature is 273 K, the pressure is 1 atmosphere, and the volume is 8 liters ($R = .0821$). Begin by solving the equation for n.

6. Write a short explanation of why the distance formula works, and include a discussion of why either point can be point 1 or point 2.

*7. Use the distance formula to find the distance between $(4, -2)$ and $(-5, 3)$.

Graph:

8. $-x - 2 \le 4, D = \{\text{negative integers}\}$

9. $-x + 2 \not> 3, D = \{\text{negative integers}\}$

Solve:

10. $\begin{cases} BT_D + 6T_D = 22 \\ BT_D - 6T_D = 10 \end{cases}$

11. $\begin{cases} x^2 + y^2 = 2 \\ x - y = 1 \end{cases}$

Simplify:

12. $\dfrac{(x^2)^{a+b}x^{-2a+b}y^a}{y^{a/4}}$

13. $\dfrac{(y^{2a+2})^2 y^{a/2}b}{y^a b^{2a}}$

14. $\dfrac{r}{m + \dfrac{r}{\dfrac{1}{r} + m}}$

15. $\dfrac{pc}{p - \dfrac{c^2}{p - \dfrac{1}{pc}}}$ 16. $\dfrac{4i - 1}{3i - 2}$ 17. $\dfrac{5i - 1}{5i - 2}$

Solve:

18. $\sqrt{z - 35} + \sqrt{z} = 7$ 19. $\begin{cases} 2x + 2y - z = 14 \\ 3x + 3y + z = 16 \\ x - 2y = 0 \end{cases}$

20. Add: $-20\underline{/-200°} + 30\underline{/-30°}$ 21. Write $4R - 12U$ in polar form.

Simplify:

22. $4i^2 - \sqrt{-9}$ 23. $4\sqrt{\dfrac{9}{3}} + 3\sqrt{\dfrac{3}{9}} - 5\sqrt{27}$

24. $\dfrac{1 - \sqrt{2}}{3 - 2\sqrt{2}}$ 25. $\dfrac{4 + \sqrt{3}}{1 - \sqrt{3}}$

26. $\dfrac{x + 2y}{c} = y\left(\dfrac{1}{x} - \dfrac{1}{r}\right)$; find r

27. Solve $-x^2 - 2x - 2 = 0$ by completing the square.

28. Convert 10 milliliters per second to cubic inches per minute.

29. Find the volume in cubic inches of a figure 8 feet high if this is the base. Dimensions are in feet.

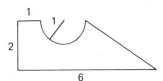

30. Solve: $\dfrac{-3x - 2}{2} - \dfrac{2x - 4}{3} = 7$

ER 87-70, 87-71

LESSON 88   *Conjunctions and disjunctions*

88.A
conjunctions

The English prefixes *com-* and *con-* come from the Latin word *cum*, which means "with" or "together." This is the reason that the English word *compose* means "to make by placing together," and the reason that the *confluence* of two rivers means "the place where the two rivers flow together." We use the prefix *con-* in algebra in the word *conjunction* to describe a situation where two or more restrictions on the variable must apply (join together) at the same time. We use the word **and** to denote conjunctions by writing **and** between the two symbolic statements of the restrictions on the variable.

example 88.A.1   Graph $-x - 3 \le -2$ **and** $x - 2 < 1$, $D = \{\text{Integers}\}$.

solution   **The word *and* means that both conditions must be met.** The additive property of inequality permits us to add the same quantity to both sides as required to find x.

$$
\begin{array}{ccc}
-x - 3 \leq -2 & \text{and} & x - 2 < 1 \\
+3 +3 & & +2 +2 \\
\hline
-x \leq 1 & \text{and} & x < 3
\end{array}
$$

Now on the left we must mentally multiply both sides by -1 and reverse the inequality symbol to solve for x.

$$x \geq -1 \qquad \text{and} \qquad x < 3$$

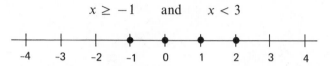

In the graph we have indicated the integers that are less than 3 **and** are greater than or equal to -1.

example 88.A.2   Graph $-2 < x - 2 \leq +4$, $D = \{\text{Reals}\}$.

solution   **This notation is an alternate way to write a conjunction.** To find the two conditions, we first cover up the $\leq +4$ on the right and read the first condition as

$$-2 < x - 2$$

Then we cover up the $-2 <$ on the left and read the second condition as

$$x - 2 \leq +4$$

We now add $+2$ to both sides of these inequalities to solve for x.

$$
\begin{array}{ccc}
-2 < x - 2 & \text{and} & x - 2 \leq +4 \\
+2 +2 & & +2 +2 \\
\hline
0 < x & \text{and} & x \leq 6
\end{array}
$$

Thus, the solution is the graph of the real numbers that are greater than 0 **and** are less than or equal to 6.

88.B
disjunctions

The Latin prefix *dis-* indicates negation or reversal. This is the reason that *disapprove* means "not to approve" and *disagree* means "not to agree." In algebra, we use the prefix *dis-* in the word *disjunction*, which describes a situation where it is not necessary to satisfy both restrictions on a variable. A number satisfies a disjunction if it satisfies either of the two restrictions stated in the disjunction. We use the word **or** to designate a disjunction.

example 88.B.1   Graph $-x \geq 3$ **or** $-x < -1$, $D = \{\text{Reals}\}$.

solution   The word **or** indicates that a number satisfies this disjunction if it satisfies the left inequality or if it satisfies the right inequality. We solve the inequalities for x by multiplying both sides by -1 and reversing the inequality symbols.

$$x \leq -3 \qquad \text{or} \qquad x > 1$$

Any real number that satisfies either of the restrictions is a solution to the disjunction.

example 88.B.2   Graph $-x - 2 \geq 0$ or $x + 3 > 6$, $D = \{$Integers$\}$.

solution   We begin by solving both inequalities for x. As the last step on the left, we mentally multiply both sides by -1 and reverse the inequality symbol.

$$\begin{array}{rclcrcl}
-x - 2 & \geq & 0 & \text{or} & x + 3 & > & 6 \\
+ 2 & & +2 & & - 3 & & - 3 \\
\hline
-x & \geq & 2 & \text{or} & x & > & 3 \\
x & \leq & -2 & \text{or} & x & > & 3
\end{array}$$

This disjunction is satisfied by all integers that are greater than 3 **or** are less than or equal to -2. A number qualifies if it satisfies either one of these conditions. Unfortunately (or fortunately), there is no concise notation for a disjunction. Both conditions must be written out, and the word **or** must be written between the conditions.

problem set 88

1. The number of altercations varied directly as the number who were belligerent by nature. If there were 500 altercations when 10 were belligerent, how many would there be if 42 were belligerent? Work once using the ratio format and once using the variation format.

2. The road was rocky so after running for a while at 10 kilometers per hour, Clytemnestra slowed to a 5-kilometer-per-hour walk. If the total trip was 65 kilometers and she made it in 9 hours, how far did she walk and how far did she run?

3. Eight hundred gallons of a 79% glycol solution was available. How much pure glycol should be extracted so that the remainder would be only 30% glycol?

4. Some were fast and the rest were slow. Ten times the number of fast was 140 less than twice the number of slow. Also, one-half the number of slow exceeded 3 times the number of fast by 10. How many were fast and how many were slow?

5. When the Grinch peered through the crack, he could see that the number of spotted ones had increased 640 percent. If he could now identify 592 spotted ones, how many spotted ones had he seen the time before?

Graph on a number line:

*6. $-x - 3 \leq -2$ and $x - 2 < 1$, $D = \{$Integers$\}$

*7. $-2 < x - 2 \leq +4$, $D = \{$Reals$\}$

*8. $-x \geq 3$ or $-x < -1$, $D = \{$Reals$\}$

*9. $-x - 2 \geq 0$ or $x + 3 > 6$, $D = \{$Integers$\}$

Solve:

10. $\begin{cases} BT_D + 5T_D = 57 \\ BT_D - 5T_D = 27 \end{cases}$

11. $\begin{cases} x^2 + y^2 = 3 \\ x - y = 2 \end{cases}$

Simplify:

12. $\dfrac{x^a y^{2b}(x^{a+2})^{1/2}}{y^{3b}}$

13. $\dfrac{(y^{a+2})^a(y^a)^a}{y^{2+a}}$

14. $\dfrac{k}{m + \dfrac{m}{a + \dfrac{1}{m}}}$

15. $\dfrac{m}{a + \dfrac{x}{b + \dfrac{d}{c}}}$

16. $\dfrac{3 - 2i}{i - 4}$

17. $\dfrac{2 - 3i}{4i - 1}$

Solve:

18. $\sqrt{s} = 3 + \sqrt{s - 21}$

19. $\begin{cases} 3x + 2y + z = 9 \\ x - 2y - 2z = -3 \\ 2x + z = 0 \end{cases}$

20. The two forces are applied to an object as indicated. Find the resultant force on the object.

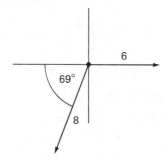

Simplify:

21. $\dfrac{1 - 3\sqrt{2}}{3 - \sqrt{2}}$

22. $\dfrac{4 + 2\sqrt{3}}{3\sqrt{3} - 2}$

23. $\dfrac{x}{a + c} = m\left(\dfrac{x}{y} + c\right)$, find y

24. $\dfrac{x}{a + c} = m\left(\dfrac{x}{y} + c\right)$; find a

25. Simplify: $3i^3 + 5i - \sqrt{-2}\sqrt{2}$

26. $3\sqrt{\dfrac{3}{8}} + 4\sqrt{\dfrac{8}{3}} - 2\sqrt{24}$

27. Solve $-7x^2 = -x - 5$ by using the quadratic formula.

28. Convert 10 cubic feet per second to cubic inches per minute.

29. Find C.

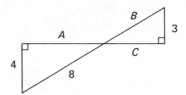

30. Find the distance between $(-2, 3)$ and $(8, 4)$.

ER 88-72, 88-73

LESSON 89   *Systems of three equations*

89.A

systems
of three
equations

The solution to a system of two linear equations in two unknowns such as

$$\text{(a)} \quad \begin{cases} 3x + 2y = -4 \\ 2x - 4y = -8 \end{cases}$$
$$\text{(b)}$$

is the value of x and the value of y that will satisfy both equations. A solution exists only if the graphs of the equations cross. The graphs of the two equations cross at the point whose coordinates are $x = -2$ and $y = 1$, so we say that the solution of this system is the **ordered pair** $(-2, 1)$. If we substitute these values for x and y in equations (a) and (b), both equations will become true numerical equations.

(a) $3(-2) + 2(1) = -4$ (b) $2(-2) - 4(1) = -8$

$-6 + 2 = -4$ $-4 - 4 = -8$

$-4 = -4$ True $-8 = -8$ True

The solution to a system of three linear equations in three unknowns, such as

$$\text{(a)} \quad \begin{cases} x + 2y + z = 4 \\ 2x - y - z = 0 \\ 2x - 2y + z = 1 \end{cases}$$
$$\text{(b)}$$
$$\text{(c)}$$

is the value of x, the value of y, and the value of z that will satisfy all three equations. The graph of a linear equation in three unknowns is a plane, and thus the three equations shown describe three planes. If the three planes meet in one point, the three coordinates of this point are the **ordered triple** (x, y, z) that will satisfy all three equations.

A graphical solution of this system is not feasible because of the difficulty in drawing accurate graphs in three dimensions. However, we may use either substitution or elimination to find the answer. **The most common procedure is to add two equations so that one variable is eliminated. Then the remaining equation is added to one of those already used so that the same variable is eliminated. The result will be two equations in two unknowns that can be solved by using either substitution or elimination.**

example 89.A.1   Solve:
$$\text{(a)} \quad \begin{cases} x + 2y + z = 4 \\ 2x - y - z = 0 \\ 2x - 2y + z = 1 \end{cases}$$
$$\text{(b)}$$
$$\text{(c)}$$

solution   We choose to eliminate z. As the first step we will add (a) to (b) so that z is eliminated. Then we can add (c) to either (a) or (b) and again eliminate z. We decide to add (c) to (b).

(a)	$x + 2y + z = 4$	(b)	$2x - y - z = 0$
(b)	$2x - y - z = 0$	(c)	$2x - 2y + z = 1$
(d)	$3x + y \quad\quad = 4$	(e)	$4x - 3y \quad\quad = 1$

Now we have the two equations (d) and (e) in the two unknowns x and y. We will use elimination and add (e) to the product of (d) and (3).

(3)(d) $9x + 3y = 12$
(e) $4x - 3y = 1$
$\overline{\ 13x = 13}$

$$x = 1$$

Now we can use $x = 1$ in either (d) or (e) to find y. We will use both to show that both will yield the same answer.

(d) $3(1) + y = 4$ (e) $4(1) - 3y = 1$

$\qquad\qquad 3 + y = 4 \qquad\qquad\qquad 4 - 3y = 1$

$\qquad\qquad\qquad\quad y = 1 \qquad\qquad\qquad\quad -3y = -3$

$\qquad\qquad\qquad\qquad\qquad\qquad\qquad\qquad\qquad\quad y = 1$

Now we can use $x = 1$ and $y = 1$ in either (a), (b), or (c) to find z. This time we will use all three equations to show that all three will produce the same result.

(a) $(1) + 2(1) + z = 4$ (b) $2(1) - (1) - z = 0$ (c) $2(1) - 2(1) + z = 1$

$\qquad 1 + 2 + z = 4 \qquad\qquad 2 - 1 - z = 0 \qquad\qquad 2 - 2 + z = 1$

$\qquad\qquad 3 + z = 4 \qquad\qquad\qquad -z = -1 \qquad\qquad\qquad 0 + z = 1$

$\qquad\qquad\qquad z = 1 \qquad\qquad\qquad\quad z = 1 \qquad\qquad\qquad\qquad\quad z = 1$

Thus, the solution is the ordered triple $(\mathbf{1}, \mathbf{1}, \mathbf{1})$.

example 89.A.2   Solve: (a) $\begin{cases} 2x - y + 3z = 9 \\ x + 2y + z = 8 \\ x - 2y + z = 0 \end{cases}$
     (b)
     (c)

solution   This time we decide to eliminate y. Thus on the left we multiply (a) by 2 and add this product to (b). On the right we add (b) and (c) in their present form.

$$
\begin{array}{llll}
\text{(2)(a)} & 4x - 2y + 6z = 18 & \text{(b)} & x + 2y + z = 8 \\
\text{(b)} & \underline{x + 2y + z = 8} & \text{(c)} & \underline{x - 2y + z = 0} \\
\text{(d)} & 5x \quad\quad + 7z = 26 & \text{(e)} & 2x \quad\quad + 2z = 8
\end{array}
$$

Now to eliminate x, we will multiply (d) by 2 and (e) by -5.

$$
\begin{array}{ll}
\text{(2)(d)} & 10x + 14z = 52 \\
(-5)\text{(e)} & \underline{-10x - 10z = -40} \\
& \qquad\quad 4z = 12 \\
& \qquad\quad\; z = 3
\end{array}
$$

Now we could use $z = 3$ in either (d) or (e) to find x. We will use (e).

(e) $2x + 2(3) = 8$

$\qquad\quad 2x + 6 = 8$

$\qquad\qquad\; 2x = 2$

$\qquad\qquad\;\; x = 1$

Now we could use $x = 1$ and $z = 3$ in either (a), (b), or (c) to find y. We decide to use (b).

(b) $(1) + 2y + (3) = 8$

$\qquad\qquad\quad 2y = 4$

$\qquad\qquad\quad\; y = 2$

Thus, the solution to this system of three linear equations is the ordered triple $(\mathbf{1}, \mathbf{2}, \mathbf{3})$.

problem set 89

1. It took Pedro 5 times as long to travel 300 miles as it took Roberto to travel 160 miles. If Roberto's speed was 50 mph greater than Pedro's speed, find the speeds of both and the times that both traveled.

2. Use the formula $PV = nRT$ to find the volume of .832 moles of gas at 3 atmospheres pressure and a temperature of 400 K. Begin by solving for V. ($R = .0821$)

3. Sergio painted some boats red and painted the rest blue. The number of red boats was 5 less than 3 times the number of blue boats. Also, 6 times the number of blue boats was 70 less than 10 times the number of red boats. How many were red and how many were blue?

4. Only 10% of the first solution was glycerine, but 40% of the second solution was glycerine. How many liters of each should Lolita use to get 800 liters of a solution that is 13% glycerine?

5. As the Grinch watched in horror, 7/16 of the spotted ones metamorphosed into striped ones. If the number that metamorphosed totaled 672, how many were spotted when the Grinch began to watch?

Solve:

*6. $\begin{cases} x + 2y + z = 4 \\ 2x - y - z = 0 \\ 2x - 2y + z = 1 \end{cases}$

*7. $\begin{cases} 2x - y + 3z = 9 \\ x + 2y + z = 8 \\ x - 2y + z = 0 \end{cases}$

Graph on a number line:

8. $-1 \le x - 1 < 4, D = \{\text{Integers}\}$

9. $x - 2 \not\ge 0$ or $x - 2 > 2, D = \{\text{Reals}\}$

Solve:

10. $\begin{cases} BT_D + 3T_D = 28 \\ BT_D - 3T_D = 16 \end{cases}$

11. $\begin{cases} x^2 + y^2 = 18 \\ y - x = 4 \end{cases}$

Simplify:

12. $\dfrac{a^2 a^{x/2}(a^2)^x}{(a^{-3})^{-x}}$

13. $\dfrac{y^c(m^{-b})^2}{m^{-b/2}}$

14. $\dfrac{x}{xm - \dfrac{m}{m - \dfrac{1}{x}}}$

15. $\dfrac{p}{x - \dfrac{xp}{1 - \dfrac{p}{x}}}$

16. $\dfrac{i - 2}{5i - 1}$

17. $\dfrac{1 + 2i}{-5 - i}$

18. Solve: $\sqrt{s} - \sqrt{s - 15} = 3$

19. Add: $4\underline{/40°} - 6\underline{/-120°}$

20. Write $-4R - 10U$ in polar form.

Simplify:

21. $\dfrac{-2 - \sqrt{3}}{2 - 2\sqrt{3}}$

22. $\dfrac{4 + \sqrt{3}}{3 - 2\sqrt{3}}$

23. $-2i^2 + \sqrt{-4}\sqrt{4} - \sqrt{-3}\sqrt{-3} - 2i^5$

24. $\dfrac{x}{y} - m = p\left(\dfrac{r}{c} - \dfrac{1}{b}\right)$; find c

25. $\dfrac{x}{y} - m = p\left(\dfrac{r}{c} - \dfrac{1}{b}\right)$; find y

26. Simplify: $\sqrt[3]{9\sqrt[4]{3}}$

27. Solve $-6x^2 - x - 5 = 0$ by completing the square.

28. Convert 42 cubic feet to cubic centimeters.

29. Find *B*.

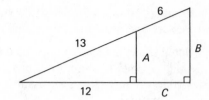

30. Find the equation of the line that passes through $(-2, 7)$ that is perpendicular to $x + 5y = 7$.

ER 89-74, 89-75

LESSON 90 *Linear inequalities*

90.A
linear inequalities

A line divides the set of all the points in the plane into three disjoint subsets. These are the set of points that lie on the line and the two sets of points that lie on either side of the line.

In the figure, we have graphed the equation $y = -\frac{1}{2}x + 1$. The coordinates of all the points that lie on the line will satisfy this equation. The coordinates of any point not on the line will satisfy one of the following inequalities.

$$\text{(a)} \quad y > -\tfrac{1}{2}x + 1 \qquad \text{(b)} \quad y < -\tfrac{1}{2}x + 1$$

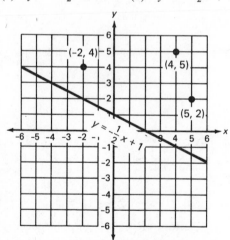

If the coordinates of a point on one side of the line will satisfy one of these inequalities, the coordinates of all the points on the same side of the line will satisfy the same inequality. We will demonstrate this by using in the left equation the coordinates of the three points graphed above. All the resulting inequalities will be true inequalities.

USING $(-2, 4)$	USING $(4, 5)$	USING $(5, 2)$
$4 > -\dfrac{1}{2}(-2) + 1$	$5 > -\dfrac{1}{2}(4) + 1$	$2 > -\dfrac{1}{2}(5) + 1$
$4 > 1 + 1$	$5 > -2 + 1$	$2 > -\dfrac{5}{2} + 1$
$4 > 2$ True	$5 > -1$ True	$2 > -\dfrac{3}{2}$ True

Thus, the coordinates of any point above the line will satisfy this inequality and will not satisfy the other inequality,

$$\text{(b)} \quad y < -\frac{1}{2}x + 1$$

because this inequality is satisfied by the coordinates of all points below the line. We can always choose a point blindly and try both inequalities to see which one the coordinates of the point will satisfy. But if we remember that y is greater than as we go up and that y is less than as we go down, beginning with a test point is unnecessary. We read

$$y > -\frac{1}{2}x + 1$$

as "y is greater than" and remember that "y is greater than" means "above," so this inequality is satisfied by all points above the line. The other inequality is read as "y is less than" and is satisfied by all points below the line.

y IS GREATER THAN (POINTS ABOVE) y IS LESS THAN (POINTS BELOW)

$$y > -\frac{1}{2}x + 1 \qquad\qquad\qquad y < -\frac{1}{2}x + 1$$

Our surmise can be checked by using a test point if we wish.

90.B
greater than or equal to

The two inequalities

$$\text{(a)} \quad y \leq \frac{1}{3}x + 2 \qquad \text{and} \qquad \text{(b)} \quad y < \frac{1}{3}x + 2$$

both designate the points that lie below the same line, for the coordinates of any point below this line will satisfy both inequalities. In addition, the coordinates of any point that lies on the line will satisfy inequality (a), for this inequality is read as "y is less than or equal to one-third x plus two."

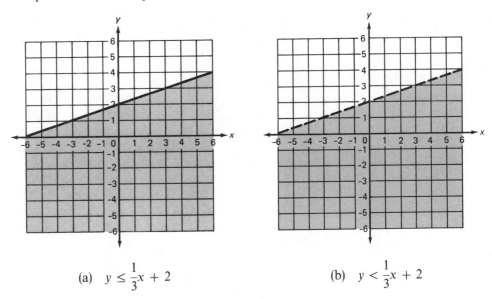

$$\text{(a)} \quad y \leq \frac{1}{3}x + 2 \qquad\qquad\qquad \text{(b)} \quad y < \frac{1}{3}x + 2$$

When we graph inequalities, we shade the regions whose coordinates satisfy the inequality. In both figures, the area below the line is shaded. **In the left-hand figure the line is drawn as a solid line to indicate that the coordinates of the points on the line also satisfy the inequality. In the right-hand figure, the line is drawn as a dashed line to indicate that the coordinates of the points on the line do not satisfy the inequality.**

90.C

**systems
of linear
inequalities**

If two linear equations are not equivalent equations, then the lines designated by them cross or the lines are parallel.

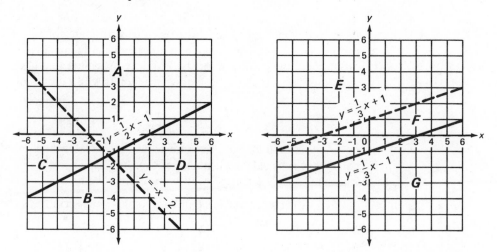

The lines cross in the left-hand figure and divide the figure into four distinct regions.

Region *A* is above the dashed line and on or above the solid line.
Region *B* is below the dashed line and on or below the solid line.
Region *C* is below the dashed line and above or on the solid line.
Region *D* is above the dashed line and on or below the solid line.

The systems of linear inequalities that define these regions are

$$A \begin{cases} y \geq \dfrac{1}{2}x - 1 \\ y > -x - 2 \end{cases} \qquad B \begin{cases} y \leq \dfrac{1}{2}x - 1 \\ y < -x - 2 \end{cases}$$

$$C \begin{cases} y \geq \dfrac{1}{2}x - 1 \\ y < -x - 2 \end{cases} \qquad D \begin{cases} y \leq \dfrac{1}{2}x - 1 \\ y > -x - 2 \end{cases}$$

In the right-hand figure,

Region *E* is above both lines.
Region *F* is on or above the solid line and below the dashed line.
Region G is below the dashed line and on or below the solid line.

The systems of inequalities that define these regions are

$$E \begin{cases} y > \dfrac{1}{3}x + 1 \\ y \geq \dfrac{1}{3}x - 1 \end{cases} \qquad F \begin{cases} y < \dfrac{1}{3}x + 1 \\ y \geq \dfrac{1}{3}x - 1 \end{cases} \qquad G \begin{cases} y < \dfrac{1}{3}x + 1 \\ y \leq \dfrac{1}{3}x - 1 \end{cases}$$

example 90.C.1 Graph the solution to: $\begin{cases} y < \dfrac{1}{2}x + 2 \\ y \geq -x - 3 \end{cases}$

solution The first step is to draw the lines. We will draw $y = -x - 3$ as a solid line and $y = \frac{1}{2}x + 2$ as a dashed line.

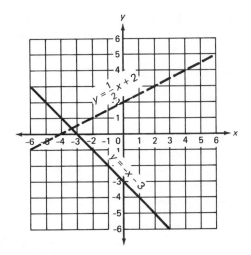

The region we wish is above or on the solid line and below the dashed line. We shade this area in the following figure.

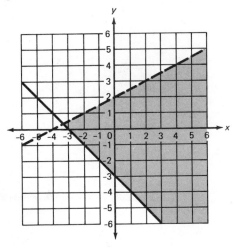

example 90.C.2   Graph the solution to: $\begin{cases} y > x - 2 \\ y \le x + 1 \end{cases}$

solution   In the left-hand figure, we draw the lines, and in the right-hand figure we shade the area between the lines because the inequalities specify the points that lie above the bottom line that are also on or below the top line.

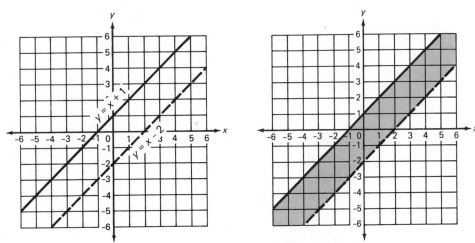

problem set 90

1. The pressure of a quantity of an ideal gas at a temperature of 400 K was 740 millimeters of mercury. If the volume was not changed, what would the pressure be if the temperature were raised to 1600 K?

2. The solution in the large container was 60% alcohol, and the solution in the small container was only 30% alcohol. How much of each should be used to get 300 ml that is 42% alcohol?

3. The container held the compound As_4O_6. What was the percentage by weight of the arsenic (As) in this compound? (As, 75; O, 16)

4. The large plane could travel 3000 miles in 1 more hour than it took the small plane to cover 800 miles. If the rate of the large plane was 3 times the rate of the small plane, find the rates and the times of both.

5. The number that were glabrous varied inversely as the average age of those present. If 35 were glabrous when the average age was 70, how many were glabrous when the average age was only 50? Work using the variation format and again using the ratio format.

Graph:

*6. $\begin{cases} y < \dfrac{1}{2}x + 2 \\ y \geq -x - 3 \end{cases}$

*7. $\begin{cases} y > x - 2 \\ y \leq x + 1 \end{cases}$

8. Solve: $\begin{cases} x + 2y - z = 1 \\ 2x - y + 2z = 9 \\ x - 2y - 3z = -9 \end{cases}$

Graph on a number line:

9. $4 < x + 4 < 6, D = \{\text{Integers}\}$

10. $x + 2 \nleq 5 \text{ or } x + 5 < 6, D = \{\text{Reals}\}$

Solve:

11. $\begin{cases} BT_D + 2T_D = 51 \\ BT_D - 2T_D = 39 \end{cases}$

12. $\begin{cases} x^2 + y^2 = 12 \\ x + y = 4 \end{cases}$

Simplify:

13. $\dfrac{(x^b)^{2-a}x^{ab}}{x^{ab/2}}$

14. $\dfrac{a}{b + \dfrac{1}{cx + \dfrac{1}{x}}}$

15. $\dfrac{m}{a + \dfrac{ma}{a + \dfrac{m}{a}}}$

16. $\dfrac{2i + 7}{i + 2}$

17. $\dfrac{3i - 6}{-2i + 1}$

18. $\dfrac{1 - \sqrt{2}}{4 - 5\sqrt{2}}$

19. $\dfrac{4 + 3\sqrt{2}}{-\sqrt{2}}$

20. $\dfrac{x}{my} = d\left(\dfrac{r}{a} + b\right)$; find y

21. $\dfrac{x}{my} = d\left(\dfrac{r}{a} + b\right)$; find a

22. The two forces are applied on the object as indicated. Find the resultant force on the object.

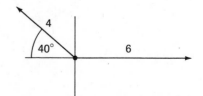

23. Draw the line suggested by the data points and write the equation that expresses output as a function of input.

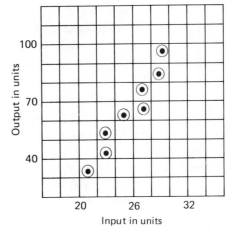

Simplify:

24. $\dfrac{-9^{3/2}}{27^{2/3}}$

25. $-2i^5 + 3i^3 - \sqrt{-9} + \sqrt{-4} - \sqrt{-2}\sqrt{-2}$

26. $2\sqrt{\dfrac{9}{5}} + 3\sqrt{\dfrac{5}{9}} + 3\sqrt{45}$

27. Begin with $ax^2 + bx + c = 0$ and derive the quadratic formula by completing the square.

28. Convert 400 milliliters per second to cubic inches per minute.

29. Find the surface area of the right circular cylinder in square centimeters. Dimensions are in meters.

30. Divide $3x^3 - 2x + 2$ by $x + 2$.

ER 90-76, 90-77

LESSON 91 *Boat-in-the-river problems*

91.A
boat-in-the-river problems

Robert has a boat that has a speed of 11 mph in still water. If the boat, with the motor turned off, is placed in a river in which the water flows at 3 mph, the boat will drift downstream at 3 mph. If the boat is headed downstream with the motor turned on, its speed (rate) downstream would be 14 mph, which is 3 mph plus 11 mph. If the boat is turned around and is headed upstream, it must go against the current; and its speed upstream would be only 8 mph, which is 11 mph minus 3 mph. Thus, in general, the downstream rate is the speed of the boat plus the speed of the water; and the upstream rate is the speed of the boat minus the speed of the water.

$$\text{Downstream rate} = B + W$$

$$\text{Upstream rate} = B - W$$

The distance downstream equals the rate downstream times the time downstream; and the distance upstream equals the rate upstream times the time upstream. These two statements lead to the following two equations which can be used to solve almost all boat-in-the-river problems.

DOWNSTREAM EQUATION   UPSTREAM EQUATION

$$(B + W)T_D = D_D \qquad\qquad (B - W)T_U = D_U$$

These two equations contain six unknowns: B, W, T_D, D_D, T_U, and D_U. Thus, boat-in-the-river problems must contain six statements of equality because six equations are required.

example 91.A.1   Robert and Clay could go 60 miles downstream in the same time it took them to go 20 miles upstream. If the speed of their boat was 8 mph in still water, what was the speed of the current and what were the times?

solution   We begin by recording the downstream equation and the upstream equation.

DOWNSTREAM EQUATION   UPSTREAM EQUATION

(a) $(B + W)T_D = D_D$ (b) $(B - W)T_U = D_U$

Rather than write the other six equations, we will just make the necessary substitutions. The times were equal, so we will use T for both T_D and T_U. The rate of the boat was 8, the distance downstream was 60, and the distance upstream was 20.

(a) $(8 + W)T = 60$ (b) $(8 - W)T = 20$

Next we multiply as required in both equations and get

(a) $8T + WT = 60$ (b) $8T - WT = 20$

We will use elimination to solve.

$$
\begin{array}{rl}
\text{(a)} & 8T + WT = 60 \\
\text{(b)} & \underline{8T - WT = 20} \\
& 16T \qquad\;\; = 80 \longrightarrow \textbf{T = 5 hours}
\end{array}
$$

We will now use equation (a) to solve for W.

$$8(5) + W(5) = 60 \qquad \text{substituted}$$
$$40 + 5W = 60 \qquad \text{multiplied}$$
$$5W = 20 \qquad \text{added } -40$$
$$\textbf{W = 4 mph} \qquad \text{divided}$$

Thus, the time for both trips was 5 hours and the speed of the current was 4 mph.

example 91.A.2   The steamboat Juby Fountain could go 70 miles downstream in 5 hours but required 6 hours to go 48 miles upstream. What was the speed of the boat in still water and what was the speed of the current?

solution   We begin by writing both equations.

DOWNSTREAM EQUATION   UPSTREAM EQUATION

(a) $(B + W)T_D = D_D$ (b) $(B - W)T_U = D_U$

Next we replace D_D, T_D, D_U, and T_U with 70, 5, 48, and 6.

(a) $(B + W)5 = 70$ (b) $(B - W)6 = 48$

Now we multiply in both equations and get

$$\text{(a)} \quad 5B + 5W = 70 \qquad \text{(b)} \quad 6B - 6W = 48$$

We choose to multiply (a) by 6 and to multiply (b) by 5 and then use elimination.

$$
\begin{array}{lll}
\text{(a)} \quad 5B + 5W = 70 & \longrightarrow \ (6) \ \longrightarrow & \text{(a}') \quad 30B + 30W = 420 \\
\text{(b)} \quad 6B - 6W = 48 & \longrightarrow \ (5) \ \longrightarrow & \text{(b}') \quad \underline{30B - 30W = 240} \\
& & \phantom{\text{(b}')} \quad 60B = 660
\end{array}
$$

$$\boldsymbol{B = 11 \text{ mph}}$$

Now we will use 11 for B in equation (a) and solve for W.

$$
\begin{array}{ll}
5(11) + 5W = 70 & \text{substituted} \\
55 + 5W = 70 & \text{multiplied} \\
5W = 15 & \text{added} -55 \\
\boldsymbol{W = 3 \text{ mph}} & \text{divided}
\end{array}
$$

Thus the speed of the boat in still water was 11 mph and the speed of the current was 3 mph.

example 91.A.3   The water in the Flint River flows at 5 kilometers per hour. A speedboat can go 15 kilometers upstream in the same time it takes to go 25 kilometers downstream. How fast can the boat go in still water?

solution   We begin by writing both equations.

$$
\begin{array}{ll}
\text{DOWNSTREAM EQUATION} & \text{UPSTREAM EQUATION} \\
\text{(a)} \quad (B + W)T_D = D_D & \text{(b)} \quad (B - W)T_U = D_U
\end{array}
$$

Next we reread the equations and make the required substitutions. We use T for both T_D and T_U since these times are equal.

$$\text{(a)} \quad (B + 5)T = 25 \qquad \text{(b)} \quad (B - 5)T = 15$$

Now we multiply.

$$\text{(a)} \quad BT + 5T = 25 \qquad \text{(b)} \quad BT - 5T = 15$$

We can eliminate the double variable BT if we multiply equation (b) by (-1) and add it to equation (a).

$$
\begin{array}{lrr}
\text{(a)} & BT + 5T = & 25 \\
(-1)\text{(b)} & \underline{-BT + 5T =} & \underline{-15} \\
& 10T = & 10
\end{array}
$$

$$T = 1 \text{ hour}$$

Now we use 1 for T in equation (a) and solve for the speed of the boat.

$$
\begin{array}{ll}
B(1) + 5 = 25 & \text{substituted} \\
\boldsymbol{B = 20 \text{ kph}} & \text{added} -5
\end{array}
$$

Thus we find that the speed of the boat in still water is 20 kph.

problem set 91   ***1.** Robert and Clay could go 60 miles downstream in the same time it took them to go 20 miles upstream. If the speed of their boat was 8 mph in still water, what was the speed of the current and what were the times?

*2. The steamboat Juby Fountain could go 70 miles downstream in 5 hours but required 6 hours to go 48 miles upstream. What was the speed of the boat in still water and what was the speed of the current?

*3. The water in the Flint River flows at 5 kilometers per hour. A speedboat can go 15 kilometers upstream in the same time it takes to go 25 kilometers downstream. How fast can the boat go in still water?

4. The gustatory score varied directly as the number of delicious comestibles offered. If the gustatory score was 500 when 20 delicious comestibles were offered, what offering was necessary for a score of 1750? Work once using the ratio format and again using the variation format.

5. Harriet gave Wilbur a 50-yard head start. How long did it take her to catch Wilbur if her speed was 8 yards per second and his was only 6 yards per second?

Graph:

6. $\begin{cases} y \geq x + 1 \\ y > -\dfrac{4}{5}x - 1 \end{cases}$

7. $\begin{cases} y \leq \dfrac{1}{2}x + 2 \\ x < 2 \end{cases}$

Solve:

8. $\begin{cases} x + y + z = 6 \\ 3x - y + z = 8 \\ x - 2y + z = 0 \end{cases}$

9. $\begin{cases} x^2 + y^2 = 10 \\ x + y = 4 \end{cases}$

Graph on a number line:

10. $-4 \leq x - 2 < 2$, $D = \{\text{Reals}\}$

11. $x - 1 \not< 2$ or $x + 2 \not\geq 2$, $D = \{\text{Integers}\}$

Simplify:

12. $\dfrac{(y^{a+2})^2 x^{2b/3}}{x^b y^{-a}}$

13. $\dfrac{x}{y + \dfrac{y}{\dfrac{1}{x} + y}}$

14. $\dfrac{m}{a + \dfrac{a}{1 + \dfrac{a}{m}}}$

15. $\dfrac{i - 5}{7 - i}$

16. $\dfrac{3i + 2}{2i - 3}$

17. Solve: $\sqrt{z} - 3 = \sqrt{z - 27}$

18. Add: $-5\underline{/20°} + 8\underline{/-150°}$

19. Write $-5R + 20U$ in polar form.

Simplify:

20. $\dfrac{3 - 5\sqrt{2}}{2 - \sqrt{2}}$

21. $\dfrac{8 - \sqrt{2}}{4 - \sqrt{8}}$

22. $\sqrt[5]{x^4 y^3}\,\sqrt[3]{xy^2}$

23. $\dfrac{a}{by} = x\left(\dfrac{1}{R_1} + \dfrac{1}{R_2}\right)$; find R_1

24. $\dfrac{a}{by} = x\left(\dfrac{1}{R_1} + \dfrac{1}{R_2}\right)$; find y

Simplify:

25. $i^3 - 2i^4 + \sqrt{-9} - \sqrt{-2}\sqrt{-2}$

26. $\sqrt{\dfrac{5}{12}} - 3\sqrt{\dfrac{12}{5}} + 2\sqrt{60}$

27. Solve $-3x^2 - x = 5$ by using the quadratic formula.

28. Convert 15 centimeters per second to yards per minute.

29. Find the surface area of the box in square centimeters. Dimensions are in meters.

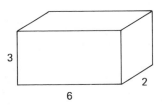

30. Add: $\dfrac{4}{x^2 - 9} - \dfrac{3x}{-3 + x}$

ER 91-78, 91-79

LESSON 92 *Discriminant*

discriminant

In Lesson 70 we completed the square on the general form of the quadratic equation $ax^2 + bx + c = 0$ to derive the quadratic formula. This formula can be used to solve any quadratic equation.

$$x = \frac{-b \pm \sqrt{b^2 - 4ac}}{2a}$$

We remember that quadratic equations can have three different types of solutions.

(a) A single real number
(b) Two different real numbers
(c) Two different complex numbers

Which type of solution an equation will have is determined by the value of the expression

$$b^2 - 4ac$$

in the quadratic formula. Because this expression determines the type of answer we will get, we say that this expression **discriminates** between the possible types of answers. This is the reason that we call this part of the quadratic formula the **discriminant.** We can see how this expression discriminates if we use the quadratic formula to solve the following quadratic equations.

(a) $x^2 + 4x + 4 = 0$ (b) $x^2 + 5x - 1 = 0$ (c) $x^2 + 5x + 7 = 0$

SOLUTION   SOLUTION   SOLUTION

$a = 1, b = 4, c = 4$ $a = 1, b = 5, c = -1$ $a = 1, b = 5, c = 7$

$x = \dfrac{-4 \pm \sqrt{(4)^2 - 4(1)(4)}}{2}$ $x = \dfrac{-5 \pm \sqrt{25 - 4(1)(-1)}}{2}$ $x = \dfrac{-5 \pm \sqrt{25 - 4(1)(7)}}{2}$

$= -2 \pm \dfrac{\sqrt{0}}{2}$ $= \dfrac{-5 \pm \sqrt{25 + 4}}{2}$ $= \dfrac{-5 \pm \sqrt{25 - 28}}{2}$

$x = -2$ $= \dfrac{-5 \pm \sqrt{29}}{2}$ $= \dfrac{-5 \pm \sqrt{-3}}{2}$

$x = -\dfrac{5}{2} \pm \dfrac{\sqrt{29}}{2}$ $x = -\dfrac{5}{2} \pm \dfrac{\sqrt{3}}{2}i$

In (a) the value of $b^2 - 4ac$ is *zero* so we get the answer

$$-2 \pm 0$$

for a solution. This has only one value, which is -2. Thus, we see that

(a)   When $b^2 - 4ac$ equals zero, the solution is one real number.

In (b) the value of $b^2 - 4ac$ is 29, **a positive number,** so we get the two solutions to this equation

$$-\frac{5}{2} + \frac{\sqrt{29}}{2} \quad \text{and} \quad -\frac{5}{2} - \frac{\sqrt{29}}{2}$$

From this we see that

(b)   When $b^2 - 4ac$ is a positive number, there are two real number solutions.

In (c) the value of $b^2 - 4ac$ is -3, **a negative number.** In this case, we get the following complex numbers as solutions

$$-\frac{5}{2} + \frac{\sqrt{3}}{2}i \quad \text{and} \quad -\frac{5}{2} - \frac{\sqrt{3}}{2}i$$

From this we see that

(c)   When $b^2 - 4ac$ is a negative number, the equation has two complex solutions which are conjugates.

example 92.A.1   What kind of solutions does the equation $x^2 = -4x + 2$ have? Do not solve.

solution   We rearrange the equation into standard form and find the values of a, b, and c.

$$x^2 + 4x - 2 = 0$$
$$a = 1, b = 4, c = -2$$

Now we find the value of $b^2 - 4ac$:

$b^2 - 4ac$	discriminant
$(4)^2 - 4(1)(-2)$	substituted
$16 + 8$	multiplied
24	added

In this equation the discriminant is a positive number, so the equation has two real, unequal solutions.

example 92.A.2   What kind of roots does the equation $-2x = -3x^2 - 8$ have? Do not solve.

solution   We first write the equation in standard form. Then we identify a, b, and c and find the value of the discriminant $b^2 - 4ac$.

$3x^2 - 2x + 8 = 0$	standard form
$a = 3, b = -2, c = 8$	values of a, b, and c
$b^2 - 4ac$	discriminant
$(-2)^2 - 4(3)(8)$	substituted
$4 - 96$	multiplied
-92	added

The discriminant in this equation has a value of -92, which is a negative number. Thus, the solution to this equation is a pair of complex numbers which are conjugates.

problem set 92

1. The Robert E. Lee could go 45 miles down the Old Muddy in the same time it took it to go 15 miles up the same stream. If the current in the Old Muddy was 5 miles per hour, what was the speed of the Robert E. Lee in still water?

2. The Memphis Belle could go 48 miles downstream in 4 hours but required 8 hours to go 64 miles upstream. What was the speed of the Memphis Belle in still water and what was the speed of the current?

3. The hydro could go 40 miles per hour on a lake. The same boat could go 210 miles down the Echeconnee River in one-half the time that it took to go 380 miles up the Echeconnee River. How fast did the Echeconnee flow?

4. Charlemagne trudged the 40 miles to the battle in 4 hours longer than it took Roland to travel 48 miles to the same battle. Roland rode a horse and thus traveled at twice the speed of Charlemagne. Find the rates and times of both.

5. The volume of a quantity of an ideal gas was held constant at 3 liters. The initial temperature and pressure were 4000 kelvins and 5 atmospheres. What would the new temperature be in kelvins if the pressure were reduced to 1 atmosphere?

Use the discriminants to determine the types of solutions that the following equations have. Do not solve.

*6. $x^2 = -4x + 2$

*7. $-2x = -3x^2 - 8$

8. Graph: $\begin{cases} x + 2y < 2 \\ y \geq -1 \end{cases}$

9. Graph on a number line: $-3 < x - 4 < 2$, $D = \{\text{Integers}\}$

Solve:

10. $\begin{cases} x + y - 2z = -3 \\ 2x + y + z = 7 \\ 3x - y - z = 13 \end{cases}$

11. $\begin{cases} x^2 + y^2 = 1 \\ x - 2y = 2 \end{cases}$

Simplify:

12. $(y^{a+2})^2 y^{a/3} y^2$

13. $\dfrac{x}{1 + \dfrac{a}{b + \dfrac{1}{c}}}$

14. $\dfrac{m}{2x + \dfrac{3}{3 + \dfrac{1}{m}}}$

15. $\dfrac{3i - 5}{i - 7}$

16. $\dfrac{2i + 4}{3i + 2}$

17. Solve: $\sqrt{p} = 5 + \sqrt{p - 35}$

18. The two forces are applied to the point as indicated. Find the resultant force.

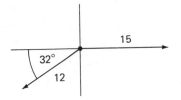

Simplify:

19. $\dfrac{3 + 4\sqrt{2}}{3\sqrt{2} - 4}$

20. $\dfrac{5 + \sqrt{5}}{5 + 2\sqrt{5}}$

21. $\sqrt[3]{4\sqrt[5]{2}}$ **22.** $\sqrt[4]{xy^6}\sqrt{x^3y^5}$

23. $\dfrac{a+c}{m} = m\left(\dfrac{1}{x} + \dfrac{b}{d}\right)$; find a **24.** $\dfrac{a+c}{m} = m\left(\dfrac{1}{x} + \dfrac{b}{d}\right)$; find x

Simplify:

25. $i^3 - 2i^2 - \sqrt{-9} + \sqrt{-4}\sqrt{-4} + 2$ **26.** $3\sqrt{\dfrac{2}{9}} + \sqrt{\dfrac{9}{2}} - 3\sqrt{18}$

27. Solve $3x^2 = -x - 2$ by completing the square.

28. Convert 10 meters per second to feet per minute.

29. Find the area of this figure in square centimeters. Dimensions are in meters.

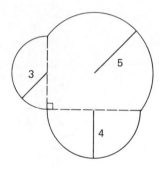

30. Divide $x^3 - 3x^2 - 2$ by $x + 3$.

ER-92-80, 92-81

LESSON 93 *Dependent and independent variables •*
Functions • Functional notation

93.A
dependent and independent variables

If we look at the equation

$$y = 2x + 4$$

we see that if we assign a value to x, then the equation will give us a value for y. For instance, if we let x equal 5, then

$$y = 2(5) + 4 \qquad \text{substituted 5 for } x$$
$$y = 10 + 4 \qquad \text{multiplied}$$
$$\mathbf{y = 14} \qquad \text{added}$$

y equals 14. It was not necessary to assign the value to x. We could have assigned a value to y and solved the equation to find the matching value of x. If we let y equal 14,

$$14 = 2x + 4 \qquad \text{substituted 14 for } y$$
$$10 = 2x \qquad \text{added } -4$$
$$\mathbf{5 = x} \qquad \text{divided}$$

we find that the matching value of x is 5.

Thus we see that when we have an equation in two variables, we can assign a value to either variable and use the equation to find the value of the other variable. **We** call the variable to which we assign the value the **independent variable, and we call the other variable the dependent variable because its value depends on the value assigned to the independent variable.** To reduce confusion as to which variable is the dependent variable, we normally use the letters x and y to represent the variables; and we let x be the variable to which we assign values. **Thus, x will always be the independent variable, and the value of y will depend on the value of x.** This is the reason that we graph a line by solving the equation for y.

$$y = 2x - 3$$

x	0	2	−2
y			

And then we assign values to x and see what values of y the equation pairs with the chosen values of x.

93.B

functions Some equations have only one answer for y for any chosen value of x. Both of the equations

   (a) $y = 2x + 4$ (b) $y = x^3 + 4x + 3$

are equations of this kind. For instance, if we give x a value of -3, each equation will give us one answer for y.

   (a) $y = 2(-3) + 4$ (b) $y = (-3)^3 + 4(-3) + 3$
        $y = -6 + 4$ $y = -27 - 12 + 3$
        $\mathbf{y = -2}$ $\mathbf{y = -36}$

We see that when we let x equal -3, equation (a) gives us an answer of -2 for y; and equation (b) gives us an answer of -36 for y. Some people prefer not to use the word **answer**, and they would say that equation (a) **pairs** or **matches** the y value of -2 with the x value of -3. They would also say that equation (b) **pairs** or **matches** the y value of -36 with the x value of -3.

Not all equations have just one answer for y for every value of x. For example, in the equation

$$y^2 = x$$

if we let x equal 4

$$y^2 = 4$$

then both $+2$ and -2 are paired values of y, for both $(+2)^2$ and $(-2)^2$ equal 4.

$$(+2)^2 = 4 \qquad (-2)^2 = 4$$

Mathematicians have found it useful to have a special name for equations that have only one answer for y for every value of x. They call these equations functional relationships and use the word *function* when discussing these relationships. Unfortunately (or fortunately), the definition of a function has been extended to cover any situation where each member of a given set of numbers or letters has only one answer.

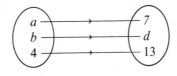

Domain   Range

Here we see that the answer for *a* is 7, the answer for *b* is *d,* and the answer for 4 is 13. This figure falls under the definition of a *function,* and we discover that an equation is not necessary. All we need is two sets. The first set is the values of *x* that we can use, and the second set is the collection of all the answers for these values of *x*. We remember that the *domain* is the word we use for the permissible values of *x*, so we call the first set the **domain of the function.** Instead of calling the answers *answers,* we call them **images. Thus, in the diagrams on the previous page, the image of *a* is 7, and the images of *b* and 4 are *d* and 13, respectively. We call the collection of all the images the** *range* **of the function.**

There are two accepted definitions for a function. One definition says that **the pairing or the matching is the function,** while the second definition says that the **ordered pairs themselves are the function.**

A **function** is a **mapping** between two sets that associates with each element of the first set a **unique** (one and only one) element of the second set. The first set is called the **domain** of the function. For each element *x* of the domain, the corresponding element *y* of the second set is called the **image** of *x* under the function. The set of all images of the elements of the domain is called the **range** of the function.

A **function** is a **set of ordered pairs** in which no two pairs have the same first element and different second elements.

Now we need to know what to think when we see the word function. We will be correct if we always think that a function is something that has for every value of *x*

---

exactly one answer

If the relationship has one or more answers, it is called a **relation.** Thus, we see that every function can also be called a relation; but every relation is not a function, for many relations have more than one answer.

We will remember that for a relationship to be called a function,

1. **The domain must be specified or implied.**
2. **A way must be designated to find every image (answer).**
3. **There is exactly one answer for every member of the domain.**

example 93.B.1   Which of the following depict functions?

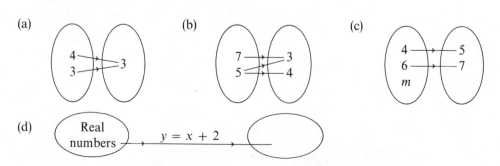

(a)   (b)   (c)

(d)

solution   Diagrams (a) and (d) depict functions because both diagrams show exactly 1 answer for each member of the domain. In (a) the domain is specified to be the numbers 4 and 3 and each member of the domain has one image. The image for 4 is 3 and the image for 3 is 3. Thus both images are the same but this is acceptable. In (d) the domain is specified and a way is given to find any image.

Diagrams (b) and (c) do not depict functions because in (b) 5 has two images and in (c) no image is designated for *m*. Thus (c) is not even a relation.

example 93.B.2   Which of the following sets of ordered pairs of *x* and *y* are functions?

   (a)   (4, 3), (7, 5), (−3, −2), (−6, −4)
   (b)   (4, 3), (5, −2), (7, 3), (6, 3)
   (c)   (5, −2), (4, −3), (7, −2), (5, 4)
   (d)   (7, −2), (3, −2), (6, 5), (4, 5)

solution   (a), (b), and (d) are functions. (c) is not a function because 5 has two images.

93.C
functional notation

If we are given the two equations

$$y = x + 2 \quad \text{and} \quad y = x - 5$$

and asked to find the value of *y* when *x* equals 2, we would want to know which equation to use. If we use functional notation, we can name the left equation the *f* equation and the right equation the *g* equation. If we do this we will use the notations $f(x)$ and g(x) instead of using *y*.

$$f(x) = x + 2 \qquad g(x) = x - 5$$

We read the left equation as "*f* of *x* equals *x* + 2" and the right equation as "*g* of *x* equals *x* − 5." Now if we are asked to find g(2), we are asked to find the value of the *g* equation when *x* equals 2.

$$g(2) = (2) - 5 \qquad \text{replaced } x \text{ with } 2$$

$$g(2) = -3 \qquad \text{simplified}$$

Since the answer is −3 when *x* equals 2, we say that "*g* of 2 equals −3."

example 93.C.1   If $h(x) = 4x - 3$ and $p(x) = x^2 - 3x$, find $p(-3)$.

solution   We are asked to find "*p* of −3," which is the value of the *p* equation when *x* equals −3.

$$p(-3) = (-3)^2 - 3(-3) \qquad \text{replaced } x \text{ with } -3$$

$$p(-3) = 9 + 9 \qquad \text{simplified}$$

$$p(-3) = 18 \qquad \text{simplified}$$

problem set 93

1.   Use the relationship $PV = nRT$ to find the number of moles in a quantity of gas when the temperature is 473 K, the pressure is 2 atmospheres, and the volume is 10 liters ($R = .0821$). Begin by solving the equation for *n*.

2.   Analysis showed that the 40 gallons in the tank was only 20% disinfectant. A 44% disinfectant was needed. How many gallons of a 60% disinfectant should be added?

3.   The Suzie Q could go 48 miles downstream in the same time that it took her to go 32 miles upstream. If her speed was 10 miles per hour in still water, what was the speed of the current?

4. The temperature of a quantity of an ideal gas was held constant at 430°C. The pressure was 700 millimeters of mercury and the volume was 1400 ml. What was the volume when the pressure was increased to 2800 millimeters of mercury?

5. The number of purples varied inversely as the square of the number of reds. When 10 reds were present, the number of purples was 4. How many purples were present when the number of reds was only 5?

*6. Which of the following sets of ordered pairs of x and y are functions?
   (a) (4, 3), (7, 5), (−3, −2), (−6, −4)
   (b) (4, 3), (5, −2), (7, 3), (6, 3)
   (c) (5, −2), (4, −3), (7, −2), (5, 4)
   (d) (7, −2), (3, −2), (6, 5), (4, 5)

*7. If $h(x) = 4x − 3$ and $p(x) = x^2 − 3x$, find $p(−3)$.

8. Use the discriminant to determine the kinds of numbers that will satisfy the equation $−x^2 = 4x + 4$. Do not solve.

9. Graph: $\begin{cases} 2x + 3y > −6 \\ x − 3y \geq −6 \end{cases}$

10. Graph on a number line: $−x + 3 \not< −2$ or $−x + 3 < −5$, $D = \{\text{Integers}\}$

Solve:

11. $\begin{cases} x + 2y − z = 2 \\ 2x − y + z = 2 \\ 3x − y + 2z = 4 \end{cases}$

12. $\begin{cases} x^2 + y^2 = 3 \\ x − y = 2 \end{cases}$

Simplify:

13. $\dfrac{x^{a/3}y^2}{x^{3a/2}(y^2)^a}$

14. $\dfrac{p}{m + \dfrac{m}{p − \dfrac{1}{mp}}}$

15. $\dfrac{x}{a + \dfrac{b}{ab − \dfrac{1}{b}}}$

16. $\dfrac{2i + 7}{−3i}$

17. $\dfrac{−5i − 2}{−2i + 5}$

18. Solve: $\sqrt{k} + \sqrt{k − 21} = 7$

19. Add: $−20/\underline{−160°} + 20/\underline{200°}$

20. Write $−4R + 8U$ in polar form.

Simplify:

21. $\dfrac{2 − 5\sqrt{2}}{3\sqrt{2} − 2}$

22. $\dfrac{2 + 3\sqrt{2}}{2\sqrt{2} − 3}$

23. $\dfrac{a + b}{x} = \dfrac{1}{R_1} + \dfrac{c}{R_2}$; find R_1

24. $\dfrac{a + b}{x} = \left(\dfrac{1}{R_1} + \dfrac{c}{R_2}\right)$; find c

Simplify:

25. $3i^2 + 2i^5 − 2i + \sqrt{−9} − \sqrt{−2}\sqrt{2}$

26. $4\sqrt{\dfrac{5}{12}} + 3\sqrt{\dfrac{12}{5}} − 2\sqrt{240}$

27. Solve $−2x^2 = −7 + x$ by using the quadratic formula.

28. Convert 20 liters per second to cubic inches per minute.

29. Draw the line suggested by the data points and write the equation that expresses the number of neutrons as a function of radiation.

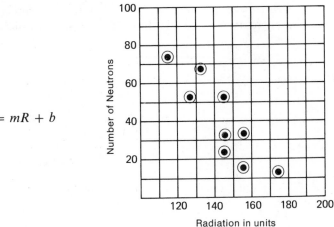

$$N = mR + b$$

30. Add: $\dfrac{4}{ax^2} + \dfrac{6x}{a^2} + \dfrac{3}{a^4x^3}$

ER 93-82, 93-83

LESSON 94 *More nonlinear systems*

94.A

more nonlinear systems

The nonlinear systems that we have investigated thus far have been like one of the following:

$$\text{(a)} \quad \begin{cases} BT_D + 5T_D = 25 \\ BT_D - 5T_D = 15 \end{cases} \qquad \text{(b)} \quad \begin{cases} x^2 + y^2 = 9 \\ 2x - y = 3 \end{cases}$$

On the left we have the equations of two hyperbolas and on the right we have the equation of a circle and the equation of a straight line. In this lesson, we will look at two other types of nonlinear systems. The first will consist of the equation of a hyperbola and the equation of a straight line.

example 94.A.1   Solve the system: (a) $\begin{cases} 6x - y = 5 \qquad \text{(straight line)} \\ xy = 4 \qquad\qquad \text{(hyperbola)} \end{cases}$
                          (b)

solution   We cannot use elimination, for the terms in both equations are not alike. Thus, some form of substitution should work. We will solve equation (b) for x and substitute this expression for x in equation (a).

$$x = \frac{4}{y} \qquad \text{equation (b)}$$

$$6\left(\frac{4}{y}\right) - y = 5 \qquad \text{substituted}$$

$$\frac{24}{y} - y = 5 \qquad \text{multiplied}$$

Now, whenever an equation has denominators, we eliminate the denominators. Thus, we multiply every term by y and cancel the denominator.

$$y \cdot \frac{24}{y} - y \cdot y = 5y \qquad \text{multiplied every term by } y$$

$$24 - y^2 = 5y \qquad \text{simplified}$$

$$y^2 + 5y - 24 = 0 \qquad \text{rearranged}$$

$$(y - 3)(y + 8) = 0 \qquad \text{factored}$$

$$y = 3, -8 \qquad \text{zero factor theorem}$$

We finish by using both 3 and -8 in equation (b) to find the paired values of x.

USING 3   USING -8

$$x(3) = 4 \qquad x(-8) = 4$$

$$x = \frac{4}{3} \qquad x = -\frac{1}{2}$$

Thus, the ordered pairs of x and y that satisfy both equations are $(\frac{4}{3}, 3)$ and $(-\frac{1}{2}, -8)$. The graphs of the two equations are shown here, and we note that the line intersects the hyperbola at the coordinates we have found.

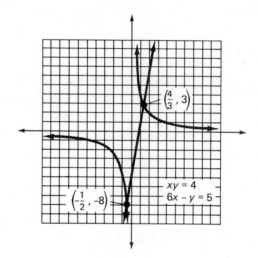

example 94.A.2   Solve the system: (a) $\begin{cases} x^2 + y^2 = 9 & \text{(circle)} \\ 2x^2 - y^2 = -6 & \text{(hyperbola)} \end{cases}$
                           (b)

solution   This system can be solved by using either substitution or elimination. We must be careful to get all the answers because this circle and hyperbola intersect at four different points. We decide to use elimination. We can eliminate the y^2 terms if we add the equations just as they are. If we do this, we get

$$3x^2 = 3 \qquad \text{added}$$

$$x^2 = 1 \qquad \text{divided by 3}$$

Here we must be careful because this equation has both $+1$ and -1 as solutions.

$$x = \pm\sqrt{1} \quad \longrightarrow \quad x = 1, -1$$

Now we must use these values of x one at a time to solve for y. We will use equation (a) and begin by letting x equal $+1$.

$$(1)^2 + y^2 = 9 \qquad \text{substituted (1) for } x$$
$$y^2 = 8 \qquad \text{added } -1$$
$$y = \pm 2\sqrt{2} \qquad \text{solved}$$

Thus, there are two points of intersection when x equals 1. So two solutions of our system are

$$(1, 2\sqrt{2}) \qquad \text{and} \qquad (1, -2\sqrt{2})$$

Next, we find the values of y that pair with a value of -1 for x. Again we use equation (a).

$$(-1)^2 + y^2 = 9 \qquad \text{substituted } (-1) \text{ for } x$$
$$y^2 = 8 \qquad \text{added } -1$$
$$y = \pm 2\sqrt{2} \qquad \text{solved}$$

Thus, our other two solutions to the system are

$$(-1, 2\sqrt{2}) \qquad \text{and} \qquad (-1, -2\sqrt{2})$$

Here we show the graphs of the two curves and note that there are four points where the curves intersect.

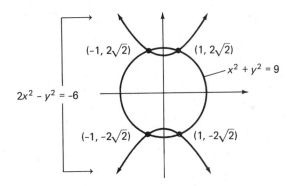

1. Weatherman Bob walked to the campsite at 6 kilometers per hour. Then he caught a ride home on an old truck at 24 kilometers per hour. If he was gone for 10 hours, how far was it to the campsite?

2. The alcohol concentration had to be exactly 52%. How many milliliters of a 60% solution should be added to 200 ml of a 20% solution to get the proper concentration?

3. The pressure of a quantity of ideal gas was held constant at 764 millimeters of mercury. The original temperature was 200 K and the original volume was 400 ml. If the temperature was increased to 600 K, what was the final volume? Begin by solving for V_2.

4. Ronald could travel 360 miles in one-fourth the time it took Jimmy to travel 480 miles. This was because Ronald's speed exceeded that of Jimmy by 60 mph. What were the speeds and times of both?

5. Detia could row 60 miles downstream in 4 hours but required 8 hours to row 72 miles upstream. What was her speed in still water and what was the speed of the current?

Solve:

*6. $\begin{cases} 6x - y = 5 \\ xy = 4 \end{cases}$

*7. $\begin{cases} x^2 + y^2 = 9 \\ 2x^2 - y^2 = -6 \end{cases}$

8. $\begin{cases} x + 2y - z = 4 \\ 2x - y + z = -3 \\ x - y + z = -4 \end{cases}$

9. Which of the following depict functions?

(a)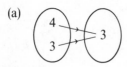

(b)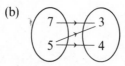

(c)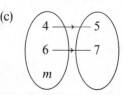

(d)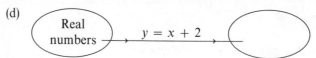

10. Use the discriminant to find the types of roots of $3x^2 - x + 5 = 0$. Do not solve.

11. Graph: $\begin{cases} x > -2 \\ 2x + 3y > -3 \end{cases}$

12. Graph on a number line: $-2 < -x - 2 < 0$, $D = \{\text{Integers}\}$

Simplify:

13. $\dfrac{(a^{x+2})^{1/2} x^{2a}}{a^x x^a}$

14. $\dfrac{x}{4 + \dfrac{4}{1 + \dfrac{x}{4}}}$

15. $\dfrac{5}{2 + \dfrac{1}{2 + \dfrac{2}{x}}}$

16. $\dfrac{3i - 5}{3 - 5i}$

17. $\dfrac{4 + i}{-i}$

18. Solve: $\sqrt{x^2 + 2x + 34} - x = 4$

19. The two forces are applied to the point as shown. Find the resultant force.

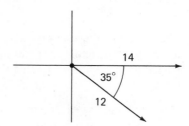

Simplify:

20. $\dfrac{2\sqrt{2} - 5}{2\sqrt{2} + 3}$

21. $\dfrac{3 - \sqrt{2}}{\sqrt{2} + 4}$

22. $\dfrac{m(a + b)}{x} = \left(\dfrac{1}{p} + \dfrac{r}{q}\right)$; find r **23.** $\dfrac{m(a + b)}{x} = \dfrac{1}{p} + \dfrac{r}{q}$; find p

Simplify:

24. $4 - 3i^2 + \sqrt{-9} + \sqrt{-3}\sqrt{-3}$ **25.** $2\sqrt{\dfrac{7}{3}} - 3\sqrt{\dfrac{3}{7}} - 2\sqrt{84}$

26. Solve $-x = x^2 - 3x - 4$ by completing the square.

27. Use the formula $PV = nRT$ to find the volume of 1.32 moles of gas at a pressure of 5 atmospheres and a temperature of 600 K ($R = .0821$).

28. Find the equation of the line that passes through $(4, 2)$ that is perpendicular to the line $3y - 2x = 5$.

29. Find the surface area of the right circular cylinder in square inches. Dimensions are in feet.

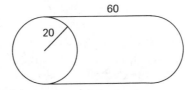

30. Convert 30 miles per hour to feet per second.

ER 94-84, 94-85

LESSON 95   *Joint and combined variation • More on irrational roots*

95.A
joint and combined variation

To review the concept of direct variation we recall that if the number of peaches varies directly as the number of apples, either of the following equations may be used.

$$\text{(a)} \quad P = kA \qquad \text{(b)} \quad \dfrac{P_1}{P_2} = \dfrac{A_1}{A_2}$$

The form on the left is the direct variation form of the relationship. On the right we show the ratio form of the same relationship. Note that P_1 is on top on the left and that A_1 is on top on the right.

The word *inverse* means inverted or turned upside down, so if we are told that peaches vary inversely as apples, we remember that the relationships are the upside-down form of the direct variation equations.

$$\text{(c)} \quad P = \dfrac{k}{A} \qquad \text{(d)} \quad \dfrac{P_1}{P_2} = \dfrac{A_2}{A_1}$$

In equation (c) note that k remains on top and that A goes below. On the right side of equation (d), note that A_2 is above and A_1 is below. This is the inverse or inverted form of equation (b).

Some statements of variation give the relationship between three or more variables. The words **varies jointly** imply a sort of double direct variation that has only one constant of proportionality. Thus, the statement that peaches vary jointly as apples and raisins implies the following relationships:

$$\text{(e)} \quad P = kAR \qquad \text{(f)} \quad \frac{P_1}{P_2} = \frac{A_1 R_1}{A_2 R_2}$$

If we are told that peaches vary inversely as apples and raisins, we must invert the variables on the right side of both equations and get

$$\text{(g)} \quad P = \frac{k}{AR} \qquad \text{(h)} \quad \frac{P_1}{P_2} = \frac{A_2 R_2}{A_1 R_1}$$

Note that in (g) we have only one constant of proportionality k and that k is not inverted but remains on top.

In some relationships we have both direct and inverse variations in the same statement. For instance, the statement that girls vary inversely as boys and directly as teachers implies the equations

$$\text{(i)} \quad G = \frac{kT}{B} \qquad \text{(j)} \quad \frac{G_1}{G_2} = \frac{B_2 T_1}{B_1 T_2}$$

In equation (i) B for boys went below because boys varied inversely and in equation (j), B_2 went above B_1 because boys varied inversely.

We note that either the variation form of the equation or the ratio form of the equation may be used. It is helpful to know how to use both forms because both approaches will be encountered in advanced courses in mathematics and science.

example 95.A.1   The number of girls varied inversely as the number of boys and directly as the number of teachers. When there were 50 girls, there were 20 teachers and 10 boys. How many boys were there when there were 10 girls and 100 teachers? Work the problem twice; use the proportion form and then use the ratio form.

solution   Since the boys varied inversely, B goes on the bottom.

$$G = \frac{kT}{B}$$

Now we solve for k.

$$50 = \frac{k(20)}{10} \quad \longrightarrow \quad k = 25$$

In the original equation, we replace k with 25 and get

$$G = \frac{25T}{B}$$

Next we use 10 for girls and 100 for teachers, and solve for boys.

$$10 = \frac{(25)(100)}{B} \quad \longrightarrow \quad B = \frac{2500}{10} \quad \longrightarrow \quad \mathbf{B = 250}$$

Now we will work the problem again, but this time we will use the ratio form of the equation. On the right side B_1 will go below because the variation is inverse for boys.

$$\frac{G_1}{G_2} = \frac{T_1 B_2}{T_2 B_1}$$

Now we replace G_1, T_1, and B_1 with 50, 20, and 10 and get

$$\frac{50}{G_2} = \frac{20 B_2}{T_2 10}$$

Now we use 10 for G_2 and 100 for T_2 and solve for B_2.

$$\frac{50}{10} = \frac{20B_2}{100(10)} \longrightarrow 5 = \frac{B_2}{50} \longrightarrow B_2 = 250$$

We could have made all the substitutions at one time, but we choose to do it in two steps because it is so easy to make a mistake when substituting for five variables.

example 95.A.2   Strawberries varied jointly as plums and tomatoes. If 500 strawberries went with 4 plums and 25 tomatoes, how many plums would go with 40 strawberries and 2 tomatoes? First work the problem using proportion and then work it again using ratio.

solution   The words **varies jointly** imply the relationship

$$S = kPT$$

Next we replace S, P, and T with 500, 4, and 25 and solve for k.

$$500 = k(4)(25) \longrightarrow k = 5$$

Now we replace k in the original equation with 5.

$$S = 5PT$$

We finish by replacing S with 40 and T with 2 and solving for P.

$$40 = 5(P)(2) \longrightarrow P = 4 \text{ plums}$$

Now we will work the problem again and use the ratio form. There is no inverse relationship so S_1, P_1 and T_1 all go on top.

$$\frac{S_1}{S_2} = \frac{P_1 T_1}{P_2 T_2}$$

We first replace S_1, P_1, and T_1 with 500, 4, and 25; and we replace S_2 and T_2 with 40 and 2.

$$\frac{500}{40} = \frac{4(25)}{P_2(2)}$$

Next we eliminate the denominators by multiplying both sides by $40P_2$.

$$40P_2 \cdot \frac{500}{40} = \frac{4(25)}{P_2(2)} \cdot 40P_2 \longrightarrow 500P_2 = 2000$$

and we finish by dividing by 500

$$\frac{500P_2}{500} = \frac{2000}{500} \longrightarrow P_2 = 4$$

Again we find that the answer is **4 plums.**

95.B
more on irrational roots

In Lesson 94 we found that the solution to the system

$$\begin{cases} 6x - y = 5 \\ xy = 4 \end{cases}$$

required the solution to the quadratic equation

$$y^2 + 5y - 24 = 0$$

This equation can be factored, and thus the solution to this system consists of ordered pairs of rational numbers (fractions). If a quadratic equation cannot be solved by factoring we can always find the solutions by using the quadratic formula. **Trying the factor method to solve quadratic equations that result from real life problems is usually a waste of time and thus many people use the formula without even trying to factor first.**

example 95.B.1 Solve: (a) $\begin{cases} x - 2y = 3 \\ (b) \quad xy = 6 \end{cases}$

solution We begin by solving equation (b) for x and substituting into equation (a).

$$x = \frac{6}{y} \qquad \text{solved for } x$$

$$\frac{6}{y} - 2y = 3 \qquad \text{substituted}$$

$$6 - 2y^2 = 3y \qquad \text{multiplied by } y$$

$$2y^2 + 3y - 6 = 0 \qquad \text{rearranged}$$

Now we use the quadratic formula to find the roots of this equation.

$$x = \frac{-b \pm \sqrt{b^2 - 4ac}}{2a} \qquad y = \frac{-3 \pm \sqrt{9 - 4(2)(-6)}}{4} \qquad y = -\frac{3}{4} \pm \frac{\sqrt{57}}{4}$$

Now we will use the linear equation $x - 2y = 3$ to find x.

$$x = 2y + 3 \qquad\qquad\qquad x = 2y + 3$$

$$x = 2\left(-\frac{3}{4} + \frac{\sqrt{57}}{4}\right) + 3 \quad \text{substituted} \qquad x = 2\left(-\frac{3}{4} - \frac{\sqrt{57}}{4}\right) + 3 \quad \text{substituted}$$

$$x = \frac{3}{2} + \frac{\sqrt{57}}{2} \qquad\qquad \text{simplified} \qquad x = \frac{3}{2} - \frac{\sqrt{57}}{2} \qquad\qquad \text{simplified}$$

Thus the solutions are the following ordered pairs of x and y:

$$\left(\frac{3}{2} + \frac{\sqrt{57}}{2}, -\frac{3}{4} + \frac{\sqrt{57}}{4}\right) \quad \text{and} \quad \left(\frac{3}{2} - \frac{\sqrt{57}}{2}, -\frac{3}{4} - \frac{\sqrt{57}}{4}\right)$$

problem set 95

*1. The number of girls varied inversely as the number of boys and directly as the number of teachers. When there were 50 girls, there were 20 teachers and 10 boys. How many boys were there when there were 10 girls and 100 teachers? Work the problem twice. Use the proportion form and then use the ratio form.

*2. Strawberries varied jointly as plums and tomatoes. If 500 strawberries went with 4 plums and 25 tomatoes, how many plums would go with 40 strawberries and 2 tomatoes? First work the problem using proportion and then work it again using ratio.

3. The current in the Bolibee River flows at 6 kilometers per hour. The boat can go 40 kilometers upstream in twice the time it takes to go 80 kilometers downstream. How fast can the boat go in still water?

4. The 300-mile trip to Aunt Lucy's took 5 hours longer than the trip home. If the speed coming back was twice as great as the speed going, find both speeds and both times.

5. The only way to mix a 32% antiseptic solution was to mix a 20% solution and a 60% solution. How much of each should be used to get 500 ml of the 32% solution?

Solve:

*6. $\begin{cases} x - 2y = 3 \\ xy = 6 \end{cases}$ 7. $\begin{cases} x^2 + y^2 = 4 \\ 4x^2 - y^2 = -4 \end{cases}$ 8. $\begin{cases} x + 3y - z = 2 \\ x + y + 2z = 6 \\ 2x + 2y - z = 2 \end{cases}$

9. Which of these sets of ordered pairs are functions?
   (a) $(4, -3), (5, -3), (-5, 2), (7, -3)$
   (b) $(6, -2), (-2, 6), (4, 6), (5, -3)$
   (c) $(4, 2), (6, 2), (5, -3), (4, 3)$

10. If $g(x) = x^2 - 4$, $D = \{\text{Integers}\}$, find $g(-2)$.

11. Graph: $\begin{cases} x - y < -2 \\ y \geq -2 \end{cases}$

12. Graph on a number line:
    $x + 4 \not> 2$ or $x - 4 > -1$, $D = \{\text{Reals}\}$

Simplify:

13. $\dfrac{(a^{x+4})^{1/2} b^{2x}}{a^{3/2} b^x}$

14. $\dfrac{x}{a + \dfrac{b}{a^2 + \dfrac{1}{ab}}}$

15. $2\sqrt{8}\sqrt[3]{2}$

16. $\dfrac{2i - 8}{4 - 6i}$

17. $\dfrac{i - i^2}{i}$

18. Solve: $\sqrt{s - 39} = 13 - \sqrt{s}$

19. The two forces act on the point as shown. Find the resultant force.

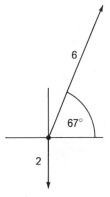

Simplify:

20. $\dfrac{3\sqrt{5} - 1}{1 - \sqrt{5}}$

21. $\dfrac{2 - \sqrt{7}}{1 + \sqrt{7}}$

22. $c = m\left(\dfrac{a + d}{p} - r\right)$; find p

23. $c = m\left(\dfrac{a + d}{p} - r\right)$; find a

24. Simplify: $2i^4 - i^2 - \sqrt{-16} - \sqrt{-4}\sqrt{-4}$

25. Use the relationship $PV = nRT$ to find the number of moles in a quantity of gas when the temperature is 673 K, the pressure is 5 atmospheres, and the volume is 20 liters ($R = .0821$).

26. Solve $4x^2 + 6 = -x$ by using the quadratic formula.

27. Solve by graphing and then find an exact solution by using either substitution or elimination:
$$\begin{cases} x + 2y = 6 \\ 2x - 5y = -10 \end{cases}$$

28. Convert 40 inches per hour to feet per minute.

29. Find C.

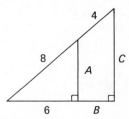

30. Find the volume of the box in cubic centimeters. Dimensions are in inches.

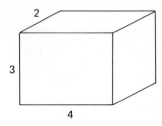

ER 95-86, 95-87

LESSON 96   *Advanced substitution*

96.A
advanced substitution

We have found that we can solve a system of two equations in two unknowns by using either the substitution method or the elimination method. We normally use elimination to solve systems such as

$$\begin{cases} 3x + 2y = -3 \\ 4x - 3y = 13 \end{cases}$$

in which every variable has a coefficient that is greater than 1. Substitution can be used to solve these systems if we remember to eliminate the denominators as the first step after we substitute.

example 96.A.1   Use substitution to solve: (a) $\begin{cases} 3x + 2y = -3 \\ (b) \quad 4x - 3y = 13 \end{cases}$

solution   We decide to solve equation (a) for y and substitute for y in equation (b). First, we solve for y

$$3x + 2y = -3 \qquad \text{equation (a)}$$

$$2y = -3 - 3x \qquad \text{added } -3x$$

$$y = \frac{-3 - 3x}{2} \qquad \text{divided by 2}$$

Now we will substitute for y in equation (b).

$$4x - 3\left(\frac{-3 - 3x}{2}\right) = 13$$

Next we eliminate the denominator 2 by multiplying every term on both sides by 2.

$$4x(2) - 3\left(\frac{-3 - 3x}{2}\right)(2) = 13(2)$$

We cancel and multiply and then solve.

$$8x + 9 + 9x = 26 \qquad \text{multiplied}$$
$$17x = 17 \qquad \text{simplified}$$
$$x = 1 \qquad \text{divided}$$

Now we use 1 for x in equation (a) and solve for y.

$$3(1) + 2y = -3 \qquad \text{substituted}$$
$$3 + 2y = -3 \qquad \text{multiplied}$$
$$2y = -6 \qquad \text{added} -3$$
$$y = -3 \qquad \text{divided}$$

Thus the solution is the ordered pair **(1, −3).**

example 96.A.2   Use substitution to solve: (a) $\begin{cases} 5x - 3y = 9 \\ 2x - 4y = -2 \end{cases}$
    (b)

solution   We decide to solve equation (a) for x and substitute for x in equation (b). First, we solve for x.

$$5x - 3y = 9 \qquad \text{equation (a)}$$
$$5x = 9 + 3y \qquad \text{added } 3y$$
$$x = \frac{9 + 3y}{5} \qquad \text{divided by 5}$$

Now we will substitute for x in equation (b):

$$2\left(\frac{9 + 3y}{5}\right) - 4y = -2 \qquad \text{substituted}$$

Next we will eliminate the denominator 5 by multiplying every term by 5.

$$(5)(2)\left(\frac{9 + 3y}{5}\right) - (5)(4y) = -2(5) \qquad \text{multiplied by 5}$$

Now we cancel, expand, and solve.

$$18 + 6y - 20y = -10 \qquad \text{canceled and expanded}$$
$$-14y = -28 \qquad \text{added } -18$$
$$y = 2 \qquad \text{divided}$$

Now we use 2 for y in equation (a) and solve for x.

$$5x - 3(2) = 9 \qquad \text{substituted}$$
$$5x - 6 = 9 \qquad \text{multiplied}$$
$$5x = 15 \qquad \text{added}$$
$$x = 3 \qquad \text{divided}$$

Thus the solution is the ordered pair **(3, 2).**

problem set 96

1. More were talented than were not. Twice the number of talented exceeded 3 times the number of untalented by 12. Also, 4 times the number of untalented was only 48 less than 3 times the number of talented. How many were talented and how many were untalented?

2. The strength of the solution had to be increased from 20% to 24%. How many milliliters of a 30% solution should be added to 300 ml of the 20% solution to get the desired result?

3. The volume of a quantity of an ideal gas was held constant at 7.4 liters. The original temperature was 500 K and the original pressure was 10 centimeters of mercury. What was the final temperature in kelvins if the pressure was increased to 30 centimeters of mercury?

4. Blues varied directly as greens and inversely as whites squared, and 4 blues and 2 whites went with 3 greens. How many greens were required for 2 blues and 4 whites? Work the problem using variation and again using ratio.

5. Cheers varied jointly as the number of fans and the square of the jubilation factor. When there were 100 fans and the jubilation factor was 4, there were 1000 cheers. How many cheers were there when there were only 10 fans whose jubilation factor was 20? Work the problem two ways.

Use substitution to solve:

*6. $\begin{cases} 3x + 2y = -3 \\ 4x - 3y = 13 \end{cases}$
    *7. $\begin{cases} 5x - 3y = 9 \\ 2x - 4y = -2 \end{cases}$

Solve:

8. $\begin{cases} y - 2x = 3 \\ xy = 4 \end{cases}$
   9. $\begin{cases} x^2 + y^2 = 11 \\ 2x^2 - y^2 = -2 \end{cases}$
   10. $\begin{cases} 3x + y + z = 2 \\ 2x - y - z = 3 \\ x + 2y - z = 8 \end{cases}$

11. If $g(x) = x^2 - 2x + 2$, $D = \{\text{Reals}\}$, find $g(5)$.

12. Graph: $\begin{cases} x - 2y < 2 \\ y \geq 0 \end{cases}$

13. Graph on a number line: $4 \not> x + 3 < 7$, $D = \{\text{Integers}\}$

Simplify:

14. $\dfrac{(x^{a+2})^{1/2} x^{3a/2} y^b}{y^{-b/2}}$

15. $\dfrac{p}{mx - \dfrac{m}{x + \dfrac{1}{mx}}}$

16. $\dfrac{5i - i^2}{2i^2 + i^3}$

17. $\sqrt[6]{8\sqrt{2}}$

18. $\dfrac{5i - 2i^2}{-i}$

19. Solve: $\sqrt{z - 33} + \sqrt{z} = 11$

20. Find the resultant of the two forces shown.

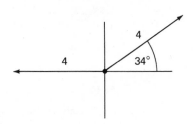

Simplify:

21. $\dfrac{3 - 2\sqrt{2}}{4 + 2\sqrt{2}}$

22. $\dfrac{3 - \sqrt{7}}{-\sqrt{7}}$

23. $-4i^2 - \sqrt{-9} - \sqrt{2}\sqrt{-2} - 5i^5$

24. $\sqrt{\dfrac{9}{3}} + 2\sqrt{\dfrac{3}{9}} - 5\sqrt{27}$

25. $\dfrac{a}{x + y} - c = \dfrac{1}{r^2}$; find y

26. $\dfrac{p}{x} = my\left(\dfrac{1}{a} + \dfrac{1}{c}\right)$; find a

27. Find B.

28. Begin with $ax^2 + bx + c = 0$ and derive the quadratic equation by completing the square.

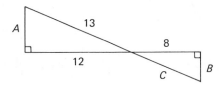

29. Convert 400 milliliters per second to cubic inches per hour.

30. Solve: $\dfrac{x - 4}{2} - \dfrac{x - 6}{3} = 7$

ER 96-88, 96-89

LESSON 97 *Relationships of numbers*

97.A
relationships
of numbers

We remember that we say that the real numbers constitute an ordered set because the real numbers are arranged in a definite order. Every real number has a definite relationship to every other real number. We use the number line to give us a visual representation of how numbers are ordered.

On this number line, we can see that 1 is greater than -3 because the graph of 1 lies to the right of the graph of -3. Also, we can see that the graph of 1 is four-sevenths of the distance from -3 to $+4$ on the number line. Furthermore, we can see that the name of a number designates its distance and direction from the origin on the number line. The graph of -3 is 3 units to the left of the origin, while the graph of $+4$ is 4 units to the right of the origin. We will use these facts to solve problems about the order relationships of numbers.

example 97.A.1   Find the number that is $\frac{7}{10}$ of the way from 30 to 40.

solution   First we locate the numbers on a number line.

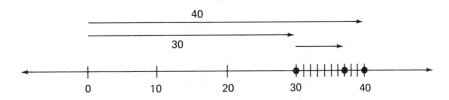

From the diagram we see that 40 is 40 units to the right of the origin and that 30 is 30 units to the right of the origin, so the distance between the numbers is $40 - 30 = 10$. Seven-tenths of 10 is 7, so the number in question lies 7 units to the right of 30 and its distance from the origin is $30 + 7 = \mathbf{37}$.

We can write the solution to this problem in a more compact form by writing

$$N = 30 + \frac{7}{10}(40 - 30) \quad \longrightarrow \quad N = \mathbf{37}$$

example 97.A.2 Find the number that is $\frac{1}{5}$ of the way from $\frac{1}{8}$ to $\frac{9}{11}$.

solution We will refer to the solution to the last problem and see that the name of a number is its distance from the origin. We will use S to represent the smaller number and L to represent the larger number.

We see that the distance between two positive numbers is the larger number minus the smaller number. A fraction F of this distance is

$$F(L - S)$$

and in this problem, the fraction is $\frac{1}{5}$ and the numbers are $\frac{1}{8}$ and $\frac{9}{11}$, so we have

$$\frac{1}{5}\left(\frac{9}{11} - \frac{1}{8}\right)$$

This is the length of the arrow shown. To this we must add the distance from the origin to the smaller number, so the number we are looking for is

$$\frac{1}{8} + \frac{1}{5}\left(\frac{9}{11} - \frac{1}{8}\right)$$

We finish by simplifying this expression.

$$\frac{1}{8} + \frac{1}{5}\left(\frac{72}{88} - \frac{11}{88}\right) \qquad \text{common denominator}$$

$$\frac{1}{8} + \frac{1}{5}\left(\frac{61}{88}\right) \qquad \text{added}$$

$$\frac{1}{8} + \frac{61}{440} \qquad \text{multiplied}$$

$$\frac{55}{440} + \frac{61}{440} \qquad \text{common denominator}$$

$$\mathbf{\frac{29}{110}} \qquad \text{added and simplified}$$

example 97.A.3 Find the number that is $\frac{2}{3}$ of the way from $2\frac{1}{4}$ to $3\frac{5}{6}$.

solution   The distance between the two numbers is

$$3\frac{5}{6} - 2\frac{1}{4}$$

and $\frac{2}{3}$ of this distance is

$$\frac{2}{3}\left(3\frac{5}{6} - 2\frac{1}{4}\right)$$

But this is not the distance from the origin, so to this we add $2\frac{1}{4}$. Thus, the number we want is

$$
\begin{aligned}
N &= 2\frac{1}{4} + \frac{2}{3}\left(3\frac{5}{6} - 2\frac{1}{4}\right) & \text{added } 2\frac{1}{4} \\[2mm]
&= \frac{9}{4} + \frac{2}{3}\left(\frac{23}{6} - \frac{9}{4}\right) & \text{improper fractions} \\[2mm]
&= \frac{9}{4} + \frac{2}{3}\left(\frac{46}{12} - \frac{27}{12}\right) & \text{common denominators} \\[2mm]
&= \frac{9}{4} + \frac{2}{3}\left(\frac{19}{12}\right) & \text{added} \\[2mm]
&= \frac{9}{4} + \frac{19}{18} & \text{multiplied} \\[2mm]
&= \frac{81}{36} + \frac{38}{36} = \mathbf{\frac{119}{36}} & \text{added}
\end{aligned}
$$

problem set 97

1. Johnny ran for a while at 12 miles per hour and then walked the rest of the way at 6 miles per hour. If he covered the 96 miles in 12 hours, how far did he run and how far did he walk?

2. The 15% alcohol solution had to be mixed by using a 10% alcohol solution and a 60% alcohol solution. How much of each should be used to get 600 ml of the 15% solution?

3. Charles and Mathew found that the 1200-mile drive to the city took 4 times as long as the 360-mile drive to the mountains. If the speed driving to the mountains was 10 mph greater than the speed driving to the city, find both times and both rates.

4. Mako could row 28 miles downstream in 4 hours but required 8 hours to go 40 miles upstream. What was the speed of the current and how fast could he row in still water?

5. The number of rabbits varied directly as the number of squirrels and inversely as the number of raccoons. When there were 10 rabbits and 40 squirrels, there were only 2 raccoons. How many raccoons went with 5 rabbits and 20 squirrels? Work the problem two ways.

*6. Find the number that is $\frac{1}{5}$ of the way from $\frac{1}{8}$ to $\frac{9}{11}$.

*7. Find the number that is $\frac{2}{3}$ of the way from $2\frac{1}{4}$ to $3\frac{5}{6}$.

8. Use substitution:
$$\begin{cases} 3x + 2y = 5 \\ 5x + 6y = 7 \end{cases}$$

Solve:

9. $$\begin{cases} y - 3x = 5 \\ xy = 6 \end{cases}$$

10. $$\begin{cases} x^2 + y^2 = 16 \\ 2x^2 - y^2 = -4 \end{cases}$$

11. Which of the following sets of ordered pairs are functions?
    (a) $(5, 7), (7, 5), (-3, -2)$
    (b) $(5, 7), (4, 7), (-3, 7)$
    (c) $(-2, 5), (4, -2), (4, -2)$

12. Graph:
    $$\begin{cases} x + 2y > 4 \\ y \geq 1 \end{cases}$$

13. Graph on a number line: $4 \not< -x + 2$ or $x + 3 < -1$, $D = \{\text{Integers}\}$

Simplify:

14. $\dfrac{(x^{2a})^{1/3} x^{2a}}{x^{a/2}}$

15. $\dfrac{xy}{x + \dfrac{xy}{x + \dfrac{1}{y}}}$

16. $\sqrt[7]{4 \sqrt[3]{2}}$

17. $\dfrac{2i - 3i^2}{-i}$

18. $\dfrac{2i^4 - i^3}{-2i}$

19. Draw the line suggested by the data points and write the equation that expresses salt as a function of pepper.

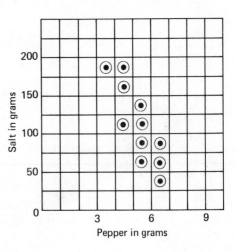

20. Find the resultant of the forces shown.

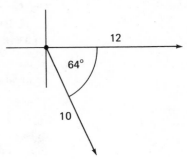

Simplify:

21. $\dfrac{5 - \sqrt{2}}{2 - 5\sqrt{2}}$

22. $\dfrac{3 - \sqrt{3}}{3 - 2\sqrt{3}}$

23. $\dfrac{a}{x + y} = z\left(\dfrac{1}{m} + \dfrac{1}{n}\right)$; find m

24. $\dfrac{a}{x + y} = z\left(\dfrac{1}{m} + \dfrac{1}{n}\right)$; find y

Simplify:

25. $-\sqrt{-9} - \sqrt{-2}\sqrt{-2} + 3i^2 - 2i^3 + 4$

26. $\sqrt{\dfrac{7}{3}} - 2\sqrt{\dfrac{3}{7}} + 5\sqrt{84}$

27. Solve $-5x^2 - x - 5 = 0$ by completing the square.

28. Convert 60 inches per second to centimeters per minute.

29. Simplify: $\dfrac{\dfrac{a}{xy} - \dfrac{x}{y^2}}{\dfrac{4}{x} - \dfrac{3}{xy^2}}$

30. Estimate: $\dfrac{(47,162)(50,312)(7 \times 10^{15})}{(.006 \times 10^{-14})(.198721)}$

ER 97-90, 97-91

LESSON 98 *Absolute value inequalities • Negative numbers and absolute value*

98.A

absolute value inequalities

Absolute value inequalities are either conjunctions or disjunctions. The reason for this is evident if we draw a number line and indicate the absolute value of each number.

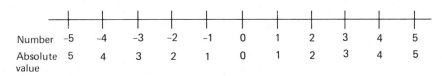

We note that except for zero, every absolute value is positive and that the numbers closest to zero have the smallest absolute values. For example, we see that every number between 3 and -3 has an absolute value that is less than 3. Further, we see that every number to the right of 3 and to the left of -3 has an absolute value that is greater than 3.

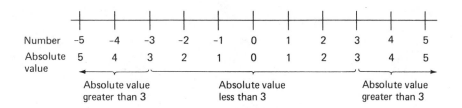

98.B

absolute value conjunctions

The graph above shows that all numbers between -3 and 3 have an absolute value that is less than 3. We can indicate these numbers in two ways. The first way is with a conjunction that does not contain absolute value

$$x > -3 \qquad \text{and} \qquad x < 3$$

and the second way is to write the absolute value inequality

$$|x| < 3$$

Thus, we see that a conjunction can be written that places the same restrictions on a variable as does an absolute value statement of less than.

example 98.B.1   Graph $|x| \leq 4$, $D = \{$Integers$\}$.

solution   This less-than absolute value inequality designates the same numbers as does the conjunction

$$x \geq -4 \quad \text{and} \quad x \leq 4$$

The domain restricts the values of x to the integers, so our graph shows the integers that are greater than or equal to -4 **and** are also less than or equal to $+4$.

example 98.B.2   Graph $-|x| + 3 > 0$, $D = \{$Integers$\}$.

solution   First we add -3 to both sides. We **do not** reverse the inequality symbol when we add negative quantities to both sides.

$$\begin{array}{rr} -|x| + 3 > & 0 \\ -3 & -3 \\ \hline -|x| > & -3 \end{array}$$

Now we mentally multiply both sides by -1 and reverse the inequality symbol

$$|x| < 3$$

Now we can replace this absolute value inequality with a conjunction that places the same restrictions on the variable.

$$x > -3 \quad \text{and} \quad x < 3$$

Thus, we are asked to indicate the integers that are less than $+3$ and that are greater than -3. The graph is

98.C
absolute value disjunctions

Since the numbers whose absolute values are greater than 3

$$|x| > 3$$

graph to the right of 3 or to the left of -3, we can make the same statement by writing the disjunction

$$x < -3 \quad \text{or} \quad x > 3$$

Thus, we see that a disjunction can be written that places the same restriction on the variable as does an absolute value statement of *greater than*.

example 98.C.1   Graph $-|x| + 2 < -2$, $D = \{$Reals$\}$.

solution   We begin by adding -2 to both sides of the inequality. We **do not** reverse the inequality symbol when we add a negative quantity.

$$-|x| + 2 < -2$$
$$\underline{\quad\;\; -\,2 \qquad -2\quad}$$
$$-|x| \qquad\; < -4$$

Now we mentally multiply both sides by -1 and reverse the inequality symbol.

$$|x| > 4$$

This absolute value statement of greater than can be replaced with the disjunction

$$x > 4 \qquad \text{or} \qquad x < -4$$

The graph of this disjunction, using the real numbers as the domain, is

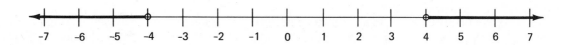

98.D
negative numbers and absolute value

The preceding absolute value inequalities stated that the absolute value of x was greater than or less than a given positive number. If we look at a number line on which the absolute values have been indicated

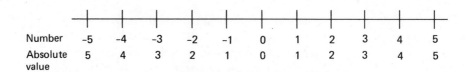

Number	-5	-4	-3	-2	-1	0	1	2	3	4	5
Absolute value	5	4	3	2	1	0	1	2	3	4	5

we note that except for zero, all the absolute values are positive numbers. **There are no absolute values that are less than zero.** Thus, a notation such as

$$|x| < -3$$

has no solution. **If the smallest absolute value is zero, then there is no absolute value less than zero and certainly no absolute value that is less than -3.** In the same way, the solution to the inequality

$$|x| > -3$$

is any number because any number (even zero) has an absolute value greater than -3.

example 98.D.1   Graph: (a) $|x| < -2$, $D = \{$Reals$\}$; (b) $|x| > -2$, $D = \{$Reals$\}$.

solution   These can be thought of as trick questions because the answers are either all the numbers or none of the numbers. The answer to (a) is none of the numbers. We write this answer by using the symbol for the null set, or the empty set.

(a) ϕ or $\{\;\}$

(b)   Every real number has an absolute value greater than -2, so the graph shows every real number.

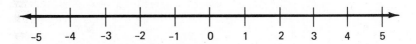

example 98.D.2   Graph $-|x| - 5 > -3, D = \{\text{Integers}\}$.

solution   We begin by adding $+5$ to both sides and we get

$$-|x| > 2$$

Now we mentally multiply both sides by -1 and reverse the inequality symbol.

$$|x| < -2$$

There are no real numbers that satisfy this inequality, and thus the original inequality has no solution.

problem set 98

1. If the silver could be separated from the sulfur, the reaction would be a success. What is the percentage by weight of the silver (Ag) in the compound Ag_2S? (Ag, 108; S, 32)

2. Some walked purposefully and some merely maundered. Twice the number of the purposeful walkers exceeded 10 times the number of maunderers by 16. Also, 13 times the number of maunderers exceeded the number of purposeful walkers by only 8. How many walked purposefully and how many just maundered?

3. The current in the river was 8 miles per hour. The boat could go 60 miles upstream in one-half the time it took to go 280 miles downstream. How fast could the boat go in still water?

4. The work accomplished varied jointly as the number of people and their average productivity factor. If 100 people with an average productivity factor of 20 could produce 8000 units on one shift, how many people whose factor was only 2 would be required to produce 16,000 units? Work the problem two ways.

5. The first container held a 5% iodine solution and the second container held a 10% iodine solution. How much of each should Nadine and Bob use to get 1200 ml of an 8% iodine solution?

Graph on a number line:

*6. $-|x| + 3 > 0, D = \{\text{Integers}\}$ *7. $-|x| + 2 < -2, D = \{\text{Reals}\}$

*8. $-|x| - 5 > -3, D = \{\text{Integers}\}$

9. Find the number that is $\frac{2}{7}$ of the way from $\frac{1}{2}$ to $2\frac{1}{3}$.

10. Use substitution: $\begin{cases} 3x + 3y = 9 \\ 4x - 6y = -8 \end{cases}$

Solve:

11. $\begin{cases} x - 3y = 2 \\ xy = 8 \end{cases}$ 12. $\begin{cases} x^2 + y^2 = 8 \\ 2x^2 - y^2 = 7 \end{cases}$ 13. $\begin{cases} 3x + 2y - z = 1 \\ x + y - z = -1 \\ 5x + 2y + 2z = 8 \end{cases}$

14. If $h(x) = x^2 - 4, D = \{\text{Negative integers}\}$, find $h(4)$

Graph:

15. $\begin{cases} x - y < -2 \\ 3x + 5y \le -5 \end{cases}$ 16. $x + 4 \not> 3$ or $x + 1 \not\le 2, \quad D = \{\text{Reals}\}$

Simplify:

17. $(x^{2-a})^2 x^{a/4}$

18. $\dfrac{ab}{a + \dfrac{b}{1 + \dfrac{a}{b^2}}}$

19. $2\sqrt[5]{16\sqrt[3]{2}}$

20. $\dfrac{2 - 3i}{2i^3}$

21. $\dfrac{2i - i^3}{-i}$

22. Draw the line suggested by the data points and write the equation that expresses output as a function of input.

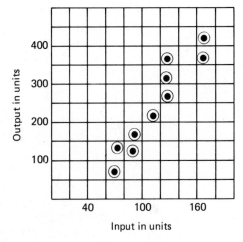

23. Find the resultant of the forces shown.

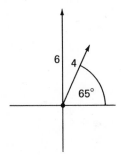

Simplify:

24. $\dfrac{3 - \sqrt{2}}{\sqrt{2} + 3}$

25. $\dfrac{4 - 2\sqrt{3}}{-\sqrt{3}}$

26. $\dfrac{m}{c} = p\left(\dfrac{1}{a} + \dfrac{1}{b}\right)$; find c

Simplify:

27. $-\sqrt{-9} - 3i^3 + 2i - \sqrt{-4}\sqrt{4} + 3\sqrt{-4}$

28. $-\sqrt{\dfrac{3}{10}} + 4\sqrt{\dfrac{10}{3}} - \sqrt{120}$

29. Use the formula $PV = nRT$ to find the volume of .0163 mole of gas at 10 atmospheres of pressure and a temperature of 870 K ($R = .0821$).

30. Convert 40 centimeters per second to feet per minute.

ER 98-92, 98-93

LESSON 99 *Graphs of parabolas*

99.A
graphs of parabolas

It can be shown that the graph of a quadratic equation in x and y, such as

$$y = x^2 + 4x + 4 \qquad \text{or} \qquad y = -x^2 + 4x - 3$$

is always symmetric about a vertical line called the **axis of symmetry** and either **opens upward** or **opens downward.** If the curve opens up, the vertex is on the axis of symmetry and the y coordinate of the vertex has a minimum value. If the curve opens down, the vertex is on the axis of symmetry and its y coordinate has a maximum value.

OPENS UPWARD

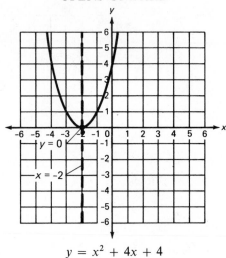

$$y = x^2 + 4x + 4$$

OPENS DOWNWARD

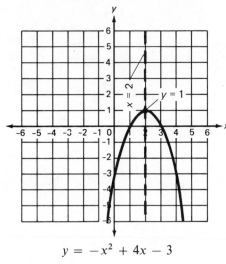

$$y = -x^2 + 4x - 3$$

To investigate the graphing of parabolas, we will look at the two quadratic equations

$$y = -x^2 - 8x - 13 \qquad \text{and} \qquad y = x^2 - 8x + 13$$

We could begin by making two tables and assigning values of x,

x	0	2	3	-2	-3
y					

x	0	2	3	-2	-3
y					

and using the equations to find the matching values of y. This method is not recommended since it is time-consuming and laborious. By completing the square, we can write the equations in a form that will allow us to determine three things.

1. The axis of symmetry
2. The y coordinate of the vertex
3. Whether the curve opens up or opens down

If we complete the square on the given equations we can change the forms of the equations to permit the following analyses.

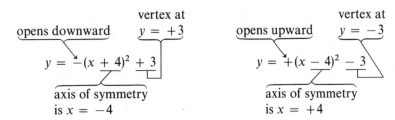

Note that when we have $(x + 4)^2$, the x value of the axis of symmetry is -4, not $+4$! Also, when we have $(x - 4)^2$, the x value of the axis of symmetry is $+4$, not -4!

The squared term is always positive, and for large values of x this term approximates the value of y. When it is preceded by a $+$ sign, y is positive for both large positive and negative values of x; and thus the curve opens upward. When the squared term is preceded by a minus sign, y is negative for both large positive and negative values of x; and the curve opens downward.

When the value of x is such that the squared term equals zero, the curve is at a maximum or minimum point; and the y coordinate of the vertex is the constant at the end of the expression.

We will show how to use completing the square as an aid to graphing in the next three examples.

example 99.A.1   Complete the square to graph $y = x^2 - 4x + 2$.

solution   We want to rearrange the right side into the form

$$y = (x + a)^2 + k$$

so we place a parentheses around the x^2 term and the x term.

$$y = (x^2 - 4x \quad\quad) + 2$$

Now to make the expression inside the parentheses a perfect square, it is necessary to add the square of one-half the coefficient of x

$$(-4 \cdot \tfrac{1}{2})^2 = 4$$

which is 4. **Thus, we add $+4$ inside the parentheses and -4 outside the parentheses. This addition of $+4$ and -4 to the same side of the equation is a net addition of zero.**

$$y = (x^2 - 4x + 4) + 2 - 4$$

Now the term in the parentheses is a perfect square and we write it as such.

$$y = (x - 2)^2 - 2$$

From this form we can determine the three things necessary to sketch the curve.

(a)   Opens upward

(b)   Axis of symmetry is $x = +2$ $y = +(x - 2)^2 - 2$

(c) y coordinate of vertex is -2

We use this information to draw the axis of symmetry and the vertex of the curve as we show on the left at the top of the next page. Now, if we get one more point on the curve, we can make a sketch. Let's let $x = 4$ in the original equation and solve for y.

$$y = (4)^2 - 4(4) + 2 \quad\quad \text{substituted}$$

$$y = 16 - 16 + 2 \quad\quad \text{multiplied}$$

$$y = 2 \qu\quad\quad \text{simplified}$$

Thus, the point (4, 2) lies on the curve. We remember that the curve is symmetric about the line $x = 2$ and complete the sketch.

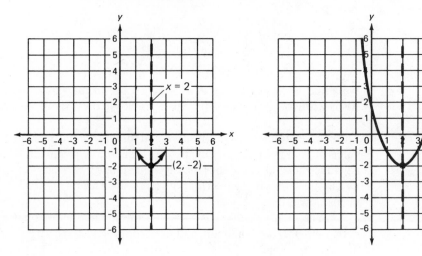

example 99.A.2 Complete the square to graph $y = -x^2 + 2x + 2$.

solution The procedure will be the same, except that as the first step, we will multiply both sides by -1 so that the coefficient of x^2 will be $+1$. As the last step, we will again multiply both sides by -1.

$$-y = x^2 - 2x - 2 \qquad \text{multiplied by } -1$$

Now to change the right side into the form

$$(x + a)^2 + k$$

we place parentheses around the $x^2 - 2x$

$$-y = (x^2 - 2x \qquad) - 2$$

and add $+1$ inside the parentheses and -1 outside the parentheses.

$$-y = (x^2 - 2x + 1) - 2 - 1$$

Now the expression in the parentheses is a perfect square.

$$-y = (x - 1)^2 - 3$$

As the last step, we multiply both sides by -1 so that y will be positive. **Note that we do not change the sign inside the parentheses!**

$$y = -(x - 1)^2 + 3$$

Now we can read the salient features of the graph.

   (a) Opens downward

   (b) Axis of symmetry is $x = +1$ $y = -(x - 1)^2 + 3$

   (c) y coordinate of vertex is $+3$

To find another point on the curve we let $x = -2$ in the original equation and solve for y.

$$y = -(-2)^2 + 2(-2) + 2 \qquad \text{substituted}$$
$$y = -4 - 4 + 2 \qquad \text{multiplied}$$
$$y = -6 \qquad \text{simplified}$$

We complete the graph by using the point $(-2, -6)$, remembering that the curve is symmetric about $x = 1$.

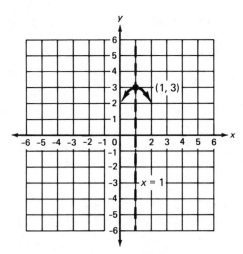

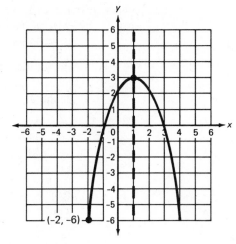

example 99.A.3   Complete the square to graph $y = -x^2 - 6x - 8$.

solution   Since the x^2 term is negative, we will begin and end by multiplying both sides of the equation by -1.

$$-y = (x^2 + 6x \quad\quad) + 8 \qquad\qquad \text{multiplied by } -1$$

$$-y = (x^2 + 6x + 9) + 8 - 9 \qquad \text{added } +9 \text{ and } -9$$

$$-y = (x + 3)^2 - 1 \qquad\qquad\quad \text{simplified}$$

$$y = -(x + 3)^2 + 1 \qquad\qquad\quad \text{multiplied by } -1$$

Now we can diagnose the salient features of the graph.

(a)   Opens downward $\rangle$

(b)   Axis of symmetry is $\underline{x = -3}$ $\qquad y = -(x + 3)^2 + 1$

(c)   y coordinate of vertex is $\underline{+1}$

On the left we use these facts to begin the curve. To find another point on the curve we replace x with -1 and find that y equals -3. Then we use the point $(-1, -3)$ and symmetry to complete the graph.

$$y = -(-1)^2 - 6(-1) - 8 \qquad \text{substituted}$$

$$y = -1 + 6 - 8 \qquad\qquad\quad \text{multiplied}$$

$$y = -3 \qquad\qquad\qquad\qquad\quad \text{simplified}$$

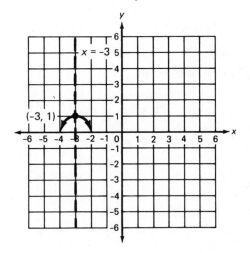

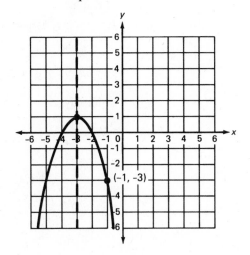

problem set 99

1. Bickford traveled twice as fast as Shawn traveled. Thus, Bickford could travel the 320 miles to the reef in only 2 hours less than it took Shawn to travel the 240 miles to Jane's house. Find the rates and times of both boys.

2. The solution was 68% alcohol and it came up to the 600-ml mark on the beaker. How much alcohol should be evaporated so that the remainder would be only 20% alcohol?

3. Some sparkled and the rest coruscated. Ten times the number of sparklers exceeded 6 times the number that coruscated by 40. But 4 times the number that coruscated exceeded twice the number that sparkled by only 160. How many were in each category?

4. Reds varied directly as yellows and inversely as greens squared; 100 reds and 40 yellows went with 10 greens. How many reds went with 20 yellows and only 5 greens?

5. Find three consecutive integers such that the product of the first and the third exceeds the product of 8 and the second by 32.

Complete the square as an aid in graphing:

*6. $y = x^2 - 4x + 2$ *7. $y = -x^2 + 2x + 2$

*8. $y = -x^2 - 6x - 8$

9. Graph on a number line: $-|x| + 5 \not\leq 3$, $D = \{\text{Reals}\}$

10. Find the number that is $\frac{5}{11}$ of the way from $2\frac{1}{3}$ to $3\frac{1}{6}$.

11. Use substitution: $\begin{cases} 2x - 2y = -1 \\ 4x + 3y = 5 \end{cases}$

Solve:

12. $\begin{cases} x - y = 5 \\ xy = 2 \end{cases}$ 13. $\begin{cases} x^2 + y^2 = 5 \\ 2x^2 - y^2 = 4 \end{cases}$

14. $\begin{cases} 2x + 2y - z = 0 \\ x + y - 2z = -12 \\ 2x - y + z = 10 \end{cases}$

15. Which of the following sets of ordered pairs are functions?
    (a) (4, 2), (5, 2), (2, 5), (2, 4)
    (b) (−7, 2), (4, 2), (2, 4), (2, −7)
    (c) (−3, 2), (3, −2), (−3, 4), (−2, 3)

Graph:

16. $\begin{cases} y \leq -3 \\ 4x + y < -2 \end{cases}$

17. $x + 2 \not\leq 5$ or $x + 3 < 3$, $D = \{\text{Integers}\}$

Simplify:

18. $\dfrac{(a^{x+2})^2 a^{b/2}}{(a^{2-b})^{1/2}}$ 19. $\dfrac{x}{x^2 y - \dfrac{1}{1 + \dfrac{1}{xy}}}$ 20. $3\sqrt{9\sqrt[2]{3}}$

21. $\dfrac{2 - i^3}{-i}$ 22. $\dfrac{3 - i}{i^2 - 3i}$

23. Determine the resultant of the force vectors shown.

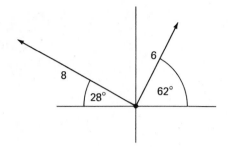

Simplify:

24. $\dfrac{5 - 2\sqrt{2}}{\sqrt{2}}$

25. $\dfrac{3 - 4\sqrt{5}}{\sqrt{5} - 1}$

26. $\sqrt{\dfrac{2}{13}} + 3\sqrt{\dfrac{13}{2}} - 2\sqrt{104}$

27. $-\sqrt{-7}\sqrt{-7} + 2\sqrt{-16} - 3i^5 + 2i^2$

28. $\dfrac{a}{k + c} = m\left(\dfrac{x}{y} + d\right)$; find y

29. Convert 400 cubic feet per minute to cubic inches per second.

ER 99-94, 99-95

LESSON 100 *Percent markups*

100.A
percent markups

The difference between the selling price of a piece of merchandise and the price paid for it is called markup. For example, if a dealer purchased an item from the factory for $40 and sold it for $50, we say that the markup was $10. Ten dollars is 25 percent of $40 and is only 20 percent of $50. Thus, the markup was 20 percent of the selling price and 25 percent of the purchase price. The store owner would say that the markup was only 20 percent, but the customer would say that the markup was 25 percent.

The selling price is the purchase price plus the markup.

Selling price = purchase price + markup

Any one of these components can be expressed as a percentage of any other component.

When we work these problems, we will use rate instead of percent. Rate is percent divided by 100 and is the decimal form of the relationship. Thus,

20 percent	equals a rate of	.2
415 percent	equals a rate of	4.15
2 percent	equals a rate of	.02

example 100.A.1 The selling price was $48. If the markup was 20 percent of the purchase price, what was the purchase price? What was the markup?

solution   First we write

$$\text{Selling price} = \text{purchase price} + \text{markup}$$

The selling price was 48 and the markup was 20 percent of the purchase price or $.2PP$. We substitute and find

$$48 = PP + .2PP \qquad \text{substituted}$$
$$48 = 1.2PP \qquad \text{simplified}$$
$$\mathbf{40 = PP} \qquad \text{divided}$$

Thus the purchase price was $40, and the markup was **$8.**

example 100.A.2   The purchase price of the item was $1800. If the markup was 40 percent of the selling price, what was the selling price?

solution   First we write

$$\text{Selling price} = \text{purchase price} + \text{markup}$$

The markup was 40 percent of the selling price or $.4SP$, and the purchase price was $1800. We make these substitutions and solve.

$$SP = 1800 + .4SP \qquad \text{substituted}$$
$$.6SP = 1800 \qquad \text{added} -.4SP$$
$$\mathbf{SP = 3000} \qquad \text{divided}$$

Thus, the markup was $1200, which is 40 percent of $3000.

example 100.A.3   The sports car retailed for $10,368. What was the purchase price if the car had been marked up 8 percent of the purchase price?

solution   First we write

$$\text{Selling price} = \text{purchase price} + \text{markup}$$

The markup was 8 percent of the purchase price, so we get

$$10{,}368 = PP + .08PP \qquad \text{substituted}$$
$$10{,}368 = 1.08PP \qquad \text{simplified}$$
$$\mathbf{9600 = PP} \qquad \text{divided}$$

Thus, the car had been purchased for $9600, and the markup was $768.

problem set 100

*1.   The selling price was $48. If the markup was 20 percent of the purchase price, what was the purchase price? What was the markup?

*2.   The purchase price of the item was $1800. If the markup was 40 percent of the selling price, what was the selling price?

*3.   The sports car retailed for $10,368. What was the purchase price if the car had been marked up 8 percent of the purchase price?

4.   Schneider had 400 liters of a solution that was 60% antifreeze. How many liters of an 80% solution should he add to make a solution that is 72% antifreeze?

5. The boat could go 104 miles downstream in the same time it took to go 56 miles upstream. If the speed of the boat was 20 miles per hour in still water, what was the speed of the current?

Complete the square as an aid in graphing:

6. $y = -x^2 - 4x - 5$

7. $y = x^2 + 2x + 2$

8. $y = -x^2 - 2x - 3$

9. Graph on a number line: $-|x| + 3 \not< 2$, $D = \{\text{Integers}\}$

10. Find the number that is $\frac{2}{5}$ of the way from $3\frac{1}{3}$ to $4\frac{5}{6}$.

11. Use substitution: $\begin{cases} 6x + 5y = 8 \\ 4x + 2y = 3 \end{cases}$

Solve:

12. $\begin{cases} 3x - y = 4 \\ xy = 5 \end{cases}$

13. $\begin{cases} x^2 + y^2 = 16 \\ 2x^2 - y^2 = 2 \end{cases}$

14. $\begin{cases} 3x + y + z = 7 \\ x - 2y - z = 2 \\ -x + y - z = -5 \end{cases}$

15. If $p(x) = x^2 - 4$, $D = \{\text{Reals}\}$, find $p(\frac{1}{2})$.

Graph:

16. $\begin{cases} 3y - 2x > -3 \\ x \geq -2 \end{cases}$

17. $-10 \not\geq x + 2 < -4$, $D = \{\text{Reals}\}$

Simplify:

18. $\dfrac{(a^{x+2})^{1/2} y}{(y^2)^a}$

19. $\dfrac{m}{my - \dfrac{1}{1 - \dfrac{1}{my}}}$

20. $7\sqrt{49\sqrt[3]{7}}$

21. $\dfrac{-2 - 3i}{3 + 2i}$

22. $\dfrac{i}{2i - 3}$

23. Determine the resultant of the two vectors.

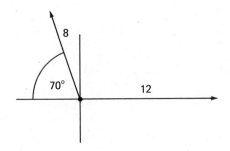

Simplify:

24. $\dfrac{2\sqrt{2} - \sqrt{3}}{\sqrt{6}}$

25. $\dfrac{3 - \sqrt{2}}{\sqrt{2} + 2}$

26. $2\sqrt{\dfrac{2}{5}} - 4\sqrt{\dfrac{5}{2}} + 2\sqrt{40}$

27. $-3i^2 - 2i^3 + 4 - i^5 + \sqrt{-9}$

28. $\dfrac{m}{x} - c = \dfrac{1}{R_1} + \dfrac{1}{R_2}$; find R_1

29. Solve $-2x^2 = x + 4$ by using the quadratic formula.

30. Multiply: $\dfrac{2x^0 y^{-2} p^{-4}}{x^2 p^{-4}} \left(\dfrac{yp^4 m}{x^{-2}} - 3xy^0 \right)$

ER 100-96, 100-97

LESSON *101* *Sums of functions • Products of functions*

101.A

sums
of functions

Recall that functional notation has several advantages. The first is that it allows us to identify the equations being considered. For instance, if we have the three equations

$$y = x + 3 \qquad y = x^2 - 6 \qquad y = 2x^2 + 5$$

we can name them from left to right as, say, equation h, equation ϕ, and equation g. If we use functional notation, we would use $h(x)$, $\phi(x)$ and $g(x)$ instead of y when we write the equations.

$$h(x) = x + 3 \qquad \phi(x) = x^2 - 6 \qquad g(x) = 2x^2 + 5$$

These notations in themselves do not completely define the function, for the domain must be specified or implied for every function. Thus, we will arbitrarily assign a domain to each equation.

$$h(x) = x + 3 \qquad \phi(x) = x^2 - 6 \qquad g(x) = 2x^2 + 5$$
$$D = \{\text{Reals}\} \qquad D = \{\text{Integers}\} \qquad D = \{\text{Negative integers}\}$$

If we add two of these equations, we get an equation for the sum. The sum of equation h and equation g is

$$
\begin{aligned}
h(x) &= \phantom{2x^2 +{}} x + 3 \\
g(x) &= 2x^2 + 5 \\
\hline
h(x) + g(x) &= 2x^2 + x + 8
\end{aligned}
$$

We see that $h(x) + g(x)$ means that we have added the h equation to the g equation. **Often we use the notation**

$$(h + g)(x)$$

to mean the same thing. The notation $(h + g)$ means that we have added the h equation to the g equation. The domain for the new equation is all numbers that were common to both of the original domains.

example 101.A.1 Given $h(x) = x + 3$, $D = \{\text{Reals}\}$ and $\phi(x) = x^2 - 6$, $D = \{\text{Integers}\}$, find $(h + \phi)(+2)$.

solution We can find the answer two ways. First we will find $h(2)$ and $\phi(2)$ and add.

$$h(2) = 2 + 3 \qquad \phi(2) = (2)^2 - 6$$
$$h(2) = 5 \qquad \phi(2) = -2$$

Thus, since $(h + \phi)(2)$ means $h(2) + \phi(2)$, we have

$$(h + \phi)(2) = (5) + (-2) = 3$$

The second way is to add the equations to find the equation $(h + \phi)(x)$.

$$h(x) = x + 3$$
$$\underline{\phi(x) = x^2 - 6}$$
$$(h + \phi)(x) = x^2 + x - 3$$

Now we use 2 for x and find the value of the $h + \phi$ equation when x equals 2.

$$(h + \phi)(2) = (2)^2 + (2) - 3$$
$$= 4 + 2 - 3$$
$$\mathbf{(h + \phi)(2) = 3}$$

We get the same answer both ways.

example 101.A.2   Given $h(x) = x + 3$, $D = \{\text{Reals}\}$ and $g(x) = 2x^2 + 5$, $D = \{\text{Negative integers}\}$, find $(h + g)(5)$.

solution   We can find the equation $(h + g)(x)$ by adding the h equation and the g equation.

$$h(x) = x + 3$$
$$\underline{g(x) = 2x^2 + 5}$$
$$(h + g)(x) = 2x^2 + x + 8$$

But we cannot use this equation to find $(h + g)(5)$ because 5 was not a member of the domain of $g(x)$ so 5 is not a member of the domain of $(h + g)(x)$. Thus, we say that the problem has no answer, or we can say that the answer is the null set ϕ or the empty set $\{\ \ \}$. Null set and empty set mean the same thing. This problem does not have an acceptable answer.

101.B
products of functions

When we multiply two functions, the product is also a function. If we have the equations

$$h(x) = x + 3 \qquad \text{and} \qquad g(x) = x^2 - 6$$
$$D = \{\text{Reals}\} \qquad\qquad D = \{\text{Negative integers}\}$$

and we multiply the h equation by the g equation, we get the product $h(x)g(x)$.

$$h(x)g(x) = (x + 3)(x^2 - 6) \qquad \text{product of functions}$$
$$h(x)g(x) = x^3 - 6x + 3x^2 - 18 \qquad \text{multiplied}$$
$$h(x)g(x) = x^3 + 3x^2 - 6x - 18 \qquad \text{rearranged}$$

Instead of writing $h(x)g(x)$ to designate the product of the two functions, we find it convenient to write

$$hg(x)$$

instead. The notation hg means that the h equation has been multiplied by the g equation in the same way that

$$(h + g)(x)$$

means that the h equation and the g equation have been added.

example 101.B.1   Find $hg(-4)$ if $h(x) = x + 3$, $D = \{\text{Reals}\}$ and $g(x) = x^2 - 6$, $D = \{\text{Negative integers}\}$.

solution   As in Example 101.A.1, we can find the answer two ways. The first is to find $h(-4)$ and $g(-4)$ and then multiply these answers.

$$h(-4) = -4 + 3 \qquad g(-4) = (-4)^2 - 6$$
$$h(-4) = -1 \qquad\qquad g(-4) = 10$$

so

$$hg(-4) = (-1)(10)$$

$$\mathbf{hg(-4) = -10}$$

The second way is to find $hg(x)$ and then find $hg(-4)$.

$$hg(x) = (x + 3)(x^2 - 6)$$

$$hg(x) = x^3 + 3x^2 - 6x - 18$$

Now to find $hg(-4)$, we use -4 for x in the equation.

$$hg(-4) = (-4)^3 + 3(-4)^2 - 6(-4) - 18$$

$$hg(-4) = -64 + 48 + 24 - 18$$

$$\mathbf{hg(-4) = -10}$$

example 101.B.2   Find $fg(-4)$ if $f(x) = x + 3$, $D = \{$Reals$\}$ and $g(x) = x - 5$, $D = \{$Positive integers$\}$.

solution   We begin by multiplying the equations to find $fg(x)$.

$$f(x)g(x) = (x + 3)(x - 5)$$

$$fg(x) = x^2 - 2x - 15$$

But we cannot use this function to find $fg(-4)$ because -4 is not a member of the domain of $g(x)$ so it is not a member of the domain of $fg(x)$. Thus, we may say that this problem has no answer, or we may say that the answer is either

$$\varnothing \quad \text{or} \quad \{ \ \}$$

problem set 101

1. Sarah had a 40-mile head start and was driving north at 46 miles per hour when James and Renée began their pursuit at 50 miles per hour. How much farther did Sarah go before James and Renée caught up?

2. The speed of the boat in still water was 10 miles per hour. The boat could go 78 miles down the Lazy River in the same time it took to go 42 miles up the Lazy River. How fast did the current flow in the Lazy River?

3. Horses varied directly as goats and inversely as pigs squared. When the barnyard contained 5 horses, there were 4 pigs and only 2 goats. How many goats went with 6 pigs and 10 horses?

4. A 70 percent markup of the purchase price made the selling price of the item $1666. What did the store owner pay for the item?

5. The selling price was $1680. This low price was possible because Audry and Sam only marked the item up 40 percent of the purchase price. What did they pay for the item?

*6.   Find $(h + \phi)(+2)$ if $h(x) = x + 3$, $D = \{$Reals$\}$ and $\phi(x) = x^2 - 6$, $D = \{$Integers$\}$.

*7.   Find $hg(-4)$ if $h(x) = x + 3$, $D = \{$Reals$\}$ and $g(x) = x^2 - 6$, $D = \{$Negative integers$\}$.

*8.   Find $fg(-4)$ if $f(x) = x + 3$, $D = \{$Reals$\}$ and $g(x) = x - 5$, $D = \{$Positive integers$\}$.

Complete the square as an aid in graphing:

9. $y = x^2 + 4x + 2$ 10. $y = -x^2 - 4x - 2$

11. Graph on a number line: $-|x| - 2 \leq -5$, $D = \{$Integers$\}$

12. Use substitution: $\begin{cases} 2x + 3y = -3 \\ 4x - 2y = 18 \end{cases}$

Solve:

13. $\begin{cases} 2x - y = 6 \\ xy = 4 \end{cases}$

14. $\begin{cases} x^2 + y^2 = 12 \\ 3x^2 - y^2 = 4 \end{cases}$

15. $\begin{cases} x + y + z = 8 \\ x + y - z = 0 \\ 2x - y + z = 3 \end{cases}$

Graph:

16. $\begin{cases} 3x - 5y < +10 \\ y \geq -2 \end{cases}$

17. $x + 2 < 0$ or $x + 3 \nleq 3$, $D = \{$Integers$\}$

18. Find the number that is $\frac{1}{8}$ of the way from $\frac{1}{5}$ to $2\frac{1}{3}$.

Simplify:

19. $\dfrac{x^{a/2}y^{2a}}{x^{3a}y^{-2a/3}}$

20. $\dfrac{kx}{x - \dfrac{kx}{k - \dfrac{1}{x}}}$

21. $\sqrt[3]{x^5 y}\ \sqrt[4]{xy^2}$

22. $\dfrac{3i + 4}{-i^2 - i^5}$

23. $\dfrac{4i - 3}{i^3 - 2i^2}$

24. Write $4R + 10U$ in polar form.

25. $\dfrac{a}{x} + \dfrac{b}{m + c} = ax$; find c

Simplify:

26. $\dfrac{2 - 2\sqrt{2}}{3\sqrt{2} - 2}$

27. $3i^5 - \sqrt{-2}\sqrt{-2} + \sqrt{3}\sqrt{-3} - \sqrt{-9}$

28. $4\sqrt{\dfrac{5}{12}} + 3\sqrt{\dfrac{12}{5}} - 3\sqrt{60}$

29. Begin with $ax^2 + bx + c = 0$ and derive the quadratic formula by completing the square.

30. The figure shows the base of a container that is 10 ft high and the dimensions are in feet. Find the volume in cubic inches.

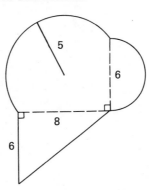

LESSON *102* *Advanced polynomial division*

102.A

advanced polynomial division

We can divide polynomials that have more than one variable by using the same method that we used when only one variable was present.

example 102.A.1   Divide $x^3 + y^3$ by $x + y$.

solution   We use the format for long division.

$$x + y \,\big|\, \overline{x^3 \qquad + y^3}$$

x^3 divided by x is x^2 so we record an x^2 above

$$\begin{array}{r} x^2 \\ x + y \,\big|\, \overline{x^3 \qquad + y^3} \end{array}$$

and multiply x^2 by $x + y$ and record.

$$\begin{array}{r} x^2 \\ x + y \,\big|\, \overline{x^3 \qquad\quad + y^3} \\ \underline{x^3 + x^2 y } \end{array}$$

Now we mentally change the signs and add.

$$\begin{array}{r} x^2 \\ x + y \,\big|\, \overline{x^3 \qquad\quad + y^3} \\ \underline{x^3 + x^2 y } \\ -x^2 y \end{array}$$

Now $-x^2 y$ divided by x equals $-xy$, so we record $-xy$ above and then multiply and add.

$$\begin{array}{r} x^2 - xy \\ x + y \,\big|\, \overline{x^3 \qquad\qquad + y^3} \\ \underline{x^3 + x^2 y } \\ -x^2 y \\ \underline{-x^2 y - xy^2 } \\ xy^2 \end{array}$$

Finally, xy^2 divided by x equals y^2. We record a y^2 above and multiply to finish.

$$\begin{array}{r} x^2 - xy + y^2 \\ x + y \,\big|\, \overline{x^3 \qquad\qquad + y^3} \\ \underline{x^3 + x^2 y } \\ -x^2 y \\ \underline{-x^2 y - xy^2 } \\ xy^2 + y^3 \\ \underline{xy^2 + y^3} \end{array}$$

example 102.A.2   Divide $x^3 - y^3$ by $x - y$.

solution   The procedure is the same and the answer is the same except that the middle sign is different.

$$x - y \overline{\smash{\big)}\ \begin{aligned} x^2 + xy + y^2 \\ x^3 \qquad\qquad - y^3 \end{aligned}}$$

```
                x² + xy + y²
          _____
   x - y ) x³              - y³
           x³ - x²y
           _____
                x²y
                x²y - xy²
                _____
                     xy² - y³
                     xy² - y³
```

problem set 102

1. A 60 percent markup of the purchase price was necessary to pay the rent, utilities, and the workers and still make a small profit. If an item sold for $1424, what did the storekeeper pay for it?

2. Sister Baby's boat could attain a speed of 18 miles per hour on a lake. If the boat took the same time to go 132 miles down the river as it took to go 84 miles up the river, how fast was the current in the river?

3. Donna took twice as long to drive 720 miles as Maple took to drive 200 miles. Find the rates and times of both if Donna's speed exceeded that of Maple by 40 miles per hour.

4. The initial pressure and temperature of a quantity of an ideal gas was 400 millimeters of mercury and 300 K. If the volume was held constant, what would the final temperature be in kelvins if the pressure was increased to 600 millimeters of mercury?

5. David and Le Van found three consecutive multiples of 11 such that 4 times the sum of the first and third was 66 less than 10 times the second. What were the numbers?

*6. Divide $x^3 + y^3$ by $x + y$. *7. Divide $x^3 - y^3$ by $x - y$.

8. Find $ab(2)$ if $a(x) = x - 5$, $D = \{\text{Reals}\}$ and $b(x) = x^2 + 4$, $D = \{\text{Negative integers}\}$.

Complete the square as an aid in graphing:

9. $y = x^2 + 4x + 6$ 10. $y = -x^2 + 4x - 6$

11. Graph on the number line: $x + 3 \geq 5$, $D = \{\text{Reals}\}$

12. Find the number that is $\frac{2}{3}$ of the way from $\frac{1}{4}$ to $2\frac{1}{2}$.

13. Use substitution: $\begin{cases} 4x + 3y = 17 \\ 2x - 3y = -5 \end{cases}$

Solve:

14. $\begin{cases} x^2 + y^2 = 6 \\ x - y = 2 \end{cases}$ 15. $\begin{cases} x^2 + y^2 = 10 \\ 2x^2 - 2y^2 = 5 \end{cases}$

16. $\begin{cases} x + 2y + z = -1 \\ 3x - y + z = 6 \\ 2x - 3y - z = 8 \end{cases}$

Graph:

17. $\begin{cases} x - 4y \leq -4 \\ x < 3 \end{cases}$ 18. $-3 \leq x - 3 \not> 4$, $D = \{\text{Integers}\}$

Simplify:

19. $\dfrac{(x^{2a-2})^b}{x^{b/2}}$ 20. $\dfrac{m}{m^2 + \dfrac{m}{m^2 + \dfrac{1}{m}}}$ 21. $\sqrt[5]{x^2y^3}\,\sqrt[4]{xy}$

22. $\dfrac{2i^2 + i^3}{i^3 + 2}$ 　　　　　 **23.** $\dfrac{2i - 5}{5i^2 - 2i}$ 　　　　　 **24.** $\dfrac{3 + 2\sqrt{5}}{5 - \sqrt{20}}$

25. The two vectors act on the point as shown. Find the resultant vector.

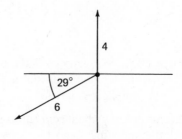

26. $a\left(\dfrac{b}{c} - \dfrac{1}{x}\right) = \dfrac{m}{p}$; find x 　　　　　 **27.** Solve: $\sqrt{z} + \sqrt{z + 33} = 11$

Simplify:

28. $\sqrt{-16} - \sqrt{-2}\sqrt{2}\sqrt{-3}\sqrt{-3} - i^5$ 　　　 **29.** $3\sqrt{\dfrac{4}{3}} - 2\sqrt{\dfrac{3}{4}} + 5\sqrt{48}$

ER 102-100, 102-101

LESSON 103 · *Decimal numerals and rational numbers*

103.A
complex numbers

We begin our review discussion of complex numbers by remembering that this number

$$\frac{4}{7} + \sqrt{2}\,i$$

is a complex number written in standard form. The real part is $\frac{4}{7}$ and is written first. The imaginary part is $\sqrt{2}\,i$ and is written after the real part. We often use the letters a and b to designate the standard form of a complex number by writing

$$a + bi$$

and say that a and b can be any real numbers. Since zero is a real number, either a or b can be zero. If b is 0, then only a is left and thus,

$$4 \qquad \frac{3}{4} \qquad -5\sqrt{2} \qquad -\frac{19}{3}$$

are all complex numbers whose imaginary parts are zero. If a is zero, then only the imaginary part b remains. Thus, the imaginary numbers

$$-\sqrt{2}i \qquad \frac{4\sqrt{2}}{3}i \qquad -3i \qquad i$$

are all complex numbers whose real parts equal zero.

103.B
subsets
of the set
of real numbers

The set of real numbers has an infinite number of members, and these can be used to form an infinite number of subsets. Normally, however, we restrict our attention to five major subsets of the set of real numbers. The first three are

The counting (natural) numbers	$\{1, 2, 3, \ldots\}$
The whole numbers	$\{0, 1, 2, 3, \ldots\}$
The integers	$\{\ldots, -3, -2, -1, 0, 1, 2, 3, \ldots\}$

These three sets account for the numbers that are designated when the number line is drawn

because we usually designate the location of the integers below a number line. All integers can be written as fractions of other integers. For example, $-4, 0,$ and 13 can be written as fractions, as shown here:

$$-4 = \frac{12}{-3} \qquad 0 = \frac{0}{2} \qquad 13 = \frac{-39}{-3}$$

We say that a number that can be written as a fraction of integers is a *rational number* because *ratio* is another name for **fraction.**

The rest of the set of real numbers is made up of all the positive numbers of arithmetic and their negative counterparts. Some of these numbers can be written as fractions of integers, and thus are rational numbers. The rest cannot be written as fractions of integers and are irrational numbers.[†] **Irrational numbers cannot be represented exactly with decimal numerals that contain a finite number of digits.** The square root of 2 is an irrational number and thus can only be approximated with a decimal numeral. A pocket calculator gives an approximation of the square root of 2 as

$$\sqrt{2} \approx 1.4142136$$

The complete representation of this number would require a numeral with an infinite number of digits, and the digits would occur in a nonrepeating pattern.

1. **If the digits in a decimal numeral terminate, the number is a rational number.**
2. **If the digits in a nonterminating decimal numeral repeat in a pattern, the number is a rational number.**

example 103.B.1 Show that .00314 is a rational number by writing it as a fraction of integers.

solution Terminating decimal numerals can be written as fractions by multiplying above and below by a judiciously chosen power of 10. If we move the decimal point five places to the right in this example, we get 314. Thus, we will multiply above and below by 10^5.

$$.00314 \times \frac{10^5}{10^5} = \frac{314}{100,000}$$

[†] At this level it is helpful to think of half the real numbers as being rational numbers and half as being irrational numbers although this is not true because an infinite set cannot be divided into halves!

We could reduce this fraction to lowest terms, but this reduction is not required because any fraction of integers will satisfy the requirement.

example 103.B.2   Show that .00000623 is a rational number by writing it as a fraction of integers.

solution   We will multiply above and below by 10^8.

$$.00000623 \times \frac{10^8}{10^8} = \frac{623}{100,000,000}$$

103.C

repeating digits

We indicate that digits in a decimal fraction repeat by drawing a bar over the repeating digits. Thus, in the numerals $0.016\overline{23}$, $1.0031\overline{543}$, the digits under the bars repeat in an endless pattern, as follows:

$$0.016\overline{23} = .01623232323232323 \cdots$$
$$1.0031\overline{543} = 1.0031543543543543543543 \cdots$$

Each of these numerals represents a rational number and can be written as a quotient of integers. To do this, we must get rid of the repeating digits. **We can eliminate these repeating digits by subtracting the given number from another number that has the same repeating digits, as shown here.**

$$
\begin{array}{r}
100N = 1.623\ 23\ 23\ 23\ 23 \cdots \\
N = .016\ 23\ 23\ 23\ 23 \cdots \\
\hline
99N = 1.607 \qquad \text{(repeating digits eliminated)}
\end{array}
$$

The equation with $100N$ is the same as the equation with N except that each side has been multiplied by 100. We multiplied by 100 because there were two repeating digits. Three repeating digits would require a multiplier of 1000; four repeating digits would require a multiplier of 10,000, etc. We will investigate this procedure in the next three examples.

example 103.C.1   Show that $.016\overline{23}$ is a rational number by writing it as a fraction of integers.

solution   We will use three steps. The first is to write the number and indicate the repeating digits. **Then we record another decimal point above the decimal point in the number.**

$$N = \overset{.}{.}016\,|23\,|23\,|23 \cdots$$

Now we mentally move the digits up and shift them two places to the left.

$$
\begin{array}{r}
100N = 1.623\ |\ 23\ |\ 23\ |\ 23\ \cdots \\
N = .016\ |\ 23\ |\ 23\ |\ 23\ \cdots
\end{array}
$$

Moving the digits two places while holding the decimal point fixed is the same as multiplying by 100, so we record $100N$ on the left side. Now we can subtract the lower equation from the upper equation.

$$
\begin{array}{r}
100N = 1.623\ 23\ 23 \cdots \\
N = .016\ 23\ 23 \cdots \\
\hline
99N = 1.607
\end{array}
$$

We have eliminated the repeating digits. Now we divide both sides of the equation by 99 to find N.

$$N = \frac{1.607}{99}$$

and, finally, we can get a fraction of integers if we multiply above and below by 1000 to get

$$N = \frac{1607}{99,000}$$

example 103.C.2 Show that $1.0031\overline{543}$ is a rational number by writing it as a quotient of integers.

solution We begin by recording the number in expanded form and writing a decimal point above the decimal point in the number.

$$N = 1\overset{.}{.}0031\big|543\big|543\big|543 \cdots$$

Now we mentally move the digits up and shift them three places to the left. Since this is the same as multiplying by 1000, we multiply N by 1000 and then subtract the lower equation from the upper equation.

$$\begin{array}{rl}
1000N = & 1003.1543 \mid 543 \mid 543 \mid 543 \cdots \\
N = & 1.0031 \mid 543 \mid 543 \mid 543 \cdots \\
\hline
999N = & 1002.1512
\end{array}$$

Next, we divide both sides by 999.

$$N = \frac{1002.1512}{999}$$

We can make this a fraction of integers if we multiply above and below by 10,000. We do this and the fraction of integers is

$$N = \frac{10,021,512}{9,990,000}$$

example 103.C.3 Show that $13.01\overline{2}$ is a rational number by writing it as a fraction of integers.

solution We record the number in expanded form, indicate the repeating digits and write the new decimal point.

$$N = 13\overset{.}{.}01\big|2\big|2\big|2\big|2\big|2\big| \cdots$$

This time only one digit repeats so when we mentally move the digits up, we shift them only one place to the left. Shifting the digits one place is equivalent to multiplying by 10, so we also multiply N by 10 and then subtract.

$$\begin{array}{rl}
10N = & 130.12 \mid 2 \mid 2 \mid 2 \mid 2 \mid 2 \mid \cdots \\
N = & 13.01 \mid 2 \mid 2 \mid 2 \mid 2 \mid 2 \mid \cdots \\
\hline
9N = & 117.11
\end{array}$$

We finish by dividing by 9 and then mentally multiplying above and below by 100; and we get

$$N = \frac{117.11}{9} \longrightarrow N = \frac{11,711}{900}$$

problem set 103

1. Twice the number of ducks was only 6 less than 24 times the number of geese. Also 10 times the number of geese was only 5 less than the number of ducks. How many ducks and geese were there?

2. Up was very far so Van and Samuel began early and traveled at 240 miles per hour. Back happened to be the same distance but the speed back was 360 miles per hour. If the time back was 4 hours less than the time up, how far was up?

3. Pamela and Gail were puzzled. They needed 400 ml of a solution that was 11% salt. They had two solutions to work with. One solution was 10% salt and the other was 20% salt. How much of each solution should they use?

4. The riverboat Emily-Eleanor could go 168 miles downstream in 12 hours, but it took her 9 hours to go 54 miles upstream. What was her speed in still water and what was the speed of the current in the river?

5. At the bazaar Deeb paid the Burk $2400 for the rug. If the Burk had only paid $400 for the rug, what was the markup as a percentage of cost, and what was the markup as a percentage of the selling price?

Show that the following numbers are rational numbers by writing them as fractions of integers.

*6. .00000623

*7. .016$\overline{23}$

*8. 1.0031$\overline{543}$

*9. 13.01$\overline{2}$

10. Divide $m^3 - p^3$ by $m - p$.

11. Which of the following sets of ordered pairs are functions?
    (a) $(7, 11), (11, 7), (4, -4), (-4, 4)$
    (b) $(5, -2), (-2, 5), (4, -2), (-2, 4)$
    (c) $(7, -3), (-3, 7), (5, 7), (3, 7)$

12. Complete the square as an aid in graphing $y = -x^2 + 4x - 2$.

13. Graph on a number line: $-|x| + 2 > -2, D = \{\text{Integers}\}$

14. Find the number that is $\frac{3}{7}$ of the way from $\frac{1}{8}$ to $3\frac{1}{4}$.

Solve:

15. $\begin{cases} \dfrac{1}{3}x - \dfrac{2}{5}y = -5 \\ .005x - .04y = -.755 \end{cases}$

16. $\begin{cases} 4x - y = 2 \\ xy = 3 \end{cases}$

17. $\begin{cases} x - y - 2z = -14 \\ 2x + y - z = 2 \\ -x + y - z = -4 \end{cases}$

Graph:

18. $\begin{cases} x - 3y \le -6 \\ x \ge -2 \end{cases}$

19. $0 < x + 2 \not> 4, D = \{\text{Reals}\}$

Simplify:

20. $\dfrac{(x^{-2})^{a+2}y^{-3a}}{y^{-a/2}}$

21. $\dfrac{k}{k^2x - \dfrac{1}{x - \dfrac{1}{k^2}}}$

22. $\sqrt[5]{4\sqrt[3]{2}}$

23. $\dfrac{6i - i^2}{-i^3 + 3}$

24. $\dfrac{2 + 3\sqrt{2}}{4 - \sqrt{18}}$

25. $-i^5 + \sqrt{-4}\sqrt{4} - 3\sqrt{-9} + 2i^4$

26. $3\sqrt{\dfrac{5}{2}} + 2\sqrt{\dfrac{2}{5}} - 4\sqrt{40}$

27. Add: $-30\underline{/135°} + 20\underline{/-20°}$

28. $m\left(\dfrac{1}{x} + \dfrac{1}{y}\right) = \dfrac{1}{c} + \dfrac{a}{b}$; find c

29. Solve $-4x^2 = x - 5$ by completing the square.

30. Change 4000 milliliters per second to cubic feet per minute.

ER 103-102, 103-103

LESSON 104   *Advanced factoring*

104.A
advanced
factoring

When two or more quantities are multiplied to form a product, we say that each of the quantities is a **factor** of the product. Thus, since

$$x(x + 3)(x + 2) = x^3 + 5x^2 + 6x$$

we can say that x and $x + 3$ and $x + 2$ are factors of $x^3 + 5x^2 + 6x$. This is the reason that we use the word *factoring* to describe the procedure of writing a sum as the product of the quantities that can be multiplied to form the sum. Thus far, our factoring of trinomials has been restricted to trinomials whose lead coefficient is 1. Now we will investigate the procedure used to factor trinomials in which the lead coefficient is a number other than 1. This type of trinomial is the product of two binomials, at least one of which has a lead coefficient that is not 1. Let's look at the pattern that evolves when binomials of this type are multiplied.

$$
\begin{array}{r}
2x + 3 \\
x - 5 \\
\hline
2x^2 + 3x \\
- 10x - 15 \\
\hline
2x^2 - 7x - 15
\end{array}
\qquad
\begin{array}{r}
3x + 2 \\
2x - 3 \\
\hline
6x^2 + 4x \\
- 9x - 6 \\
\hline
6x^2 - 5x - 6
\end{array}
$$

Here we note that the first term of both trinomials is the product of the first terms of the binomials and that the last terms of the trinomials are the products of the last terms of the binomials. But, alas, the coefficient of the middle terms of the trinomials is no longer the sum of the last two terms of the binomials. **In both examples the middle term is the sum of the product of the first term of the first binomial and the last term of the second binomial and the product of the last term of the first binomial and the first term of the second binomial.** It is easier to see this if we write the original indicated multiplication in horizontal form and note that the middle term is the sum of the products of the means and the extremes.[†]

$$
\underbrace{(2x + 3)(x - 5)}_{\text{means}}^{\text{extremes}} = 2x^2 - 7x - 15
\qquad
\underbrace{(3x + 2)(2x - 3)}_{\text{means}}^{\text{extremes}} = 6x^2 - 5x - 6
$$

Product of means $= 3x$
Product of extremes $= -10x$
Sum $= -7x$

Product of means $= 4x$
Product of extremes $= -9x$
Sum $= -5x$

[†] Mathematicians sometimes use the word *mean* to mean *middle* and the word *extreme* to mean *end.* Thus the mean terms in the multiplication shown are the middle terms and the extreme terms are the end terms.

example 104.A.1   Factor $-7x - 15 + 2x^2$.

solution   We begin by writing the trinomial in descending powers of the variable.

$$2x^2 - 7x - 15$$

Now to factor $2x^2 - 7x - 15$, we note that the product of the first terms of the binomials is $2x^2$, the product of the last terms of the binomials is -15, and the middle term is the sum of the products of the means and the extremes. Since the term $2x^2$ is the product of the first terms of the binomials, we write

$$(2x \quad)(x \quad)$$

Now, the pairs of integral factors of -15 are 15 and -1, -15 and 1, 5 and -3, and -5 and 3. We must try each pair *twice* and see what middle term will result in each case.

For (15, -1): $(2x + 15)(x - 1)$ middle term is $13x$
                   $(2x - 1)(x + 15)$ middle term is $29x$

For (5, -3): $(2x + 5)(x - 3)$ middle term is $-x$
               $(2x - 3)(x + 5)$ middle term is $7x$

For (-15, 1): $(2x - 15)(x + 1)$ middle term is $-13x$
              $(2x + 1)(x - 15)$ middle term is $-29x$

For (-5, 3): $(2x - 5)(x + 3)$ middle term is x
              $(2x + 3)(x - 5)$ middle term is $-7x$

We will use the last entry because the sum of the products of the means and the extremes is $-7x$. Thus we see that $2x^2 - 7x - 15$ can be factored over the integers as $(2x + 3)(x - 5)$. If our original problem had been to find the roots of $2x^2 - 7x - 15 = 0$, we would begin by factoring

$$2x^2 - 7x - 15 = 0 \quad \longrightarrow \quad (2x + 3)(x - 5) = 0$$

and complete the solution by using the zero factor theorem.

If $2x + 3 = 0$ and if $x - 5 = 0$

$$2x = -3 \qquad\qquad\qquad \boldsymbol{x = 5}$$

$$x = -\frac{3}{2}$$

Check:

$$2\left(-\frac{3}{2}\right)^2 - 7\left(-\frac{3}{2}\right) - 15 = 0 \qquad\qquad 2(5)^2 - 7(5) - 15 = 0$$

$$2\left(\frac{9}{4}\right) + \frac{21}{2} - 15 = 0 \qquad\qquad 2(25) - 35 - 15 = 0$$

$$\frac{9}{2} + \frac{21}{2} - 15 = 0 \qquad\qquad\qquad 50 - 35 - 15 = 0$$

$$\qquad\qquad\qquad\qquad\qquad\qquad\qquad\qquad 0 = 0 \quad \text{Check}$$

$$15 - 15 = 0$$

$$0 = 0 \quad \text{Check}$$

example 104.A.2   Factor to find the solutions of $-5x - 6 = -6x^2$.

solution   We begin by writing the equation in standard form as

$$6x^2 - 5x - 6 = 0$$

The pairs of factors of $6x^2$ are $6x$ and x and also $3x$ and $2x$. The pairs of factors of -6 are 3 and -2, -3 and 2, 6 and -1, and -6 and 1. We must try each of the last four pairs *twice* with each of the first two pairs to find out which combination yields the proper middle term.

	MIDDLE		MIDDLE
using $6x$ and x	TERM	using $3x$ and $2x$	TERM
$(6x - 3)(x + 2) \longrightarrow$	$+9x$	$(3x + 6)(2x - 1) \longrightarrow$	$9x$
$(6x + 2)(x - 3) \longrightarrow$	$-16x$	$(3x - 1)(2x + 6) \longrightarrow$	$16x$
$(6x + 3)(x - 2) \longrightarrow$	$-9x$	$(3x - 6)(2x + 1) \longrightarrow$	$-9x$
$(6x - 2)(x + 3) \longrightarrow$	$+16x$	$(3x + 1)(2x - 6) \longrightarrow$	$-16x$
$(6x + 6)(x - 1) \longrightarrow$	0	$(3x + 3)(2x - 2) \longrightarrow$	0
$(6x - 1)(x + 6) \longrightarrow$	$35x$	$(3x - 2)(2x + 3) \longrightarrow$	$5x$
$(6x - 6)(x + 1) \longrightarrow$	0	$(3x - 3)(2x + 2) \longrightarrow$	0
$(6x + 1)(x - 6) \longrightarrow$	$-35x$	$(3x + 2)(2x - 3) \longrightarrow$	$-5x$

Since the last combination gives us a middle term of $-5x$, this is the combination we want. Now we factor.

$$6x^2 - 5x - 6 = 0 \quad \longrightarrow \quad (3x + 2)(2x - 3) = 0$$

And next we complete the solution by using the zero factor theorem.

$$\text{If} \quad 3x + 2 = 0 \quad \text{and if} \quad 2x - 3 = 0$$
$$3x = -2 \qquad\qquad 2x = 3$$
$$x = -\frac{2}{3} \qquad\qquad x = \frac{3}{2}$$

Check:

When $x = -\dfrac{2}{3}$:

$$-5x - 6 = -6x^2$$
$$-5\left(-\frac{2}{3}\right) - 6 = -6\left(-\frac{2}{3}\right)^2$$
$$\frac{10}{3} - 6 = -\frac{24}{9}$$
$$\frac{30}{9} - \frac{54}{9} = -\frac{24}{9}$$
$$-\frac{24}{9} = -\frac{24}{9} \quad \text{Check}$$

When $x = \dfrac{3}{2}$:

$$-5x - 6 = -6x^2$$
$$-5\left(\frac{3}{2}\right) - 6 = -6\left(\frac{3}{2}\right)^2$$
$$-\frac{15}{2} - 6 = -\frac{54}{4}$$
$$-\frac{30}{4} - \frac{24}{4} = -\frac{54}{4}$$
$$-\frac{54}{4} = -\frac{54}{4} \quad \text{Check}$$

1. The weight of the chlorine (Cl) in a quantity of the compound $C_3H_3Cl_5$ was 1050 grams. What was the total weight of the compound? (C, 12; H, 1; Cl, 35)

2. Weasel headed for Table Rock Lake in a 10 mile per hour trot. But soon short wind forced her to slow to a 6 mile per hour walk. If she covered 64 miles in 8 hours, how far did she walk and how far did she trot?

3. In the flask was 140 ml of a 20% alcohol solution. How much pure alcohol should be added to get a final solution that is 44% alcohol?

4. The number of potatoes varied jointly as the number of mules and the number of farmers squared. If 5 mules and 5 farmers went with 750 potatoes, how many potatoes went with 10 mules and 10 farmers?

5. Eight hundred dollars was the markup and the cost was $2400. What was the markup as a percentage of selling price and what was the markup as a percentage of the cost?

Solve by factoring:

*6. $-7x - 15 + 2x^2 = 0$ $\qquad\qquad$ *7. $-5x - 6 = -6x^2$

Show that the following numbers are rational numbers by writing them as fractions of integers:

8. $.000\overline{1234}$ $\qquad\qquad\qquad\qquad$ 9. $.01\overline{651}$

10. Divide $m^3 + p^3$ by $m + p$. $\qquad\qquad$ 11. Divide $x^3 + y^3$ by $x + y$.

12. Complete the square as an aid in graphing: $y = x^2 + 6x + 8$

13. Graph on a number line: $-|x| + 2 > 1$, $D = \{\text{Integers}\}$

14. Find the number that is $\frac{3}{5}$ of the way from $2\frac{1}{4}$ to $4\frac{1}{2}$.

Solve:

15. $\begin{cases} \frac{2}{7}x - \frac{1}{4}y = -6 \\ .07x + .14y = 6.58 \end{cases}$ $\qquad$ 16. $\begin{cases} 5x - y = 2 \\ xy = 6 \end{cases}$

17. $\begin{cases} x^2 + y^2 = 8 \\ x - y = 2 \end{cases}$ $\qquad\qquad$ 18. $\begin{cases} 2x - y - 2z = 2 \\ x + y - z = 7 \\ 2x - y - z = 0 \end{cases}$

19. Use substitution: $\begin{cases} 3x + 5y = 4 \\ 10x - 15y = -50 \end{cases}$ $\quad$ 20. Graph: $\begin{cases} x - 3y \geq 6 \\ x \geq -3 \end{cases}$

Solve by factoring. Rearrange if necessary.

21. $-14x - 30 + 4x^2 = 0$ $\qquad\qquad$ 22. $10x = 3x^3 - 13x^2$

23. $-5x^2 - 2x + 3x^3 = 0$ $\qquad\qquad$ 24. $-10x - 4 + 6x^2 = 0$

25. $8x + 4 + 3x^2 = 0$ $\qquad\qquad\qquad$ 26. $24x^2 + 9x^3 + 12x = 0$

27. $2p^2 - 3p - 5 = 0$ $\qquad\qquad\qquad$ 28. $8 + 18x + 4x^2 = 0$

Simplify:

29. $\dfrac{x^a(x^{\frac{a}{2}+4})^2 y^b}{y^{b/3} x^{a/6}}$ $\qquad$ 30. $\dfrac{2 - 3i^3}{i + 2i^2 + 3i^3}$ $\qquad$ 31. $\dfrac{4 + 2\sqrt{5}}{5 - 3\sqrt{5}}$

ER 104-104, 104-105

LESSON 105 *More on systems of three equations*

105.A
more on systems of three equations

Some systems of three equations in three unknowns do not have all three variables in each of the equations. These equations can also be solved by using the substitution method or the elimination method.

example 105.A.1   Solve: (a)
$$\begin{cases} 2x + 3y = -4 \\ x - 2z = -3 \\ 2y - z = -6 \end{cases}$$
(b)
(c)

solution   One variable is missing in each equation. We can see this better if we write the equations in expanded form.

$$\text{(a)} \quad 2x + 3y \qquad\quad = -4$$
$$\text{(b)} \quad\; x \qquad\quad - 2z = -3$$
$$\text{(c)} \qquad\quad 2y - \; z = -6$$

The first step is to add any two of the equations so that one variable is eliminated. We could add (a) and (b) to eliminate x; or add (b) and (c) to eliminate z; or add (a) and (c) to eliminate y. We choose to eliminate x so we add equation (a) to the product of equation (b) and (-2).

$$\begin{array}{ll} \text{(a)} & 2x + 3y \qquad\quad = -4 \\ (-2)\text{(b)} & \underline{-2x \qquad\quad + 4z = \;\; 6} \\ \text{(d)} & \qquad\quad 3y + 4z = 2 \end{array}$$

The resulting equation, (d), has y and z as variables. So does equation (c). We use these equations to eliminate z by adding equation (d) to the product of (4) and equation (c).

$$\begin{array}{ll} \text{(4)(c)} & 8y - 4z = -24 \\ \text{(d)} & \underline{3y + 4z = \;\;\; 2} \\ & 11y \qquad\;\; = -22 \\ & \qquad\quad y = -2 \end{array}$$

Now we replace y with -2 in equation (a) and solve for x. Then we replace y with -2 in equation (c) and solve for z.

$$\begin{array}{ll} \text{(a)} \quad 2x + 3(-2) = -4 & \text{(c)} \quad 2(-2) - z = -6 \\ \qquad\quad 2x - 6 = -4 & \qquad\quad -4 - z = -6 \\ \qquad\qquad\;\; 2x = 2 & \qquad\qquad\;\; -z = -2 \\ \qquad\qquad\;\;\; x = 1 & \qquad\qquad\;\;\;\; z = 2 \end{array}$$

Thus, our solution is the ordered triple $(1, -2, 2)$.

example 105.A.2   Solve: (a)
$$\begin{cases} 3y - 2z = -12 \\ 2x - 3z = -5 \\ x - 2y = 6 \end{cases}$$
(b)
(c)

solution Although it is not necessary, we will begin by writing the equations in expanded form.

$$\text{(a)} \qquad 3y - 2z = -12$$
$$\text{(b)} \quad 2x \qquad - 3z = -5$$
$$\text{(c)} \qquad x - 2y \qquad = 6$$

This time we decide to eliminate y so we will add the product of 2 and equation (a) to the product of 3 and equation (c).

$$\begin{array}{ll} (2)(a) & 6y - 4z = -24 \\ (3)(c) & \underline{3x - 6y \qquad = 18} \\ (d) & 3x \qquad - 4z = -6 \end{array}$$

Now equation (b) also is an equation in x and z, so we decide to add the product of -3 and equation (b) to the product of 2 and equation (d)

$$\begin{array}{ll} (-3)(b) & -6x + 9z = 15 \\ (2)(d) & \underline{6x - 8z = -12} \\ & z = 3 \end{array}$$

Now we will replace z with 3 in equations (a) and (b) and solve for x and y.

$$\begin{array}{ll} \text{(a)} \quad 3y - 2(3) = -12 & \text{(b)} \quad 2x - 3(3) = -5 \\ \qquad 3y - 6 = -12 & \qquad 2x - 9 = -5 \\ \qquad 3y = -6 & \qquad 2x = 4 \\ \qquad \boldsymbol{y = -2} & \qquad \boldsymbol{x = 2} \end{array}$$

So the solution to this system is the ordered triple **(2, −2, 3)**.

It was not necessary to use elimination as the first step. For example, we could have solved equation (c) for x.

$$x = 6 + 2y$$

and substituted $6 + 2y$ for x in equation (b).

$$\text{(b)} \quad 2(6 + 2y) - 3z = -5$$
$$12 + 4y - 3z = -5$$
$$\text{(e)} \qquad 4y - 3z = -17$$

Now equation (e) could be used with equation (a) to solve for y and z.

problem set 105

1. Jerry started a used parts establishment. He marked up the items 30 percent of the purchase price. If one item sold for $715, what did Jerry pay for it?

2. The small plane could go 6 times as fast as the small car. Thus, the plane could go 1200 miles in only 1 hour less than it took the car to go 250 miles. Find the rate and the time of the small plane and the rate and the time of the small car.

3. The near-stagnant river flowed at only 2 miles per hour. Harold's boat could go 56 miles down the river in one-half the time it took to go 80 miles up the river. What was the speed of his boat in still water?

4. The temperature of a quantity of an ideal gas was held constant at 500°C. The initial pressure and volume were 700 torr and 500 ml. What would the final pressure be if the volume were increased to 1000 ml?

Solve:

*5. $\begin{cases} 2x + 3y = -4 \\ x - 2z = -3 \\ 2y - z = -6 \end{cases}$ ***6.** $\begin{cases} 3y - 2z = -12 \\ 2x - 3z = -5 \\ x - 2y = 6 \end{cases}$

7. $\begin{cases} 2x - 2y - z = 9 \\ 3x + 3y - z = 6 \\ x + y + z = -2 \end{cases}$

Show that these numbers are rational numbers by writing them as quotients of integers:

8. $.0007\overline{013}$ **9.** $4.10\overline{26}$

10. Divide $m^3 - p^3$ by $m - p$.

11. Complete the square as an aid in graphing: $y = -x^2 - 2x - 3$

12. Graph on a number line: $-|x| - 4 \not> 0$, $D = \{\text{Integers}\}$

13. Find the number that is $\frac{1}{10}$ of the way from $3\frac{1}{3}$ to $6\frac{1}{2}$.

Solve:

14. $\begin{cases} \dfrac{3}{5}x - \dfrac{1}{4}y = 5 \\ .012x + .07y = 2.20 \end{cases}$ **15.** $\begin{cases} 5x - y = 3 \\ xy = 4 \end{cases}$

16. $\begin{cases} x^2 + y^2 = 7 \\ 2x - y = 2 \end{cases}$ **17.** Use substitution: $\begin{cases} 5x - 3y = 27 \\ 2x - 5y = 26 \end{cases}$

18. Graph: $\begin{cases} 3x - 4y \geq 8 \\ y > 2 \end{cases}$ **19.** Simplify: $\dfrac{3 - 2i^2 - i}{3i^3 + 3i + 2}$

20. Solve $3x^2 - x - 7 = 0$ by completing the square.

21. Find the resultant of the vectors shown.

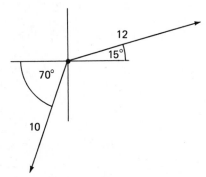

Solve by factoring. Rearrange as necessary. Always look for a common factor.

22. $3x^2 + 7x + 2 = 0$ **23.** $3x^2 + x - 2 = 0$

24. $2z^2 + 13z + 15 = 0$ **25.** $33p^2 + 45p + 6p^3 = 0$

26. $3p^2 - 13p - 10 = 0$ **27.** $-11a + 15 = -2a^2$

28. $14x^2 + 12x + 4x^3 = 0$ **29.** $3x^2 + 4x = -1$

ER 105-106, 105-107

LESSON *106* *Number word problems*

106.A

numbers, numerals, and value

We remember that a numeral is the symbol that we use to represent a particular number. We say that the value of each of the following numerals is three because each one represents the number 3.

$$3 \qquad 1+1+1 \qquad \frac{81}{27} \qquad 2^3 - 5 \qquad 2^2 - 1$$

Thus, we see that **number** and **value** mean the same thing. Also, we see that we should not speak of the value of a number because that is the same thing as saying the number of a number. But we can speak of the value of a numeral because this is the number represented by the numeral. Since paying excessive attention to the difference between a number and a numeral is often counterproductive, we sometimes use the word number when we should use the word numeral. See if you can find where this mistake is made in this lesson. The mistake is not serious.

106.B

number word problems

The value of a digit in a decimal numeral depends on the position of the digit with respect to the decimal point. The digit 6 in the numeral

$$496.0$$

has a value of 6 times 1, or 6, because it is in the units' place which is the first place to the left of the decimal point. The digit 9 has a value of 9 times 10, or 90, because it is in the tens' place, which is two places to the left of the decimal point. The digit 4 has a value of 400 because it is in the hundreds' place, which is three places to the left of the decimal point. We can use the fact that the value of a digit depends on its position to solve some rather interesting word problems.

To solve these problems, we will use U to represent the units' digit; T to represent the tens' digit, and H to represent the hundreds' digit. Also, we will say that

The value of the units' digit is	$1U$
The value of the tens' digit is	$10T$
The value of the hundreds' digit is	$100H$

example 106.B.1 The sum of the digits in a two-digit counting number is 11. If the digits are reversed, the new number is 27 greater than the original number. What was the original number?

solution The original number is written with the tens' digit followed by the units digit, as

$$TU$$

and the sum of T and U is 11. This gives us

$$\text{(a)} \quad U + T = 11$$

The value of the original number is

$$\text{(b)} \quad 10T + U$$

The value of the number with the digits reversed is

$$\text{(c)} \quad 10U + T$$

Since (c) is 27 greater than (b), we add 27 to (b) when we write the equation

$$10T + U + 27 = 10U + T$$

Now from (a) we substitute $11 - T$ for U and solve.

$$10T + (11 - T) + 27 = 10(11 - T) + T$$
$$10T + 11 - T + 27 = 110 - 10T + T$$
$$9T + 38 = 110 - 9T$$
$$18T = 72$$
$$T = 4$$

and since $T + U = 11$, then $U = 7$ and thus the original number was **47**.

example 106.B.2   The sum of the digits of a two-digit counting number was 9. When the digits were reversed, the new number was 45 less than the original number. What was the original number?

solution   The original number is written as TU and the sum of the digits is 9.

$$\text{(a)} \quad T + U = 9$$

The original number had a value of

$$\text{(b)} \quad 10T + U$$

and when the digits were reversed, the value was

$$\text{(c)} \quad 10U + T$$

Now (c) is 45 less than (b) so we add 45 to (c)

$$10U + T + 45 = 10T + U$$

and substitute $9 - U$ for T and get

$$10U + (9 - U) + 45 = 10(9 - U) + U$$

which we solve.

$$9U + 54 = 90 - 10U + U$$
$$9U + 54 = 90 - 9U$$
$$18U = 36$$
$$U = 2$$

Thus, since $T + U = 9$, T equals 7 and the original number was **72**.

problem set 106

1.   Lynn's mixture was 30% formaldehyde while Lucy had a mixture that was 60% formaldehyde. How much of each should they use to get a 400 ml of a mixture that is 36% formaldehyde?

2.   The beaker was filled with methyl bromide, CH_3Br. What percent by weight of this compound is bromine (Br)? (C, 12; H, 1; Br, 80)

*3.   The sum of the digits in a two-digit counting number is 11. If the digits are reversed, the new number is 27 greater than the original number. What was the original number?

*4.   The sum of the digits of a two-digit counting number was 9. When the digits were reversed, the new number was 45 less than the original number. What was the original number?

5.   A 70% markup on the selling price brought the selling price to $1400. What did the shopkeeper pay for the item and what was the markup?

Solve:

6. $\begin{cases} 2x - y = -6 \\ 3y + 2z = 12 \\ x - 3z = -11 \end{cases}$ **7.** $\begin{cases} 5x - y - z = 2 \\ x - 5y + z = -2 \\ -x + y - z = -2 \end{cases}$

8. Show that $.001\overline{213}$ is a rational number by writing it as a fraction of integers.

9. Divide $x^3 + y^3$ by $x + y$.

10. Complete the square as an aid in graphing: $y = -x^2 + 2x + 1$

11. Graph on a number line: $x - 5 \not< -4$, $D = \{\text{Reals}\}$

12. Find the number that is $\frac{3}{4}$ of the way from 2 to $6\frac{2}{3}$.

Solve:

13. $\begin{cases} \frac{2}{3}x - \frac{1}{3}y = 5 \\ -.006x - .04y = -.432 \end{cases}$ **14.** $\begin{cases} 4y - x = 2 \\ xy = 5 \end{cases}$

15. $\begin{cases} x^2 + y^2 = 4 \\ x - 2y = 1 \end{cases}$

16. Use substitution: $\begin{cases} 5x - 3y = 32 \\ 2x - 2y = 16 \end{cases}$

17. Graph: $\begin{cases} 3x - 8y > -x \\ x \le y \end{cases}$

Simplify:

18. $\dfrac{2i^3 - i + 2}{3 + 4i}$ **19.** $\sqrt[3]{x^5 y^2} \sqrt[4]{xy^3}$

20. Solve: $\sqrt{s - 48} = 8 - \sqrt{s}$

21. Write $-4R + 8U$ in polar form.

22. Find the equation of the line that passes through $(-2, 3)$ that is perpendicular to $5x - 3y = 4$.

23. $mx = \dfrac{1}{m}\left(\dfrac{1}{r} + \dfrac{1}{p}\right)$; find r

24. Convert 800 liters per minute to cubic feet per second.

25. Find the surface area of the right circular cylinder in square inches. Dimensions are in inches.

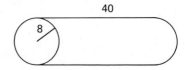

Solve by factoring. Rearrange as necessary. Always look for a common factor.

26. $2x^2 - 30x + 4x^3 = 0$ **27.** $4 + 12x + 8x^2 = 0$

28. $11m^2 + 2m + 5m^3 = 0$ **29.** $16s^2 + 6s^3 + 10s = 0$

30. $2x^3 + x^2 = 3x$

ER 106-108, 106-109

LESSON 107   *Sum and difference of two cubes*

107.A
sum and difference of two cubes

Expressions that are the sum of two squares such as

$$x^2 + y^2 \qquad \text{and} \qquad 9x^2y^2 + 4p^2$$

cannot be factored, but we can factor expressions that are the difference of two squares such as

$$x^2 - y^2 \qquad \text{and} \qquad 9x^2y^2 - 4p^2$$

To factor these expressions, we must recognize that each one is the difference of two squares. **There is no procedure to follow.** We just write down the factored forms by inspection.

$$x^2 - y^2 = (x + y)(x - y) \qquad \text{and} \qquad 9x^2y^2 - 4p^2 = (3xy + 2p)(3xy - 2p)$$

If we had not recognized the forms, we could not have factored the expressions. Two other forms that require recognition for factoring are the sum and difference of two cubes. Both of these can be factored.

$$a^3y^3 - p^3 \qquad \text{and} \qquad a^3y^3 + p^3$$

Unfortunately, the factored forms of these expressions are sometimes difficult to remember. However, they can be easily derived when needed by using simple expressions and polynomial division. To find the factored forms, we must remember that $a^3 + b^3$ is evenly divisible by $a + b$ and that $a^3 - b^3$ is evenly divisible by $a - b$. We will do both of these divisions here.

$$
\begin{array}{r}
a^2 - ab + b^2 \\
a + b \overline{\smash{\big)}\ a^3 + b^3} \\
\underline{a^3 + a^2b } \\
-a^2b \\
\underline{-a^2b - ab^2 } \\
ab^2 + b^3 \\
\underline{ab^2 + b^3}
\end{array}
\qquad
\begin{array}{r}
a^2 + ab + b^2 \\
a - b \overline{\smash{\big)}\ a^3 - b^3} \\
\underline{a^3 - a^2b } \\
a^2b \\
\underline{a^2b - ab^2 } \\
ab^2 - b^3 \\
\underline{ab^2 - b^3}
\end{array}
$$

Thus, we see that we can factor as follows:

$$(1) \quad a^3 + b^3 = (a + b)(a^2 - ab + b^2)$$

$$(2) \quad a^3 - b^3 = (a - b)(a^2 + ab + b^2)$$

To extend these forms to more complicated expressions, some people find that it is helpful to think F_T for *first thing* and S_T for *second thing* instead of a and b. If we do this in the above, we get

$$(1') \quad F_T{}^3 + S_T{}^3 = (F_T + S_T)(F_T{}^2 - F_TS_T + S_T{}^2)$$

$$(2') \quad F_T{}^3 - S_T{}^3 = (F_T - S_T)(F_T{}^2 + F_TS_T + S_T{}^2)$$

example 107.A.1   Factor $x^3y^3 - p^3$.

solution   We recognize that this expression can be written as the difference of two cubes:

$$(xy)^3 - (p)^3$$

The first thing that is cubed is xy and the second thing that is cubed is p. If we use form (2') above,

$$(2') \quad F_T{}^3 - S_T{}^3 = (F_T - S_T)(F_T{}^2 + F_TS_T + S_T{}^2)$$

and replace F_T with xy and S_T with p, we can write the given expression in factored form.

$$x^3y^3 - p^3 = (xy - p)(x^2y^2 + xyp + p^2)$$

example 107.A.2   Factor $8m^3y^6 + x^3$.

solution   We recognize this as the sum of two cubes.

$$(2my^2)^3 + (x)^3$$

and note that the first thing cubed is $2my^2$ and that the second thing cubed is x. Thus if we use form (1′),

$$(1') \quad F_T{}^3 + S_T{}^3 = (F_T + S_T)(F_T{}^2 - F_TS_T + S_T{}^2)$$

and replace F_T with $2my^2$ and S_T with x, we get

$$8m^3y^6 + x^3 = (2my^2 + x)(4m^2y^4 - 2my^2x + x^2)$$

example 107.A.3   Factor $a^{12} + b^{12}$.

solution   We can write this expression as the sum of two cubes as

$$(a^4)^3 + (b^4)^3$$

and we see that the first thing that is cubed is a^4, and the second thing that is cubed is b^4. Thus if we use the form

$$(1') \quad F_T{}^3 + S_T{}^3 = (F_T + S_T)(F_T{}^2 - F_TS_T + S_T{}^2)$$

and use a^4 for F_T and b^4 for S_T, we get

$$a^{12} + b^{12} = (a^4 + b^4)(a^8 - a^4b^4 + b^8)$$

**problem
set 107**

1. The break-even markup was 40 percent of the selling price. If the total income for a day was $6400, what did the store owner pay for these items?

2. Ross and Thais had a two-digit counting number. They saw that the sum of the digits was 6, and if the digits were reversed, the new number was 18 less than the original number. What was the original number?

3. Jeff and Cindy noted that the sum of the digits of a two-digit counting number was 15. If the digits were reversed, they found that the new number was 27 less than the original number. What was the original number?

4. Yellows varied directly as greens squared and inversely as blues. When there were 100 yellows, there were 5 blues but only 1 green. How many blues went with 10 yellows and 10 greens? Solve two ways.

5. The current in the river was only 3 miles per hour so the fast boat could go 92 miles downstream in the same time it took to go 68 miles upstream. How long did each trip take and what was the speed of the boat in still water?

Factor:

*6. $8m^3y^6 + x^3$ *7. $a^{12} + b^{12}$

*8. $x^3y^3 - p^3$ 9. $8x^{12}y^6 - m^3y^9$

10.   Show that $4.12\overline{3}$ is a rational number by writing it as a fraction of integers.

11.   Complete the square as an aid in graphing: $y = x^2 - 2x - 1$

12. Graph on a number line: $-|x| - 3 < -5, D = \{\text{Reals}\}$

Solve:

13. $\begin{cases} \dfrac{2}{5}x - \dfrac{1}{4}y = 2 \\ -.008x - .2y = -1.68 \end{cases}$

14. $\begin{cases} x^2 + y^2 = 5 \\ y - 2x = 2 \end{cases}$

15. $\begin{cases} 4x + 2y = 8 \\ 3x - 3z = -9 \\ -3x + z = 1 \end{cases}$

16. $\begin{cases} x - y - 3z = -2 \\ 3x + y + z = 12 \\ 2x - y + z = 5 \end{cases}$

17. Graph: $\begin{cases} -x - 3y \geq -9 \\ y < 2x \end{cases}$

Simplify:

18. $\dfrac{-i^3 - \sqrt{-2}\sqrt{-2}}{i^2 - 2i}$

19. $\dfrac{4 + 3\sqrt{5}}{2 + \sqrt{5}}$

20. $\sqrt[6]{9\sqrt[3]{3}}$

21. $5\sqrt{\dfrac{3}{5}} + 2\sqrt{\dfrac{5}{3}} - \sqrt{60}$

22. Convert $4R - 14U$ to polar form.

23. Convert 40 feet per second to miles per hour.

24. Find the distance between $(-3, 7)$ and $(5, 3)$.

25. Draw the line suggested by the data points and write the equation that expresses work as a function of energy.

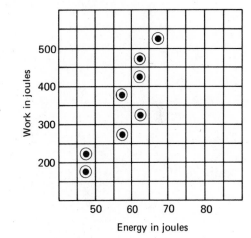

26. Simplify: $\dfrac{ab}{a^2 + \dfrac{b^2}{a^2 + \dfrac{1}{a}}}$

27. Begin with $ax^2 + bx + c = 0$ and derive the quadratic formula by completing the square.

Solve by factoring. Rearrange as necessary. Always look for a common factor.

28. $9x + 2x^2 + 9 = 0$

29. $6b + 20b^2 + 6b^3 = 0$

30. $-14k + 6k^2 - 12 = 0$

ER 107-110, 107-111

LESSON *108* *More on fractional exponents*

108.A
more on fractional exponents

In Lesson 38 we noted that the power theorem for exponents cannot be used when the base is a sum. This is very important so we will review the reason here. We note that when we write

$$x^2$$

we indicate that x is to be multiplied by itself, so

$$x^2 = x \cdot x$$

In the same way if we write

$$(2x^2y^3z^{-4})^2$$

we indicate that $2x^2y^3z^{-4}$ is to be multiplied by itself, and we get

$$(2x^2y^3z^{-4})^2 = (2x^2y^3z^{-4})(2x^2y^3z^{-4}) = 4x^4y^6z^{-8}$$

The power theorem permits this expansion directly if we multiply the exponents to get

$$(2x^2y^3z^{-4})^2 = \mathbf{4x^4y^6z^{-8}}$$

The power theorem cannot be used when we have a sum. Thus, $(x^2 + y^2)^2$ does not equal $x^4 + y^4$.

$$(x^2 + y^2)^2 \neq x^4 + y^4$$

To expand $(x^2 + y^2)^2$, we must multiply $x^2 + y^2$ by itself as follows.

$$
\begin{array}{r}
x^2 + y^2 \\
x^2 + y^2 \\
\hline
x^4 + \quad x^2y^2 \\
x^2y^2 + y^4 \\
\hline
\mathbf{x^4 + 2x^2y^2 + y^4}
\end{array}
$$

The power theorem can be used only when the expression raised to a power is a product of factors.

example 108.A.1   Expand $(2x^{1/2}y^{1/4}z)^3$.

solution   We apply the power theorem and get

$$\mathbf{8x^{3/2}y^{3/4}z^3}$$

example 108.A.2   Expand $(x^{1/2} + y^{1/2})^2$.

solution   The power theorem cannot be used for a sum. We must multiply $x^{1/2} + y^{1/2}$ by $x^{1/2} + y^{1/2}$.

$$
\begin{array}{r}
x^{1/2} + y^{1/2} \\
x^{1/2} + y^{1/2} \\
\hline
x + \quad x^{1/2}y^{1/2} \\
x^{1/2}y^{1/2} + y \\
\hline
\mathbf{x + 2x^{1/2}y^{1/2} + y}
\end{array}
$$

example 108.A.3   Expand $(x^{1/2} + y^{-1/2})^2$.

solution   We must multiply $x^{1/2} + y^{-1/2}$ by $x^{1/2} + y^{-1/2}$.

$$\begin{array}{r} x^{1/2} + y^{-1/2} \\ x^{1/2} + y^{-1/2} \\ \hline x + x^{1/2}y^{-1/2} \\ x^{1/2}y^{-1/2} + y^{-1} \\ \hline \mathbf{x + 2x^{1/2}y^{-1/2} + y^{-1}} \end{array}$$

problem set 108

1. A two-digit counting number has a value that is 8 times the sum of its digits. If 6 times the units' digit is 5 more than the tens' digit, what is the number?

2. The sum of the digits of a two-digit counting number was 11. If the digits were reversed the new number would be 27 less than the original number. What was the original number?

3. Lucretius could cover the 63 miles to Pompeii in 11 more hours than it took Cassius to drive the 60 miles to Rome. Find the rates and times of both if Lucretius traveled at half the speed of Cassius.

4. Doctor Steve held the volume of a quantity of an ideal gas constant at 400 ml. The original pressure and temperature were 800 torr and 400 K. What would be the final pressure if the temperature were raised to 1200 K?

5. Tom and Zollie had six hundred milliliters of a 46% alcohol solution that had to be reduced to a 40 percent solution by extracting pure alcohol. How much pure alcohol should be extracted?

Expand:

*6. $(2x^{1/2}y^{1/4}z)^3$ *7. $(x^{1/2} + y^{1/2})^2$ *8. $(x^{1/2} + y^{-1/2})^2$

Factor:

9. $x^3y^6 - 27m^3$ 10. $64x^9y^6 + p^{12}z^3$

11. Show that $1.023\overline{42}$ is a rational number by writing it as a fraction of integers.

12. Complete the square as an aid in graphing: $y = -x^2 + 4x - 1$

13. Graph on a number line: $-2 \not< x + 2 > -4, D = \{\text{Reals}\}$

Solve:

14. $\begin{cases} \dfrac{2}{5}x - \dfrac{1}{3}y = 1 \\ .3x - .05y = 2.55 \end{cases}$ 15. $\begin{cases} x - 2y = 5 \\ xy = 3 \end{cases}$

16. $\begin{cases} x + y - z = 7 \\ 4x + y + z = 4 \\ 3x + y - z = 9 \end{cases}$ 17. $\begin{cases} 2x + 3y = 15 \\ x - 2z = -3 \\ 3y - z = 6 \end{cases}$

18. Graph: $\begin{cases} 2x - 5y \geq 15 \\ x \leq -y \end{cases}$

Simplify:

19. $\dfrac{2i^2 - \sqrt{-9} + 2}{3 - \sqrt{-2}\sqrt{2}}$ 20. $\dfrac{3 + 2\sqrt{2}}{5\sqrt{2} - 2}$

21. $\sqrt[3]{27}\sqrt[3]{3}$ 22. $\sqrt{x^5y}\sqrt{x^2y}$

23. $\dfrac{(x^{a-2})^{1/3}(y^b)^{1/2}}{x^{2a}y^{-b}}$ 24. $\sqrt{\dfrac{2}{3}} + 4\sqrt{\dfrac{3}{2}} - 6\sqrt{24}$

25. $\dfrac{ka^2}{ka - \dfrac{a^2}{k^2 - \dfrac{1}{a}}}$

26. Find the resultant of the force vectors shown.

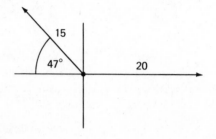

27. Convert 1000 liters per minute to milliliters per second.

Solve by factoring. Rearrange as necessary. Always look for a common factor.

28. $-3p + 6p^2 - 30 = 0$ **29.** $-8x - 14x^2 + 4x^3 = 0$

30. $3x^2 - 7x - 6 = 0$

ER 108-112, 108-113

LESSON 109   *Quadratic inequalities (greater than)*

109.A

quadratic inequalities

If we wish to say that x is a positive number, we can write

x is a positive number

or we can use the greater than/less than symbol and write

$x > 0$

This is the symbolic way to designate a positive number because all the numbers that are greater than zero are positive numbers.

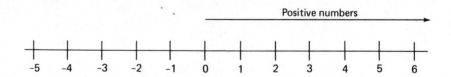

If the product of two quantities is greater than zero,

$(\quad)(\quad) > 0$

then the product is a positive number. There are two ways that a product of two factors can be positive. Either both factors are positive or both factors are negative.

(Pos.)(pos.) > 0 and   (Neg.)(neg.) > 0

We will use this fact to graph the solution to the following inequality.

example 109.A.1   Graph the solution to $(x + 2)(x - 3) > 0$, $D = \{\text{Reals}\}$.

solution   The product of the factors is greater than zero. This means that the product is a positive number. For a product of two factors to be positive, both factors must be positive *or* both factors must be negative.

(POS)	**and**	(POS)	**or**	(NEG)	**and**	(NEG)
$x + 2$		$x - 3$		$x + 2$		$x - 3$

Since each of these factors is positive, each must be greater than zero.

$$x + 2 > 0 \quad \text{and} \quad x - 3 > 0$$

Since each of these factors is negative, each must be less than zero.

$$x + 2 < 0 \quad \text{and} \quad x - 3 < 0$$

$$\underline{-2 \quad -2} \qquad \underline{3 \quad 3}$$

$$x \quad > -2 \quad \text{and} \quad x \quad > 3 \quad \textbf{or} \quad x \quad < -2 \quad \text{and} \quad x \quad < 3$$

The numbers that satisfy the conjunction on the left are the numbers that are greater than -2 and that are also greater than 3. **Of course, all numbers that are greater than 3 are also greater than -2, so the numbers that are greater than 3 satisfy both conditions on the left.**

The numbers that satisfy both conditions of the right conjunction are the numbers that are less than -2 and that are also less than 3. **Of course, all numbers that are less than -2 are also less than 3, so the numbers that are less than -2 satisfy both conditions.**

Thus, our solution is

$$x > 3 \quad \textbf{or} \quad x < -2$$

Now we graph all the real numbers that satisfy either one of these conditions.

example 109.A.2   Graph $x^2 - 2x \geq 3$, $D = \{\text{Integers}\}$.

solution   We begin the solution of this quadratic inequality by writing it in standard form.

$$x^2 - 2x - 3 \geq 0$$

Next we factor and get

$$(x - 3)(x + 1) \geq 0$$

This notation uses symbols to tell us that the product is equal to or is greater than zero. This is another way of saying that the product is zero or is a positive number. If the product is a positive number, both factors must be positive or both factors must be negative.

(POS)		(POS)		(NEG)		(NEG)
$x - 3 \geq 0$	and	$x + 1 \geq 0$	or	$x - 3 \leq 0$	and	$x + 1 \leq 0$

$$x \quad \geq 3 \quad \text{and} \quad x \quad \geq -1 \quad \textbf{or} \quad x \quad \leq 3 \quad \text{and} \quad x \quad \leq -1$$

Now both of the inequalities on the left must be true **or** both the inequalities on the right must be true. On the left, if $x \geq 3$ **and** $x \geq -1$, then

$$x \geq 3$$

On the right, if $x \le 3$ **and** $x \le -1$, then

$$x \le -1$$

Thus, the solution is the graph of the disjunction

$$x \ge 3 \quad \textbf{or} \quad x \le -1$$

When we graph, we remember that the domain is the set of integers.

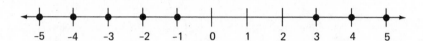

problem set 109

1. RJ and Cheng noted that the sum of the digits of their two-digit counting number was 9. If the digits were reversed, they found that the new number was 27 less than the original number. What was the original number?

2. Suzane had a two-digit counting number and the sum of the digits was 7. If she reversed the digits, she found that the new number was 45 greater than the original number. What was the original number?

3. The weight of the oxygen in a quantity of the compound KCr_2O_7 was 336 grams. What was the total weight of the compound? (K, 39; Cr, 52; O, 16)

4. The Delta Queen could go 120 miles downstream in 8 hours, but it took her 9 hours to go 63 miles upstream. What was her speed in still water, and what was the speed of the current in the river?

5. Steve and Susu found that the cost of the machine was \$1400 and that it was marked up \$700 over what it had cost. What was the markup as a percentage of the selling price and what was the markup as a percentage of the cost?

Graph the solution on a number line:

*6. $(x + 2)(x - 3) > 0$, $D = \{\text{Reals}\}$

*7. $x^2 - 2x \ge 3$, $D = \{\text{Integers}\}$

Expand:

8. $(-2x^3y^2z^3)^2$

9. $(x^{1/2} - y^{1/4})^2$

10. $(x^{1/2} + y^{1/2})(x^{1/2} - y^{1/2})$

Factor:

11. $8x^9 - y^6p^3$

12. $27x^{12}y^9 + p^6m^{15}$

13. Show that $.01\overline{362}$ is a rational number by writing it as a fraction of integers.

14. Complete the square as an aid in graphing: $y = x^2 - 4x + 3$

15. Graph on a number line: $3 \not\le x + 2$ or $x + 5 \not< 8$, $D = \{\text{Reals}\}$

Solve:

16. $\begin{cases} \dfrac{2}{7}x - \dfrac{1}{5}y = 2 \\ .03x + .07y = 1.12 \end{cases}$

17. $\begin{cases} x + y + z = 1 \\ 4x - 2y - z = 6 \\ 3x - y + z = -1 \end{cases}$

18. $\begin{cases} x^2 + y^2 = 6 \\ x - y = 1 \end{cases}$

19. $\begin{cases} x - z = 3 \\ x + 2y = 5 \\ y + z = 0 \end{cases}$

20. Graph: $\begin{cases} 2x \ge 6 \\ x + y < 3 \end{cases}$

Simplify:

21. $\dfrac{2i^2 - 2i + 2}{\sqrt{-9} - \sqrt{-3}\sqrt{-3}}$ **22.** $\dfrac{3 + 2\sqrt{20}}{1 - \sqrt{5}}$ **23.** $\dfrac{(a^{2-b})^2 x^2}{x^{b/2} a^{b/2}}$

24. $\sqrt[4]{8\sqrt{2}}$ **25.** $\sqrt{\dfrac{2}{3}} - 5\sqrt{\dfrac{3}{2}} + 3\sqrt{24}$

26. Solve: $\sqrt{k} = 6 + \sqrt{k - 48}$ **27.** Write $4R - 15U$ in polar form.

Solve by factoring:

28. $11s + 3s^2 + 10 = 0$ **29.** $-2x + 8x^3 + 6x^2 = 0$

30. $12n + 11n^2 + 2n^3 = 0$

ER 109-114, 109-115

LESSON 110 *Three statements of equality*

110.A
three statements of equality

Many word problems contain three statements of equality. We solve these problems by writing an equation for each statement of equality. If we have to use only three unknowns and if a solution exists, it can be found by using substitution or elimination, or by using both substitution and elimination.

example 110.A.1   There were 26 nickels, dimes, and quarters in all, and their value was \$2.25. How many coins of each type were there if there were 10 times as many nickels as quarters?

solution   We have three statements of equality, and each one can be written as an equation.

(a)   Number of nickels + number of dimes + number of quarters = 26

$$N_N + N_D + N_Q = 26$$

(b)   Value of the nickels + value of dimes + value of quarters = 225 pennies

$$5N_N + 10N_D + 25N_Q = 225$$

(c)   ten times the number of quarters equals the number of nickels

$$10N_Q = N_N$$

We begin by substituting $10N_Q$ for N_N in equations (a) and (b).

(a) $(10N_Q) + N_D + N_Q = 26$ $\longrightarrow$ $N_D + 11N_Q = 26$ (a')

(b) $5(10N_Q) + 10N_D + 25N_Q = 225$ $\longrightarrow$ $10N_D + 75N_Q = 225$ (b')

To solve, we will multiply equation (a') by -10 and add the equations.

$$
\begin{array}{rl}
-10(a') & -10N_D - 110N_Q = -260 \\
(b') & \underline{10N_D + 75N_Q = 225} \\
& \, -35N_Q = -35 \\
& \, N_Q = 1
\end{array}
$$

Now we can state that there are 10 nickels because $N_N = 10N_Q$, and this means 15 dimes because there are 26 coins in all. So

$$N_D = 15 \qquad N_N = 10 \qquad N_Q = 1$$

example 110.A.2   The total number of blues, greens, and yellows in the pot was 7. The blues weighed 1 pound, the greens weighed 4 pounds, and the yellows weighed 5 pounds. The total weight was 25 pounds. If there was 1 more yellow than green, how many of each color were there?

solution   Again we get three statements of equality that we can write as equations.

(a)   Number of blues + number of greens + number of yellows = 7

$$N_B + N_G + N_Y = 7$$

(b)   Weight of blues + weight of greens + weight of yellows = 25

$$N_B + 4N_G + 5N_Y = 25$$

(c)   There was 1 more yellow than there were greens

$$N_G + 1 = N_Y$$

To begin we will replace N_Y in (a) and (b) with $N_G + 1$ and then simplify.

(a) $N_B + N_G + (N_G + 1) = 7 \qquad \longrightarrow \qquad N_B + 2N_G = 6 \qquad$ (a')

(b) $N_B + 4N_G + 5(N_G + 1) = 25 \quad \longrightarrow \quad N_B + 9N_G = 20 \qquad$ (b')

Now we multiply (a') by -1 and add the result to (b').

$$\begin{array}{rl}
-1(a') & -N_B - 2N_G = -6 \\
(b') & \underline{N_B + 9N_G = 20} \\
& 7N_G = 14 \\
& N_G = 2
\end{array}$$

Now since $N_Y = N_G + 1$, there were 3 yellows.

$$N_Y = 3$$

and there must have been 2 blues because the total was 7.

$$N_B = 2 \qquad N_Y = 3 \qquad N_G = 2$$

problem
set 110

*1.   There were 26 nickels, dimes, and quarters in all and their value was $2.25. How many coins of each type were there if there were 10 times as many nickels as quarters?

*2.   The total number of blues, greens, and yellows in the pot was 7. The blues weighed 1 pound, the greens weighed 4 pounds, and the yellows weighed 5 pounds. The total weight was 25 pounds. If there was 1 more yellow than green, how many of each color were there?

3.   A two-digit counting number has a value that equals 4 times the sum of its digits. If the units' digit is 1 greater than the tens' digit, what is the number?

4.   Find three consecutive integers such that the product of the first and the third is 35 greater than the product of the second and 5.

5.   The number of students varied directly as the number of teachers and as the number of administrators squared. One thousand students were present when there were 5 teachers and 2 administrators. How many students were there when there were 8 teachers and only 1 administrator? Solve two ways.

Graph the solution on a number line:

6. $(x + 4)(x - 2) > 0$, $D = \{\text{Integers}\}$ **7.** $x^2 > -6 + 5x$, $D = \{\text{Integers}\}$

Expand:

8. $(x^{1/2} + y^{1/4})^2$ **9.** $(x^{1/2} - y^{-1/2})^2$ **<u>10.</u>** $(x^{1/2}y^{-1/2})^2$

Factor:

11. $x^3 - m^6y^6$ **12.** $8x^6y^3 - 27m^3p^{12}$

13. Show that $1.02\overline{13}$ is a rational number by writing it as a quotient of integers.

14. Complete the square as an aid in graphing: $y = -x^2 - 4x - 1$

Graph on a number line:

15. $-|x| - 3 \not< -7$, $D = \{\text{Reals}\}$ **16.** $-2 \not> x + 5 < 4$, $D = \{\text{Integers}\}$

Solve:

17. $\begin{cases} \dfrac{3}{5}x - \dfrac{2}{5}y = -10 \\ .003x + .2y = 1.97 \end{cases}$ **18.** $\begin{cases} x + 2y = 10 \\ x - 3z = -16 \\ y + 2z = 16 \end{cases}$ **<u>19.</u>** $\begin{cases} x^2 + y^2 = 4 \\ x - y = 1 \end{cases}$

20. Solve $-2x^2 + 3x + 5 = 0$ by completing the square.

21. Change 40 inches per second to feet per hour.

Simplify:

<u>22.</u> $\dfrac{2i^3 - \sqrt{-3}\sqrt{-3}}{4 - 3i^2}$ **23.** $\dfrac{2\sqrt{3} + 2}{3 - \sqrt{3}}$ **24.** $\dfrac{a^{x/2}(y^{2-x})^{1/2}}{a^{3x}y^{-2x}}$

25. $\sqrt{xy}\sqrt{x^2y}$ **<u>26.</u>** $\sqrt{\dfrac{2}{7}} - 3\sqrt{\dfrac{7}{2}} + 2\sqrt{126}$

27. Graph: $\begin{cases} -y < 3 \\ 3x + y \le 3 \end{cases}$

28. Find the resultant vector of the two vectors shown.

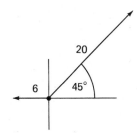

Solve by factoring.

29. $2x^2 = x + 10$ **30.** $-15x = 7x^2 - 2x^3$

ER 110-116, 110-117

Additional

12 Lessons

Topics

LESSON 111 *Quadratic inequalities (less than)*

111.A
quadratic inequalities

In Lesson 109 we discovered that the solution to a greater-than quadratic inequality is a disjunction. We found that the solution to the inequality

$$(x + 2)(x - 3) > 0$$

is the disjunction

$$x < -2 \quad \textbf{or} \quad x > 3$$

If we reverse the inequality symbol in the original inequality, we get

$$(x + 2)(x - 3) < 0$$

This product is a negative number because all numbers that are less than zero are negative numbers. We will find that the solution to a less-than inequality such as this one will be a conjunction.

example 111.A.1 Solve the inequality $(x + 2)(x - 3) < 0$, $D = \{\text{Reals}\}$.

solution For a product to be less than zero, the product must be negative; for all real numbers that are less than zero are negative numbers. **Thus the first factor must be negative and the second factor positive or the first factor must be positive and the second factor negative.**

(NEG)	**and**	(POS)	**or**	(POS)	**and**	(NEG)
$x + 2$		$x - 3$		$x + 2$		$x - 3$

For $x + 2$ to be negative, $x + 2$ must be less than zero; and for $x - 3$ to be positive, $x - 3$ must be greater than zero. Thus,	For $x + 2$ to be positive, $x + 2$ must be greater than zero; and for $x - 3$ to be negative, $x - 3$ must be less than zero. Thus,

$$
\begin{array}{lll}
x + 2 < \;\;\; 0 & \textbf{and} & x - 3 > 0 \\
\underline{-2 \quad -2} & & \underline{\;\;\;\; 3 \quad\;\; 3} \\
x \qquad < -2 & \textbf{and} & x \qquad > 3
\end{array}
\qquad \textbf{or} \qquad
\begin{array}{lll}
x + 2 > \;\;\; 0 & \textbf{and} & x - 3 < 0 \\
\underline{-2 \quad -2} & & \underline{\;\;\;\; 3 \quad\;\; 3} \\
x \qquad > -2 & \textbf{and} & x \qquad < 3
\end{array}
$$

The numbers that satisfy the conjunction on the left are the real numbers that are less than -2 and that are also greater than 3. **There are no numbers that are less than -2 and that are also greater than 3. Thus, there are no numbers that satisfy the left side, and the total** solution must come from the right side. The numbers that satisfy the inequalities on the right are the real numbers greater than -2 that are also less than 3.

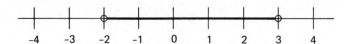

example 111.A.2   Solve the inequality $x^2 + 2x - 8 < 0$, $D = \{\text{Integers}\}$.

solution   We begin by factoring and we get

$$(x + 4)(x - 2) < 0$$

The product of these two factors is less than zero. **This means that the product is a negative number.** Thus, the first factor must be negative and the second factor must be positive *or* the first factor must be positive and the second factor must be negative.

(NEG)	**and**	(POS)	**or**	(POS)	**and**	(NEG)
$x + 4$		$x - 2$		$x + 4$		$x - 2$

For $x + 4$ to be negative, $x + 4$ must be less than zero; and for $x - 2$ to be positive, $x - 2$ must be greater than zero.

For $x + 4$ to be positive, $x + 4$ must be greater than zero; and for $x - 2$ to be negative, $x - 2$ must be less than zero.

$$\begin{array}{rcl} x + 4 < & 0 \\ -4 & -4 \\ \hline x & < -4 \end{array} \quad \textbf{and} \quad \begin{array}{rcl} x - 2 > 0 \\ 2 \quad 2 \\ \hline x \quad > 2 \end{array} \quad \textbf{or} \quad \begin{array}{rcl} x + 4 > & 0 \\ -4 & -4 \\ \hline x & > -4 \end{array} \quad \textbf{and} \quad \begin{array}{rcl} x - 2 < 0 \\ 2 \quad 2 \\ \hline x \quad < 2 \end{array}$$

There are no integers that satisfy the conjunction on the left because there are no integers that are less than -4 that are also greater than 2. Thus, the total solution must come from the conjunction on the right and consists of the integers that are greater than -4 and that are also less than 2.

problem set 111

1. There were 20 nickels, dimes, and quarters whose value was \$3.25. If there were twice as many quarters as dimes, how many coins of each kind were there?

2. The sum of the digits of a two-digit counting number is 7. If the digits are reversed, the new number is 9 less than the original number. What was the original number?

3. The bookstore went out of business because the markup they used was only 20 percent of the selling price of the books. If they sold one shipment for a total price of \$1800, what did they pay for these books?

4. The Ochlochnee River meandered at 3 miles per hour. A fast boat could go 230 miles down the river in twice the time it took to go 85 miles up the river. What was the speed of the boat in still water and what were the times?

5. Fabian took 3 times as long to fly 1800 miles as it took Hamilcar to fly 1200 miles. Find the rates and times of both if Hamilcar's speed was 200 miles per hour greater than that of Fabian.

Graph the solution on a number line:

*6. $(x + 2)(x - 3) < 0, D = \{\text{Reals}\}$

*7. $x^2 + 2x - 8 < 0, D = \{\text{Integers}\}$

8. $(x + 2)(x - 3) > 0, D = \{\text{Reals}\}$

9. Multiply: $(x^{1/2} + y^{1/2})(x^{1/2} - y^{-1/4})$

10. Factor: $p^6 x^6 - k^3$

11. Show that $4.01\overline{43}$ is a rational number by writing it as a fraction of integers.

12. Complete the square as an aid in graphing: $y = x^2 + 2x + 3$

Graph on a number line:

13. $-|x| + 3 \le 0, D = \{\text{Reals}\}$

14. $x - 2 \not> 0$ or $x + 4 > 8, D = \{\text{Integers}\}$

15. Which of the following sets of ordered pairs are functions?
    (a) $(4, 2), (2, 4), (5, 7), (7, 5)$
    (b) $(4, -2), (-2, 6), (5, 7), (7, -5)$
    (c) $(-4, 2), (5, 3), (-4, 7), (3, 5)$

16. Find the number $\frac{1}{3}$ of the way from $2\frac{1}{8}$ to 5.

<u>17.</u> Solve $-5x^2 - x = 4$ by using the quadratic formula.

<u>18.</u> Find the equation of the line through $(5, -7)$ that is perpendicular to $x + 3y - 4 = 0$.

Solve:

19. $\begin{cases} \dfrac{1}{2}x + \dfrac{1}{4}y = 7 \\ .7x - .02y = 5.36 \end{cases}$

20. $\begin{cases} x - 2y = 10 \\ 3x - z = 11 \\ 2y - 3z = -9 \end{cases}$

21. $\begin{cases} 2x - y = 7 \\ xy = 4 \end{cases}$

22. $\begin{cases} x + y + z = 1 \\ 2x - y + z = -5 \\ 3x + y + z = 5 \end{cases}$

Simplify:

23. $\dfrac{-3i^2 - 2i^3}{\sqrt{-3}\sqrt{-3} - \sqrt{-9}}$

24. $\dfrac{(a^{x-2})^{1/2}y^2}{y^{x/2}a^{x/2}}$

25. $\sqrt[3]{x^5 y}\sqrt{x^5 y^2}$

26. $\sqrt{\dfrac{2}{7}} + \sqrt{\dfrac{7}{2}} - 3\sqrt{56}$

27. $\dfrac{3\sqrt{2} - 2}{7\sqrt{2} - 3}$

Solve by factoring:

28. $3x^3 + 5x^2 = -2x$

29. $2 = 2x^2 - 3x$

<u>30.</u> $3x^2 + 8x + 4 = 0$

ER 111-118, 111-119

LESSON *112* *Logarithms*

A **logarithm** is an **exponent,** and the rules of exponents are used when we work with logarithms. We know that 10 squared equals 100.

$$\text{(a)} \quad 10^2 = 100$$

The base is 10, the exponent is 2, and the number represented by the exponential is 100. We write this same statement in the language of logarithms by writing

$$\text{(a}') \quad \log_{10} 100 = 2$$

This is read as "the logarithm of 100 to the base 10 is 2." Beginners sometimes become confused because they forget that the word **logarithm** means **exponent. To emphasize that logarithm and exponent have the same meaning, we will use the words interchangeably in this book.** Thus, we can read (a′) as

The exponent of 100 to the base 10 is 2.

Any positive real number except 1 can be used as the base of a logarithm.

$$\text{(b)} \quad 3^4 = 81 \qquad \text{(b}') \quad \log_3 81 = 4$$

These two expressions make the same statement. We read (b) by saying "3 to the fourth power is 81"; and we read (b′) by saying that "the exponent of 81 to the base 3 is 4." Both forms tell us that the base is 3, that the exponent is 4, and that the number represented is 81.

We will restrict our investigation in this book to logarithms (exponents) whose base is 10. On the left we show several exponentials whose base is 10, and on the right we show the logarithmic notation that makes the same statement.

EXPONENTIAL NOTATION	LOGARITHMIC NOTATION
$10^3 = 1000$	$\log_{10} 1000 = 3$
$10^2 = 100$	$\log_{10} 100 = 2$
$10^1 = 10$	$\log_{10} 10 = 1$
$10^{-1} = .1$	$\log_{10} .1 = -1$
$10^{-2} = .01$	$\log_{10} .01 = -2$

In the past, logarithms were used to do difficult and involved calculations. Now we do these calculations electronically by using either computers or small, handheld calculators. Logarithms today are useful primarily because of the utility of the logarithmic function as an algebraic tool. We will do calculations in this book to enhance our understanding of the logarithmic function and not with the aim of learning to calculate by using logarithms.

To find the logarithm (exponent) of a number that is not an integral power of 10, we will use scientific notation and a table of logarithms. We could get the answer directly from a handheld calculator, but this approach would deny us some of the understanding we seek. Thus, we will use a more involved approach.

A table of logarithms (exponents) matches positive numbers between 1 and 10 with the proper positive decimal exponent (logarithm) necessary to express the number in exponential form with a base of 10.

A portion of a table of logarithms (exponents) is shown here. Note that all exponents have been rounded off to four decimal places. Tables that have more than four decimal places are available, but four places are sufficient for our purposes.

Positive numbers between 1 and 10

n	0	1	2	3	4	5	6	7	8	9
1.0	.0000	.0043	.0086	.0128	.0170	.0212	.0253	.0294	.0334	.0374
1.1	.0414	.0453	.0492	.0531	.0569	.0607	.0645	.0682	.0719	.0755
1.2	.0792	.0828	.0864	.0899	.0934	.0969	.1004	.1038	.1072	.1106
1.3	.1139	.1173	.1206	.1239	.1271	.1303	.1335	.1367	.1399	.1430
1.4	.1461	.1492	.1523	.1553	.1584	.1614	.1644	.1673	.1703	.1732
1.5	.1761	.1790	.1818	.1847	.1875	.1903	.1931	.1959	.1987	.2014
1.6	.2041	.2068	.2095	.2122	.2148	.2175	.2201	.2227	.2253	.2279
1.7	.2304	.2330	.2355	.2380	.2405	.2430	.2455	.2480	.2504	.2529
1.8	.2553	.2577	.2601	.2625	.2648	.2672	.2695	.2718	.2742	.2765
1.9	.2788	.2810	.2833	.2856	.2878	.2900	.2923	.2945	.2967	.2989

Positive decimal numbers between 0 and 1

The first two digits of the positive number between 1 and 10 are written in the far left column under *n*. The third digit of the positive number between 1 and 10 is written in the top row. The proper positive decimal exponent (logarithm) so that 10 to that power will equal the given number appears in the body of the table at the intersection of the row of the first two digits of the number and the column of the last digit of the number.

example 112.B.1   Use the portion of the table below to write the number 5.8437912 in exponential notation with 10 as the base. Then write the expression using logarithmic notation.

n	0	1	2	3	4	5	6	7	8	9
5.5	.7404	.7412	.7419	.7427	.7435	.7443	.7451	.7459	.7466	.7474
5.6	.7482	.7490	.7497	.7505	.7513	.7520	.7528	.7536	.7543	.7551
5.7	.7559	.7566	.7574	.7582	.7589	.7597	.7604	.7612	.7619	.7627
5.8	.7634	.7642	.7649	.7657	.7664	.7672	.7679	.7686	.7694	.7701
5.9	.7709	.7716	.7723	.7731	.7738	.7745	.7752	.7760	.7767	.7774
6.0	.7782	.7789	.7796	.7803	.7810	.7818	.7825	.7832	.7839	.7846
6.1	.7853	.7860	.7868	.7875	.7882	.7889	.7896	.7903	.7910	.7917
6.2	.7924	.7931	.7938	.7945	.7952	.7959	.7966	.7973	.7980	.7987

solution   We begin by rounding off the number to three digits and get 5.84, the positive number between 1 and 10 for which we need the matching exponent. On the left side of the table, an arrow points to 5.8, and above the table an arrow points to 4. At the intersection of the row and column indicated by the arrows, we find the positive decimal number .7664. Using this number as an exponent, 5.84 can be written in exponential form as

$$5.84 = 10^{.7664}$$

Now we write this same expression in logarithmic form as

$$\log_{10} 5.84 = .7664$$

example 112.B.2   Write (a) $584{,}312 \times 10^4$ and (b) $.00058423 \times 10^{-5}$ in exponential notation with 10 as the base. Then write each of the expressions in logarithmic notation.

solution (a) We begin by rounding off the number to three digits and writing it in scientific notation as

$$5.84 \times 10^9$$

From the last problem, we know that 5.84 can be written as 10 with an exponent of .7664. So we have

$$\mathbf{5.84 \times 10^9} = 10^{.7664} \times 10^9 = \mathbf{10^{9.7664}}$$

We write the same information in logarithmic form as

$$\mathbf{\log_{10} (5.84 \times 10^9) = 9.7664}$$

(b) We again begin by rounding off and using scientific notation to write

$$5.84 \times 10^{-9}$$

Now the exponential form of 5.84 is the same, and we simplify by adding .7664 and −9 algebraically.

$$\mathbf{5.84 \times 10^{-9}} = 10^{.7664} \times 10^{-9} = \mathbf{10^{-8.2336}}$$

We can write the same information in logarithmic form by writing

$$\mathbf{\log_{10} (5.84 \times 10^{-9}) = -8.2336}$$

**problem
set 112**

1. There were 19 nickels, dimes, and quarters in the pot. James noted that their value was $2. How many coins of each type were there if there were twice as many nickels as dimes?

2. Reds varied directly as blues and inversely as mauves squared. When there were 10 reds, there were 2 mauves and 4 blues. How many blues went with 20 mauves and 3 reds?

3. It was necessary to mix 200 ml of a solution in which the key ingredient made up exactly 63 percent of the total. One container held a solution that contained 70 percent key ingredient and the solution in other container was only 60 percent key ingredient. How much of each should be used?

4. Carlene made the trip in only 8 hours while Dan took 12 hours to make the same trip. This was because Dan dawdled and drove 20 miles per hour slower than Carlene. How fast did each drive and how long was the trip?

5. The pressure of a quantity of an ideal gas was held constant at 1400 torr. The initial volume and temperature were 1000 ml and 1700 K. If the volume were increased to 2000 ml, what would the final temperature be in kelvins?

Use the table of logarithms to write the following numbers in exponential notation with 10 as the base. Then write the number using logarithmic notation.

*6. 5.8437912   *7. $584{,}312 \times 10^4$ *8. $.00058423 \times 10^{-5}$

Graph the solution on a number line:

9. $(x + 2)(x - 3) < 0, D = \{\text{Reals}\}$ 10. $x^2 - x - 6 \geq 0, D = \{\text{Integers}\}$

11. $x^2 - 8 \leq 2x, D = \{\text{Integers}\}$ 12. $|x| - 1 \not\leq 0, D = \{\text{Integers}\}$

13. $7 \not> x - 2 < 10, D = \{\text{Reals}\}$ 14. Multiply: $(x^{1/3} + y^{2/3})(x^{2/3} + y^{1/3})$

15. Factor: $8p^6k^{15} - x^3m^6$

16. Show that $.003\overline{16}$ is a rational number by writing it as a fraction of integers.

17. Complete the square as an aid in graphing: $y = -x^2 + 4x - 1$

18. Find the number $\frac{2}{11}$ of the way from 2 to $4\frac{1}{6}$.

19. Solve $3x^2 - x - 7 = 0$ by completing the square.

Solve:

20. $\begin{cases} 1\frac{1}{5}x + \frac{2}{3}y = 30 \\ -.18x - .02y = -3.78 \end{cases}$

21. $\begin{cases} x^2 + y^2 = 4 \\ 3x - y = 2 \end{cases}$

22. $\begin{cases} x - 4y = -15 \\ 3x + z = 20 \\ 2y - z = 5 \end{cases}$

23. $\begin{cases} x - 2y - z = -8 \\ 3x - y - 2z = -5 \\ x + y + z = 9 \end{cases}$

Simplify:

24. $\dfrac{2i^3 - i}{-\sqrt{-3}\sqrt{-3} + 3}$

25. $\dfrac{\sqrt{2} - 5}{2\sqrt{2} - 4}$

26. $\sqrt[6]{xy^3}\sqrt[3]{xy^2}$

27. $\sqrt{\dfrac{4}{3}} + 2\sqrt{\dfrac{3}{4}} - 3\sqrt{48}$

Solve by factoring:

28. $2x^3 = -x^2 + 3x$ **29.** $3x^2 - 2 = x$ **30.** $-7x^2 - 2x = 3x^3$

ER 112-120, 112-121

LESSON *113* *Nonlinear inequalities*

113.A
nonlinear inequalities

We remember that a straight line divides the set of all points in the plane into three disjoint subsets. These are the set of points that lie on the line and the sets of points that lie on either side of the line. In the figure we show the graph of the equation $y = -\frac{1}{2}x - 1$. All points whose coordinates satisfy this equation lie on this line. The rest of the points lie on one side of the line or the other side of the line and will satisfy one of the following inequalities.

(a) $y > -\dfrac{1}{2}x - 1$

(b) $y < -\dfrac{1}{2}x - 1$

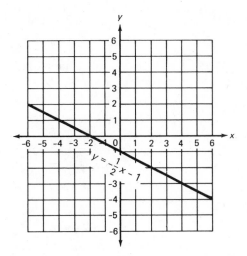

If the coordinates of a point satisfy one of these inequalities, the coordinates of all the points on the same side of the line will satisfy the same inequality. Inequality (a) is read as "y is greater than negative one-half x minus 1" and since y is "greater than" as we move up, we surmise that this inequality designates the points that lie above the line. We can always use a test point to check our surmise.

The same thoughts apply to curved lines, as will be demonstrated in the following examples.

example 113.A.1 The line and the parabola graphed in the figure divide the coordinate system into four distinct regions, which have been labeled A, B, C, and D. The coordinates of the points in which region satisfy the given system of inequalities?

(a) $\begin{cases} y \geq \frac{1}{2}x - 4 \end{cases}$ (line)

(b) $\begin{cases} y \geq -x^2 + 4x - 1 \end{cases}$ (parabola)

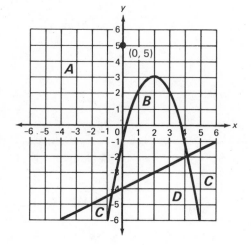

solution Inequality (a) is read as "y is greater than or equal to $\frac{1}{2}x - 4$." Since y is greater than as we go up, we suspect that this inequality designates the points on or above the line. Inequality (b) reads "y is greater than or equal to $-x^2 + 4x - 1$." Thus we suspect that this inequality designates the points on or above the parabola. The area above both the line and the parabola is area A. We will use the point $(0, 5)$ as a test point.

$$\text{Line:} \qquad y \geq \frac{1}{2}x - 4 \quad \longrightarrow \quad 5 \geq \frac{1}{2}(0) - 4 \quad \longrightarrow \quad 5 \geq -4 \qquad \text{True}$$

$$\text{Parabola:} \qquad y \geq -x^2 + 4x - 1 \quad \longrightarrow \quad 5 \geq -(0)^2 + 4(0) - 1$$
$$\longrightarrow \quad 5 \geq -1 \qquad \text{True}$$

Thus the solution is area A and includes the bordering points that lie on the line or on the parabola. The coordinates of any point in area A or on the border will satisfy both of the inequalities given.

example 113.A.2 In the figure we show the graphs of the given line and parabola. The coordinates of which of the areas will satisfy both inequalities?

(a) $\begin{cases} y \geq x^2 + 4x + 2 \end{cases}$ (parabola)

(b) $\begin{cases} y < -x + 1 \end{cases}$ (line)

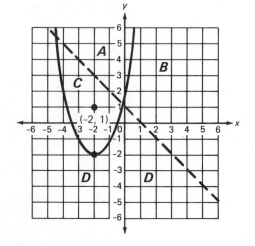

solution   The quadratic inequality reads ''y is greater than or equal to,'' which indicates the points above or on the parabola; and the linear inequality reads ''y is less than,'' which indicates the points below the line. We will use the point $(-2, 1)$ as a test point.

$$\text{Parabola:} \quad y \geq x^2 + 4x + 2 \quad \longrightarrow \quad 1 \geq (-2)^2 + 4(-2) + 2$$
$$\longrightarrow \quad 1 \geq 4 - 8 + 2 \quad \longrightarrow \quad 1 \geq -2 \quad \text{True}$$
$$\text{Line:} \quad y < -x + 1 \quad \longrightarrow \quad 1 < -(-2) + 1 \quad \longrightarrow \quad 1 < 2 + 1$$
$$\longrightarrow \quad 1 < 3 \quad \text{True}$$

Thus the coordinates of all points that lie below the line and on or above the parabola satisfy this system of inequalities and the answer is region C.

example 113.A.3   Which area of the graph satisfies this system of nonlinear inequalities?

(a) $\quad \begin{cases} x^2 + y^2 \leq 16 & \text{(circle)} \\ y \geq x & \text{(line)} \end{cases}$
(b)

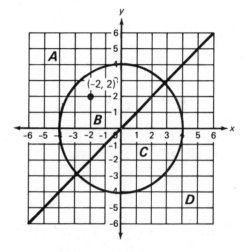

solution   The inequality $x^2 + y^2 \leq 16$ designates the points that lie on or inside the circle (try a few points to check this out), and the linear inequality designates the points on or above the line. We will use the point $(-2, 2)$ as a test point.

$$\text{Circle:} \quad x^2 + y^2 \leq 16 \quad \longrightarrow \quad (-2)^2 + (2)^2 \leq 16$$
$$\longrightarrow \quad 4 + 4 \leq 16 \quad \text{True}$$
$$\text{Line:} \quad y \geq x \quad \longrightarrow \quad 2 \geq -2 \quad \text{True}$$

This verifies our surmise, and thus the area designated is B and includes the points on the boundary of this area.

problem set 113

1. Find three consecutive even integers such that the product of the first and the third exceeds the product of the second and 8 by 16.

2. The druggist began with 240 ml of a 20% antiseptic solution. How much pure antiseptic should be added to get a solution that is 52% antiseptic?

3. The long trip was 4800 miles, and for this trip Ken used fast transportation, which took 1 hour more than the slower conveyance took to cover 2000 miles. Find the rates and times of both methods of transportation if the faster moved at twice the speed of the slower.

4. The current in the river flowed at 4 miles per hour. The steamboat could go 34 miles downstream in one-third the time it took to go 54 miles upstream. What was the speed of the boat in still water?

5. The sum of the digits of a two-digit counting number was 8. If the digits were reversed, the new number would be 54 greater than the original number. What was the original number?

In the following figures, designate the areas in which the coordinates of the points satisfy the given systems of inequalities.

*6. $\begin{cases} y \ge \frac{1}{2}x - 4 \\ y \ge -x^2 + 4x - 1 \end{cases}$ *7. $\begin{cases} y \ge x^2 + 4x + 2 \\ y < -x + 1 \end{cases}$ *8. $\begin{cases} x^2 + y^2 \le 16 \\ y \ge x \end{cases}$

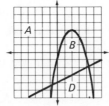   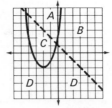   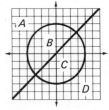

Use a table of logarithms to write the following numbers as exponentials whose base is 10. Then write the same thing using logarithmic notation. Show your work.

9. $47{,}830 \times 10^5$ 10. $.000715 \times 10^{-4}$ 11. $38{,}000 \times 10^{-15}$

12. Expand: $(x^{1/2} - y^{-1/2})^2$ 13. Factor: $m^3 p^3 + 27 x^6 y^9$

Graph the solutions on a number line:

14. $x^2 \ge -3x + 4$, $D = \{\text{Integers}\}$ 15. $x^2 - 4 \le 3x$, $D = \{\text{Integers}\}$

16. $-|x| - 3 > -7$, $D = \{\text{Integers}\}$ 17. $6 \not> x - 4 < 8$, $D = \{\text{Integers}\}$

18. Show that $.001\overline{056}$ is a rational number by writing it as a quotient of integers.

19. Complete the square as an aid in graphing: $y = x^2 - 4x + 7$

20. Find the number $\frac{3}{8}$ of the way from $4\frac{1}{2}$ to $6\frac{1}{4}$.

Solve:

21. $\begin{cases} 2\frac{1}{3}x + \frac{1}{5}y = 10 \\ .03x - .03y = -.36 \end{cases}$ 22. $\begin{cases} 3x - z = 8 \\ 2x - 2y = -4 \\ 2y + 3z = 2 \end{cases}$

23. $\begin{cases} x + y = 6 \\ xy = -1 \end{cases}$ 24. $\begin{cases} x - y + z = 3 \\ 2x - y + 2z = 9 \\ -x + y + z = 1 \end{cases}$

25. Draw the line suggested by the data points and write the equation that expresses output as a function of input.

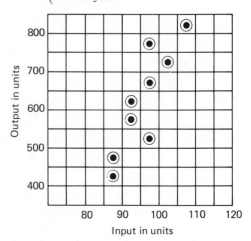

Simplify:

26. $\dfrac{i^3 - i^2}{i^5 + 2}$

27. $\sqrt{\dfrac{5}{8}} + 2\sqrt{\dfrac{8}{5}} - 3\sqrt{40}$

28. Solve: $\sqrt{x - 15} + \sqrt{x} = 5$

Solve by factoring:

29. $5x^3 + 2x = -7x^2$

30. $2 - 5x^2 = 3x$

ER 113-122, 113-123

LESSON *114* *Antilogarithms*

114.A
**exponents
to numbers**

In Lesson 112 we used tables of logarithms to find the proper exponents so that **positive numbers between 1 and 10** could be written as 10 raised **to a positive decimal power.** In this lesson, we will reverse the procedure and use the table to match **a positive decimal exponent with a positive number between 1 and 10.** Often the table will not contain the exact number we are looking for. **When this happens, we will use the number in the table that is closest to the number we are looking for.** (In later courses, we will use a process called linear interpolation when the exact number cannot be found in the table.) It is advisable to use a hand held calculator to help with the arithmetic of these problems. We advise that the log tables in the calculator not be used. The use of these tables would deny us some of the understanding that we seek. The calculator tables may be used to check the answers.

example 114.A.1 Use the portion of the table shown here to find the number represented by $10^{.7592}$.

n	0	1	2	3	4	5	6	7	8	9
5.5	.7404	.7412	.7419	.7427	.7435	.7443	.7451	.7459	.7466	.7474
5.6	.7482	.7490	.7497	.7505	.7513	.7520	.7528	.7536	.7543	.7551
5.7	.7559	.7566	.7574	.7582	.7589	.7597	.7604	.7612	.7619	.7627
5.8	.7634	.7642	.7649	.7657	.7664	.7672	.7679	.7686	.7694	.7701
5.9	.7709	.7716	.7723	.7731	.7738	.7745	.7752	.7760	.7767	.7774

solution We look for the number .7592 in the body of the table but do not find it. The closest numbers to .7592 are .7589 and .7597. We will use .7589 because this number is closer to .7592. We read the table backwards and find that the number represented is 5.74.

$$10^{.7592} = 5.74$$

example 114.A.2 Use the same table to find the number represented by $10^{6.7592}$

solution **Since the body of the table contains only positive decimal numbers, the exponential must be rewritten as a product of two exponentials—one with a positive decimal exponent and the other with an integral exponent.**

$$10^{.7592} \times 10^6$$

In the last problem, we found that the first factor represented 5.74 so our answer is

5.74×10^6

example 114.A.3   Use the following table of logarithms to find the number represented by $10^{-5.2393}$

solution   As the first step, we write the exponential as a product of two exponentials. One must have a positive decimal exponent and the other must have an integral exponent. **We always go to the next negative-integer exponent, which in this case is −6**, and find

$$10^{-5.2393} = 10^{.7607} \times 10^{-6}$$

We enter the table with .7607 and find that this exponent pairs with the number 5.76.

n	0	1	2	3	4	5	6	7	8	9
5.5	.7404	.7412	.7419	.7427	.7435	.7443	.7451	.7459	.7466	.7474
5.6	.7482	.7490	.7497	.7505	.7513	.7520	.7528	.7536	.7543	.7551
5.7	.7559	.7566	.7574	.7582	.7589	.7597	.7604	.7612	.7619	.7627
5.8	.7634	.7642	.7649	.7657	.7664	.7672	.7679	.7686	.7694	.7701
5.9	.7709	.7716	.7723	.7731	.7738	.7745	.7752	.7760	.7767	.7774

So our answer is

$$5.76 \times 10^{-6}$$

114.B
antilogarithms   In the last example, we discovered that the exponential $10^{-5.2393}$ can be written as 5.76×10^{-6}.

$$\text{exponent or logarithm} \diagup 10^{-5.2393} = 5.76 \times 10^{-6} \quad \text{number}$$

We say that −5.2393 is the exponent (logarithm) of the number 5.76×10^{-6}, and we say that the number 5.76×10^{-6} is the antilogarithm of the exponent −5.2393.

example 114.B.1   Use the following table to find the antilogarithm of −6.8717.

n	0	1	2	3	4	5	6	7	8	9
1.0	.0000	.0043	.0086	.0128	.0170	.0212	.0253	.0294	.0334	.0374
1.1	.0414	.0453	.0492	.0531	.0569	.0607	.0645	.0682	.0719	.0755
1.2	.0792	.0828	.0864	.0899	.0934	.0969	.1004	.1038	.1072	.1106
1.3	.1139	.1173	.1206	.1239	.1271	.1303	.1335	.1367	.1399	.1430
1.4	.1461	.1492	.1523	.1553	.1584	.1614	.1644	.1673	.1703	.1732

solution   If we are asked for the antilogarithm of −6.8717, then −6.8717 must be a logarithm. Thus, we have

$$10^{-6.8717}$$

which we can write as a product, one of whose exponents is a positive decimal number, as

$$10^{.1283} \times 10^{-7}$$

We enter the table with the positive decimal number .1283 and find that the closest number to this is .1271, which is paired with 1.34. Thus, our answer is

$$1.34 \times 10^{-7}$$

problem set 114

1. A chemist had a container of the compound $KMnO_4$ that weighed 790 grams. What was the weight of the oxygen in the container? (K, 39; Mn, 55; O, 16)

2. Marsha rode for a while at 20 miles per hour with Bertha and completed the 280-mile trip by riding with Sherri at 45 miles per hour. If the total trip took 9 hours, how far did she ride with each?

3. Hathaway flew in the fast airplane, traveling 1200 miles in only 1 more hour than it took Beauregard to travel 480 miles. If Hathaway's speed was twice that of Beauregard, find the rates and times of both.

4. There were 14 nickels, dimes, and quarters whose total value equaled $1.05. How many coins of each kind were there if there were 3 times as many dimes as quarters?

5. A two-digit counting number has a value that is 4 times the sum of its digits. If 4 times the units' digit is 14 greater than the tens' digit, what is the number?

Use the table of logarithms to write the following numbers as exponentials whose base is 10. Then write the same thing using logarithmic notation. Show your work.

6. .00071623   7. $482,517 \times 10^4$ 8. $.007 \times 10^{-2}$

Use the table of logarithms to write the following exponentials as standard base 10 numerals. Show your work.

*9. $10^{.7592}$ *10. $10^{-5.2393}$ 11. $10^{-7.0523}$

Find the antilogarithms. Show your work.

*12. -6.8717 13. 3.0260

14. The coordinates of the points in which of the areas designated in the figure satisfy the given system of inequalities?
$$\begin{cases} x^2 + y^2 \le 16 & \text{(circle)} \\ y \le x^2 + 4x + 2 & \text{(parabola)} \end{cases}$$

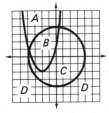

Graph the solutions on a number line:

15. $(x + 5)(x - 2) > 0, D = \{\text{Reals}\}$ 16. $x^2 < -7x - 10, D = \{\text{Reals}\}$

17. $|x| + 2 \not> 4, D = \{\text{Integers}\}$ 18. $6 \not\ge x - 5 \not\ge 10, D = \{\text{Integers}\}$

19. Expand: $(x^{1/4} - y^{-1/4})^2$ 20. Factor: $x^3m^{15} - 8p^3y^6$

21. Show that $1.0\overline{47}$ is a rational number by writing it as a fraction of integers.

22. Complete the square as an aid in graphing: $y = x^2 + 2x + 1$

23. Find the number $\frac{1}{10}$ of the way from $5\frac{2}{3}$ to $6\frac{1}{2}$.

Solve:

24. $\begin{cases} 1\frac{2}{3}x - 2\frac{1}{4}y = -7 \\ -.2x + .05y = -1.8 \end{cases}$ 25. $\begin{cases} 4x - z = 12 \\ x - 3y = -10 \\ 3y + z = 8 \end{cases}$

26. $\begin{cases} x + y - z = 4 \\ 2x + y + z = 10 \\ 3x + y + z = 14 \end{cases}$

Simplify:

27. $3i^2 - 2i^5 - \sqrt{-2}\sqrt{-2} + (i + 2)(i - 3)$

28. $\dfrac{4\sqrt{2} + 1}{1 - 3\sqrt{2}}$

29. Solve $-3x^2 = -x - 6$ by completing the square.

30. Solve $3x^2 + 2x = 8$ by factoring.

ER 114-124, 114-125

LESSON 115 *Power of the hydrogen*

115.A
Power of the hydrogen

A logarithmic function is used in chemistry to determine the relative acidity of a liquid. The relative acidity is called pH, is written with a small p and a large H. This symbol combines the letter H for hydrogen and the letter p from the Latin word *potentia* which means power or strength. We use pH as a measure of the relative concentration of hydrogen ions in a liquid. The formula can be written in exponential notation or in logarithmic notation.

<table>
<tr><td>EXPONENTIAL NOTATION</td><td>LOGARITHMIC NOTATION</td></tr>
<tr><td>$10^{-\text{pH}} = \text{H}^+$</td><td>$\text{pH} = -\log \text{H}^+$</td></tr>
</table>

The pH of liquids ranges from about 0 to $+14$, and the pH of water, considered to be neutral, is 7. Since investigation of the theory of pH is a topic for chemistry, we will limit our discussion and concentrate on the mathematical problem of finding H^+ if given pH and of finding pH if given H^+.

example 115.A.1

Find the pH of a solution when the concentration of hydrogen ions is .00204 moles per liter.

solution

We will not concern ourselves with the units. We assume that they are correct. We will use the relationship

$$10^{-\text{pH}} = \text{H}^+$$

and replace H^+ with .00204

$$10^{-\text{pH}} = .00204 = 2.04 \times 10^{-3}$$

Next we use the table to write 2.04 as 10 raised to a power.

$\downarrow$

n	0	1	2	3	4	5	6	7	8	9
2.0	.3010	.3032	.3054	.3075	.3096	.3118	.3139	.3160	.3181	.3201
2.1	.3222	.3243	.3263	.3284	.3304	.3324	.3345	.3365	.3385	.3404
2.2	.3424	.3444	.3464	.3483	.3502	.3522	.3541	.3560	.3579	.3598

$$10^{-\text{pH}} = 10^{.3096} \times 10^{-3} = 10^{-2.6904}$$

Now if these quantities are equal, then the exponents must be equal, so

$$-pH = -2.6904$$

and we solve for pH by multiplying both sides by (-1) and we get

$$pH = 2.6904$$

example 115.A.2   Find the pH of a solution when the concentration of hydrogen ions is 4.36×10^{-3} moles per liter.

solution   Again we do not concern ourselves with units but replace H^+ in the formula with 4.36×10^{-3}.

$$10^{-pH} = 4.36 \times 10^{-3}$$

To solve, we write 4.36×10^{-3} as 10 raised to a power.

n	0	1	2	3	4	5	6	7	8	9
4.0	.6021	.6031	.6042	.6053	.6064	.6075	.6085	.6096	.6107	.6117
4.1	.6128	.6138	.6149	.6160	.6170	.6180	.6191	.6201	.6212	.6222
4.2	.6232	.6243	.6253	.6263	.6274	.6284	.6294	.6304	.6314	.6325
4.3	.6335	.6345	.6355	.6365	.6375	.6385	.6395	.6405	.6415	.6425
4.4	.6435	.6444	.6454	.6464	.6474	.6484	.6493	.6503	.6513	.6522

$$4.36 \times 10^{-3} = 10^{.6395} \times 10^{-3} = 10^{-2.3605}$$

So now we can write

$$10^{-pH} = 10^{-2.3605}$$

Thus, $-pH$ must equal -2.3605, and therefore

$$pH = 2.3605$$

example 115.A.3   Find the concentration of hydrogen ions in a solution whose pH is 7.05.

solution   We use the same formula, but this time we replace pH with the number 7.05

$$10^{-7.05} = H^+$$

To find the number represented by the exponential, we write it as a product of exponentials,

$$10^{.95} \times 10^{-8} = H^+$$

and we can find the approximate value of $10^{.9500}$ by using .9499 as the exponent.

n	0	1	2	3	4	5	6	7	8	9
8.7	.9395	.9400	.9405	.9410	.9415	.9420	.9425	.9430	.9435	.9440
8.8	.9445	.9450	.9455	.9460	.9465	.9469	.9474	.9479	.9484	.9489
8.9	.9494	.9499	.9504	.9509	.9513	.9518	.9523	.9528	.9533	.9538

And we get **$8.91 \times 10^{-8} = H^+$**.

example 115.A.4   Find the concentration of hydrogen ions in a solution whose pH is 3.4.

solution   We replace pH in the formula with 3.4 and get

$$10^{-3.4} = H^+$$

Now we write the exponential as a product as follows.

$$10^{.6000} \times 10^{-4} = H^+$$

and we read out from .5999 in the table

n	0	1	2	3	4	5	6	7	8	9
3.7	.5682	.5694	.5705	.5717	.5729	.5740	.5752	.5763	.5775	.5786
3.8	.5798	.5809	.5821	.5832	.5843	.5855	.5866	.5877	.5888	.5899
⟶ 3.9	.5911	.5922	.5933	.5944	.5955	.5966	.5977	.5988	.5999	.6010

and get

$$3.98 \times 10^{-4} = H^+$$

problem set 115

1. Find three consecutive multiples of 6 such that 6 times the sum of the first and the third is 84 less than 10 times the second.

2. The pressure of a quantity of an ideal gas was held constant at 600 torr. The initial volume and temperature were 400 ml and 800 K. What was the final temperature in kelvins when the volume was decreased to 200 ml?

3. The Natchez Belle could go 65 miles downstream in 5 hours, but it took her 8 hours to go 56 miles upstream. What was her speed in still water and what was the speed of the current in the river?

4. The number of tomatoes varied jointly as the rain and as the fertilizer squared. If 1000 tomatoes resulted from 2 inches of rain and 1 ton of fertilizer, how many tomatoes would be caused by 2 tons of fertilizer and 1 inch of rain? Solve two ways.

5. A two-digit counting number has a value that is 7 greater than twice the sum of its digits. If the units' digit is 3 greater than 3 times the tens' digit, what is the number?

Find the pH (pH $= -\log H^+$ or $10^{-pH} = H^+$) of the solution when the concentration of hydrogen ions (H^+) in moles per liter is as follows. Show your work.

*6. .00204   7. .00142   8. .0032

Find the concentration of hydrogen ions (H^+) in moles per liter when the pH of the liquid is as follows. Show your work.

*9. 7.05   *10. 3.4   11. 1.060

12. The coordinates of the points in which of the areas designated in the figure satisfy the given system of inequalities?

$$\begin{cases} y \ge \frac{2}{3}x + 2 & \text{(line)} \\ y \ge x^2 + 2x & \text{(parabola)} \end{cases}$$

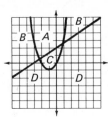

13. The figure shown is the base of a container 6 feet high. Find the volume of the container in cubic inches. Dimensions are in feet.

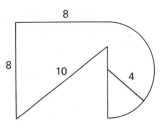

14. Find B.

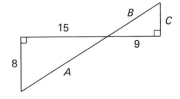

Graph the solutions on a number line:

15. $x^2 - 3 \geq -2x$, $D = \{\text{Integers}\}$

16. $x^2 + 2x < 3$, $D = \{\text{Integers}\}$

17. $-|x| + 2 \nleq -1$, $D = \{\text{Integers}\}$

18. Show that $1.04\overline{76}$ is a rational number by writing it as a fraction of integers.

19. Complete the square as an aid in graphing: $y = -x^2 - 4x - 6$

20. Expand: $(x^{1/2} - y^{3/4})^2$

21. Factor: $27m^9p^3 - x^{12}y^3$

22. Find the number that is $\frac{2}{9}$ of the way from $\frac{1}{4}$ to $3\frac{1}{2}$.

Solve:

23. $\begin{cases} \dfrac{3}{5}x - \dfrac{1}{7}y = 6 \\ -.21x + .02y = -2.73 \end{cases}$

24. $\begin{cases} x - 2z = 7 \\ y + 2z = -9 \\ -x + 2y = -7 \end{cases}$

25. $\begin{cases} 2x - y + 2z = -9 \\ 2x + 2y + z = -15 \\ x - 2y + z = 0 \end{cases}$

Simplify:

26. $\dfrac{2i^3 - 3}{1 - \sqrt{-4}\sqrt{4}}$

27. $\dfrac{3 - \sqrt{24}}{2 - \sqrt{6}}$

28. $\dfrac{(a^{\frac{x}{2} - 2})^2 m^x}{(m^2 a^2)^{x/4}}$

29. Solve: $\sqrt{p} = 9 - \sqrt{p - 45}$

30. Solve $3x^3 = x^2 + 2x$ by factoring.

ER 115-126, 115-127

LESSON 116   *Set-builder notation*

116.A
letter symbols for sets

The most commonly encountered sets of numbers are the sets of natural or counting numbers, whole numbers, integers, rational numbers, irrational numbers, real numbers, and complex numbers. While most mathematicians have agreed on the definitions of and composition of these sets of numbers, they have not been able to agree totally on a capital letter to assign to each set. The capital letters shown below are those that seem to be used most frequently.

$$N = \{1, 2, 3, \ldots\} \qquad \text{Natural or counting numbers}$$

The set of natural numbers is a subset of the set of complex numbers, has an infinite number of members, and is a subset of all the sets listed above except the set of irrational numbers.

$$W = \{0, 1, 2, 3, \ldots\} \qquad \text{Whole numbers}$$

We note that the set of whole numbers contains the set of natural numbers and also contains the number zero.

$$J = \{\ldots, -3, -2, -1, 0, 1, 2, 3, \ldots\} \qquad \text{Integers}$$

The set of integers contains the set of whole numbers and also contains the negative of each member of the set of natural numbers

$$Q = \{\text{Rational numbers}\}$$

The set of rational numbers contains the set of integers and also contains all other numbers that can be written as a fraction (quotient) of integers. Of course, any integer can be written as a fraction of integers. For example, -4 can be written as $\frac{40}{-10}$ or $\frac{240}{-60}$. Both represent the number -4. The rational numbers can also be described as real numbers that can be represented by a terminating or repeating decimal numeral.

$$P = \{\text{Irrational numbers}\}$$

The set of irrational numbers consists of all numbers that cannot be written as fractions of integers. Numbers such as $\sqrt{2}$, π, $\sqrt[3]{4}$, $\sqrt[5]{3}$ are examples of the infinite number of numbers that cannot be written as fractions of integers and are therefore irrational numbers. The irrational numbers can also be described as real numbers that cannot be represented by repeating or terminating decimal numerals. The decimal representation of these numbers consists of nonrepeating decimal numerals of infinite length.

$$R = \{\text{Real numbers}\}$$

The set of real numbers consists of all members of the set of rational numbers and all members of the set of irrational numbers. A real number can be thought of as any number that does not have i as a factor. Any real number can be paired with a unique point on the number line and any point on the number line can be paired with a unique real number. **To use simpler, nonrigorous language, we may say that any number on the number line is a real number.**

$$C = \{\text{Complex numbers}\}$$

The set of complex numbers consists of all numbers of the form $a + bi$, where a and b are real numbers. Any real number is also a complex number. For instance, the number 3 may be thought of as being in the form $a + bi$ where the value of b is zero; thus $3 + 0i$ equals 3.

116.B
set-builder notation

Use of the **set-builder notation** allows us to describe a set completely and exactly. Its compactness appeals to mathematicians, and it is used extensively in more advanced courses in mathematics. The first component is the leading half of the set of braces. This is read as *the set whose members are*. The second component is a statement about the variable and is followed by the third component, a vertical line that is read *such that*. This is followed by one or more restrictions on the variable and the last component is the terminal half of the set of braces.

$$A = \{x \in J \mid x + 2 > 4\}$$

This would be read, as: "A is the set whose members x are integers, such that x plus 2 is greater than 4." Previously in this book we would have described this set by writing:

$$x + 2 > 4, \qquad D = \{\text{Integers}\}$$

and the graph of the solution set would be

The set-builder notation can also be used to designate ordered pairs and to state the restrictions on the ordered pairs.

example 116.B.1 Graph the solution of $\{(x, y) \in R \mid y > x + 2 \text{ and } y < -x\}$.

solution This is read as: "the set whose members are ordered pairs of x and y where x and y are real numbers such that $y > x + 2$ and $y < -x$." Previously this problem would have been stated as

Graph the solution set to this system of linear inequalities:

$$\begin{cases} y > x + 2 \\ y < -x \end{cases} \qquad D = \{\text{Reals}\}$$

Thus, we see that the set-builder notation is just another way to describe a particular set, and the reader should not be confused by the notation. The graph of the solution set of this problem is shown here.

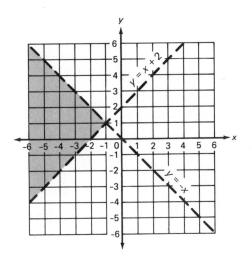

1. There were 16 nickels, dimes, and quarters whose total value was $1.50. How many coins of each kind were there if there were 3 times as many dimes as quarters?

2. The dealer paid $4000 for the car and sold it for $5000. What was the percent markup of the purchase price and what was the percent markup of the selling price?

3. Gabriel could drive the 200 miles to the seashore in one-half the time it took Martha to drive the 650 miles to the mountains. Find the speeds and times of both if Martha's speed was 25 miles per hour greater than that of Gabriel.

4. The two containers were side by side on the shelf. The first one was 30% medicine and the second one was 60% medicine. How much of each sould have been used to mix 600 ml of a solution that was 39% medicine?

5. Kenneth and David rebelled at paying \$4320 because this was a markup of 80 percent of the selling price. What was the markup and what did the store pay for the item?

Find the pH of the solution when the concentration of hydrogen ions (H^+) in moles per liter is as follows: ($pH = -\log H^+$ or $10^{-pH} = H^+$)

6. .00263   7. 11.7×10^{-5} 8. .0046

Find the concentration of hydrogen ions (H^+) in moles per liter when the pH of the liquid is as follows:

9. 5.34   10. 2.016   11. 4.0424

12. The coordinates of the points in which of the areas designated in the figure satisfy the given system of inequalities?
$$\begin{cases} x^2 + y^2 \geq 16 & \text{(circle)} \\ y \leq -x^2 + 2 & \text{(parabola)} \end{cases}$$

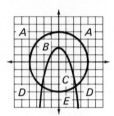

13. The figure shown is the base of a container 10 meters high. The dimensions are in meters. Find the volume in cubic centimeters.

14. Find C.

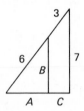

Graph the solutions:

*15. $A = \{x \in J \mid x + 2 > 4\}$ 16. $B = \{x \in R \mid x^2 \geq -4x - 3\}$

17. $C = \{x \in R \mid x^2 + 4x < -3\}$ 18. $D = \{x \in J \mid |x| + 2 \leq 3\}$

*19. $\{(x, y) \in R \mid y > x + 2 \text{ and } y < -x\}$

20. Show that $2.04\overline{25}$ is a rational number by writing it as a fraction of integers.

21. Complete the square as an aid in graphing: $y = -x^2 + 4x - 2$

22. Expand: $(x^{1/2} - y^{-1/2})^2$ 23. Factor: $m^3 - 8p^6k^9$

24. Find the number that is $\frac{4}{9}$ of the way from $\frac{3}{5}$ to $1\frac{2}{5}$.

Solve:

25. $\begin{cases} \dfrac{5}{8}x - \dfrac{2}{3}y = 4 \\ -.15x + .2y = -.6 \end{cases}$ 26. $\begin{cases} 3x + y = 2 \\ 2x - z = 0 \\ 2y + z = -4 \end{cases}$

Simplify:

27. $\dfrac{\sqrt{-2}\sqrt{-2} - 3i^3}{i + 2}$ 28. $\dfrac{2\sqrt{2} - 4}{2\sqrt{4} - \sqrt{2}}$ 29. $\sqrt[3]{x^5} \ \sqrt[4]{x^3y}$

30. Solve $6 = 2x^2 - 11x$ by factoring.

ER 116-128, 116-129

LESSON *117* *Operations with exponentials*

117.A
operations with exponentials

The exponential function and the logarithmic function relate the same variables and give the same information. We know that ten squared equals one hundred.

$$10^2 = 100$$

We make the same statement using the language of logarithms if we say that the exponent (logarithm) of 100 to the base 10 is 2.

$$\log_{10} 100 = 2$$

The exponential function and the logarithmic function are used extensively in calculus and other advanced mathematics courses. Again we note that in this book we will do calculations with these functions with the aim of increasing our understanding of exponentials and logarithms, and not with the aim of learning a new way of calculating. We will continue to use calculators to help speed up arithmetic operations, but for the time being will avoid their use for logarithms. We will use the tables instead.

We multiply exponentials with like bases by adding the exponents.

$$10^5 \cdot 10^2 = 10^7$$

and we divide exponentials with like bases by subtracting the exponent of the divisor.

$$\frac{10^5}{10^2} = 10^{5-2} = 10^3$$

We raise exponentials to powers by multiplying the exponents.

$$(10^5)^7 = 10^{35}$$

We perform calculations with logarithms by first writing the numbers in exponential notation and then using these three rules for exponents as indicated.

example 117.A.1   Use logarithms (exponents) as required to multiply $(146{,}321)(12{,}780 \times 10^{-14})$. Begin by writing both numbers as exponentials whose base is 10.

solution   The table has only three digits, so as the first step we write each number in scientific notation and round off to three digits.

$$146{,}321 = 1.46 \times 10^5$$

$$12{,}780 \times 10^{-14} = 1.28 \times 10^{-10}$$

Next we use the table to write each number as an exponential whose base is 10.

n	0	1	2	3	4	5	6	7	8	9
1.0	.0000	.0043	.0086	.0128	.0170	.0212	.0253	.0294	.0334	.0374
1.1	.0414	.0453	.0492	.0531	.0569	.0607	.0645	.0682	.0719	.0755
⟶ 1.2	.0792	.0828	.0864	.0899	.0934	.0969	.1004	.1038	.1072	.1106
1.3	.1139	.1173	.1206	.1239	.1271	.1303	.1335	.1367	.1399	.1430
⟶ 1.4	.1461	.1492	.1523	.1553	.1584	.1614	.1644	.1673	.1703	.1732

$$1.46 \times 10^5 = 10^{.1644} \times 10^5 = 10^{5.1644}$$

$$1.28 \times 10^{-10} = 10^{.1072} \times 10^{-10} = 10^{-9.8928}$$

Now we multiply the exponentials by adding the exponents (logarithms).

$$10^{5.1644} \times 10^{-9.8928} = 10^{-4.7284}$$

We write this exponential as a product by writing

$$10^{.2716} \times 10^{-5}$$

And we finish by using the table to write $10^{.2716}$ as 1.87

n	0	1	2	3	4	5	6	7	8	9
1.0	.0000	.0043	.0086	.0128	.0170	.0212	.0253	.0294	.0334	.0374
1.1	.0414	.0453	.0492	.0531	.0569	.0607	.0645	.0682	.0719	.0755
1.2	.0792	.0828	.0864	.0899	.0934	.0969	.1004	.1038	.1072	.1106
1.3	.1139	.1173	.1206	.1239	.1271	.1303	.1335	.1367	.1399	.1430
1.4	.1461	.1492	.1523	.1553	.1584	.1614	.1644	.1673	.1703	.1732
1.5	.1761	.1790	.1818	.1847	.1875	.1903	.1931	.1959	.1987	.2014
1.6	.2041	.2068	.2095	.2122	.2148	.2175	.2201	.2227	.2253	.2279
1.7	.2304	.2330	.2355	.2380	.2405	.2430	.2455	.2480	.2504	.2529
⟶ 1.8	.2553	.2577	.2601	.2625	.2648	.2672	.2695	.2718	.2742	.2765
1.9	.2788	.2810	.2833	.2856	.2878	.2900	.2923	.2945	.2967	.2989

and we get **1.87×10^{-5}**.

example 117.A.2   Use logarithms as required to simplify:

$$\frac{(21,482 \times 10^{-7})(56,541 \times 10^{5})}{57,143}$$

solution   We begin by rounding off and writing each of the numbers as an exponential whose base is 10.

n	0	1	2	3	4	5	6	7	8	9
2.0	.3010	.3032	.3054	.3075	.3096	.3118	.3139	.3160	.3181	.3201
⟶ 2.1	.3222	.3243	.3263	.3284	.3304	.3324	.3345	.3365	.3385	.3404
2.2	.3424	.3444	.3464	.3483	.3502	.3522	.3541	.3560	.3579	.3598

n	0	1	2	3	4	5	6	7	8	9
5.5	.7404	.7412	.7419	.7427	.7435	.7443	.7451	.7459	.7466	.7474
⟶ 5.6	.7482	.7490	.7497	.7505	.7513	.7520	.7528	.7536	.7543	.7551
⟶ 5.7	.7559	.7566	.7574	.7582	.7589	.7597	.7604	.7612	.7619	.7627

$$21,482 \times 10^{-7} \approx 2.15 \times 10^{-3} = 10^{.3324} \times 10^{-3} = 10^{-2.6676}$$

$$56,541 \times 10^{5} \approx 5.65 \times 10^{9} = 10^{.7520} \times 10^{9} = 10^{9.7520}$$

$$57,143 \approx 5.71 \times 10^{4} = 10^{.7566} \times 10^{4} = 10^{4.7566}$$

Now we replace each of the numbers with its equivalent exponential.

$$\frac{10^{-2.6676} \times 10^{9.7520}}{10^{4.7566}}$$

and simplify by using the rules for exponents.

$$10^{-2.6676+9.7520-4.7566} = 10^{2.3278}$$

We write this exponential in two parts as

$$10^{.3278} \times 10^2$$

and use the table to finish, and we get

n	0	1	2	3	4	5	6	7	8	9
2.0	.3010	.3032	.3054	.3075	.3096	.3118	.3139	.3160	.3181	.3201
2.1	.3222	.3243	.3263	.3284	.3304	.3324	.3345	.3365	.3385	.3404
2.2	.3424	.3444	.3464	.3483	.3502	.3522	.3541	.3560	.3579	.3598
2.3	.3617	.3636	.3655	.3674	.3692	.3711	.3729	.3747	.3766	.3784
2.4	.3802	.3820	.3838	.3856	.3874	.3892	.3909	.3927	.3945	.3962

$$\mathbf{2.13 \times 10^2}$$

example 117.A.3   Use logarithms to simplify $(21{,}482 \times 10^{-7})^{2/5}$.

solution   We begin by writing $21{,}482 \times 10^{-7}$ as an exponential, which we did in the last problem.

$$(10^{-2.6676})^{2/5}$$

Now we multiply the exponents and get

$$10^{-1.067}$$

which we write as a product

$$10^{.9330} \times 10^{-2}$$

and find .9330 in the table corresponds to 8.57.

n	0	1	2	3	4	5	6	7	8	9
8.5	.9294	.9299	.9304	.9309	.9315	.9320	.9325	.9330	.9335	.9340
8.6	.9345	.9350	.9355	.9360	.9365	.9370	.9375	.9380	.9385	.9390
8.7	.9395	.9400	.9405	.9410	.9415	.9420	.9425	.9430	.9435	.9440
8.8	.9445	.9450	.9455	.9460	.9465	.9469	.9474	.9479	.9484	.9489
8.9	.9494	.9499	.9504	.9509	.9513	.9518	.9523	.9528	.9533	.9538

Our answer is

$$\mathbf{8.57 \times 10^{-2}}$$

problem set 117

1. Seven-sixteenths of the assembled throng squatted in place. If 6399 did not squat in place, how many did squat in place?

2. The jar was half full of the compound C_2H_4O. The total weight of the compound was 396 grams. What was the weight of the hydrogen (H) in the jar? (C, 12; H, 1; O, 16)

3. The class had a collection of nickels and quarters that totaled 18 coins and that had a value of $2.70. How many were nickels and how many were quarters?

4. The current in the Ogeechee River flowed at 5 miles per hour. The boat could go 160 miles downstream in twice the time it took to go 40 miles upstream. What was the speed of the boat in still water and what were the times?

5. A two-digit counting number has a value that is 1 greater than 8 times the sum of its digits. If 3 times the tens' digit is 11 greater than the units' digit, what is the number?

Use logarithms as required to perform the indicated operations. Begin by writing each number as an exponential whose base is 10. Show your work.

*6. $(146,321)(12,780 \times 10^{-14})$

*7. $\dfrac{(21,482 \times 10^{-7})(56,541 \times 10^{5})}{(57,143)}$

*8. $(21,482 \times 10^{-7})^{2/5}$

9. $(.00168 \times 10^{-5})^{.213}$

Find the pH ($pH = -\log H^+$ or $10^{-pH} = H^+$) of the solution when the concentration of hydrogen ions (H^+) in moles per liter is:

10. 3.142×10^{-3}

11. 10.08×10^{-8}

12. $.0037$

Find the concentration of hydrogen ions (H^+) in moles per liter when the pH of the liquid is:

13. 5.042

14. 2.43

15. 8.9508

16. Find the surface area of this prismatic storage bin in square inches. Dimensions are in feet.

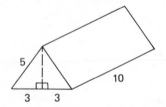

17. Show that $.001\overline{63}$ is a rational number by writing it as a fraction of integers.

18. Complete the square as an aid in graphing: $y = -x^2 - 6x - 10$

Graph the solutions.

19. $\{x \in J \mid |x| - 3 > -7\}$

20. $\{(x, y) \in R \mid x + y \geq -3 \text{ and } x - 2y < -4\}$

21. $\{x \in J \mid x^2 \geq 2x + 3\}$

22. $\{x \in J \mid x^2 - 3 < 2x\}$

23. Solve the equation $6x + x^2 = -10$ by completing the square.

24. Expand: $(x^{1/4} - y^{1/4})^2$

25. Factor: $x^6 y^3 - 27 p^6 m^9$

26. Find the number that is $\frac{3}{8}$ of the way from $4\frac{1}{5}$ to $6\frac{1}{10}$.

27. Solve: $\begin{cases} 3x - y - z = 9 \\ 2x + y - z = 12 \\ 2x - y + z = 0 \end{cases}$

Simplify:

28. $\dfrac{-2i^3 + 2}{i - i^2}$

29. $\sqrt[5]{27\sqrt{3}}$

30. $\dfrac{\sqrt{2} - 5}{3 - 2\sqrt{2}}$

ER 117-130, 117-131

LESSON 118 *Absolute value inequalities*

118.A
absolute value inequalities

The absolute value notation with a variable usually implies two answers.[†] For instance,

$$|\text{Something}| = 4, \qquad D = \{\text{Reals}\}$$

has two values that satisfy, for both $+4$ and -4 have an absolute value of 4.

An absolute value statement of **less than** tells us that the value of the variable lies between a positive number and a negative number. For instance,

$$|\text{Something}| < 4, \qquad D = \{\text{Reals}\}$$

tells us that the value of something is between 4 and -4 and is described by the conjunction

$$\text{Something} > -4 \qquad \textbf{and} \qquad \text{Something} < 4$$

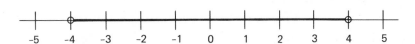

Thus, we see that an absolute value statement of *less than* can be replaced with two statements that do not contain absolute value but place the same restrictions on the variable.

example 118.A.1   Graph $\{x \in R \,\big|\, |x - 2| < 4\}$.

solution   We know from the discussion above if the absolute value of something is less than 4, then something is greater than -4 **and** something is less than 4.

$$\text{Something} > -4 \qquad \textbf{and} \qquad \text{Something} < 4$$

But in this problem, something is $x - 2$. So we replace something with $x - 2$ and solve.

$$
\begin{array}{ccc}
x - 2 > -4 & \textbf{and} & x - 2 < 4 \\
\underline{2\quad\ \ 2} & & \underline{2\quad\ 2} \\
x > -2 & \textbf{and} & x < 6
\end{array}
$$

Thus, all real numbers greater than -2 and less than 6 will satisfy the given inequality.

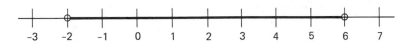

Sometimes it helps understanding to check one of the solutions. From the graph, we see that $+3$ satisfies the condition. If we replace x in the original problem with $+3$, we get

$$|(3) - 2| < 4$$

$$|1| < 4$$

$$1 < 4 \qquad \text{True}$$

[†] The absolute value of zero is zero so $|x| = 0$ has only one answer.

example 118.A.2   Graph $\{x \in J \mid |x + 2| \leq 3\}$.

solution   This problem in the old notation would have been written as

$$\text{Graph} \qquad |x + 2| \leq 3, \qquad D = \{\text{Integers}\}$$

Here our "something" is less than or equal to 3, so the absolute value inequality can be replaced with the conjunction.

$$\text{Something} \geq -3 \qquad \textbf{and} \qquad \text{something} \leq 3$$

and since our something is $x + 2$, we get

$$
\begin{array}{ccc}
x + 2 \geq -3 & \textbf{and} & x + 2 \leq 3 \\
\underline{-2 -2} & & \underline{-2 -2} \\
x \geq -5 & \textbf{and} & x \leq 1
\end{array}
$$

Thus, our solution consists of the integers that are greater than or equal to -5 and that are also less than or equal to 1.

118.B

greater than   **When the absolute value is greater than a given positive number, then a disjunction is implied.** The disjunction consists of two statements of greater than, neither of which use the absolute value notation. For instance, if

$$|\text{Something}| > 3, \qquad D = \{\text{Reals}\}$$

then something is less than -3 *or* is greater than 3.

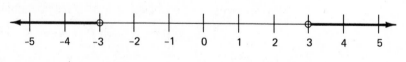

$$\text{Something} < -3 \qquad \textbf{or} \qquad \text{Something} > 3$$

example 118.B.1   Graph $|x - 2| > 3$, $D = \{\text{Reals}\}$.

solution   This is the same statement as the statement above except here something is $x - 2$. If we replace something with $x - 2$, we get

$$
\begin{array}{ccc}
x - 2 < -3 & \textbf{or} & x - 2 > 3 \\
\underline{2 2} & & \underline{2 2} \\
x < -1 & \textbf{or} & x > 5
\end{array}
$$

The graph of the solution is

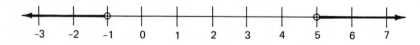

All of the numbers shown used for x satisfy the original inequality

$$|x - 2| > 3$$

example 118.B.2   Graph $\{x \in J \mid |x + 2| > 4\}$.

solution   If the absolute value of something is greater than 4, then something is less than -4 *or* is greater than 4.

$$\text{Something} < -4 \quad \textbf{or} \quad \text{Something} > 4$$

Since our something is $x + 2$, we get

$$
\begin{array}{ccc}
x + 2 < -4 & \textbf{or} & x + 2 > \quad 4 \\
\underline{-2 \quad -2} & & \underline{-2 \quad -2} \\
x < -6 & \textbf{or} & x \quad > \quad 2
\end{array}
$$

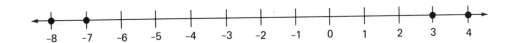

Thus, the graph indicates all integers that are less than -6 **or** are greater than 2.

problem set 118

1. The pressure of a quantity of an ideal gas was held constant at 475 torr. The initial volume and temperature were 500 ml and 700 K. If the temperature had been raised to 2100 K, what would the final volume have been?

2. Gomez could drive the 250 miles to the mountains in one-half the time it took de la Tore to drive the 400 miles to the seashore. Gomez drove 10 miles per hour faster than de la Tore. Find the rates and times of both.

3. Pinks varied inversely as blacks squared and directly as whites. Two pinks and 10 blacks went with 4 whites. How many pinks went with 1 black and 20 whites? Use both ratio and variation formats.

4. There were 24 nickels, dimes, and quarters whose total value equaled $4.25. How many coins of each kind were there if the number of nickels equaled the number of dimes?

5. The sum of the digits of a two-digit counting number was 5. If the digits were reversed, the new number would be 27 less than the original number. What was the original number?

Graph the solutions on a number line:

*6. $\{x \in R \mid |x - 2| < 4\}$

*7. $\{x \in J \mid |x + 2| \le 3\}$

*8. $\{x \in R \mid |x - 2| > 3\}$

*9. $\{x \in J \mid |x + 2| > 4\}$

10. $\{x \in R \mid x^2 - 5 > 4x\}$

11. $\{x \in R \mid x^2 - 4x < 5\}$

12. $\{x \in J \mid -10 \not> x + 3 < -6\}$

Use logarithms as required to perform the following operations. Begin by writing each number as an exponential whose base is 10. Show your work.

13. $\dfrac{4165 \times 10^5}{.0725 \times 10^{-4}}$

14. $(4166 \times 10^5)(.0725 \times 10^{-4})$

15. $(4166 \times 10^7)^{1/7}$

Find the pH ($\text{pH} = -\log H^+$ or $10^{-\text{pH}} = H^+$) of the solution when the concentration of hydrogens ions (H^+) in moles per liter is:

16. 2.07×10^{-10}

17. 9.52×10^{-12}

18. $.017$

Find the concentration of hydrogen ions (H^+) in moles per liter when the pH of the liquid is:

19. 6.041   **20.** 2.23   **21.** 1.48

22. Convert 5000 liters per minute to cubic inches per second.

23. Find the surface area of this right circular cylinder in square centimeters. Dimensions are in meters.

24. Show that $.01\overline{3}$ is a rational number by writing it as a fraction of integers.

25. Complete the square as an aid in graphing: $y = -x^2 - 2x + 1$

26. Begin with $ax^2 + bx + c = 0$ and derive the quadratic formula by completing the square.

27. Find the number that is $\frac{1}{5}$ of the way from $3\frac{1}{3}$ to $5\frac{2}{5}$.

Simplify:

28. $\dfrac{2i^3 - 2}{3 - \sqrt{-2\sqrt{2}}}$ **29.** $\dfrac{(a^2)^{b/2} x^b}{(ax)^{b/2}}$

30. Solve $3x^3 - 3x^2 = 6x$ by factoring.

ER 118-132 118-133

LESSON 119 *Age word problems*

119.A

age word problems

Most age word problems discuss the present ages of two or more people and their ages at some given time in the past and/or in the future. **The key to these problems is the proper choice of unknowns. Subscripted variables are most helpful.**

example 119.A.1   A man is 13 times as old as his son. In 10 years he will be 3 times as old as his son will be then. How old are they now?

solution   We will use these variables.

	AGE NOW	AGE 10 YEARS IN THE FUTURE
Man	M_N	$M_N + 10$
Son	S_N	$S_N + 10$

The first statement is that the man is 13 times as old as his son. This gives the equation

$$\text{(a)} \quad M_N = 13 S_N$$

And we use the same variables to say that in 10 years he will be 3 times as old as his son will be then.

$$\text{(b)} \quad M_N + 10 = 3(S_N + 10)$$

Substitute for M_N in equation (b) its value from equation (a) and solve.

$$13S_N + 10 = 3(S_N + 10) \qquad \text{substituted}$$

$$13S_N + 10 = 3S_N + 30 \qquad \text{multiplied}$$

$$10S_N = 20 \qquad \text{simplified}$$

$$S_N = 2, \text{ so } M_N = 26 \qquad \text{solved}$$

example 119.A.2   Five years ago Brenda was $\frac{4}{5}$ as old as Layton. Ten years from now she will be $\frac{7}{8}$ as old as Layton. How old is each now?

solution

	NOW	5 YEARS AGO	10 YEARS FROM NOW
Brenda	B_N	$B_N - 5$	$B_N + 10$
Layton	L_N	$L_N - 5$	$L_N + 10$

(a)   Brenda's age 5 years ago equaled $\frac{4}{5}$ Layton's age 5 years ago.

$$\text{(a)} \quad B_N - 5 = (L_N - 5)\frac{4}{5}$$

Note that Layton's age then is multiplied by $\frac{4}{5}$, but not Brenda's.

(b)   Brenda's age 10 years from now equals $\frac{7}{8}$ of Layton's age 10 years from now.

$$\text{(b)} \quad (B_N + 10) = (L_N + 10)\frac{7}{8}$$

To clear fractions, multiply every term in equation (a) by 5 and every term in equation (b) by 8 to get equations (a′) and (b′):

$$\text{(a′)} \quad (5)(B_N - 5) = (L_N - 5)\frac{4}{5}(5)$$

$$\text{(b′)} \quad (8)(B_N + 10) = (L_N + 10)\frac{7}{8}(8)$$

Simplify and get equations (a″) and (b″)

$$\text{(a″)} \quad 5B_N - 4L_N = 5 \qquad \text{(b″)} \quad 8B_N - 7L_N = -10$$

which can be solved by using elimination:

$$
\begin{array}{lllll}
\text{(a″)} & 5B_N - 4L_N = 5 & \longrightarrow & (7) & \longrightarrow & 35B_N - 28L_N = 35 \\
\text{(b″)} & 8B_N - 7L_N = -10 & \longrightarrow & (-4) & \longrightarrow & -32B_N + 28L_N = 40 \\
\hline
& & & & & 3B_N \qquad\quad = 75
\end{array}
$$

$$B_N = 25$$

Now we will replace B_N with 25 in equation (a″) and solve for L_N.

$$
\begin{aligned}
\text{(a″)} \quad 5(25) - 4L_N &= 5 \\
125 - 4L_N &= 5 \\
-4L_N &= -120 \\
L_N &= 30
\end{aligned}
$$

example 119.A.3   Thirty years ago Barbie was 1 year older than twice Mary's age then. Twenty years ago Mary was $\frac{4}{5}$ as old as Barbie was then. How old is each girl now?

solution   We will use the variables B_N for Barbie now and M_N for Mary now.

	NOW	30 YEARS AGO	20 YEARS AGO
	B_N	$B_N - 30$	$B_N - 20$
	M_N	$M_N - 30$	$M_N - 20$

The first sentence gives us the equation

$$\text{(a)} \quad (B_N - 30) - 1 = 2(M_N - 30)$$

which simplifies to

$$B_N - 2M_N = -29$$

The second sentence gives us the equation

$$\text{(b)} \quad M_N - 20 = \frac{4}{5}(B_N - 20)$$

which simplifies to

$$4B_N - 5M_N = -20$$

Next we will use elimination to solve the two equations for M_N.

$$
\begin{array}{lcccl}
B_N - 2M_N = -29 & \longrightarrow & (-4) & \longrightarrow & -4B_N + 8M_N = 116 \\
4B_N - 5M_N = -20 & \longrightarrow & (1) & \longrightarrow & 4B_N - 5M_N = -20 \\
\hline
& & & & 3M_N = 96 \\
& & & & M_N = 32
\end{array}
$$

Now we replace M_N in equation (a) with 32 to find B_N.

$$(B_N - 30) - 1 = 2(32 - 30) \quad \longrightarrow \quad B_N = 35$$

problem set 119

1. Three hundred liters of a 76% antifreeze solution had to be reduced to a 20% solution. How many liters of pure antifreeze should be extracted?

2. The sum of the digits of a two-digit counting number is 11. If the digits are reversed, the new number is 5 greater than 3 times the original number. What was the original number?

*3. A man is 13 times as old as his son. In 10 years, he will be 3 times as old as his son will be then. How old are they now?

*4. Five years ago, Brenda was $\frac{4}{5}$ as old as Layton. Ten years from now, she will be $\frac{7}{8}$ as old as Layton. How old are they now?

*5. Thirty years ago, Barbie was 1 year older than twice Mary's age then. Twenty years ago, Mary was $\frac{4}{5}$ as old as Barbie was then. How old is each girl now?

Graph the solution on a number line:

6. $\{x \in J \mid |x - 3| < 2\}$

7. $\{x \in R \mid |x + 3| \le 2\}$

8. $\{x \in R \mid x^2 - 6x > -8\}$

9. $\{x \in J \mid |x + 1| > 4\}$

<u>10.</u> $\{x \in R \mid x + 1 > 5\}$

11. $\{x \in J \mid x^2 + x - 12 \le 0\}$

12. $\{x \in R \mid x + 2 \not> -2 \text{ or } x - 3 > -5\}$

Use logarithms as required to perform the following operations. Begin by writing each number as an exponential whose base is 10. Show your work.

13. $\dfrac{.0123 \times 10^{-5}}{375,000}$

14. $(.0123 \times 10^{-5})^{1/5}$

15. $(.5712 \times 10^{-2})(.0123 \times 10^{-5})$

Find the pH ($\text{pH} = -\log \text{H}^+$ or $10^{-\text{pH}} = \text{H}^+$) of the solution when the concentration of hydrogen ions (H^+) in moles per liter is:

16. 3.26×10^{-12}

17. 9.83×10^{-7}

18. $.062$

Find the concentration of hydrogen ions (H^+) in moles per liter when the pH of the liquid is:

19. 5.92   **20.** 3.13   **21.** 9.2306

22. Convert 40 cubic feet per minute to cubic inches per second.

23. Show that $.02\overline{163}$ is a rational number by writing it as a fraction of integers.

24. Find the volume of this prism in cubic inches. Dimensions are in inches.

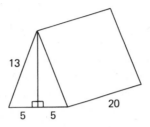

25. Complete the square as an aid in graphing: $y = -x^2 + 2x - 3$

26. Solve $-5x^2 = 2x - 1$ by completing the square.

27. Solve: $\begin{cases} \dfrac{3}{2}x + y = 13 \\ .2x - .02y = 1.12 \end{cases}$

28. Write $4R - 6U$ in polar form.

29. Add: $4\underline{/20°} - 6\underline{/230°}$

30. Solve $5x^3 + 9x^2 - 2x = 0$ by factoring.

ER 119-134, 119-135

LESSON 120 *Rational inequalities*

120.A
rational inequalities

We have learned to graph the solution to quadratic inequalities such as

$$x^2 - x - 6 > 0 \quad \text{and} \quad x^2 - x - 6 < 0$$

by factoring first.

$$(x - 3)(x + 2) > 0 \quad \text{and} \quad (x - 3)(x + 2) < 0$$

In the left hand inequality we note that the product is greater than zero and thus is a positive number. For this to be true, both factors must represent positive numbers *or* both factors must represent negative numbers.

(POS)	and	(POS)	**or**	(NEG)	and	(NEG)
$(x - 3)$		$(x + 2)$		$(x - 3)$		$(x + 2)$

In the right hand inequality we note that the product is less than zero and thus is a negative number. For this to be true, one factor must represent a positive number while the other factor must represent a negative number.

	(POS)	and	(NEG)	**or**	(NEG)	and	(POS)
	$(x - 3)$		$(x + 2)$		$(x - 3)$		$(x + 2)$

We will use the same thought to solve quadratic inequalities that are written in fractional form such as

$$1 \le \frac{-1}{x - 4}$$

As the first step, we must eliminate the denominator. **If we multiply both sides by $x - 4$ to do this**

$$(x - 4)1 \le \frac{-1}{x - 4}(x - 4) \qquad \text{incorrect}$$

we have a problem. We must reverse the inequality symbol if $x - 4$ represents a negative quantity, but the value of $x - 4$ can be either positive or negative as determined by the replacement value of x. We can resolve our dilemma if instead of multiplying by $x - 4$, we multiply by $(x - 4)^2$ because the square of any nonzero quantity always represents a positive number.

example 120.A.1   Graph $1 \le \dfrac{-1}{x - 4}$, $D = \{\text{Reals}\}$.

solution   **We begin by noting that $x = 4$ cannot be a solution because division by zero is not defined.** To eliminate the denominator, we will multiply both sides by $(x-4)^2$ and cancel the denominator.

$$(x - 4)^2 1 \le \frac{-1}{x - 4}(x - 4)(x - 4)$$

For clarity on the right side, we used $(x - 4)(x - 4)$ to represent $(x - 4)^2$. Now we simplify and get

$$x^2 - 8x + 16 \le -x + 4$$

and add $+x - 4$ to both sides and get

$$x^2 - 7x + 12 \le 0$$

which we factor as

$$(x - 4)(x - 3) \le 0$$

This product is less than or equal to zero. If the product is less than zero, it must be a negative number because all real numbers that are less than zero are negative. Thus, if the product is negative, the first factor is positive **and** the second factor is negative; **or** the first factor is negative **and** the second factor is positive. So we get

(POS)		(NEG)	**or**	(NEG)		(POS)
$(x - 4)$	**and**	$(x-3)$		$(x - 4)$	**and**	$(x - 3)$
$x - 4 \ge 0$	**and**	$x - 3 \le 0$		$x - 4 \le 0$	**and**	$x - 3 \ge 0$
$x - 4 \ge 0$		$x - 3 \le 0$		$x - 4 \le 0$		$x - 3 \ge 0$
$\underline{+4 \quad +4}$	**and**	$\underline{+3 \quad +3}$	**or**	$\underline{+4 \quad +4}$	**and**	$\underline{+3 \quad +3}$
$x \ge 4$		$x \le 3$		$x \le 4$		$x \ge 3$

Impossible. There is no real number that is greater than 4 and also less than 3.

Thus the total solution comes from the conjunction stated on this side.

$$x \le 4 \quad \text{**and**} \quad x \ge 3$$

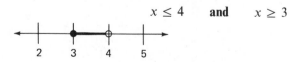

Note that the graph excludes 4, an answer that we rejected at the outset.

example 120.A.2   Graph $\dfrac{m-2}{m+2} \le 2$, $D = \{\text{Reals}\}$.

solution   **As the first step we note that -2 cannot be a solution, for this would make the denominator equal to zero.** Next we multiply both sides by $(m+2)^2$ and do not reverse the inequality symbol because $(m+2)^2$ is a positive quantity.

$$(m+2)(m+2)\left(\frac{m-2}{m+2}\right) \le 2(m+2)(m+2)$$

$$m^2 - 4 \le 2m^2 + 8m + 8$$

Now we add $-m^2 + 4$ to both sides and get

$$m^2 + 8m + 12 \ge 0$$

which factors as

$$(m+2)(m+6) \ge 0$$

Now for the product to be positive, both factors must be positive **or** both factors must be negative.

(POS)		(POS)	**or**	(NEG)		(NEG)
$m+2 \ge 0$	**and**	$m+6 \ge 0$		$m+2 \le 0$	**and**	$m+6 \le 0$
$m \ge -2$	**and**	$m \ge -6$		$m \le -2$	**and**	$m \le -6$

Any number greater than -2 is certainly greater than -6, so the values of m that satisfy this conjunction are the values of m such that

Any number less than -6 is certainly less than -2, so the values of m that satisfy this conjunction are the values of m such that

$$m \ge -2 \qquad \text{or} \qquad m \le -6$$

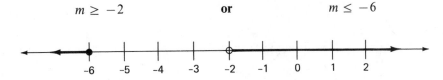

In the graph we have an open circle at -2 because at the beginning of the problem we noted that this number could not be a part of the solution.

problem set 120

1. Eighty liters of a 40% antifreeze solution had to be strengthened so that it contained 52% antifreeze. How many liters of pure antifreeze should be added?

2. The volume of a quantity of an ideal gas was held constant at 500 ml. The initial pressure and temperature were 10 atmospheres and 600 K. What would the final temperature be in kelvins if the pressure were increased to 20 atmospheres?

3. The Bayou Belle could go 84 miles downstream in 6 hours, but it took her 7 hours to go 42 miles upstream. What was her speed in still water and what was the speed of the current in the river?

4. Garfunkel was twice as old as his dog, Spot. Ten years later, he found that 4 times his age exceeded 3 times his dog's age by only 15 years. How old were both in the beginning?

5. Rover Boy was 5 years older than Yolanda. In 10 years he found that 4 times his age exceeded twice Yolanda's age by only 50. How old were Rover Boy and Yolanda in the beginning?

Graph the solutions on a number line:

*6. $1 \le \dfrac{-1}{x-4}, D = \{\text{Reals}\}$ ***7.** $\dfrac{m-2}{m+2} \le 2, D = \{\text{Reals}\}$

8. $\{x \in J \,|\, |x - 4| < 2\}$ **9.** $\{x \in R \,|\, |x + 2| \le 3\}$

10. $\{x \in J \,|\, |x + 2| \ge 1\}$

Use logarithms as required to perform the following operations. Begin by writing each number as an exponential whose base is 10.

11. $\dfrac{4.163 \times 10^5}{.007 \times 10^{-4}}$ **12.** $(417,\!000)^{3/8}$

13. $(1.2 \times 10^5)(.007 \times 10^{-4})$

Find the pH ($\text{pH} = -\log \text{H}^+$ or $10^{-\text{pH}} = \text{H}^+$) of the solution when the concentration of hydrogen ions (H^+) in moles per liter is:

14. 4.18×10^{-11} **15.** 7.24×10^{-5} **16.** $.053$

Find the concentration of hydrogen ions (H^+) in moles per liter when the pH of the liquid is:

17. 3.41 **18.** 7.25 **19.** 3.13

20. Convert 4000 centimeters per minute to yards per second.

21. Show that $.0\overline{4234}$ is a rational number by writing it as a fraction of integers.

22. The figure is the base of a solid 10 yards high. Dimensions are in feet. Find the volume in cubic feet.

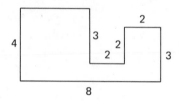

23. Complete the square as an aid in graphing: $y = -x^2 + 4x - 7$

24. Find the resultant vector of the two vectors shown.

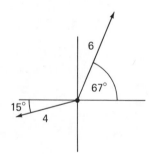

Solve:

25. $\begin{cases} 3x - y + z = 1 \\ x - y - z = 1 \\ x - 2y - z = -2 \end{cases}$ **26.** $\begin{cases} \dfrac{1}{2}x + \dfrac{1}{3}y = 5 \\ .4x - .2y = -.2 \end{cases}$ **27.** $\begin{cases} 5x + y = 7 \\ 2x - z = -1 \\ y + z = 5 \end{cases}$

28. Simplify: $\dfrac{4 - \sqrt{5}}{\sqrt{5} + 2}$

29. Solve $3x^3 = 4x^2 + 7x$ by factoring.

ER 120-136, 120-137

LESSON 121   *Laws of logarithms*

121.A
**rules
of logarithms—
product rule**

A logarithm is the exponent of an exponential. Thus, since these two exponentials

$$\text{(a)} \quad 10^2 \qquad \text{(b)} \quad 10^4$$

represent 100 and 10,000, we say that the logarithm of 100 is 2, and the logarithm of 10,000 is 4. Since logarithms are exponents, we use the rules for exponents when we use logarithms. To multiply 10^2 by 10^4, we add the exponents

$$10^2 \cdot 10^4 = 10^{2+4}$$

Thus, we see that when we multiply numbers in exponential form, we add their logarithms. If we have two numbers m and n whose exponential forms are b^a and b^c,

$$m = b^a \qquad n = b^c$$

we see that the product of the numbers is

$$b^a \cdot b^c = b^{a+c}$$

Since b to the a represents m and b to the c represents n, the product of m and n can be represented by b to the $a + c$ power. Thus, the first rule of logarithms is

(1)   **The exponent of a product of exponentials is the sum of the exponents; and since an exponent is also a logarithm we can also say that the logarithm of a product is the sum of the logarithms, which we say in the language of logarithms by writing**

$$\log mn = \log m + \log n$$

Note that we have omitted the base 10 here. Anytime log is written by itself, it is understood that the base is 10. Thus, log 32 is understood to mean $\log_{10} 32$.

example 121.A.1   Find the logarithm of $(10{,}461 \times 10^{14})(1378 \times 10^{-25})$.

solution   The logarithm of a product is the sum of the individual logarithms, so first we find the logarithm (exponent) of each number.

n	0	1	2	3	4	5	6	7	8	9
1.0	.0000	.0043	.0086	.0128	.0170	.0212	.0253	.0294	.0334	.0374
1.1	.0414	.0453	.0492	.0531	.0569	.0607	.0645	.0682	.0719	.0755
1.2	.0792	.0828	.0864	.0899	.0934	.0969	.1004	.1038	.1072	.1106
1.3	.1139	.1173	.1206	.1239	.1271	.1303	.1335	.1367	.1399	.1430
1.4	.1461	.1492	.1523	.1553	.1584	.1614	.1644	.1673	.1703	.1732

$$10{,}461 \times 10^{14} \approx 1.05 \times 10^{18} = 10^{.0212} \times 10^{18} = 10^{18.0212}$$

Thus

$$\log (10{,}461 \times 10^{14}) = 18.0212$$

$$1378 \times 10^{-25} \approx 1.38 \times 10^{-22} = 10^{.1399} \times 10^{-22} = 10^{-21.8601}$$

Thus

$$\log (1378 \times 10^{-25}) = -21.8601$$

and the sum of these logs is $(18.0212) + (-21.8601) = -3.8389$. If we use logarithmic notation, we say

$$\log (10{,}461 \times 10^{14})(1378 \times 10^{-25}) = \mathbf{-3.8389}$$

121.B
quotient rule

When we divide numbers written in exponential notation, we use the quotient theorem for exponents. Thus, we perform this indicated division

$$\frac{10^4}{10^2}$$

by writing

$$10^{4-2}$$

This is the reason that when we divide numbers in exponential form, we subtract the logarithm of the denominator from the logarithm of the numerator. Again if we use b^a and b^c as the exponential forms of m and n

$$m = b^a \qquad n = b^c$$

we see that the quotient of the numbers is

$$\frac{m}{n} = \frac{b^a}{b^c} = b^{a-c}$$

Thus, the exponent of the quotient is $a - c$, and the second rule for logarithms is the quotient rule which indicates the difference of the logarithms.

(2) **The exponent of a quotient of exponentials is the difference of the exponents; or the logarithm of a quotient is the difference of the logarithms, which we say in the language of logarithms by writing**

$$\log \frac{m}{n} = \log m - \log n$$

example 121.B.1 Find the logarithm of $\dfrac{10{,}461 \times 10^{14}}{1378 \times 10^{-25}}$.

solution We found the logarithm of each number in the last example as

$$\log (10{,}461 \times 10^{14}) = 18.0212$$
$$\log (1378 \times 10^{-25}) = -21.8601$$

Using the quotient rule for logarithms, we get

$$\log \frac{10{,}461 \times 10^{14}}{1378 \times 10^{-25}} = 18.0212 - (-21.8601) = \mathbf{39.8813}$$

121.C
power rule

We raise exponentials to a power by multiplying the exponents. Thus

$$(10^3)^2 = 10^6$$

If we let b^a represent the exponential form of some number m

$$m = b^a$$

we raise this number to the c power by multiplying a and c

$$m^c = b^{ac}$$

Thus the power rule for logarithms can be stated as

(3) **The exponent of a power of an exponential is the product of the power and the exponent of the exponential, which we say in the language of logarithms by writing**

$$\log n^c = c \log n$$

example 121.C.1 Find the logarithm of $(10{,}461 \times 10^{14})^5$.

solution We have found that

$$\log (10{,}461 \times 10^{14}) = 18.0212$$

and if we use the power rule for logarithms, we get

$$\log (10{,}461 \times 10^{14})^5 = 5(18.0212) = 90.106$$

problem set 121

1. Citronella dropped the last coin into her bank and smiled. She had kept a tally and knew that her bank contained $4.55 in nickels and dimes and that there were 25 more nickels than dimes. How many coins of each type did Citronella have?

2. Monkeys varied jointly as apes and edible vines squared. If 400 monkeys went with 2 apes and 2 pounds of edible vines, how many apes were there when there were 1600 monkeys and $\frac{1}{2}$ pound of edible vines? Use both ratio and variation formats.

3. A two-digit counting number has a value that is 8 more than twice the sum of its digits. If 4 times the units' digit is 30 greater than the tens' digit, what is the number?

4. Yehudi is 4 years older than his brother Mohab. In 10 years, twice his age will exceed his brother's age by 24. How old are the boys now?

5. Petunia was proud because she was twice as old as Daisy was. Ten years later she was chagrined to realize that twice her age exceeded Daisy's age by 25 years. How old were the girls at the outset?

The three rules for logarithms are the product rule, the quotient rule, and the power rule.

6. Write the product rule for logarithms and explain this rule.

7. Write the quotient rule for logarithms and explain this rule.

8. Write the power rule for logarithms and explain this rule.

Graph the solutions on a number line:

9. $2 \leq \dfrac{-2}{x - 2}$, $D = \{\text{Reals}\}$

10. $\dfrac{m + 3}{m - 3} \leq 1$, $D = \{\text{Integers}\}$

11. $\{x \in R \mid |x - 2| < 1\}$

12. $\{x \in J \mid |x + 3| \leq 4\}$

13. $\{x \in J \mid -8 \not> x + 2 < -4\}$

Use logarithms as required to perform the following operations. Begin by writing each number as an exponential whose base is 10. Show your work.

14. $\dfrac{516 \times 10^7}{.00713 \times 10^{-5}}$

15. $(321{,}000)^{2/7}$

16. $(4.6 \times 10^{14})(3.02 \times 10^{-20})$

Find the pH (pH $= -\log H^+$ or $10^{-pH} = H^+$) of the solution when the concentration of hydrogen ions (H^+) in moles per liter is:

17. 3.26×10^{-9} **18.** 7.04×10^{-5} **19.** .0016

Find the concentration of hydrogen ions (H^+) in moles per liter when the pH of the liquid is:

20. 4.02   **21.** 8.23   **22.** $+ 10.13$

23. Find the area of the figure in square inches. Dimensions are in yards.

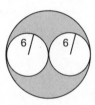

24. Convert 1000 inches per second to kilometers per hour.

25. Show that $.001\overline{68}$ is a rational number by writing it as a fraction of integers.

26. Solve $-5x^2 - x = 4$ by completing the square.

Simplify:

27. $\dfrac{\sqrt{-3}\sqrt{-3} - i^3}{2 - \sqrt{-2}\sqrt{2}}$ **28.** $\dfrac{4\sqrt{2} - 5}{1 - \sqrt{2}}$ **29.** $\sqrt[4]{xy^5}\sqrt[6]{x^3y}$

ER 121-138, 121-139

LESSON 122 *Intersection, union, Venn diagrams*

122.A
intersection   If we have the two sets

$$A = \{1, 2, 3, 6, 7\} \qquad B = \{1, 3, 8, 9\}$$

we see that the numbers 1 and 3 are members of both sets. We say the set $\{1, 3\}$ is the **intersection** of sets A and B. If we use the symbol $\cap$ to represent the word intersection, we can write

$$A \cap B = \{1, 3\}$$

This is read as the intersection of sets A and B and often as

$$A \text{ intersection } B$$

We see that the intersection of two sets is the set whose members are members of both of the given sets.

example 122.A.1   Given $P = \{1, 2, 3, 4, 7, 9, 13\}$ and $K = \{2, 5, 7, 8, 10, 13, 15\}$, find $P \cap K$.

solution   We are asked to find P intersection K, which is the set whose members are members of set P and are also members of set K.

$$P \cap K = \{2, 7, 13\}$$

122.B

union   If we look at sets A and B

$$A = \{1, 2, 3, 7\} \qquad B = \{1, 3, 8, 9\}$$

and list all the members of both sets, we would write

$$1, 2, 3, 7, 1, 3, 8, 9$$

If we list these numbers using set notation, we would write

$$\{1, 2, 3, 7, 8, 9\}$$

for we only write a number once when we use set notation. **This set is called the union of sets A and B and consists of all the members of set A and all the members of set B, none listed more than once.** We use the symbol $\cup$ to represent the word **union**.

$$A \cup B = \{1, 2, 3, 7, 8, 9\}$$

example 122.B.1   Given $P = \{1, 2, 3, 4, 7, 9, 13\}$ and $K = \{2, 5, 7, 8, 10, 13, 15\}$, find $P \cup K$.

solution   The union of the sets consists of all the members of both sets, none listed more than once. Thus,

$$P \cup K = \{1, 2, 3, 4, 5, 7, 8, 9, 10, 13, 15\}$$

122.C

Venn diagrams   Diagrams can be used to enhance our understanding of intersection and union. In the diagram we have designated set A by drawing a circle around the members of this set. We have also circled the members of set B. We can see that the numbers 15 and 17 are members of both set A and set B, so these numbers are the intersection of sets A and B.

$$A \cap B = \{15, 17\}$$

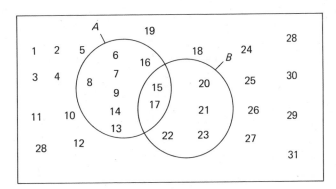

The union of sets A and B consists of all the members of both sets. Thus, we can write

$$A \cup B = \{6, 7, 8, 9, 13, 14, 15, 16, 17, 20, 21, 22, 23\}$$

example 122.C.1   The circles contain the members of sets A, B, and C as indicated. Designate the areas that represent: (a) $A \cap B$ (b) $B \cup C$ (c) $B \cap C$

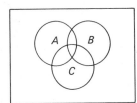

solution

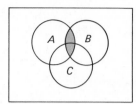

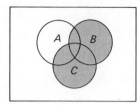

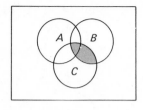

(a) $A \cap B$
The points that are
members of both
A and B

(b) $B \cup C$
The points that are
members of either
B or C or both

(c) $B \cap C$
The points that are
members of both
B and C

**problem
set 122**

1. Find three consecutive odd integers such that the product of the first and third is 13 less than the product of the second and 10.

2. Mulberry was exasperated because his plane took twice as long to cover 800 miles as it took the other plane to cover 650 miles. If the other plane was only 50 miles per hour faster, find the rates and times of both.

3. Tommy and Sarah huffed off in a hurry at 16 miles per hour. After a time they broke down and had to walk home at 10 miles per hour. If they were gone for 13 hours, how far did they get before they broke down?

4. A two-digit counting number has a value that is 13 greater than 3 times the sum of the digits. If the units' digit is 1 greater than the tens' digit, what is the number?

5. A man is 6 times as old as his son. In 5 years he will be 2 years older than 3 times his son's age then. How old are both now?

*6. Given that $P = \{1, 2, 3, 4, 7, 9, 13\}$ and $K = \{2, 5, 7, 8, 10, 13, 15\}$, find $P \cap K$.

*7. Given that $A = \{1, 2, 3, 7\}$ and $B = \{1, 3, 8, 9\}$, find $A \cup B$.

*8. Given P and K as defined in Problem 6, find $P \cup K$.

The circles contain the members of sets A, B, and C as indicated. Designate the areas that represent:

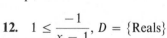

*9. $A \cap B$

*10. $B \cup C$

*11. $B \cap C$

Graph the solutions on a number line.

12. $1 \le \dfrac{-1}{x - 1}$, $D = \{\text{Reals}\}$

13. $\dfrac{m + 1}{m - 1} \le 1$, $D = \{\text{Reals}\}$ **14.** $\{x \in J \mid |x - 1| < 2\}$

15. $\{x \in R \mid |x + 3| \le 3\}$ 16. $\{x \in J \mid -6 \not> x - 3 < -2\}$

Use logarithms as required to perform the following operations. Begin by writing each number as an exponential whose base is 10. Show your work.

17. $\dfrac{302 \times 10^5}{.0013 \times 10^{-7}}$ 18. $(416{,}000)^{1/9}$

19. $(416 \times 10^3)(.0013 \times 10^{-7})$

Find the pH (pH $= -\log H^+$ or $10^{-pH} = H^+$) of the solution when the concentration of hydrogen ions (H^+) in moles per liter is:

20. 1.42×10^{-11} **21.** 6.63×10^{-4} **22.** $.025$

Find the concentration of hydrogen ions (H^+) in moles per liter when the pH of the liquid is:

23. 3.97 **24.** 2.82 **25.** 9.031

26. Show that $.0012\overline{352}$ is a rational number by writing it as a fraction of integers.

27. Complete the square as an aid in graphing: $y = -x^2 + 4x - 1$

28. Solve $3x^2 = -5x - 2$ by using the quadratic formula.

29. Simplify: $-3i^2 - 2\sqrt{-9} + \sqrt{-4} - \sqrt{-2}\sqrt{2} + 2i^5$

ER 122-140, 122-141

Appendix

Appendix

ENRICHMENT LESSON 1

Scientific calculator: addition and subtraction

ER1.A
the calculator

The calculator frees the user from mundane arithmetic chores and permits more emphasis to be placed on understanding. However, a calculator cannot be used to replace understanding itself. For instance, if the user does not know that the simplification of

$$-3^2$$

is -9, the calculator would be of no help in simplifying

$$-3.0165^2$$

because there would be no way of telling whether the calculator answer made sense or not.

Calculator instruction manuals often seem to concentrate on exotic manipulations that invariably lead to errors when tried by the beginner. The four enrichment lessons in this book will emphasize fundamental operations and are designed as an introduction to the use of the calculator. We will try to show how easy mistakes are to make and will try to point out ways to avoid mistakes. The only sure way to avoid mistakes on a calculator is never to accept an answer that does not agree with an estimated answer. This rule is so important that it will be restated in bold print.

> **NEVER ACCEPT A CALCULATOR RESULT THAT IS NOT BACKED UP BY AN ESTIMATE.**

There is almost never an excuse for accepting a wrong answer from a calculator because wrong answers usually differ greatly from any reasonable estimate of the correct answer.

The reader should read the instruction manual and be able to turn the calculator on and off and be able to add, subtract, multiply, and divide arithmetically before going further.

The instructions in this book are for calculators that use algebraic operating systems. If your calculator uses reverse Polish or other programs, it will be necessary to consult your instruction manual.

ER1.B
add, subtract, and sign change keys

In algebra the plus sign can indicate addition, or it can indicate a positive number. The minus sign can indicate subtraction, or it can indicate a negative number.

On the calculator, these keys

$$\boxed{+} \qquad \boxed{-}$$

are the add and subtract keys. **These keys instruct the calculator to add or to subtract**

algebraically. The sign change key

$$\boxed{+/-}$$

is used to designate negative numbers. **This key will change the sign of the number in the display from + to − or from − to +.** To enter a negative number, a positive number is entered and then the $\boxed{+/-}$ key is used to make the number negative. Thus, to enter −5, we enter 5 and use the $\boxed{+/-}$ key to change the sign.

DEPRESS	DISPLAY
5	5
$\boxed{+/-}$	−5

example ER1.B.1   Use the calculator to simplify 47 − 5.

solution   We enter 47, designate subtract (depress $\boxed{-}$), enter 5, and then depress $\boxed{=}$.

DEPRESS	DISPLAY
47	47
$\boxed{-}$	47
5	5
$\boxed{=}$	42

Note that a minus sign did not appear in the display when the $\boxed{-}$ key was depressed.

example ER1.B.2   Use the calculator to simplify 47 − (−5).

solution   We know the answer will be 52, and we note that this operation contains both algebraic subtraction and a negative number. We could change the signs mentally but do not because we are learning how the calculator handles subtraction. First we enter 47 and designate algebraic subtraction.

DEPRESS	DISPLAY
47	47
$\boxed{-}$	47

Note that a minus sign does not appear in the display when we use the $\boxed{-}$ key. Now we enter 5 and use $\boxed{+/-}$ to change it to −5 and then depress the equals key. **When we use the $\boxed{+/-}$ key, a − sign appears in the display.**

DEPRESS	DISPLAY
5	5
$\boxed{+/-}$	−5
$\boxed{=}$	52

This demonstrates that the calculator knows how to subtract negative numbers.

example ER1.B.3   Use the calculator to simplify −3.026 + (−91.062) − 47.0163.

solution   **We always make an estimate as the first step.** We estimate the answer as −3 − 91 − 47 = −141. Next we use the calculator as follows

DEPRESS	DISPLAY
3.026	3.026
$\boxed{+/-}$	−3.026

This enters the −3.026. We disregard the parentheses and proceed.

DEPRESS	DISPLAY
$+$	−3.026
91.062	91.062
$+/-$	−91.062
$-$	−94.088
47.0163	47.0163
$=$	−141.1043

Note that in the last step, we subtracted 47.0163. We could have added −47.0163, but we decided to use subtraction instead. **We accept this answer because it is close to our estimate. A mistake will normally cause an answer that differs greatly from the estimated answer.**

example ER1.B.4   Use the calculator to find the exact answer to $-3.0163 - .216 + (-13.152 - 1.056)$

solution   **We begin by making an estimate:** $-3 - .2 - 14 = -17.2$. Calculator instruction manuals give a procedure for using the parentheses keys. These and other unusual procedures often lead to errors so we will avoid them until we become more proficient. We want to be sure we get the right answer. Thus, we will simplify within the parentheses and rewrite the problem.

DEPRESS	DISPLAY
13.152	13.152
$+/-$	−13.152
$-$	−13.152
1.056	1.056
$=$	−14.208

Now we rewrite the problem as

$$-3.0163 - .216 - 14.208$$

DEPRESS	DISPLAY
3.0163	3.0163
$+/-$	−3.0163
$-$	−3.0163
.216	.216
$-$	−3.2323
14.208	14.208
$=$	−17.4403

We accept this answer. Note that twice we subtracted rather than change the sign and add. Either procedure would have produced the same result.

problem set ER-1

1. Nine-seventeenths of the tomatoes had blemishes. If 7600 did not have blemishes, how many had blemishes?

2. The noise was deafening, as 38 percent of the four-year-olds vied to be the cynosure. If 248 did not vie, how many four-year-olds were there in all?

3. The weight of the oxygen (O) in a quantity of water, H_2O, was 176 grams. What was the total weight of the H_2O? (H, 1; O, 16)

4. When the crowd had finally left, Kaydo counted the nickels and dimes. He found that they were worth $6.50 and that there were 10 more nickels than dimes. How many coins of each kind were there?

5. Seventy miles per hour was Timothy's average speed, and he made the trip in 4 hours less than Mary because she drove at a sedate 50 miles per hour. How long was the trip?

First estimate the answer, and then use a calculator to get the exact answer. Do not use mental shortcuts but practice calculator skills.

*6. $-3.026 + (-91.062) - 47.0163$

*7. $-3.0163 - .216 + (-13.152 - 1.056)$

8. $-.00718 + (-2.0612) + (-3.0162) - (2.03)$

Simplify:

9. $4i^3 - 3i^4 + 2i^2 - 5$ 10. $-3i^2 - 6i^3 + 2i - 4$

Solve by completing the square:

11. $x^2 = -x + 1$ 12. $-5 = -x^2 - 4x$

13. Find the equation of this line.   14. Find sides M and P.

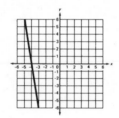
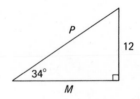

15. Solve: $\sqrt{x - 4} - 2 = 0$

16. Convert 2 feet per second to miles per hour.

Simplify:

17. $\sqrt[4]{2\sqrt[3]{2}}$ **18.** $\sqrt{25\sqrt[3]{5}}$ 19. $\sqrt{x^3y^2m}\sqrt[5]{xy^2m^2}$

20. $4\sqrt{\dfrac{2}{13}} + 3\sqrt{\dfrac{13}{2}} - 2\sqrt{104}$ 21. $-(81)^{-3/4}$

22. Estimate: $\dfrac{(6,049,213)(7,521,800)}{(.00713 \times 10^{40})}$ 23. $\dfrac{m}{c} - k = \dfrac{a}{b}$; find c

24. How many 1-inch-square floor tiles would it take to cover this figure? Dimensions are in feet.

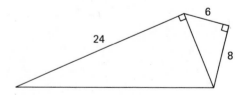

25. Find the equation of the line through $(-7, 2)$ that is perpendicular to the line $4y - 3x = 2$.

26. Find the distance between $(-4, -1)$ and $(-3, -6)$.

27. Multiply: $\dfrac{3x^2y^0x}{y^3p}\left(\dfrac{px^3}{y^{-3}} - \dfrac{4xy^{-2}}{p^2}\right)$ **28.** Divide $2x^3 - x^2 - 1$ by $x + 2$.

Solve:

29. $\dfrac{-x - 3}{3} - \dfrac{x + 1}{2} = 5$

30. $-3(x^0 - x - 3) - |-4| + 20(-x - x^0) = -4x$

ENRICHMENT LESSON 2

Scientific calculator: multiplication and division

ER2.A
multiplication and division

Multiplication and division on the calculator are straightforward. Care must be used when multiplying and dividing signed numbers to be sure that the answer has the proper sign. **We remember that we must always estimate the answer before the calculator is used.**

example ER2.A.1 Simplify: $\dfrac{(-3.02)(2.006)}{(4.13)(-1.015)}$

solution First we estimate the answer. There is a minus sign below and one above so the answer will be a positive number. We round off to estimate.

$$\frac{3 \times 2}{4} = +1.5$$

To get the exact answer, we decide to multiply in the numerator and then divide twice.

DEPRESS	DISPLAY
3.02	3.02
$+/-$	-3.02
$\times$	-3.02
2.006	2.006
$=$	-6.05812

Now we will divide by 4.13 and then by -1.015.

DEPRESS	DISPLAY
$\div$	-6.05812
4.13	4.13

Next we add an unnecessary step and depress the equals sign to check our work thus far. It should be about -1.5.

DEPRESS	DISPLAY
$=$	-1.4668571

Now we do the other division

DEPRESS	DISPLAY
$\div$	-1.4668571
1.015	1.015
$+/-$	-1.015
$=$	1.4451795

We accept this answer as being reasonably close to our estimate of $+1.5$.

example ER2.A.2   Simplify: $\dfrac{(-6.42)(-3.06)}{(1.31)(-2.28)}$

solution   There are three negative signs so the answer will be a negative number, and we estimate it by rounding off as

$$-\frac{6 \times 3}{2} = -9$$

To find the exact value, we will divide -6.42 by 1.31, multiply the result by -3.06, and finish by dividing by -2.28.

DEPRESS	DISPLAY
6.42	6.42
$+/-$	-6.42
$\div$	-6.42
1.31	1.31
$\times$	-4.9007634
3.06	3.06
$+/-$	-3.06
$\div$	14.996336
2.28	2.28
$+/-$	-2.28
$=$	-6.5773403

This answer is close to our estimate, which was -9. If necessary, we could redo our estimate. Note that in two places the calculator completed and gave the answer to one operation when the next operation was designated.

In simple problems such as this one, it is usually helpful to decide on the sign of the answer and then use only positive numbers to obtain the absolute value of the answer.

problem set ER-2

1. Find three consecutive odd integers such that the product of the first and the third is 4 less than the product of the second and -7.

2. Find four consecutive even integers such that the product of the first and the fourth is 40 less than the product of the second and -10.

3. One solution was 50% key ingredient while the other was only 20% key ingredient. How many milliliters of each should be used to get 800 ml that is 32% key ingredient?

4. The formula for methane is CH_4. What is the weight of the carbon in 160 grams of methane? (C, 12; H, 1)

5. Find three consecutive multiples of 4 such that 6 times the first is only 8 greater than twice the sum of the second and the third.

First estimate the answer, and then use a calculator to get an exact answer. Do not use mental shortcuts but practice calculator skills.

*6. $\dfrac{(-3.02)(2.006)}{(4.13)(-1.015)}$

*7. $\dfrac{(-6.42)(-3.06)}{(1.31)(-2.28)}$

8. $\dfrac{m + x}{b} + \dfrac{a}{c} = f$; find c

9. $\dfrac{cy}{d} - \dfrac{c + e}{b} = k$; find b

10. $\dfrac{ab}{c} + \dfrac{m}{x + y} = f$; find x

11. Change $20\underline{/250°}$ to rectangular form.

12. Change $20\underline{/340°}$ to rectangular form.

13. Change 500 meters per second to yards per minute.

Simplify:

14. $3i^5 - 2\sqrt{-4} + 3\sqrt{-9} - 2i^4 + i^3$

15. $5 - 2i^3 + 2i^2 + 5i - \sqrt{-16}$

16. $5i^3 - 2\sqrt{-9} + 3i^4 + 3 - 3i$

Solve by completing the square:

17. $x^2 = 6x + 6$

18. $x^2 - 2 = 2x$

Solve:

19. $\sqrt{x - 3} + 2 = 1$

20. $\sqrt{x^2 - 3x + 8} + 1 = x$

Simplify:

21. $\sqrt[5]{8\sqrt[5]{2}}$

22. $\sqrt[4]{16\sqrt{2}}$

23. $\sqrt[6]{x^2y}\sqrt{x^3y^2}$

24. $-27^{5/3}$

25. $3\sqrt{\dfrac{3}{11}} - 3\sqrt{\dfrac{11}{3}} - 2\sqrt{132}$

26. Estimate: $\dfrac{(360{,}182)(71{,}462 \times 10^{14})}{.0062 \times 10^{-6}}$

27. Simplify: $\dfrac{-35 - 2x + x^2}{x^2 - 14 - 5x} \div \dfrac{x^2 + x - 20}{x^2 + 6 + 5x}$

28. Find B.

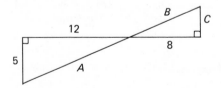

29. Find the volume of this right circular cylinder in cubic feet. Dimensions are in feet.

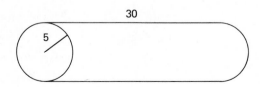

30. Find the distance between $(-3, 5)$ and $(-2, -7)$.

---

*ENRICHMENT
LESSON 3*

ER3.A

**raising
to powers**

Scientific calculator: exponents and roots

We use the

$$\boxed{y^x} \quad \text{and} \quad \boxed{\sqrt[x]{y}}$$

keys to raise positive numbers to powers and to take roots of positive numbers.[†] Thus the calculator will simplify

$$2^3 \quad \text{and} \quad 2^{-3}$$

because the bases are positive numbers but will not simplify

$$(-2)^3 \quad \text{and} \quad (-2)^{-3}$$

because the bases are negative numbers.

We use the

$$\boxed{y^x}$$

key to raise a positive number to a power. First we enter the base y and depress the y^x key. Next we enter the value of the exponent x. We finish by depressing the equals key. **It is a good policy always to check a procedure by working a problem to which the answer is known.** We will verify our procedure for the y^x key by computing

$$2^3$$

and getting 8 for a result.

DEPRESS	DISPLAY
2	2
$\boxed{y^x}$	2
3	3
$\boxed{=}$	**8**

example ER3.A.1   Simplify $2^{3.64}$.

solution   **We always begin by estimating the answer.** Now we know that

$$2^3 = 8 \quad \text{and} \quad 2^4 = 16$$

so our answer should be between 8 and 16 and should be closer to 16.

[†] Taking the root of is the inverse operation of raising to a power. Thus, some calculators require the use of the two keys $\boxed{\text{INV}}$ $\boxed{y^x}$ in sequence instead of using the $\sqrt[x]{y}$ key.

DEPRESS	DISPLAY
2	2
$\boxed{y^x}$	2
3.64	3.64
$\boxed{=}$	**12.47**

We rounded the answer off to 12.47. We accept this answer, as it agrees with our estimate.

example ER3.A.2   Simplify $4.32^{2.16}$.

solution   **First we estimate the answer.** We know that

$$4^2 = 16 \quad \text{and} \quad 4^3 = 64$$

Thus our answer should lie between these numbers.

DEPRESS	DISPLAY
4.32	4.32
$\boxed{y^x}$	4.32
2.16	2.16
$\boxed{=}$	**23.59**

We accept 23.59 as probably correct because it is within our estimated range of answers.

ER3.B
roots
of numbers

The inverse operation of raising to a power is taking the root of and the key[†]

$$\boxed{\sqrt[x]{y}}$$

instructs the calculator to take the xth root of y. **In this operation, the calculator will only accept positive values of y.** Thus, it will calculate

$$\sqrt[3]{8} \quad \text{or} \quad {}^-\sqrt[3]{8}$$

but will not attempt to calculate

$$\sqrt[3]{-8} \quad \text{nor} \quad {}^-\sqrt[3]{-8}$$

We enter y first and depress both keys in the order indicated. Then we enter x and finish by depressing the equals key. **We always check our procedure by making a calculation to which we know the answer.** Thus, we will calculate

$$\sqrt[3]{8}$$

which we know equals 2.

DEPRESS	DISPLAY
8	8
	8
3	3
$\boxed{=}$	**2**

[†] Or $\boxed{\text{INV}}$ $\boxed{y^x}$ used in sequence.

example ER3.B.1   Calculate $\sqrt[4]{204}$.

solution   **We begin by estimating the answer.** We can use the calculator to help us with the estimate. We need some fourth powers that are near 200. We calculate that

$$3^4 = 81 \quad \text{and} \quad 4^4 = 256$$

Thus, we guess that the number whose fourth power is 204 is between 3 and 4 and is closer to 4.

DEPRESS	DISPLAY
204	204
$\boxed{\sqrt[x]{y}}$	204
4	4
$\boxed{=}$	**3.78**

We accept this answer as probably correct because it is within our range of acceptable answers.

example ER.3.B.2   Calculate $\sqrt[3.26]{489}$.

solution   To estimate the answer we use the calculator and get

$$8^3 = 512 \quad \text{and} \quad 5^3 = 125$$

Thus, it would seem reasonable that our answer would be somewhere near 5 and 8. Let's see.

DEPRESS	DISPLAY
489	489
$\boxed{\sqrt[x]{y}}$	489
3.26	3.26
$\boxed{=}$	**6.68**

Since our estimate was so haphazard, we will check this answer by calculating

$$6.68^{3.26}$$

DEPRESS	DISPLAY
6.68	6.68
$\boxed{y^x}$	6.68
3.26	3.26
$\boxed{=}$	488.40

The difference must have been caused by our rounding off the first answer.

example ER3.B.3   Calculate $(3.06)^{2.1} \sqrt[4.3]{284}$.

solution   This could be calculated in one continuous procedure, and the answer would most likely be incorrect. We will break the problem up into parts and work each part. **We want the correct answer.**

$$(3.06)^{2.1} \qquad\qquad\qquad {}^{4.3}\!\sqrt{284}$$

$3^2 = 9$ so we estimate 9 $\qquad\qquad$ $4^4 = 256$ so we estimate 4

DEPRESS	DISPLAY	DEPRESS	DISPLAY
3.06	3.06	284	284
$\boxed{y^x}$	3.06	$\boxed{\sqrt[x]{y}}$	284
2.1	2.1	4.3	4.3
$\boxed{=}$	10.47	$\boxed{=}$	3.72

We finish by multiplying:

$$\text{Estimate: } (10)(4) = 40$$

$$(10.47)(3.72) = \mathbf{38.95}$$

We accept this answer, as we checked our work all the way through. We note that we rounded off in four different places in this problem. In the problem sets your answers to problems like this one might be slightly different from the answer in the book because of differences in the way the problem was rounded off.

problem set ER-3

1. The number of dastards varied directly as the number of poltroons. When there were 800 dastards, the poltroons numbered 9600. When there were 24,000 poltroons, how many dastards were there?

2. The temperature of a quantity of an ideal gas was held constant at 1400 K. If the pressure of 1200 torr was increased to 1600 torr, what was the final volume if the volume was initially 200 ml?

3. The original mixture weighed 800 pounds and was 40% slaked lime. How much of a 20% slaked lime mixture should be added to reduce the lime concentration to 36%?

4. The fugacious numbered 14 more than twice the number of ephemeral. Also, twice the number of fugacious was 100 less than 20 times the number of ephemeral. How many of each were there?

5. Twenty percent wanted to storm the fortress. If the rest totalled 6720 and just wanted to sleep in the shade, how many were in the advancing army?

First estimate the answer and then use a calculator to get a more exact answer.

*6. $2^{3.64}$ $\qquad$ *7. $4.32^{2.16}$ $\qquad$ *8. $\sqrt[4]{204}$

*9. ${}^{3.26}\!\sqrt{489}$ $\qquad$ *10. $(3.06)^{2.1}({}^{4.3}\!\sqrt{284})$

Simplify:

11. $\dfrac{1}{-2-\sqrt{2}}$ $\qquad$ 12. $\dfrac{4}{3\sqrt{2}-1}$ $\qquad$ 13. $\dfrac{2}{3\sqrt{3}-5}$

Add:

14. $\dfrac{-7}{-x+3} - \dfrac{2x}{x^2-9}$ $\qquad$ 15. $\dfrac{2}{4-x} - \dfrac{3}{x^2-16}$

16. Solve: $R_G T_G = 140$, $R_B T_B = 140$, $R_B = 2R_G$, $T_B = T_G - 7$

Simplify:

17. $4r + \dfrac{m}{x + \dfrac{m}{x}}$ $\qquad\qquad$ 18. $3a + \dfrac{ax}{a + \dfrac{x}{a}}$

19. $(5i - 2)(3i + 4)$

20. $\sqrt{-2}\sqrt{2} - 3i^2 - \sqrt{-9} + 2i^4 + 4$

21. Add: $40\underline{/315°} + 10\underline{/24°}$

22. Solve: $\begin{cases} \dfrac{2}{5}x - \dfrac{2}{3}y = -2 \\ -.06x - .4y = -7.2 \end{cases}$

Solve by completing the square:

23. $-2x = -1 - 4x^2$

24. $-3x^2 + 5 = -2x$

25. Convert 700 cubic centimeters per minute to cubic inches per hour.

Simplify:

26. $\sqrt[7]{3\sqrt[3]{3}}$

27. $\sqrt[3]{xy^5}\sqrt{x^5y}$

28. $3\sqrt{\dfrac{2}{3}} - 5\sqrt{\dfrac{3}{2}} + 2\sqrt{24}$

29. $\dfrac{m}{x(a + b)} - \dfrac{p}{y} = c$; find b

30. Find the equation of the line that passes through $(-8, 2)$ and $(5, -7)$.

ENRICHMENT LESSON 4

Scientific calculator: scientific notation

ER4.A
scientific notation

The **enter exponent** key on the calculator

$$\boxed{\text{EE}}$$

is used for scientific notation. If this key is depressed and then the equals key is depressed, the number in the display will be converted to scientific notation. If we enter the number

$$12345678$$

and depress the enter exponent key and then the equals key, the displayed number will change to

$$1.2346 \qquad 07$$

which is the calculator notation for 1.2346×10^7. The last two digits represent the power of 10. Note that the calculator rounded off the number to five digits. We can reverse this procedure by depressing the **inverse operation key**

$$\boxed{\text{INV}}$$

and then depressing the enter exponent key. If a number is too large to be displayed in standard notation, the calculator will ignore the command.

To enter the product of a number and a power of 10, the number is entered first, the $\boxed{\text{EE}}$ key depressed, and then the exponent entered. If the number is a negative number, the sign must be changed before the $\boxed{\text{EE}}$ key is depressed. A negative exponent is entered by depressing the $\boxed{+/-}$ key after depressing $\boxed{\text{EE}}$.

example ER4.A.1 Enter -4062×10^{13} in scientific notation.

solution We could move the decimal point now, but we decide to enter the number as written and use the $=$ key to move the decimal point.

DEPRESS	DISPLAY	
4062	4062	
$\boxed{+/-}$	-4062	
$\boxed{\text{EE}}$	-4062	00
13	-4062	13
$\boxed{=}$	-4.062	16

example ER4.A.2   Enter -41.36×10^{-15}.

solution   The first $-$ sign is entered before the $\boxed{EE}$ key is used. The sign of the exponent is changed as the last step.

DEPRESS	DISPLAY	
41.36	41.36	
$\boxed{+/-}$	-41.36	
$\boxed{EE}$	-41.36	00
15	-41.36	15
$\boxed{+/-}$	-41.36	-15
$\boxed{=}$	-4.136	-14

ER4.B

calculations   **It is absolutely necessary to estimate the answer before a calculation is made using scientific notation on the calculator.**

example ER4.B.1   Simplify: $(.0036 \times 10^{-14})(-232 \times 10^{8})$

solution   **As the first step we estimate.**

$$(4 \times 10^{-17})(-2 \times 10^{10}) = -8 \times 10^{-7}$$

Now we use the calculator to get a more exact answer.

DEPRESS	DISPLAY	
.0036	.0036	
$\boxed{EE}$	.0036	00
14	.0036	14
$\boxed{+/-}$	.0036	-14
$\boxed{\times}$	3.6	-17
232	232	
$\boxed{+/-}$	-232	
$\boxed{EE}$	-232	00
8	-232	08
$\boxed{=}$	-8.352	-07

We accept this answer because

$$\mathbf{-8.352 \times 10^{-7}}$$

is very close to our estimate of -8×10^{-7}.

example ER4.B.2   Simplify: $\dfrac{-13.216 \times 10^{-4}}{203.4 \times 10^{11}}$

solution   **We always begin by estimating the answer.**

$$\frac{-1 \times 10^{-3}}{2 \times 10^{13}} = -.5 \times 10^{-16} = -5 \times 10^{-17}$$

Now we use the calculator to get a more exact answer.

DEPRESS	DISPLAY	
13.216	13.216	
$\boxed{+/-}$	-13.216	
$\boxed{EE}$	-13.216	00
4	-13.216	04
$\boxed{+/-}$	-13.216	-04
$\div$	-1.3216	-03
203.4	203.4	
$\boxed{EE}$	203.4	00
11	203.4	11
$\boxed{=}$	-6.4975	-17

We accept the answer

$$-6.4975 \times 10^{-17}$$

as being reasonably close to our estimate of

$$-5 \times 10^{-17}$$

problem set ER-4

1. Muhara could drive 360 miles in 2 hours less than it took Habib to drive 240 miles. This was because Muhara exceeded the speed limit and drove twice as fast as Habib drove. Find the rates of both and the times of both.

2. A quantity of an ideal gas was confined in a container whose volume was constant. The initial pressure and temperature were 600 torr and 700 K. If the temperature was changed to 2800 K, what was the final pressure? Begin by solving for P_2.

3. The number that shied away varied directly with the volume of calumny directed toward them. If 400 shied away when the calumnations numbered 20, how many shied away when there were only 8 calumnations?

4. The blue car headed east at noon. At 2 p.m. the red car headed west. The rate of the blue car was 55 miles per hour and the rate of the red car was only 30 miles per hour. What time was it when they were 790 miles apart?

5. Gwinn bought hams for $12 each and Dale bought tenderloins for $20 each. If Dale bought 2 fewer tenderloins than Gwinn bought hams, and they spent a total of $600, how many hams did they buy and how many tenderloins did they buy?

First estimate the answer, and then use a calculator to get a more exact answer.

*6. $(.0036 \times 10^{-14})(-232 \times 10^8)$

*7. $\dfrac{-13.216 \times 10^{-4}}{203.4 \times 10^{11}}$

Add:

8. $\dfrac{x + 3}{x^2 + 2x + 1} - \dfrac{3}{-1 - x}$

9. $\dfrac{x + 7}{x^2 - 3x - 10} + \dfrac{x - 2}{-2 - x}$

Simplify:

10. $\dfrac{3 - \sqrt{3}}{-5 + \sqrt{3}}$

11. $\dfrac{5 - 2\sqrt{2}}{3 + \sqrt{2}}$

12. $\dfrac{1 - \sqrt{3}}{2 - 3\sqrt{27}}$

13. Add: $20\underline{/40°} + 10\underline{/-370°}$

14. Write $5R - 2U$ in polar form.

15. Find the roots of $-3x - 3 = 5x^2$ by completing the square.

16. $\dfrac{a}{b} = y\left(m - \dfrac{d}{f}\right)$; find f

17. $\dfrac{a}{b} = -c\left(\dfrac{1}{m} + \dfrac{a}{f}\right)$; find m

Simplify:

18. $m^2p + \dfrac{m^2}{p + \dfrac{m}{p}}$

19. $(3 + i)(i - 2) + \sqrt{-16}$

20. Solve $2x^2 - 2x = -5$ by using the quadratic formula.

21. Solve: $\begin{cases} \dfrac{1}{8}x + \dfrac{2}{5}y = 3 \\ -.05x + .07y = -.05 \end{cases}$

22. Find the surface area of the box in square inches. Dimensions are in feet.

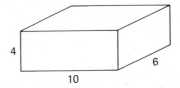

Simplify:

23. $2\sqrt{2\sqrt[3]{2}}$

24. $\dfrac{-3^0(-3^0)}{-4^{-5/2}}$

25. $\sqrt[5]{m^2p^3}\sqrt[3]{mp^3}$

26. $4\sqrt{\dfrac{7}{5}} - 3\sqrt{\dfrac{5}{7}} + \sqrt{35}$

27. Estimate: $\dfrac{(4860)(7,142,397)}{(800 \times 10^{14})(.00007)}$

28. Find the equation of the line through $(2, -3)$ that is perpendicular to the line that passes through $(-5, 2)$ and $(4, 3)$.

Solve:

29. $\sqrt{2x + 4} - 3 = 3$

30. $\dfrac{4x + 9}{3} - \dfrac{7x}{2} = 5$

Enrichment problems

The following problems are provided to permit a continuing review of the enrichment lessons on the scientific calculator. The first part of the number for each problem designates the problem set to which the problem should be appended.

Always estimate the answer before using the calculator. A calculator answer that is not backed up by an estimated answer is not reliable and should never be used.
First estimate and then use the calculator to simplify.

52.1. $-316.2 - (-2.06) - 104.62$

52.2. $-3.06 - .032 - (3.16 - 2.05)$

53.3. $-.004 - 3.06 + (-2.017 - 3.026)$

53.4. $-.13162 + .016 - 4.006 - (3.02 - 5.06)$

54.5. $-1.423 - (-8 + 6.02) - (-4.07)$

54.6. $-3.825 - (-8.015 - 3.06) + 50.42$

55.7. $-123.085 - (1.086 + 2.731) - 88.32$

55.8. $-14.016 - (-3.062 + 7.843) - 5$

56.9. $-.0816 - (-2.75 - 463.05) + (-5.32)$

56.10. $\dfrac{(4.16)(5.02)}{(2.12)(1.02)}$ **57.11.** $\dfrac{(6.21)(5.03)}{(2.01)(3.12)}$

57.12. $-7.0421 - (-5.0816 + 742) - .006$ **58.13.** $\dfrac{(7.98)(6.02)}{(2.01)(3.12)}$

58.14. $-5.06214 - (-3.06) - (-2.07 + 5.3)$

59.15. $\dfrac{(74618)(513.06)}{(50.1)(32161.1)}$

59.16. $-7.016 - 3.72(-2.04) - 2.105 + (-8.32)$

60.17. $\dfrac{(-.004)(746,012)}{-3.0172 + 8.014}$ **61.18.** $\dfrac{-3.016(-2.4) - 52.6}{(-4.02)(3.91) - 2.6}$

62.19. $\dfrac{(-8.016) - 3.016(2.413)}{(4785.1)(-.00123)}$ **63.20.** $\dfrac{(-5.026) - 7.063(-2.43)}{(7.93)(-.0632)}$

64.21. $\dfrac{-80.5 - 9.316(-.0735)}{.012(328)}$ **65.22.** $\dfrac{-918 + 4.016(5.062)}{(78,421)(.000523)}$

66.23. $\dfrac{-9.031 - 4.02(7.016 - 2.07)}{(4786.2)(.0015)}$

67.24. $\dfrac{(-924.6)(71,852)}{(602.14)(41,738)}$

68.25. $(4.03)^{2.14}$

68.26. $\sqrt[6.1]{5.42}$

68.27. $\dfrac{(4016)(-5.021)}{(-433)}$

69.28. $\sqrt[5.02]{91}$

69.29. $(4.06)^{4.03}$

69.30. $\dfrac{(417.6)(-.0052)(-572)}{(7319)(.0073)}$

70.31. $\sqrt[4.06]{80.6}$

70.32. $(3.12)^{4.06}$

70.33. $\dfrac{(5132)(-.01782)}{(4183)(.006413)}$

71.34. $\sqrt[3.1]{4.06}$

71.35. $(7.42)^{2.05}$

71.36. $\dfrac{(-401)(74.3) - 5176}{(572)}$

72.37. $\sqrt[3.15]{49.3}$

72.38. $(5.2)^{3.07}$

72.39. $\dfrac{(-.032)(4761)}{49.3 - .762}$

73.40. $\sqrt[5.2]{490.6}$

73.41. $(4.52)^{3.07}$

73.42. $\dfrac{(-.614)(5934)}{(487)}$

74.43. $\sqrt[4.07]{93.62}$ **74.44.** $(5.16)^{2.43}$ **74.45.** $\dfrac{(-521)(4173)}{(.016)(97842)}$

75.46. $\dfrac{(4732 \times 10^{42})(-9.08 \times 10^{-15})}{(51,876 \times 10^{5})(.00716)}$

75.47. $\sqrt[5.06]{4720}$

76.48. $42^{2.03}$

76.49. $\dfrac{(572 \times 10^{12})(7015 \times 10^{-4})}{(-5837)(40,162 \times 10^{-10})}$

77.50. $\sqrt[3.72]{4.03}$

77.51. $\dfrac{(3.06 \times 10^{-5})(74,618 \times 10^{7})}{(4193)(9387 \times 10^{-14})}$

78.52. $(5.72)^{3.32}$

78.53. $\dfrac{(4937)(.00163 \times 10^{-5})}{(903)(517 \times 10^{-14})}$

79.54. $\sqrt[4.06]{12.732}$

79.55. $\dfrac{(-598.61 \times 10^{-14})(2031 \times 10^{15})}{(7062 \times 10^{8})(.004 \times 10^{-5})}$

80.56. $(4.072)^{.43}$

80.57. $\dfrac{(.5061 \times 10^{5})(67,240)}{(3621 \times 10^{3})(.0071643)}$

81.58. $\sqrt[4.06]{3.21}$

81.59. $\dfrac{(9315 \times 10^{3})(-2.065 \times 10^{4})}{(-300.62)(500 \times 10^{6})}$

82.60. $42.3^{1.97}$

82.61. $\dfrac{(4316 \times 10^{7})(3.152 \times 10^{-2})}{(758 \times 10^{-8})(5713 \times 10^{4})}$

83.62. $\sqrt[3.06]{942}$

83.63. $\dfrac{(80416 \times 10^{4})(5712 \times 10^{-2})}{(.0416 \times 10^{3})(.00192)}$

84.64. $(.032)^{4.32}$

84.65. $\dfrac{(785 \times 10^{-8})(50,000)}{(-7406 \times 10^{8})(9012)}$

85.66. $-.019^{5.04}$

85.67. $\dfrac{(918 \times 10^{5})(.001852 \times 10^{-3})}{(-472.3 \times 10^{5})(801 \times 10^{4})}$

86.68. $-4.06^{2.3}$

86.69. $\dfrac{(914.62 \times 10^{-4})(.001847)}{(3016 \times 10^{5})(947 \times 10^{-8})}$

87.70. $\sqrt[2.3]{8.61}$

87.71. $\dfrac{(5205.6 \times 10^{-4})(3.015 \times 10^5)}{(9871)(946 \times 10^{-15})}$

88.72. $(4.86)^{4.82}$

88.73. $\dfrac{(-4063)(718) + (293 \times 10^4)}{7182 \times 10^{-3}}$

89.74. $\sqrt[3.4]{56.21}$

89.75. $\dfrac{(402 \times 10^5)(3074 \times 10^{-5})}{7034 \times 10^{14}}$

90.76. $(30.3)^{1.92}$

90.77. $\dfrac{(.0461)(-3.24) - .043}{432 \times 10^{-14}}$

91.78. $\sqrt[4.2]{420}$

91.79. $\dfrac{(416 \times 10^{-5})(3275 \times 10^{14})}{5219 \times 10^{23}}$

92.80. $(.032)^{2.72}$

92.81. $\dfrac{(-5063)(7429 \times 10^5)}{4783 \times 10^{10}}$

93.82. $\sqrt[1.8]{4060}$

93.83. $\dfrac{(.00183 \times 10^5)(796 \times 10^{14})}{7032 \times 10^{-18}}$

94.84. $-4.62^{3.8}$

94.85. $\dfrac{(3751 \times 10^5)(.0315 \times 10^{-14})}{78916 \times 10^5}$

95.86. $\sqrt[2.38]{9600}$

95.87. $\dfrac{(-503 \times 10^4)(.00712 \times 10^8)}{.001327 \times 10^7}$

96.88. $-30^{2.42}$

96.89. $\dfrac{(.0413)(12{,}642)}{.000153 \times 10^{-20}}$

97.90. $\sqrt[5.3]{85.62}$

97.91. $\dfrac{(5762)(1437 \times 10^{15})}{783 \times 10^{-10}}$

98.92. $3.062^{4.32}$

98.93. $\dfrac{(9052 \times 10^5)(70{,}218 \times 10^{-4})}{5062 \times 10^5}$

99.94. $\sqrt[4.2]{.046}$

99.95. $\dfrac{(764 \times 10^{-14})(.00316 \times 10^{20})}{73{,}214 \times 10^{-15}}$

100.96. $-7.08^{2.13}$

100.97. $\dfrac{(5184 \times 10^5)(.00718 \times 10^{10})}{472 \times 10^{-10}}$

101.98. $15.06^{1.75}$

101.99. $\dfrac{(50{,}372)(715 \times 10^{-10})}{-40.61 \times 10^{-4}}$

102.100. $\sqrt[3.04]{22.5}$

102.101. $\dfrac{(9502 \times 10^4)(.75 \times 10^{-5})}{-.00135 \times 10^{10}}$

103.102. $5.06^{3.52}$

103.103. $\dfrac{(78{,}102)(573 \times 10^{-14})}{5061 \times 10^5}$

104.104. $\sqrt[4.2]{90.31}$

104.105. $\dfrac{(9302)(7.06 \times 10^{-4})}{7429 \times 10^{-8}}$

105.106. $4.3^{4.05}$

105.107. $\dfrac{(-23.06 \times 10^4)(3.02 \times 10^8)}{14 \times 10^{-5}}$

106.108. $\sqrt[2.65]{31.3}$

106.109. $\dfrac{(5703 \times 10^5)(71{,}602 \times 10^{-2})}{405 \times 10^{-5}}$

107.110. $90.21^{2.22}$

107.111. $\dfrac{(8704 \times 10^5)(.0073 \times 10^{15})}{612 \times 10^{-4}}$

108.112. $\sqrt[2.08]{54.32}$

108.113. $\dfrac{(51,205 \times 10^{-4})(703 \times 10^{15})}{716.12 \times 10^{-8}}$

109.114. $3.062^{2.15}$

109.115. $\dfrac{(372 \times 10^4)(7091 \times 10^{-3})}{4.06 \times 10^5}$

110.116. $\sqrt[4.62]{500}$

110.117. $\dfrac{(5091 \times 10^5)(478 \times 10^5)}{6012 \times 10^{-12}}$

111.118. $40.31^{3.2}$

111.119. $\dfrac{(7418 \times 10^{-2})(3.05)}{7.16 \times 10^{-14}}$

112.120. $\sqrt[3.21]{42.36}$

112.121. $\dfrac{(5.03 \times 10^5)(7489 \times 10^{-5})}{473,026}$

113.122. $95.21^{1.82}$

113.123. $\dfrac{(903 \times 10^{-14})(.0071 \times 10^5)}{483 \times 10^{10}}$

114.124. $\sqrt[1.93]{54.63}$

114.125. $\dfrac{(903.62 \times 10^5)(7184 \times 10^{10})}{784 \times 10^{-8}}$

115.126. $-6.32^{3.052}$

115.127. $\dfrac{(4032 \times 10^5)(7081 \times 10^{14})}{90,812 \times 10^4}$

116.128. $\sqrt[6.2]{527.4}$

116.129. $\dfrac{(7642 \times 10^{14})(7108 \times 10^{-5})}{409 \times 10^{15}}$

117.130. $-4.06^{4.32}$

117.131. $\dfrac{(-3628 \times 10^{-4})(.00315 \times 10^{-4})}{78,516 \times 10^{-14}}$

118.132. $\sqrt[3.08]{42.5}$

118.133. $\dfrac{(5063 \times 10^5)(7018 \times 10^{-7})}{.016 \times 10^{15}}$

119.134. $-.042^{-3.2}$

119.135. $\dfrac{(6148 \times 10^{10})(7032 \times 10^{11})}{.475 \times 10^{22}}$

120.136. $\sqrt[5.72]{63.42}$

120.137. $\dfrac{(.9063 \times 10^4)(7.015 \times 10^{10})}{9417 \times 10^6}$

121.138. $-2.06^{-.93}$

121.139. $\dfrac{(765.423 \times 10^{-4})(8017 \times 10^5)}{9,063,275}$

122.140. $\sqrt[4.13]{506}$

122.141. $\dfrac{(70,215 \times 10^{-3})(4106 \times 10^5)}{90,317 \times 10^{10}}$

TABLE 1 **Values of Trigonometric Functions**
TABLE 2 **Common Logarithms**

TABLE 1 Values of Trigonometric Functions

deg	sin	cos	tan	deg	sin	cos	tan	deg	sin	cos	tan
0	.0000	1.000	.0000	6.0	.1045	.9945	.1051	12.0	.2079	.9781	.2126
0.1	.0017	1.000	.0017	6.1	.1063	.9943	.1069	12.1	.2096	.9778	.2144
0.2	.0035	1.000	.0035	6.2	.1080	.9942	.1086	12.2	.2113	.9774	.2162
0.3	.0052	1.000	.0052	6.3	.1097	.9940	.1104	12.3	.2130	.9770	.2180
0.4	.0070	1.000	.0070	6.4	.1115	.9938	.1122	12.4	.2147	.9767	.2199
0.5	.0087	1.000	.0087	6.5	.1132	.9936	.1139	12.5	.2164	.9763	.2217
0.6	.0105	.9999	.0105	6.6	.1149	.9934	.1157	12.6	.2181	.9759	.2235
0.7	.0122	.9999	.0122	6.7	.1167	.9932	.1175	12.7	.2198	.9755	.2254
0.8	.0140	.9999	.0140	6.8	.1184	.9930	.1192	12.8	.2215	.9751	.2272
0.9	.0157	.9999	.0157	6.9	.1201	.9928	.1210	12.9	.2233	.9748	.2290
1.0	.0175	.9998	.0175	7.0	.1219	.9925	.1228	13.0	.2250	.9744	.2309
1.1	.0192	.9998	.0192	7.1	.1236	.9923	.1246	13.1	.2267	.9740	.2327
1.2	.0209	.9998	.0209	7.2	.1253	.9921	.1263	13.2	.2284	.9736	.2345
1.3	.0227	.9997	.0227	7.3	.1271	.9919	.1281	13.3	.2300	.9732	.2364
1.4	.0244	.9997	.0244	7.4	.1288	.9917	.1299	13.4	.2317	.9728	.2382
1.5	.0262	.9997	.0262	7.5	.1305	.9914	.1317	13.5	.2334	.9724	.2401
1.6	.0279	.9996	.0279	7.6	.1323	.9912	.1334	13.6	.2351	.9720	.2419
1.7	.0297	.9996	.0297	7.7	.1340	.9910	.1352	13.7	.2368	.9715	.2438
1.8	.0314	.9995	.0314	7.8	.1357	.9907	.1370	13.8	.2385	.9711	.2456
1.9	.0332	.9995	.0332	7.9	.1374	.9905	.1388	13.9	.2402	.9707	.2475
2.0	.0349	.9994	.0349	8.0	.1392	.9903	.1405	14.0	.2419	.9703	.2493
2.1	.0366	.9993	.0367	8.1	.1409	.9900	.1423	14.1	.2436	.9699	.2512
2.2	.0384	.9993	.0384	8.2	.1426	.9898	.1441	14.2	.2453	.9694	.2530
2.3	.0401	.9992	.0402	8.3	.1444	.9895	.1459	14.3	.2470	.9690	.2549
2.4	.0419	.9991	.0419	8.4	.1461	.9893	.1477	14.4	.2487	.9686	.2568
2.5	.0436	.9990	.0437	8.5	.1478	.9890	.1495	14.5	.2504	.9681	.2586
2.6	.0454	.9990	.0454	8.6	.1495	.9888	.1512	14.6	.2521	.9677	.2605
2.7	.0471	.9989	.0472	8.7	.1513	.9885	.1530	14.7	.2538	.9673	.2623
2.8	.0488	.9988	.0489	8.8	.1530	.9882	.1548	14.8	.2554	.9668	.2642
2.9	.0506	.9987	.0507	8.9	.1547	.9880	.1566	14.9	.2571	.9664	.2661
3.0	.0523	.9986	.0524	9.0	.1564	.9877	.1584	15.0	.2588	.9659	.2679
3.1	.0541	.9985	.0542	9.1	.1582	.9874	.1602	15.1	.2605	.9655	.2698
3.2	.0558	.9984	.0559	9.2	.1599	.9871	.1620	15.2	.2622	.9650	.2717
3.3	.0576	.9983	.0577	9.3	.1616	.9869	.1638	15.3	.2639	.9646	.2736
3.4	.0593	.9982	.0594	9.4	.1633	.9866	.1655	15.4	.2656	.9641	.2754
3.5	.0610	.9981	.0612	9.5	.1650	.9863	.1673	15.5	.2672	.9636	.2773
3.6	.0628	.9980	.0629	9.6	.1668	.9860	.1691	15.6	.2689	.9632	.2792
3.7	.0645	.9979	.0647	9.7	.1685	.9857	.1709	15.7	.2706	.9627	.2811
3.8	.0663	.9978	.0664	9.8	.1702	.9854	.1727	15.8	.2723	.9622	.2830
3.9	.0680	.9977	.0682	9.9	.1719	.9851	.1745	15.9	.2740	.9617	.2849
4.0	.0698	.9976	.0699	10.0	.1736	.9848	.1763	16.0	.2756	.9613	.2867
4.1	.0715	.9974	.0717	10.1	.1754	.9845	.1781	16.1	.2773	.9608	.2886
4.2	.0732	.9973	.0734	10.2	.1771	.9842	.1799	16.2	.2790	.9603	.2905
4.3	.0750	.9972	.0752	10.3	.1788	.9839	.1817	16.3	.2807	.9598	.2924
4.4	.0767	.9971	.0769	10.4	.1805	.9836	.1835	16.4	.2823	.9593	.2943
4.5	.0785	.9969	.0787	10.5	.1822	.9833	.1853	16.5	.2840	.9588	.2962
4.6	.0802	.9968	.0805	10.6	.1840	.9829	.1871	16.6	.2857	.9583	.2981
4.7	.0819	.9966	.0822	10.7	.1857	.9826	.1890	16.7	.2874	.9578	.3000
4.8	.0837	.9965	.0840	10.8	.1874	.9823	.1908	16.8	.2890	.9573	.3019
4.9	.0854	.9963	.0857	10.9	.1891	.9820	.1926	16.9	.2907	.9568	.3038
5.0	.0872	.9962	.0875	11.0	.1908	.9816	.1944	17.0	.2924	.9563	.3057
5.1	.0889	.9960	.0892	11.1	.1925	.9813	.1962	17.1	.2940	.9558	.3076
5.2	.0906	.9959	.0910	11.2	.1942	.9810	.1980	17.2	.2957	.9553	.3096
5.3	.0924	.9957	.0928	11.3	.1959	.9806	.1998	17.3	.2974	.9548	.3115
5.4	.0941	.9956	.0945	11.4	.1977	.9803	.2016	17.4	.2990	.9542	.3134
5.5	.0958	.9954	.0963	11.5	.1994	.9799	.2035	17.5	.3007	.9537	.3153
5.6	.0976	.9952	.0981	11.6	.2011	.9796	.2053	17.6	.3024	.9532	.3172
5.7	.0993	.9951	.0998	11.7	.2028	.9792	.2071	17.7	.3040	.9527	.3191
5.8	.1011	.9949	.1016	11.8	.2045	.9789	.2089	17.8	.3057	.9521	.3211
5.9	.1028	.9947	.1033	11.9	.2062	.9785	.2107	17.9	.3074	.9516	.3230

TABLE 1 **Values of Trigonometric Functions** (*continued*)

deg	sin	cos	tan	deg	sin	cos	tan	deg	sin	cos	tan
18.0	.3090	.9511	.3249	24.0	.4067	.9135	.4452	30.0	.5000	.8660	.5774
18.1	.3107	.9505	.3269	24.1	.4083	.9128	.4473	30.1	.5015	.8652	.5797
18.2	.3123	.9500	.3288	24.2	.4099	.9121	.4494	30.2	.5030	.8643	.5820
18.3	.3140	.9494	.3307	24.3	.4115	.9114	.4515	30.3	.5045	.8634	.5844
18.4	.3156	.9489	.3327	24.4	.4131	.9107	.4536	30.4	.5060	.8625	.5867
18.5	.3173	.9483	.3346	24.5	.4147	.9100	.4557	30.5	.5075	.8616	.5890
18.6	.3190	.9478	.3365	24.6	.4163	.9092	.4578	30.6	.5090	.8607	.5914
18.7	.3206	.9472	.3385	24.7	.4179	.9085	.4599	30.7	.5105	.8599	.5938
18.8	.3223	.9466	.3404	24.8	.4195	.9078	.4621	30.8	.5120	.8590	.5961
18.9	.3239	.9461	.3424	24.9	.4210	.9070	.4642	30.9	.5135	.8581	.5985
19.0	.3256	.9455	.3443	25.0	.4226	.9063	.4663	31.0	.5150	.8572	.6009
19.1	.3272	.9449	.3463	25.1	.4242	.9056	.4684	31.1	.5165	.8563	.6032
19.2	.3289	.9444	.3482	25.2	.4258	.9048	.4706	31.2	.5180	.8554	.6056
19.3	.3305	.9438	.3502	25.3	.4274	.9041	.4727	31.3	.5195	.8545	.6080
19.4	.3322	.9432	.3522	25.4	.4289	.9033	.4748	31.4	.5210	.8536	.6104
19.5	.3338	.9426	.3541	25.5	.4305	.9026	.4770	31.5	.5225	.8526	.6128
19.6	.3355	.9421	.3561	25.6	.4321	.9018	.4791	31.6	.5240	.8517	.6152
19.7	.3371	.9415	.3581	25.7	.4337	.9011	.4813	31.7	.5255	.8508	.6176
19.8	.3387	.9409	.3600	25.8	.4352	.9003	.4834	31.8	.5270	.8499	.6200
19.9	.3404	.9403	.3620	25.9	.4368	.8996	.4856	31.9	.5284	.8490	.6224
20.0	.3420	.9397	.3640	26.0	.4384	.8988	.4877	32.0	.5299	.8480	.6249
20.1	.3437	.9391	.3659	26.1	.4399	.8980	.4899	32.1	.5314	.8471	.6273
20.2	.3453	.9385	.3679	26.2	.4415	.8973	.4921	32.2	.5329	.8462	.6297
20.3	.3469	.9379	.3699	26.3	.4431	.8965	.4942	32.3	.5344	.8453	.6322
20.4	.3486	.9373	.3719	26.4	.4446	.8957	.4964	32.4	.5358	.8443	.6346
20.5	.3502	.9367	.3739	26.5	.4462	.8949	.4986	32.5	.5373	.8434	.6371
20.6	.3518	.9361	.3759	26.6	.4478	.8942	.5008	32.6	.5388	.8425	.6395
20.7	.3535	.9354	.3779	26.7	.4493	.8934	.5029	32.7	.5402	.8415	.6420
20.8	.3551	.9348	.3799	26.8	.4509	.8926	.5051	32.8	.5417	.8406	.6445
20.9	.3567	.9342	.3819	26.9	.4524	.8918	.5073	32.9	.5432	.8396	.6469
21.0	.3584	.9336	.3839	27.0	.4540	.8910	.5095	33.0	.5446	.8387	.6494
21.1	.3600	.9330	.3859	27.1	.4555	.8902	.5117	33.1	.5461	.8377	.6519
21.2	.3616	.9323	.3879	27.2	.4571	.8894	.5139	33.2	.5476	.8368	.6544
21.3	.3633	.9317	.3899	27.3	.4586	.8886	.5161	33.3	.5490	.8358	.6569
21.4	.3649	.9311	.3919	27.4	.4602	.8878	.5184	33.4	.5505	.8348	.6594
21.5	.3665	.9304	.3939	27.5	.4617	.8870	.5206	33.5	.5519	.8339	.6619
21.6	.3681	.9298	.3959	27.6	.4633	.8862	.5228	33.6	.5534	.8329	.6644
21.7	.3697	.9291	.3979	27.7	.4648	.8854	.5250	33.7	.5548	.8320	.6669
21.8	.3714	.9285	.4000	27.8	.4664	.8846	.5272	33.8	.5563	.8310	.6694
21.9	.3730	.9278	.4020	27.9	.4679	.8838	.5295	33.9	.5577	.8300	.6720
22.0	.3746	.9272	.4040	28.0	.4695	.8829	.5317	34.0	.5592	.8290	.6745
22.1	.3762	.9265	.4061	28.1	.4710	.8821	.5340	34.1	.5606	.8281	.6771
22.2	.3778	.9259	.4081	28.2	.4726	.8813	.5362	34.2	.5621	.8271	.6796
22.3	.3795	.9252	.4101	28.3	.4741	.8805	.5384	34.3	.5635	.8261	.6822
22.4	.3811	.9245	.4122	28.4	.4756	.8796	.5407	34.4	.5650	.8251	.6847
22.5	.3827	.9239	.4142	28.5	.4772	.8788	.5430	34.5	.5664	.8241	.6873
22.6	.3843	.9232	.4163	28.6	.4787	.8780	.5452	34.6	.5678	.8231	.6899
22.7	.3859	.9225	.4183	28.7	.4802	.8771	.5475	34.7	.5693	.8221	.6924
22.8	.3875	.9219	.4204	28.8	.4818	.8763	.5498	34.8	.5707	.8211	.6950
22.9	.3891	.9212	.4224	28.9	.4833	.8755	.5520	34.9	.5721	.8202	.6976
23.0	.3907	.9205	.4245	29.0	.4848	.8746	.5543	35.0	.5736	.8192	.7002
23.1	.3923	.9198	.4265	29.1	.4863	.8738	.5566	35.1	.5750	.8181	.7028
23.2	.3939	.9191	.4286	29.2	.4879	.8729	.5589	35.2	.5764	.8171	.7054
23.3	.3955	.9184	.4307	29.3	.4894	.8721	.5612	35.3	.5779	.8161	.7080
23.4	.3971	.9178	.4327	29.4	.4909	.8712	.5635	35.4	.5793	.8151	.7107
23.5	.3987	.9171	.4348	29.5	.4924	.8704	.5658	35.5	.5807	.8141	.7133
23.6	.4003	.9164	.4369	29.6	.4939	.8695	.5681	35.6	.5821	.8131	.7159
23.7	.4019	.9157	.4390	29.7	.4955	.8686	.5704	35.7	.5835	.8121	.7186
23.8	.4035	.9150	.4411	29.8	.4970	.8678	.5727	35.8	.5850	.8111	.7212
23.9	.4051	.9143	.4431	29.9	.4985	.8669	.5750	35.9	.5864	.8100	.7239

TABLE 1 **Values of Trigonometric Functions** (*continued*)

deg	sin	cos	tan	deg	sin	cos	tan	deg	sin	cos	tan
36.0	.5878	.8090	.7265	42.0	.6691	.7431	.9004	48.0	.7431	.6691	1.1106
36.1	.5892	.8080	.7292	42.1	.6704	.7420	.9036	48.1	.7443	.6678	1.1145
36.2	.5906	.8070	.7319	42.2	.6717	.7408	.9067	48.2	.7455	.6665	1.1184
36.3	.5920	.8059	.7346	42.3	.6730	.7396	.9099	48.3	.7466	.6652	1.1224
36.4	.5934	.8049	.7373	42.4	.6743	.7385	.9131	48.4	.7478	.6639	1.1263
36.5	.5948	.8039	.7400	42.5	.6756	.7373	.9163	48.5	.7490	.6626	1.1303
36.6	.5962	.8028	.7427	42.6	.6769	.7361	.9195	48.6	.7501	.6613	1.1343
36.7	.5976	.8018	.7454	42.7	.6782	.7349	.9228	48.7	.7513	.6600	1.1383
36.8	.5990	.8007	.7481	42.8	.6794	.7337	.9260	48.8	.7524	.6587	1.1423
36.9	.6004	.7997	.7508	42.9	.6807	.7325	.9293	48.9	.7536	.6574	1.1463
37.0	.6018	.7986	.7536	43.0	.6820	.7314	.9325	49.0	.7547	.6561	1.1504
37.1	.6032	.7976	.7563	43.1	.6833	.7302	.9358	49.1	.7559	.6547	1.1544
37.2	.6046	.7965	.7590	43.2	.6845	.7290	.9391	49.2	.7570	.6534	1.1585
37.3	.6060	.7955	.7618	43.3	.6858	.7278	.9424	49.3	.7581	.6521	1.1626
37.4	.6074	.7944	.7646	43.4	.6871	.7266	.9457	49.4	.7593	.6508	1.1667
37.5	.6088	.7934	.7673	43.5	.6884	.7254	.9490	49.5	.7604	.6494	1.1708
37.6	.6101	.7923	.7701	43.6	.6896	.7242	.9523	49.6	.7615	.6481	1.1750
37.7	.6115	.7912	.7729	43.7	.6909	.7230	.9556	49.7	.7627	.6468	1.1792
37.8	.6129	.7902	.7757	43.8	.6921	.7218	.9590	49.8	.7638	.6455	1.1833
37.9	.6143	.7891	.7785	43.9	.6934	.7206	.9623	49.9	.7649	.6441	1.1875
38.0	.6157	.7880	.7813	44.0	.6947	.7193	.9657	50.0	.7660	.6428	1.1918
38.1	.6170	.7869	.7841	44.1	.6959	.7181	.9691	50.1	.7672	.6414	1.1960
38.2	.6184	.7859	.7869	44.2	.6972	.7169	.9725	50.2	.7683	.6401	1.2002
38.3	.6198	.7848	.7898	44.3	.6984	.7157	.9759	50.3	.7694	.6388	1.2045
38.4	.6211	.7837	.7926	44.4	.6997	.7145	.9793	50.4	.7705	.6374	1.2088
38.5	.6225	.7826	.7954	44.5	.7009	.7133	.9827	50.5	.7716	.6361	1.2131
38.6	.6239	.7815	.7983	44.6	.7022	.7120	.9861	50.6	.7727	.6347	1.2174
38.7	.6252	.7804	.8012	44.7	.7034	.7108	.9896	50.7	.7738	.6334	1.2218
38.8	.6266	.7793	.8040	44.8	.7046	.7096	.9930	50.8	.7749	.6320	1.2261
38.9	.6280	.7782	.8069	44.9	.7059	.7083	.9965	50.9	.7760	.6307	1.2305
39.0	.6293	.7771	.8098	45.0	.7071	.7071	1.0000	51.0	.7771	.6293	1.2349
39.1	.6307	.7760	.8127	45.1	.7083	.7059	1.0035	51.1	.7782	.6280	1.2393
39.2	.6320	.7749	.8156	45.2	.7096	.7046	1.0070	51.2	.7793	.6266	1.2437
39.3	.6334	.7738	.8185	45.3	.7108	.7034	1.0105	51.3	.7804	.6252	1.2482
39.4	.6347	.7727	.8214	45.4	.7120	.7022	1.0141	51.4	.7815	.6239	1.2527
39.5	.6361	.7716	.8243	45.5	.7133	.7009	1.0176	51.5	.7826	.6225	1.2572
39.6	.6374	.7705	.8273	45.6	.7145	.6997	1.0212	51.6	.7837	.6211	1.2617
39.7	.6388	.7694	.8302	45.7	.7157	.6984	1.0247	51.7	.7848	.6198	1.2662
39.8	.6401	.7683	.8332	45.8	.7169	.6972	1.0283	51.8	.7859	.6184	1.2708
39.9	.6414	.7672	.8361	45.9	.7181	.6959	1.0319	51.9	.7869	.6170	1.2753
40.0	.6428	.7660	.8391	46.0	.7193	.6947	1.0355	52.0	.7880	.6157	1.2799
40.1	.6441	.7649	.8421	46.1	.7206	.6934	1.0392	52.1	.7891	.6143	1.2846
40.2	.6455	.7638	.8451	46.2	.7218	.6921	1.0428	52.2	.7902	.6129	1.2892
40.3	.6468	.7627	.8481	46.3	.7230	.6909	1.0464	52.3	.7912	.6115	1.2938
40.4	.6481	.7615	.8511	46.4	.7242	.6896	1.0501	52.4	.7923	.6101	1.2985
40.5	.6494	.7604	.8541	46.5	.7254	.6884	1.0538	52.5	.7934	.6088	1.3032
40.6	.6508	.7593	.8571	46.6	.7266	.6871	1.0575	52.6	.7944	.6074	1.3079
40.7	.6521	.7581	.8601	46.7	.7278	.6858	1.0612	52.7	.7955	.6060	1.3127
40.8	.6534	.7570	.8632	46.8	.7290	.6845	1.0649	52.8	.7965	.6046	1.3175
40.9	.6547	.7559	.8662	46.9	.7302	.6833	1.0686	52.9	.7976	.6032	1.3222
41.0	.6561	.7547	.8693	47.0	.7314	.6820	1.0724	53.0	.7986	.6018	1.3270
41.1	.6574	.7536	.8724	47.1	.7325	.6807	1.0761	53.1	.7997	.6004	1.3319
41.2	.6587	.7524	.8754	47.2	.7337	.6794	1.0799	53.2	.8007	.5990	1.3367
41.3	.6600	.7513	.8785	47.3	.7349	.6782	1.0837	53.3	.8018	.5976	1.3416
41.4	.6613	.7501	.8816	47.4	.7361	.6769	1.0875	53.4	.8028	.5962	1.3465
41.5	.6626	.7490	.8847	47.5	.7373	.6756	1.0913	53.5	.8039	.5948	1.3514
41.6	.6639	.7478	.8878	47.6	.7385	.6743	1.0951	53.6	.8049	.5934	1.3564
41.7	.6652	.7466	.8910	47.7	.7396	.6730	1.0990	53.7	.8059	.5920	1.3613
41.8	.6665	.7455	.8941	47.8	.7408	.6717	1.1028	53.8	.8070	.5906	1.3663
41.9	.6678	.7443	.8972	47.9	.7420	.6704	1.1067	53.9	.8080	.5892	1.3713

TABLE 1 Values of Trigonometric Functions (*continued*)

deg	sin	cos	tan	deg	sin	cos	tan	deg	sin	cos	tan
54.0	.8090	.5878	1.3764	60.0	.8660	.5000	1.7321	66.0	.9135	.4067	2.2460
54.1	.8100	.5864	1.3814	60.1	.8669	.4985	1.7391	66.1	.9143	.4051	2.2566
54.2	.8111	.5850	1.3865	60.2	.8678	.4970	1.7461	66.2	.9150	.4035	2.2673
54.3	.8121	.5835	1.3916	60.3	.8686	.4955	1.7532	66.3	.9157	.4019	2.2781
54.4	.8131	.5821	1.3968	60.4	.8695	.4939	1.7603	66.4	.9164	.4003	2.2889
54.5	.8141	.5807	1.4019	60.5	.8704	.4924	1.7675	66.5	.9171	.3987	2.2998
54.6	.8151	.5793	1.4071	60.6	.8712	.4909	1.7747	66.6	.9178	.3971	2.3109
54.7	.8161	.5779	1.4124	60.7	.8721	.4894	1.7820	66.7	.9184	.3955	2.3220
54.8	.8171	.5764	1.4176	60.8	.8729	.4879	1.7893	66.8	.9191	.3939	2.3332
54.9	.8181	.5750	1.4229	60.9	.8738	.4863	1.7966	66.9	.9198	.3923	2.3445
55.0	.8192	.5736	1.4281	61.0	.8746	.4848	1.8040	67.0	.9205	.3907	2.3559
55.1	.8202	.5721	1.4335	61.1	.8755	.4833	1.8115	67.1	.9212	.3891	2.3673
55.2	.8211	.5707	1.4388	61.2	.8763	.4818	1.8190	67.2	.9219	.3875	2.3789
55.3	.8221	.5693	1.4442	61.3	.8771	.4802	1.8265	67.3	.9225	.3859	2.3906
55.4	.8231	.5678	1.4496	61.4	.8780	.4787	1.8341	67.4	.9232	.3843	2.4023
55.5	.8241	.5664	1.4550	61.5	.8788	.4772	1.8418	67.5	.9239	.3827	2.4142
55.6	.8251	.5650	1.4605	61.6	.8796	.4756	1.8495	67.6	.9245	.3811	2.4262
55.7	.8261	.5635	1.4659	61.7	.8805	.4741	1.8572	67.7	.9252	.3795	2.4383
55.8	.8271	.5621	1.4715	61.8	.8813	.4726	1.8650	67.8	.9259	.3778	2.4504
55.9	.8281	.5606	1.4770	61.9	.8821	.4710	1.8728	67.9	.9265	.3762	2.4627
56.0	.8290	.5592	1.4826	62.0	.8829	.4695	1.8807	68.0	.9272	.3746	2.4751
56.1	.8300	.5577	1.4882	62.1	.8838	.4679	1.8887	68.1	.9278	.3730	2.4876
56.2	.8310	.5563	1.4938	62.2	.8846	.4664	1.8967	68.2	.9285	.3714	2.5002
56.3	.8320	.5548	1.4994	62.3	.8854	.4648	1.9047	68.3	.9291	.3697	2.5129
56.4	.8329	.5534	1.5051	62.4	.8862	.4633	1.9128	68.4	.9298	.3681	2.5257
56.5	.8339	.5519	1.5108	62.5	.8870	.4617	1.9210	68.5	.9304	.3665	2.5386
56.6	.8348	.5505	1.5166	62.6	.8878	.4602	1.9292	68.6	.9311	.3649	2.5517
56.7	.8358	.5490	1.5224	62.7	.8886	.4586	1.9375	68.7	.9317	.3633	2.5649
56.8	.8368	.5476	1.5282	62.8	.8894	.4571	1.9458	68.8	.9323	.3616	2.5782
56.9	.8377	.5461	1.5340	62.9	.8902	.4555	1.9542	68.9	.9330	.3600	2.5916
57.0	.8387	.5446	1.5399	63.0	.8910	.4540	1.9626	69.0	.9336	.3584	2.6051
57.1	.8396	.5432	1.5458	63.1	.8918	.4524	1.9711	69.1	.9342	.3567	2.6187
57.2	.8406	.5417	1.5517	63.2	.8926	.4509	1.9797	69.2	.9348	.3551	2.6325
57.3	.8415	.5402	1.5577	63.3	.8934	.4493	1.9883	69.3	.9354	.3535	2.6464
57.4	.8425	.5388	1.5637	63.4	.8942	.4478	1.9970	69.4	.9361	.3518	2.6605
57.5	.8434	.5373	1.5697	63.5	.8949	.4462	2.0057	69.5	.9367	.3502	2.6746
57.6	.8443	.5358	1.5757	63.6	.8957	.4446	2.0145	69.6	.9373	.3486	2.6889
57.7	.8453	.5344	1.5818	63.7	.8965	.4431	2.0233	69.7	.9379	.3469	2.7034
57.8	.8462	.5329	1.5880	63.8	.8973	.4415	2.0323	69.8	.9385	.3453	2.7179
57.9	.8471	.5314	1.5941	63.9	.8980	.4399	2.0413	69.9	.9391	.3437	2.7326
58.0	.8480	.5299	1.6003	64.0	.8988	.4384	2.0503	70.0	.9397	.3420	2.7475
58.1	.8490	.5284	1.6066	64.1	.8996	.4368	2.0594	70.1	.9403	.3404	2.7625
58.2	.8499	.5270	1.6128	64.2	.9003	.4352	2.0686	70.2	.9409	.3387	2.7776
58.3	.8508	.5255	1.6191	64.3	.9011	.4337	2.0778	70.3	.9415	.3371	2.7929
58.4	.8517	.5240	1.6255	64.4	.9018	.4321	2.0872	70.4	.9421	.3355	2.8083
58.5	.8526	.5225	1.6319	64.5	.9026	.4305	2.0965	70.5	.9426	.3338	2.8239
58.6	.8536	.5210	1.6383	64.6	.9033	.4289	2.1060	70.6	.9432	.3322	2.8397
58.7	.8545	.5195	1.6447	64.7	.9041	.4274	2.1155	70.7	.9438	.3305	2.8556
58.8	.8554	.5180	1.6512	64.8	.9048	.4258	2.1251	70.8	.9444	.3289	2.8716
58.9	.8563	.5165	1.6577	64.9	.9056	.4242	2.1348	70.9	.9449	.3272	2.8878
59.0	.8572	.5150	1.6643	65.0	.9063	.4226	2.1445	71.0	.9455	.3256	2.9042
59.1	.8581	.5135	1.6709	65.1	.9070	.4210	2.1543	71.1	.9461	.3239	2.9208
59.2	.8590	.5120	1.6775	65.2	.9078	.4195	2.1642	71.2	.9466	.3223	2.9375
59.3	.8599	.5105	1.6842	65.3	.9085	.4179	2.1742	71.3	.9472	.3206	2.9544
59.4	.8607	.5090	1.6909	65.4	.9092	.4163	2.1842	71.4	.9478	.3190	2.9714
59.5	.8616	.5075	1.6977	65.5	.9100	.4147	2.1943	71.5	.9483	.3173	2.9887
59.6	.8625	.5060	1.7045	65.6	.9107	.4131	2.2045	71.6	.9489	.3156	3.0061
59.7	.8634	.5045	1.7113	65.7	.9114	.4115	2.2148	71.7	.9494	.3140	3.0237
59.8	.8643	.5030	1.7182	65.8	.9121	.4099	2.2251	71.8	.9500	.3123	3.0415
59.9	.8652	.5015	1.7251	65.9	.9128	.4083	2.2355	71.9	.9505	.3107	3.0595

TABLE 1 **Values of Trigonometric Functions** (*continued*)

deg	sin	cos	tan	deg	sin	cos	tan	deg	sin	cos	tan
72.0	.9511	.3090	3.0777	78.0	.9781	.2079	4.7046	84.0	.9945	.1045	9.5141
72.1	.9516	.3074	3.0961	78.1	.9785	.2062	4.7453	84.1	.9947	.1028	9.6768
72.2	.9521	.3057	3.1146	78.2	.9789	.2045	4.7867	84.2	.9949	.1011	9.8448
72.3	.9527	.3040	3.1334	78.3	.9792	.2028	4.8288	84.3	.9951	.0993	10.0187
72.4	.9532	.3024	3.1524	78.4	.9796	.2011	4.8716	84.4	.9952	.0976	10.1988
72.5	.9537	.3007	3.1716	78.5	.9799	.1994	4.9152	84.5	.9954	.0958	10.3854
72.6	.9542	.2990	3.1910	78.6	.9803	.1977	4.9594	84.6	.9956	.0941	10.5789
72.7	.9548	.2974	3.2106	78.7	.9806	.1959	5.0045	84.7	.9957	.0924	10.7797
72.8	.9553	.2957	3.2305	78.8	.9810	.1942	5.0504	84.8	.9959	.0906	10.9882
72.9	.9558	.2940	3.2506	78.9	.9813	.1925	5.0970	84.9	.9960	.0889	11.2048
73.0	.9563	.2924	3.2709	79.0	.9816	.1908	5.1446	85.0	.9962	.0872	11.4301
73.1	.9568	.2907	3.2914	79.1	.9820	.1891	5.1929	85.1	.9963	.0854	11.6645
73.2	.9573	.2890	3.3122	79.2	.9823	.1874	5.2422	85.2	.9965	.0837	11.9087
73.3	.9578	.2874	3.3332	79.3	.9826	.1857	5.2924	85.3	.9966	.0819	12.1632
73.4	.9583	.2857	3.3544	79.4	.9829	.1840	5.3435	85.4	.9968	.0802	12.4288
73.5	.9588	.2840	3.3759	79.5	.9833	.1822	5.3955	85.5	.9969	.0785	12.7062
73.6	.9593	.2823	3.3977	79.6	.9836	.1805	5.4486	85.6	.9971	.0767	12.9962
73.7	.9598	.2807	3.4197	79.7	.9839	.1788	5.5026	85.7	.9972	.0750	13.2996
73.8	.9603	.2790	3.4420	79.8	.9842	.1771	5.5578	85.8	.9973	.0732	13.6174
73.9	.9608	.2773	3.4646	79.9	.9845	.1754	5.6140	85.9	.9974	.0715	13.9507
74.0	.9613	.2756	3.4874	80.0	.9848	.1736	5.6713	86.0	.9976	.0698	14.3007
74.1	.9617	.2740	3.5105	80.1	.9851	.1719	5.7297	86.1	.9977	.0680	14.6685
74.2	.9622	.2723	3.5339	80.2	.9854	.1702	5.7894	86.2	.9978	.0663	15.0557
74.3	.9627	.2706	3.5576	80.3	.9857	.1685	5.8502	86.3	.9979	.0645	15.4638
74.4	.9632	.2689	3.5816	80.4	.9860	.1668	5.9124	86.4	.9980	.0628	15.8945
74.5	.9636	.2672	3.6059	80.5	.9863	.1650	5.9758	86.5	.9981	.0610	16.3499
74.6	.9641	.2656	3.6305	80.6	.9866	.1633	6.0405	86.6	.9982	.0593	16.8319
74.7	.9646	.2639	3.6554	80.7	.9869	.1616	6.1066	86.7	.9983	.0576	17.3432
74.8	.9650	.2622	3.6806	80.8	.9871	.1599	6.1742	86.8	.9984	.0558	17.8863
74.9	.9655	.2605	3.7062	80.9	.9874	.1582	6.2432	86.9	.9985	.0541	18.4645
75.0	.9659	.2588	3.7321	81.0	.9877	.1564	6.3138	87.0	.9986	.0523	19.0811
75.1	.9664	.2571	3.7583	81.1	.9880	.1547	6.3859	87.1	.9987	.0506	19.7403
75.2	.9668	.2554	3.7848	81.2	.9882	.1530	6.4596	87.2	.9988	.0488	20.4465
75.3	.9673	.2538	3.8118	81.3	.9885	.1513	6.5350	87.3	.9989	.0471	21.2049
75.4	.9677	.2521	3.8391	81.4	.9888	.1495	6.6122	87.4	.9990	.0454	22.0217
75.5	.9681	.2504	3.8667	81.5	.9890	.1478	6.6912	87.5	.9990	.0436	22.9038
75.6	.9686	.2487	3.8947	81.6	.9893	.1461	6.7720	87.6	.9991	.0419	23.8593
75.7	.9690	.2470	3.9232	81.7	.9895	.1444	6.8548	87.7	.9992	.0401	24.8978
75.8	.9694	.2453	3.9520	81.8	.9898	.1426	6.9395	87.8	.9993	.0384	26.0307
75.9	.9699	.2436	3.9812	81.9	.9900	.1409	7.0264	87.9	.9993	.0366	27.2715
76.0	.9703	.2419	4.0108	82.0	.9903	.1392	7.1154	88.0	.9994	.0349	28.6363
76.1	.9707	.2402	4.0408	82.1	.9905	.1374	7.2066	88.1	.9995	.0332	30.1446
76.2	.9711	.2385	4.0713	82.2	.9907	.1357	7.3002	88.2	.9995	.0314	31.8205
76.3	.9715	.2368	4.1022	82.3	.9910	.1340	7.3962	88.3	.9996	.0297	33.6935
76.4	.9720	.2351	4.1335	82.4	.9912	.1323	7.4947	88.4	.9996	.0279	35.8006
76.5	.9724	.2334	4.1653	82.5	.9914	.1305	7.5958	88.5	.9997	.0262	38.1885
76.6	.9728	.2317	4.1976	82.6	.9917	.1288	7.6996	88.6	.9997	.0244	40.9174
76.7	.9732	.2300	4.2303	82.7	.9919	.1271	7.8062	88.7	.9997	.0227	44.0661
76.8	.9736	.2284	4.2635	82.8	.9921	.1253	7.9158	88.8	.9998	.0209	47.7395
76.9	.9740	.2267	4.2972	82.9	.9923	.1236	8.0285	88.9	.9998	.0192	52.0807
77.0	.9744	.2250	4.3315	83.0	.9925	.1219	8.1443	89.0	.9998	.0175	57.2900
77.1	.9748	.2233	4.3662	83.1	.9928	.1201	8.2636	89.1	.9999	.0157	63.6567
77.2	.9751	.2215	4.4015	83.2	.9930	.1184	8.3863	89.2	.9999	.0140	71.6151
77.3	.9755	.2198	4.4373	83.3	.9932	.1167	8.5126	89.3	.9999	.0122	81.8470
77.4	.9759	.2181	4.4737	83.4	.9934	.1149	8.6427	89.4	.9999	.0105	95.4895
77.5	.9763	.2164	4.5107	83.5	.9936	.1132	8.7769	89.5	1.000	.0087	114.5887
77.6	.9767	.2147	4.5483	83.6	.9938	.1115	8.9152	89.6	1.000	.0070	143.2371
77.7	.9770	.2130	4.5864	83.7	.9940	.1097	9.0579	89.7	1.000	.0052	190.9842
77.8	.9774	.2113	4.6252	83.8	.9942	.1080	9.2052	89.8	1.000	.0035	286.4777
77.9	.9778	.2096	4.6646	83.9	.9943	.1063	9.3572	89.9	1.000	.0017	572.9571
								90.0	1.000	.0000	

TABLE 2 Common Logarithms

n	0	1	2	3	4	5	6	7	8	9
1.0	.0000	.0043	.0086	.0128	.0170	.0212	.0253	.0294	.0334	.0374
1.1	.0414	.0453	.0492	.0531	.0569	.0607	.0645	.0682	.0719	.0755
1.2	.0792	.0828	.0864	.0899	.0934	.0969	.1004	.1038	.1072	.1106
1.3	.1139	.1173	.1206	.1239	.1271	.1303	.1335	.1367	.1399	.1430
1.4	.1461	.1492	.1523	.1553	.1584	.1614	.1644	.1673	.1703	.1732
1.5	.1761	.1790	.1818	.1847	.1875	.1903	.1931	.1959	.1987	.2014
1.6	.2041	.2068	.2095	.2122	.2148	.2175	.2201	.2227	.2253	.2279
1.7	.2304	.2330	.2355	.2380	.2405	.2430	.2455	.2480	.2504	.2529
1.8	.2553	.2577	.2601	.2625	.2648	.2672	.2695	.2718	.2742	.2765
1.9	.2788	.2810	.2833	.2856	.2878	.2900	.2923	.2945	.2967	.2989
2.0	.3010	.3032	.3054	.3075	.3096	.3118	.3139	.3160	.3181	.3201
2.1	.3222	.3243	.3263	.3284	.3304	.3324	.3345	.3365	.3385	.3404
2.2	.3424	.3444	.3464	.3483	.3502	.3522	.3541	.3560	.3579	.3598
2.3	.3617	.3636	.3655	.3674	.3692	.3711	.3729	.3747	.3766	.3784
2.4	.3802	.3820	.3838	.3856	.3874	.3892	.3909	.3927	.3945	.3962
2.5	.3979	.3997	.4014	.4031	.4048	.4065	.4082	.4099	.4116	.4133
2.6	.4150	.4166	.4183	.4200	.4216	.4232	.4249	.4265	.4281	.4298
2.7	.4314	.4330	.4346	.4362	.4378	.4393	.4409	.4425	.4440	.4456
2.8	.4472	.4487	.4502	.4518	.4533	.4548	.4564	.4579	.4594	.4609
2.9	.4624	.4639	.4654	.4669	.4683	.4698	.4713	.4728	.4742	.4757
3.0	.4771	.4786	.4800	.4814	.4829	.4843	.4857	.4871	.4886	.4900
3.1	.4914	.4928	.4942	.4955	.4969	.4983	.4997	.5011	.5024	.5038
3.2	.5051	.5065	.5079	.5092	.5105	.5119	.5132	.5145	.5159	.5172
3.3	.5185	.5198	.5211	.5224	.5237	.5250	.5263	.5276	.5289	.5302
3.4	.5315	.5328	.5340	.5353	.5366	.5378	.5391	.5403	.5416	.5428
3.5	.5441	.5453	.5465	.5478	.5490	.5502	.5514	.5527	.5539	.5551
3.6	.5563	.5575	.5587	.5599	.5611	.5623	.5635	.5647	.5658	.5670
3.7	.5682	.5694	.5705	.5717	.5729	.5740	.5752	.5763	.5775	.5786
3.8	.5798	.5809	.5821	.5832	.5843	.5855	.5866	.5877	.5888	.5899
3.9	.5911	.5922	.5933	.5944	.5955	.5966	.5977	.5988	.5999	.6010
4.0	.6021	.6031	.6042	.6053	.6064	.6075	.6085	.6096	.6107	.6117
4.1	.6128	.6138	.6149	.6160	.6170	.6180	.6191	.6201	.6212	.6222
4.2	.6232	.6243	.6253	.6263	.6274	.6284	.6294	.6304	.6314	.6325
4.3	.6335	.6345	.6355	.6365	.6375	.6385	.6395	.6405	.6415	.6425
4.4	.6435	.6444	.6454	.6464	.6474	.6484	.6493	.6503	.6513	.6522
4.5	.6532	.6542	.6551	.6561	.6571	.6580	.6590	.6599	.6609	.6618
4.6	.6628	.6637	.6646	.6656	.6665	.6675	.6684	.6693	.6702	.6712
4.7	.6721	.6730	.6739	.6749	.6758	.6767	.6776	.6785	.6794	.6803
4.8	.6812	.6821	.6830	.6839	.6848	.6857	.6866	.6875	.6884	.6893
4.9	.6902	.6911	.6920	.6928	.6937	.6946	.6955	.6964	.6972	.6981
5.0	.6990	.6998	.7007	.7016	.7024	.7033	.7042	.7050	.7059	.7067
5.1	.7076	.7084	.7093	.7101	.7110	.7118	.7126	.7135	.7143	.7152
5.2	.7160	.7168	.7177	.7185	.7193	.7202	.7210	.7218	.7226	.7235
5.3	.7243	.7251	.7259	.7267	.7275	.7284	.7292	.7300	.7308	.7316
5.4	.7324	.7332	.7340	.7348	.7356	.7364	.7372	.7380	.7388	.7396
n	0	1	2	3	4	5	6	7	8	9

TABLE 2 **Common Logarithms** (*continued*)

n	0	1	2	3	4	5	6	7	8	9
5.5	.7404	.7412	.7419	.7427	.7435	.7443	.7451	.7459	.7466	.7474
5.6	.7482	.7490	.7497	.7505	.7513	.7520	.7528	.7536	.7543	.7551
5.7	.7559	.7566	.7574	.7582	.7589	.7597	.7604	.7612	.7619	.7627
5.8	.7634	.7642	.7649	.7657	.7664	.7672	.7679	.7686	.7694	.7701
5.9	.7709	.7716	.7723	.7731	.7738	.7745	.7752	.7760	.7767	.7774
6.0	.7782	.7789	.7796	.7803	.7810	.7818	.7825	.7832	.7839	.7846
6.1	.7853	.7860	.7868	.7875	.7882	.7889	.7896	.7903	.7910	.7917
6.2	.7924	.7931	.7938	.7945	.7952	.7959	.7966	.7973	.7980	.7987
6.3	.7993	.8000	.8007	.8014	.8021	.8028	.8035	.8041	.8048	.8055
6.4	.8062	.8069	.8075	.8082	.8089	.8096	.8102	.8109	.8116	.8122
6.5	.8129	.8136	.8142	.8149	.8156	.8162	.8169	.8176	.8182	.8189
6.6	.8195	.8202	.8209	.8215	.8222	.8228	.8235	.8241	.8248	.8254
6.7	.8261	.8267	.8274	.8280	.8287	.8293	.8299	.8306	.8312	.8319
6.8	.8325	.8331	.8338	.8344	.8351	.8357	.8363	.8370	.8376	.8382
6.9	.8388	.8395	.8401	.8407	.8414	.8420	.8426	.8432	.8439	.8445
7.0	.8451	.8457	.8463	.8470	.8476	.8482	.8488	.8494	.8500	.8506
7.1	.8513	.8519	.8525	.8531	.8537	.8543	.8549	.8555	.8561	.8567
7.2	.8573	.8579	.8585	.8591	.8597	.8603	.8609	.8615	.8621	.8627
7.3	.8633	.8639	.8645	.8651	.8657	.8663	.8669	.8675	.8681	.8686
7.4	.8692	.8698	.8704	.8710	.8716	.8722	.8727	.8733	.8739	.8745
7.5	.8751	.8756	.8762	.8768	.8774	.8779	.8785	.8791	.8797	.8802
7.6	.8808	.8814	.8820	.8825	.8831	.8837	.8842	.8848	.8854	.8859
7.7	.8865	.8871	.8876	.8882	.8887	.8893	.8899	.8904	.8910	.8915
7.8	.8921	.8927	.8932	.8938	.8943	.8949	.8954	.8960	.8965	.8971
7.9	.8976	.8982	.8987	.8993	.8998	.9004	.9009	.9015	.9020	.9025
8.0	.9031	.9036	.9042	.9047	.9053	.9058	.9063	.9069	.9074	.9079
8.1	.9085	.9090	.9096	.9101	.9106	.9112	.9117	.9122	.9128	.9133
8.2	.9138	.9143	.9149	.9154	.9159	.9165	.9170	.9175	.9180	.9186
8.3	.9191	.9196	.9201	.9206	.9212	.9217	.9222	.9227	.9232	.9238
8.4	.9243	.9248	.9253	.9258	.9263	.9269	.9274	.9279	.9284	.9289
8.5	.9294	.9299	.9304	.9309	.9315	.9320	.9325	.9330	.9335	.9340
8.6	.9345	.9350	.9355	.9360	.9365	.9370	.9375	.9380	.9385	.9390
8.7	.9395	.9400	.9405	.9410	.9415	.9420	.9425	.9430	.9435	.9440
8.8	.9445	.9450	.9455	.9460	.9465	.9469	.9474	.9479	.9484	.9489
8.9	.9494	.9499	.9504	.9509	.9513	.9518	.9523	.9528	.9533	.9538
9.0	.9542	.9547	.9552	.9557	.9562	.9566	.9571	.9576	.9581	.9586
9.1	.9590	.9595	.9600	.9605	.9609	.9614	.9619	.9624	.9628	.9633
9.2	.9638	.9643	.9647	.9652	.9657	.9661	.9666	.9671	.9675	.9680
9.3	.9685	.9689	.9694	.9699	.9703	.9708	.9713	.9717	.9722	.9727
9.4	.9731	.9736	.9741	.9745	.9750	.9754	.9759	.9763	.9768	.9773
9.5	.9777	.9782	.9786	.9791	.9795	.9800	.9805	.9809	.9814	.9818
9.6	.9823	.9827	.9832	.9836	.9841	.9845	.9850	.9854	.9859	.9863
9.7	.9868	.9872	.9877	.9881	.9886	.9890	.9894	.9899	.9903	.9908
9.8	.9912	.9917	.9921	.9926	.9930	.9934	.9939	.9943	.9948	.9952
9.9	.9956	.9961	.9965	.9969	.9974	.9978	.9983	.9987	.9991	.9996
n	0	1	2	3	4	5	6	7	8	9

Index

Answers to problem sets

problem set 1

1. -33 2. -2 3. 0 4. -5 5. -15 6. -100 7. -12 8. 52

9. 11 10. -35 11. -13 12. -21 13. 36 14. 87 15. 1

16. -12 17. -17 18. 46 19. -34 20. -13 21. -87 22. 6

23. -69 24. 35 25. -12 26. $-\dfrac{6}{13}$ 27. -214 28. -15 29. 216

30. 888

problem set 2

1. $x^{-1}y^5$ 2. $x^{-1}y^7$ 3. x^6y^{-12} 4. m^9p^{-3} 5. $k^{-3}a^3$ 6. ab^6

7. $\dfrac{m^5y^3}{x^2}$ 8. $\dfrac{1}{a^2b^5}$ 9. $\dfrac{1}{c^7d^6}$ 10. n^4m 11. $\dfrac{y^{19}}{x^5}$ 12. $\dfrac{c^3}{b^6}$ 13. $b^{-3}c^{-3}$

14. $\dfrac{k^{-2}}{L^{-3}}$ 15. $\dfrac{p^{-2}}{r^{-4}}$ 16. $\dfrac{1}{s^{-1}y^{-1}t^{-5}}$ 17. $\dfrac{x^{-3}z^{-12}}{y^{-1}}$ 18. $\dfrac{1}{x^{-3}y^{-6}}$ 19. $-\dfrac{1}{9}$

20. 8 21. -8 22. -5 23. -8 24. -52 25. -38 26. -30

27. -9 28. -12 29. -27 30. -9

problem set 3

1. -12 2. 12 3. -26 4. -23 5. 72 6. 895 7. -24

8. 2677 9. 1 10. $\dfrac{6b}{a^2c} - \dfrac{4ba^2}{c}$ 11. $\dfrac{-2p^2x^4}{m^5} + \dfrac{5p^4m^5}{x^{-4}}$ 12. $\dfrac{m^4x^5}{k^5} - \dfrac{3m^7}{xk^3}$

13. $x^5y^4 + 4xy$ 14. $-xy^2m + 2ym^2$ 15. $\dfrac{x^8p^6}{y^8}$ 16. $\dfrac{m^4x^5y^{10}}{p^{-2}}$ 17. $\dfrac{p^2}{x^7m^7}$

18. $\dfrac{p^6}{m^3y^4}$ 19. $\dfrac{p^2k^3}{x^{10}}$ 20. x^3 21. $y^{16}x^{-8}p^{-2}$ 22. $\dfrac{y^{12}}{x^2}$ 23. $\dfrac{71}{9}$ 24. -1

25. -15 26. -1 27. 18 28. -15 29. 100 30. -12

problem set 4

1. 4 2. $\dfrac{2}{3}$ 3. $\dfrac{1}{2}$ 4. $\dfrac{1}{20}$ 5. $-\dfrac{16}{9}$ 6. -3 7. $\dfrac{9}{10}$ 8. $-\dfrac{1145}{168}$

9. $\dfrac{4}{b^3a^2} - 12a$ 10. $\dfrac{a^{-1}b}{c^2} - \dfrac{3a^{-5}}{cb^{-2}}$ 11. $x^4 - 3$ 12. $y^{-6}p^{-3} - 3y^{-3}p$

13. y^{-9} **14.** $9x^{-1}y^4$ **15.** $\dfrac{y^2x^4}{4}$ **16.** $2x^2y^{-2}$ **17.** $\dfrac{x^{-8}y^{-3}}{4}$ **18.** $12x^3y^4$

19. $6x^2y^2$ **20.** $-4xy^{-1}+2yx^{-1}$ **21.** $7ay^2x^{-1}+2xya^{-1}$ **22.** 16 **23.** -18

24. 2 **25.** -89 **26.** -1 **27.** 48 **28.** -19 **29.** 4 **30.** $-\dfrac{31}{8}$

problem set 5

1. -3 **2.** 20 **3.** -11 **4.** 150 **5.** $17,500$ **6.** $\dfrac{9}{8}$ **7.** $-\dfrac{13}{6}$

8. $-\dfrac{19}{28}$ **9.** -6 **10.** $\dfrac{41}{15}$ **11.** $+6$ **12.** $\dfrac{71}{14}$ **13.** $2-\dfrac{3x^{-5}}{y^{-2}}$

14. $2-\dfrac{6a^2}{c}$ **15.** $-x^5+\dfrac{3a^2x^3}{b}$ **16.** a^6b^{10} **17.** $\dfrac{4}{x}$ **18.** d^3c^4 **19.** $\dfrac{m^2}{8p^7}$

20. $-2x^3y^3+7x^3y^{-3}$ **21.** $-8x^3a^2+2a^2x$ **22.** -18 **23.** 2 **24.** $\dfrac{7}{16}$

25. $-\dfrac{9}{16}$ **26.** 10 **27.** $\dfrac{73}{4}$ **28.** -35 **29.** 6 **30.** -16

problem set 6

1. $20,000$ **2.** 136 **3.** -3 **4.** 70 **5.** $14, 16,$ and 18 **6.** $1, 2, 3,$ and 4

7. 550 **8.** $\dfrac{365}{204}$ **9.** 5 **10.** $\dfrac{31}{4}$ **11.** 28 **12.** $-3p^{-2}+2x^2$

13. $-\dfrac{k}{x}+2k$ **14.** $\dfrac{y^3}{4x}$ **15.** $9y^{-4}x^{-3}$ **16.** $a^{-5}b^{-3}c^{-1}$ **17.** $\dfrac{1}{16x^{12}y^2}$

18. $3xy-5x$ **19.** $-3x^3y^{-2}-x^3y^2$ **20.** 28 **21.** -14 **22.** $\dfrac{5}{8}$ **23.** $\dfrac{35}{324}$

24. $-\dfrac{9}{8000}$ **25.** 1 **26.** 6 **27.** 0 **28.** -20 **29.** $\dfrac{134}{9}$ **30.** $-\dfrac{215}{72}$

problem set 7

1. 90

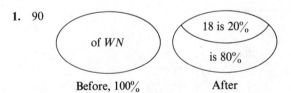

of WN

18 is 20%

is 80%

Before, 100%   After

2. 600%

of 250

1500 is WP

Before, 100%   After

3. 2300

of WN

460 is 20%

is 80%

Before, 100%   After

4. 3400%

of 20

680 is WP

Before, 100%   After

5. 20

of WN   380 is 1900%

Before, 100%   After

6. $-7, -5, -3$ **7.** 2, 4, 6

8. 5   **9.** $-\dfrac{1}{18}$ **10.** -4.88 **11.** $-\dfrac{44}{51}$ **12.** $-\dfrac{47}{100}$ **13.** $-\dfrac{13}{7}$

14. $2 - \dfrac{6x^2y}{p}$ **15.** $8 - \dfrac{12x^{-1}y^2}{k^2}$ **16.** $x^4y^4p^8$ **17.** $\dfrac{2x^2}{y^{10}}$ **18.** $4y^9x^{-6}$

19. $\dfrac{2x^3y}{p}$ **20.** $-6xp^2 + 3p^2$ **21.** $-\dfrac{5}{12}$ **22.** $-\dfrac{1}{16}$ **23.** $\dfrac{1}{8}$ **24.** 16

25. $\dfrac{9}{4}$ **26.** -50 **27.** 7   **28.** 38   **29.** 10   **30.** -4

problem set 8

1. 2300   **2.** 18, 20, 22, and 24   **3.** -13 **4.** 960   **5.** 7, 9, 11

6. 430

of WN

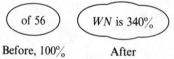
86 is 20%

is 80%

Before, 100%   After

7. 190.4

of 56   WN is 340%

Before, 100%   After

8.

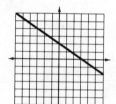

9.

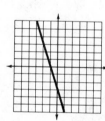

10. 9   **11.** $\dfrac{87}{35}$ **12.** $\dfrac{3}{2}$ **13.** $\dfrac{151}{56}$ **14.** $-3 - \dfrac{4y^3}{xp^3}$

15. $3x - 5xyp^2$ **16.** $\dfrac{2}{x^2y^5}$ **17.** $2x^{-4}$ **18.** $20x^{-12}y^{-8}$ **19.** $5x^2y^{-1}$

20. $4xy^{-3} - 7xy$ **21.** $-\dfrac{3}{32}$ **22.** $\dfrac{1}{27}$ **23.** $\dfrac{11}{40}$ **24.** $-\dfrac{13}{45}$ **25.** $\dfrac{2}{9}$ **26.** 24

27. 2   **28.** $-\dfrac{3}{2}$ **29.** -2 **30.** $-\dfrac{11}{8}$

problem set 9

1. 280   **2.** 64,000   **3.** $97,500   **4.** 3, 4, and 5   **5.** 250,000   **6.** 11,200

7.

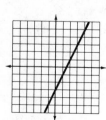

8.

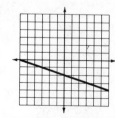

9. 5   **10.** $\dfrac{259}{288}$ **11.** $\dfrac{6}{5}$

12. $\dfrac{8}{7}$ **13.** 0   **14.** 0   **15.** $2k^2 - \dfrac{3a^4}{k}$ **16.** $\dfrac{a^2}{8y^2x^8}$ **17.** $\dfrac{1}{9a^3y^5}$

18. $-\dfrac{x^3}{8z^{12}y^3}$ **19.** $7x - 2xy$ **20.** $5xyp^{-1} - 5xy$ **21.** $\dfrac{7}{16}$ **22.** $\dfrac{9}{32}$ **23.** -90

24. 112   **25.** 5   **26.** 14   **27.** $-15\dfrac{26}{27}$ **28.** -15 **29.** 8   **30.** 7

problem set 10

1. 1750   **2.** 8000   **3.** 5310   **4.** $-5, -3, -1$ **5.** 4000   **6.** 25,800

7. 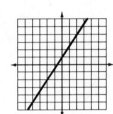   **8.** 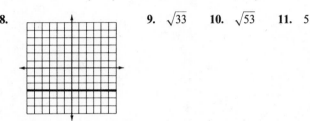   **9.** $\sqrt{33}$ **10.** $\sqrt{53}$ **11.** 5

12. 2   **13.** $-\dfrac{12}{5}$ **14.** $12 - 8x^2m$ **15.** $4 - \dfrac{6xy^2}{p^2}$ **16.** $3x^7y^4$ **17.** $\dfrac{a^2}{4x^5}$

18. $x^{-1}p^{-2}$ **19.** $5x - 1$ **20.** $\dfrac{9x^2y}{z}$ **21.** $\dfrac{5}{18}$ **22.** $\dfrac{7}{54}$ **23.** $-\dfrac{1}{3}$ **24.** $\dfrac{11}{16}$

25. $\dfrac{3}{32}$ **26.** -16 **27.** 2   **28.** 15   **29.** 8   **30.** $-\dfrac{9}{8}$

problem set 11

1. 780   **2.** 4, 6, 8, 10   **3.** \$3054   **4.** 600   **5.** 8800   **6.** $-3, -1, 1, 3$

7. $\dfrac{akm - bak^2 + cx}{ak^2}$ **8.** $\dfrac{kx^3 + 2xbc - 2m}{2ax^3}$ **9.** $\dfrac{4p - 4cak + 3a^2}{4ak}$

10. $\dfrac{4c^2xm^2 - 12p^2c - 5p}{4c^2xp}$ **11.** $\sqrt{105}$ **12.** $\sqrt{89}$ **13.**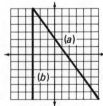

14. $\dfrac{3}{28}$ **15.** 3   **16.** $\dfrac{3}{7}$ **17.** $3 - \dfrac{2}{a^4}$ **18.** $6 - 4y^2$ **19.** $\dfrac{a^7x^7}{4}$

20. $16a^{-8}p^9$ **21.** $8x^{10}m^{-1}y^3$ **22.** 0   **23.** $-5ak$ **24.** -1 **25.** $-\dfrac{1}{2}$

26. $-\dfrac{1}{4}$ **27.** $\dfrac{51}{200}$ **28.** -72 **29.** $\dfrac{21}{4}$ **30.** 3

problem set 12

1. 1600   **2.** $-6, -5, -4, -3$ **3.** 1200   **4.** $\dfrac{5}{11}$ **5.** -5 **6.** 9000

7. $\dfrac{mcx^2b + x^3b + c^2}{cx^2b}$ **8.** $\dfrac{a^3c - 3b^2c - 2a}{a^2bc}$ **9.** $\dfrac{m + k^2m + k^3}{km}$ **10.** $\dfrac{b + a}{b}$

11. 12 **12.** $3\sqrt{13}$ **13.**

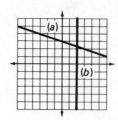

14. (a) $y = \dfrac{1}{3}x + 2$; (b) $y = -2$ **15.** $-\dfrac{13}{4}$ **16.** 5.5 **17.** $-\dfrac{20}{11}$

18. $\dfrac{-2a^{-3}x}{y^{-3}}$ **19.** $1 - \dfrac{2}{x^4}$ **20.** $2x^{-4}y^8$ **21.** $\dfrac{a^8}{b^5}$ **22.** $\dfrac{2a}{x}$ **23.** $-7abc$

24. $\dfrac{11}{32}$ **25.** $-\dfrac{1}{16}$ **26.** 0 **27.** 102 **28.** -2 **29.** -21 **30.** 0

problem set 13

1. 450 **2.** 500,000 **3.** 16, 18, 20, 22 **4.** 5 **5.** $(3, -2)$ **6.** $(2, -5)$

7. $(10, 10)$ **8.** $(13, 7)$ **9.** $\dfrac{p^2c^2 + pck + m}{pc^2}$ **10.** $\dfrac{4a + 2}{a}$ **11.** $\dfrac{4a^2 + 5k^2}{4k}$

12. $\dfrac{m^2ap^2 + map + m}{ap^2}$ **13.** $4\sqrt{2}$ **14.** $2\sqrt{13}$ **15.**

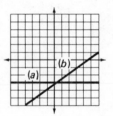

16. (a) $x = 5$; (b) $y = -\dfrac{2}{3}x - 2$ **17.** $\dfrac{52}{63}$ **18.** $-.9$ **19.** $-\dfrac{8}{5}$

20. $3 - \dfrac{2x^2a^2}{3}$ **21.** $-2p^2a^2 + \dfrac{4a^{-3}}{pm^2}$ **22.** $4x^{-3}a^{-6}$ **23.** $\dfrac{9m^2}{p^4x^9}$ **24.** 0

25. $-\dfrac{2am^2}{p} + \dfrac{5pa}{m^2}$ **26.** $\dfrac{9}{8}$ **27.** 124 **28.** 2 **29.** 11 **30.** -1

problem set 14

1. 2530 **2.** 17,150 **3.** -10 **4.** 11, 13, 15 **5.** $(5, -2)$ **6.** $(5, -2)$

7. $(10, 18)$ **8.** $(20, 2)$ **9.** $\dfrac{xcy^2 + ycx^2 - 3x}{cy^2}$ **10.** $\dfrac{m + 4x}{x}$ **11.** $\dfrac{x + a}{x}$

12. $\dfrac{4x + c - cx^2y}{x}$ **13.** $5\sqrt{3}$ **14.** 8 **15.**

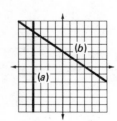

16. (a) $y = \dfrac{2}{3}x + 4$; (b) $y = -4$ **17.** $y = -\dfrac{5}{6}x - \dfrac{1}{2}$ **18.** $y = -\dfrac{3}{5}x + \dfrac{23}{5}$

19. $-\dfrac{1}{5}$ **20.** 110 **21.** $\dfrac{16}{5}$ **22.** $\dfrac{x}{y} + \dfrac{x^2}{3y^2}$ **23.** $\dfrac{y^5}{4x}$ **24.** $1 - 4m^2$

25. $-\dfrac{11xy}{m} + \dfrac{7x}{my}$ **26.** $-\dfrac{11}{36}$ **27.** 1 **28.** -5 **29.** -4 **30.** -6

problem set 15

1. 7000 **2.** 20% **3.** 4 **4.** $-14, -12, -10$ **5.** $(4, 3)$ **6.** $(5, 4)$

7. $(4, 4)$ **8.** $(3, -4)$ **9.** $\dfrac{3y^3a^2 + 3xy^2 - mxa^2}{3y^2a^2}$ **10.** $\dfrac{4x - 3a}{x}$ **11.** $\dfrac{7 - 3p}{p}$

12. $\dfrac{cx + c^2 + ac^2x}{x}$ **13.** $4\sqrt{5}$ **14.** $\sqrt{85}$ **15.**

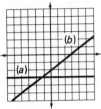

16. (a) $y = 4\dfrac{1}{2}$; (b) $y = -\dfrac{2}{3}x - 1$ **17.** $y = \dfrac{5}{6}x - \dfrac{1}{3}$ **18.** $y = -\dfrac{1}{7}x + \dfrac{33}{7}$

19. $3 + 9xy^3p^2$ **20.** $\dfrac{81}{34}$ **21.** 6 **22.** $\dfrac{3}{8}$ **23.** y^{-5} **24.** $p^2x^8y^{-2}$

25. $6y^2m^2$ **26.** $-\dfrac{1}{4}$ **27.** -9 **28.** 11 **29.** -25 **30.** -5

problem set 16

1. 420 **2.** 6720 **3.** 1050 **4.** 12,900 **5.** $(2, 3)$ **6.** $(3, 4)$ **7.** $(1, -1)$

8. $(4, 0)$ **9.** $6x^3 - 4x^2 - 6x + 4$ **10.** $5x^2 + 18x + 77 + \dfrac{310}{x - 4}$

11. $x^2 + 2x + 4 + \dfrac{2}{x - 2}$ **12.** $\dfrac{4x^2 + a}{2x^2}$ **13.** $\dfrac{16c + 4c^3x - 3}{4c^2x}$ **14.** $\dfrac{-3xyp + 2x^2}{yp}$

15. $4\sqrt{10}$ **16.** $\sqrt{53}$ **17.**

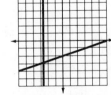

18. (a) $y = -4$; (b) $y = -\dfrac{1}{2}x + 3$ **19.** $y = -\dfrac{2}{7}x - \dfrac{32}{7}$ **20.** $y = \dfrac{3}{5}x + \dfrac{17}{5}$

21. $\dfrac{67}{70}$ **22.** 11 **23.** $-\dfrac{24}{5}$ **24.** $2 - \dfrac{y^2z}{3x^2}$ **25.** $\dfrac{x^4}{9y^2}$ **26.** $-6x^4y^5$

27. $\dfrac{25}{4}$ **28.** -22 **29.** -6 **30.** $-\dfrac{53}{27}$

problem set 17

1. 1000 **2.** 390 **3.** 1440 **4.** 22, 23, 24 **5.** $N_D = 10, N_N = 30$

6. $N_D = 10, N_P = 40$ **7.** $4x^3 + 10x^2 + 10x + 6$ **8.** $3x^2 - 3x + 3 - \dfrac{5}{x + 1}$

9. $T_W = 3, T_M = 2$ **10.** $T_E = 5, T_W = 4$ **11.** $T_M = 1, T_R = 4$ **12.** $\dfrac{4xa + 3x}{a}$

13. $\dfrac{4x + p}{x}$ **14.** $\dfrac{-2xnp^2 - cxnp^2y + 7x^2y^2}{np^2y}$ **15.** $2\sqrt{11}$ **16.** $2\sqrt{13}$

17. 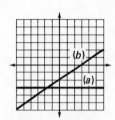   **18.** (a) $y = -2x - 2$; (b) $x = 4$ **19.** $y = -\dfrac{5}{2}x + 12$

20. $y = -\dfrac{2}{7}x + \dfrac{29}{7}$ **21.** $\dfrac{18}{11}$ **22.** 11.09   **23.** 13   **24.** $\dfrac{1}{my^2} - \dfrac{3x^2m^2}{y}$

25. $\dfrac{x^4y^2}{8}$ **26.** $-2xy^3 + 7y$ **27.** -3 **28.** -11 **29.** -30 **30.** 0

problem set 18

1. 40   **2.** 3200   **3.** 200,000   **4.** 1,000,000   **5.** $N_D = 150, N_Q = 50$

6. $N_P = 1, N_D = 29$ **7.** $6x^3 + 8x^2 - 28x - 40$ **8.** $5x^2 + 10x + 20 + \dfrac{39}{x - 2}$

9. $R_F = 96, R_S = 80$ **10.** $T_M = 1, T_R = 4$ **11.** $T_G = 5, T_B = 8$

12. $\dfrac{7x^2y^2z^2 + 1}{xyz}$ **13.** $\dfrac{x + a}{x}$ **14.** $\dfrac{-3x^2y^2 - cxy^3 + 7c}{xy^3}$ **15.** $2\sqrt{29}$ **16.** $7\sqrt{2}$

17. 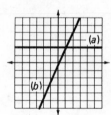   **18.** (a) $y = -3$; (b) $y = -3x$ **19.** $y = -x + 2$

20. $y = \dfrac{5}{3}x - \dfrac{26}{3}$ **21.** $\dfrac{9}{20}$ **22.** 4   **23.** $-\dfrac{3}{2}$ **24.** $\dfrac{6x}{yz^3p} - \dfrac{27x^2}{z^2}$ **25.** $\dfrac{y^2x^3}{a^5}$

26. 0   **27.** $-\dfrac{1}{2}$ **28.** 21   **29.** -10 **30.** -19

problem set 19

1. $N_D = 30, N_N = 20$ **2.** $N_C = 12, N_M = 14$ **3.** 198   **4.** 1400 liters, 266 liters

5. -2 **6.** $(3, -7)$ **7.** $x^5 - 2x^4 - 4x^3 + 8x^2 + 4x - 8$

8. $-3x^2 - 6x - 12 - \dfrac{26}{x - 2}$ **9.** $T_H = 2, T_S = 2$ **10.** $R_F = 60, R_S = 50$

11. $T_M = 1, T_R = 4$ **12.** $\dfrac{28y^2z + 3x^2}{7y^2z}$ **13.** $\dfrac{ay - 2b - 2cx^2y}{2x^2y}$

14. $\dfrac{1 + m^3p}{m^2p}$ **15.** 8   **16.** $\sqrt{106}$ **17.**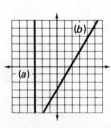

18. (a) $y = 2$; (b) $y = 2x$ **19.** $y = \dfrac{9}{5}x + \dfrac{2}{5}$ **20.** $y = \dfrac{2}{7}x - \dfrac{29}{7}$ **21.** $\dfrac{23}{39}$

22. -100 **23.** $-\dfrac{1}{2}$ **24.** $-3x + \dfrac{9y^3 p}{x}$ **25.** $\dfrac{y^4}{x}$ **26.** $\dfrac{3x^2 a}{y}$ **27.** $-\dfrac{35}{108}$

28. 1   **29.** 8   **30.** -16

problem set 20

1. 560   **2.** 40, 12   **3.** $N_W = 10, N_E = 13$ **4.** 360   **5.** $-4, -2, 0, 2$

6. $(4, 16)$ **7.** $-2x^2 - 2x - 3 - \dfrac{1}{x - 1}$ **8.** $T_G = 5, T_B = 8$

9. $-35\sqrt{2}$ **10.** $144\sqrt{2}$ **11.** $24 - 12\sqrt{2}$ **12.** $50 - 75\sqrt{2}$ **13.** $\dfrac{7y + x}{y}$

14. $\dfrac{m^2 + 5x - mx^2}{ax^2}$ **15.** $\dfrac{2a^5 - 2a^5 x^2 - 3x^3}{2a^4 x^2}$ **16.** $4\sqrt{6}$

17.

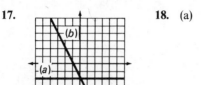

18. (a) $y = 1$; (b) $y = -x$ **19.** $y = \dfrac{1}{2}x - \dfrac{5}{2}$

20. $y = -\dfrac{1}{12}x + \dfrac{31}{6}$ **21.** $\dfrac{25}{4}$ **22.** -5 **23.** 0   **24.** $2x$ **25.** $-\dfrac{y^8}{2x^2}$

26. $-\dfrac{2x}{a} - \dfrac{5}{ax}$ **27.** $-\dfrac{129}{512}$ **28.** 19   **29.** 3   **30.** 27

problem set 21

1. 36, 48   **2.** 86, 42   **3.** 2250   **4.** 185 kg   **5.** 25   **6.** $-3, -4, -5$

7. $(6, 5)$ **8.** $8x^3 - 16x^2 + 10x - 6$ **9.** $T_K = 8, T_N = 16$ **10.** $50\sqrt{2}$

11. 240   **12.** $24\sqrt{3} - 12$ **13.** $\dfrac{4xp + 1}{p}$ **14.** $\dfrac{m^2 - 3ax - ma^2 x}{a^2 x^2}$

15. 1×10^{-26} **16.** 7   **17.**

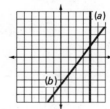

18. (a) $y = -2$; (b) $y = -2x$ **19.** $y = -5$ **20.** $y = -\dfrac{3}{7}x + \dfrac{20}{7}$ **21.** $\dfrac{39}{128}$

22. 24   **23.** 14   **24.** $-\dfrac{10}{x^4} - \dfrac{5}{x^3}$ **25.** $4x^6 y^2$ **26.** $5xa$ **27.** $-\dfrac{1}{12}$

28. 0   **29.** 8   **30.** 5

problem set 22

1. 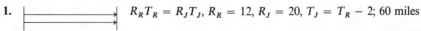 $R_R T_R = R_J T_J, R_R = 12, R_J = 20, T_J = T_R - 2$; 60 miles

2. $R_R T_R = R_W T_W, R_R = 10, T_R = 4, T_W = 2$; 20 kph   **3.** 84, 60

4. $3700   **5.** 8   **6.** 4750   **7.** (4, 4)   **8.** $x^2 + x - 3 - \dfrac{1}{x-1}$

9. $144\sqrt{2}$ **10.** $-9\sqrt{3}$ **11.** $12 - 6\sqrt{3}$ **12.** $30 - 12\sqrt{2}$ **13.** $\dfrac{2x+1}{x}$

14. $\dfrac{5x^2p + p^3y - 3x}{py}$ **15.** 2×10^{-2} **16.**

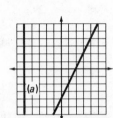

17. (a) $x = -4$; (b) $y = -\dfrac{3}{2}x + 3$ **18.** $y = \dfrac{1}{3}x - 2$ **19.** $3\sqrt{10}$

20. $y = \dfrac{2}{5}x - \dfrac{31}{5}$ **21.** $-\dfrac{45}{14}$ **22.** $\dfrac{7}{2}$ **23.** 4   **24.** $1 - \dfrac{3x^2}{p^2y^7}$

25. $4x^{10}y^{-1}p^{-1}$ **26.** $-6y$ **27.** $-\dfrac{7}{200}$ **28.** 14   **29.** 3   **30.** 43

problem set 23

1. $R_M T_M = R_J T_J$, $R_M = 600$, $R_J = 800$, $T_M = T_J + 4$; 12 min

2. $R_F T_F = R_S T_S$, $T_F = 10$, $T_S = 12$, $R_F = R_S + 10$; 600 km

3. $\dfrac{110}{120}$ **4.** $N_V = 175$, $N_S = 125$ **5.** $N_Q = 10$, $N_N = 50$ **6.** 320 tons

7. (10, 10)   **8.** $6x^5 - 3x^4 - 16x^3 + 2x^2 + 8x$ **9.** $144\sqrt{3}$ **10.** $-\sqrt{3}$

11. $20\sqrt{6} - 24$ **12.** $50 - 30\sqrt{3}$ **13.** $\dfrac{3x^2y^2m + 4}{x}$ **14.** $\dfrac{5x^2p - 4p^2m + c}{p^2m}$

15. 1×10^{-36} **16.** $\left(-\dfrac{9}{5}, \dfrac{4}{5}\right)$

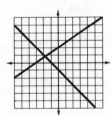

17. $\left(-\dfrac{5}{2}, -3\right)$ **18.** (a) $y = 1$; (b) $y = x$ **19.** $y = \dfrac{1}{2}x$

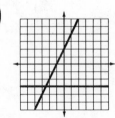

20. $2\sqrt{5}$ **21.** $y = -\dfrac{3}{8}x + \dfrac{11}{2}$ **22.** $-\dfrac{38}{39}$ **23.** 3   **24.** 12   **25.** $1 - \dfrac{3}{x}$

26. $\dfrac{x^4}{9y}$ **27.** $\dfrac{6m}{x}$ **28.** $\dfrac{3}{2}$ **29.** 6   **30.** $\dfrac{33}{4}$

problem set 24

1. $R_F T_F = R_S T_S$, $T_F = 6$, $T_S = 8$, $R_F = 60$; 45 mph

2. $R_1 T_1 = R_2 T_2$, $R_1 = 4$, $R_2 = 5$, $T_2 = T_1 - 1$; 20 miles   3. $\dfrac{50}{70}$

4. $N_N = 81$, $N_C = 92$ 5. 20   6. 4140   7. (8, 7)

8. $-3x^2 + 6x - 12 + \dfrac{27}{x + 2}$ 9. $720\sqrt{2}$ 10. $6\sqrt{7}$ 11. $30 - 36\sqrt{6}$

12. $20\sqrt{5} - 12$ 13. $\dfrac{4m^4 y^2 p + 6}{m^2 y}$ 14. $\dfrac{k^2 pc + 2p^2 c^2 - 8}{2p^2 c}$ 15. 1×10^{-19}

16. $\left(\dfrac{32}{11}, \dfrac{18}{11}\right)$ 17. (a) $y = \dfrac{5}{6}x + 2$;   (b) $y = -4$

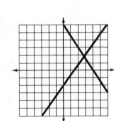

18. $y = 2x + 9$ 19. $4\sqrt{5}$ 20. $y = \dfrac{2}{9}x - \dfrac{37}{9}$ 21. $\dfrac{43}{24}$ 22. 2   23. $-\dfrac{13}{90}$

24. -250 25. $\dfrac{3}{2}$ 26. $1 + 5x^3 y$ 27. $-32x^3 y^{-2}$ 28. $\dfrac{-2p^2 x - 5p^2 x^7}{y}$

29. 12   30. $-\dfrac{1}{36}$

problem set 25

1. 95 days   2. 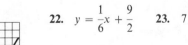 $R_B T_B = R_F T_F$, $T_B = 40$, $T_F = 30$, $R_F = R_B + 6$; 720 km

3. $N_B = 50$, $N_G = 10$ 4. 700   5. $N_Q = 100$, $N_H = 100$ 6. 2200 grams

7. $x^2 + 5x + 25 + \dfrac{123}{x - 5}$ 8. $2xp^2 k(2x - 3kp^2)$ 9. $2xy(2x - 1 + 5y)$

10. $x^2 ym^2(y^2 m^3 + 12xm^2 - 3y)$ 11. $4mp^2 y(4mp - 2py^3 + my)$

12. $x^2 yz(xyz^2 + z - 3x)$ 13. $p^3 x(p^2 x^2 + px - 1)$ 14. $180\sqrt{6}$ 15. $13\sqrt{2}$

16. $30\sqrt{3} - 20$ 17. $162 - 54\sqrt{2}$ 18. $\dfrac{ab + a}{b}$ 19. $\dfrac{ax^2 - cm^2 p + 2mp}{m^2 p}$

20. 6×10^{14} 21. $(3, -\frac{1}{2})$ 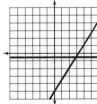   22. $y = \dfrac{1}{6}x + \dfrac{9}{2}$ 23. 7

24. 7   25. $\dfrac{17}{7}$ 26. $\dfrac{7}{10}$ 27. $6x + 1$ 28. $\dfrac{2}{y^4} - \dfrac{4x^4}{y^2}$ 29. 16   30. $\dfrac{5}{192}$

problem set 26

1. $R_A T_A = R_B T_B$, $T_A = 20$, $T_B = 8$, $R_B = R_A + 60$; 800 mi

2. $N_R = 40$, $N_B = 250$ 3. $N_T = 200$, $N_F = 1000$ 4. 700   5. 2080

6. 8100 **7.** $x^3 - x^2 + x - 1 - \dfrac{1}{x + 1}$ **8.** $7x^4y^2m(5x^3y^3 - xm + 2y^5m)$

9. $2xym(3xm^4 - x + 2)$ **10.** $2xy^4p^5(2xp - y^3 + 4x^3y)$ **11.** $(x - 3)(x - 2)$

12. $(x - 7)(-x - 2)$ **13.** $-3(x - 4)(x + 2)$ **14.** $-2x^2(x + 2)(x - 8)$

15. $ab(x + 2)(x - 1)$ **16.** $6x - 1$ **17.** $108\sqrt{2}$ **18.** $9\sqrt{3} - 40$

19. $15\sqrt{6} - 12$ **20.** $\dfrac{m + x^2y}{m}$ **21.** $\dfrac{a^2mc - x^4c - x^4}{x^2c}$ **22.** 6×10^{-22}

23. $(4, 5)$

24. $\sqrt{58}$ **25.** -66 **26.** $\dfrac{26}{3}$ **27.** $-\dfrac{21}{2}$

28. $-\dfrac{37}{19}$ **29.** $-8x^8y$ **30.** 1

problem set 27

1. $R_HT_H = R_RT_R$, $R_H = 4$, $R_R = 20$, $T_R + T_H = 18$; 60 miles

2. $R_HT_H = R_RT_R$, $R_H = 5$, $R_R = 20$, $T_R + T_H = 10$; 40 km

3. $N_B = 10$, $N_G = 26$ **4.** 14, 21, 28 **5.** 1800 grams **6.** 600 grams

7. $x^2 - 2x + 4 - \dfrac{14}{x + 2}$ **8.** $3x^2m^2p^2(3x^2 + m^2p^4 - 2x^2m)$ **9.** $mx^2y(x^2 - y^2 - 4)$

10. $a^2x^3p(1 - 4a - x)$ **11.** $a(x + 5)(x - 1)$ **12.** $-x(x - 5)(x - 3)$

13. $-ax(x + 8)(x - 3)$ **14.** $-ax^2(x - 5)(x + 1)$ **15.** $p(x - 8)(x - 7)$

16. $a + 1$ **17.** $3\sqrt{2} - 36$ **18.** $19\sqrt{5}$ **19.** $6\sqrt{6} - 18$ **20.** $\dfrac{10x + 4a}{x(x + a)}$

21. $\dfrac{7x^3 + 26x^2 + 2x + 8}{x^2(x + 4)}$ **22.** $\dfrac{x^2 + x - 3}{x(x + 3)(x + 1)}$ **23.** 40

24. $(2, 3)$

25. $y = -\dfrac{3}{8}x - \dfrac{33}{8}$ **26.** $\dfrac{97}{3}$ **27.** -65

28. $-\dfrac{19}{6}$ **29.** $5a^{-2}yb^{-1}$ **30.** -55

problem set 28

1. $R_MT_M = R_PT_P$, $R_M = 60$, $R_P = 3$, $T_M + T_P = 21$; 60 miles

2. $-9, -6, -3, 0$ **3.** $N_B = 4$, $N_G = 11$ **4.** $N_D = 100$, $N_N = 400$ **5.** 142,500

6. 1400 kg, 8750 kg **7.** $x^2 + 5x + 25 + \dfrac{118}{x - 5}$ **8.** $2x^2y(1 - 4x^2y^3)$

9. $x^2y^3p(4p^2 - 16 - x^2p^3)$ 10. $xy(x + 7)(x - 5)$ 11. $a(x - 8)(x + 1)$

12. $m^2(x + 1)(x + 2)$ 13. $-a^2(x + 1)(x + 1)$ 14. $4x + 1$ 15. $10\sqrt{3}$

16. $15\sqrt{3} - 30$ 17. 2×10^{-5} 18. $\dfrac{a}{x + y}$ 19. $\dfrac{a}{c}$ 20. $\dfrac{3\sqrt{5}}{10}$ 21. $\dfrac{\sqrt{3}}{9}$

22. $\dfrac{4a^2 + 6a + 6x}{a(a + x)}$ 23. $\dfrac{5x + 3}{x^2 + 2x + 1}$ 24. $(1, 1/2)$

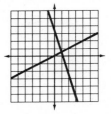

25. (a) $y = -4$; (b) $y = -\dfrac{1}{3}x + 2$ 26. $\dfrac{24}{13}$ 27. $\dfrac{139}{66}$ 28. $4\sqrt{5}$

29. $1 - \dfrac{12z^2y^3}{x^4}$ 30. $\dfrac{29}{36}$

**problem
set 29**

1. $\longmapsto\!\!\longrightarrow\!\!\longmapsto$ $R_WT_W + R_RT_R = 60,\ R_W = 3,\ R_R = 9,\ T_W + T_R = 8;\ 2\ \text{hrs}$

2. $\longmapsto\!\!\longrightarrow\!\!\mapsfrom\!\!\longmapsto$ $R_LT_L + R_QT_Q = 54,\ R_L = 3,\ T_L = 8,\ T_Q = 6;\ 5\ \text{mph}$

3. $\longleftarrow\!\!\bullet\!\!\longrightarrow$ $R_RT_R + R_MT_M = 11{,}800;\ T_R = 5,\ T_M = 4,\ R_M = R_R - 200;$

$R_M = 1200\ \text{kph},\ R_R = 1400\ \text{kph}$

4. 175   5. 3,900,000   6. 140,000   7. $x^2 + 2x + 5$ 8. $8x^2y^2z^2(2xz - 1)$

9. $2x^2yp^2(p^2 - 3xp - 1)$ 10. $a^2(x + 7)(x + 5)$ 11. $-p(x + 5)(x - 3)$

12. $-m^2(x + 1)(x + 1)$ 13. $k(x + 8)(x - 5)$ 14. $1 + a$ 15. $-6\sqrt{3}$

16. $18\sqrt{2} - 24$ 17. 1×10^{23} 18. $\dfrac{m}{m + x}$ 19. $\dfrac{a}{b}$ 20. $\dfrac{\sqrt{3}}{10}$ 21. $\dfrac{14\sqrt{3}}{45}$

22. $\dfrac{4x^2 + 14x + 24}{x^2 + 6x + 8}$ 23. $\dfrac{-5m^2 - 7m}{m^2 + 3m + 2}$ 24. $(1, 4)$

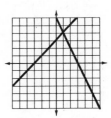

25. $y = -\dfrac{3}{8}x - \dfrac{9}{4}$

26. $\dfrac{58}{49}$ 27. 20   28. $-\dfrac{33}{35}$ 29. $x^{-6}y^{15}z^{-9}$ 30. $\dfrac{25}{108}$

**problem
set 30**

1. 480 grams   2. 500 grams

3. $\longleftarrow\!\!\bullet\!\!\longrightarrow$ $R_FT_F + R_BT_B = 68,\ R_F = 6,\qquad T_F = 6,\ T_B = 4;\ 8\ \text{mph}$

4. $N_D = 10,\ N_P = 25$ 5. $N_Q = 5,\ N_D = 10$ 6. (a) $400\ \text{ft}^2$; (b) 100 ft

7. 30 ft^2 **8.** (a) 31.4 ft; (b) 78.5 ft^2 **9.** $x^2 - x + 1 + \dfrac{1}{x + 1}$

10. $-a^2(x - 7)(x + 5)$ **11.** $3(x - 5)(x - 2)$ **12.** $-ab(x - 5)(x - 5)$

13. $a^2b^2(x + 7)(x + 2)$ **14.** $1 - 4x$ **15.** $-17\sqrt{2}$ **16.** $24\sqrt{6} - 36$

17. 7×10^{49} **18.** $\dfrac{x}{x + y}$ **19.** $\dfrac{a}{p}$ **20.** $\dfrac{\sqrt{6}}{9}$ **21.** $\dfrac{\sqrt{2}}{15}$ **22.** $\dfrac{9a^2 + 6a + 8}{2a^2 + 8a}$

23. $\dfrac{6x + 6}{x^2 + 5x + 6}$ **24.** $\left(\dfrac{12}{5}, \dfrac{2}{5}\right)$ **25.** $y = -\dfrac{1}{3}x + \dfrac{10}{3}$

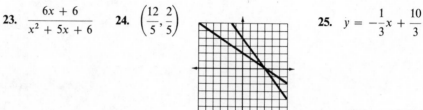

26. $-\dfrac{63}{5}$ **27.** -18 **28.** $\dfrac{11}{10}$ **29.** $\dfrac{2xa^2y}{p} - \dfrac{a^2x^2y}{p}$ **30.** $\dfrac{7}{16}$

problem set 31

1. $40, 45, 50, 55$

2. $R_BT_B + R_ST_S = 960, R_B = 40, R_S = 70, T_B - 2 = T_S$; 8 p.m.

3. $R_JT_J = R_RT_R, R_J = 6, R_R = 30, T_J + T_R = 12$; 60 miles

4. $N_T = 40, N_P = 30$ **5.** $12{,}000 \text{ kg}$ **6.** 120 g **7.** $y = -\dfrac{5}{2}x + \dfrac{19}{2}$

8. $A = 100 \text{ ft}^2, P = 48 \text{ ft}$ **9.** $P = 38.56 \text{ in}, 89.12 \text{ tiles}$ **10.** 3×10^{-31}

11. $\dfrac{a}{a + b}$ **12.** $\dfrac{4}{m}$ **13.** $\dfrac{3\sqrt{5}}{10}$ **14.** $\dfrac{7\sqrt{2}}{6}$ **15.** $\sqrt{3}$ **16.** $30 - 24\sqrt{3}$

17. $1 + 4x$ **18.** $\dfrac{x^2 + 3x + 3}{x^2 + 4x + 4}$ **19.** $\dfrac{x + 2}{x - 3}$ **20.** $-x(x - 3)(x - 2)$

21. $2ax(x - 5)(x - 4)$ **22.** $pa(x - 2)(x - 1)$ **23.** $mc(x + 5)(x - 2)$ **24.** -3

25. $\dfrac{2}{25}$ **26.** $\left(\dfrac{27}{8}, \dfrac{15}{4}\right)$ **27.** $x^3 - x^2 + x - 1 - \dfrac{1}{x + 1}$

28. $(\tfrac{1}{2}, 2)$ **29.** 17 **30.** $-\dfrac{1}{2}$

problem set 32

1. $R_WT_W + R_JT_J = 56, R_W = 4, R_J = 8, T_W + T_J = 10$; $D_W = 24 \text{ miles}, D_J = 32 \text{ miles}$ **2.** $N_S = 10, N_L = 60$ **3.** $N_N = 40, N_D = 70$

4. 500 kg **5.** 1200 g **6.** 2200 **7.** $\dfrac{7\sqrt{10}}{10}$ **8.** $-\dfrac{26\sqrt{21}}{21}$ **9.** $-\dfrac{31\sqrt{14}}{14}$

10. 35 ft^2 **11.** 92 ft^2 **12.** $y = -\frac{1}{3}x + \frac{8}{3}$ **13.** 1×10^{-10} **14.** $\dfrac{x + 4y}{x + y}$

15. $1 + 4xy$ **16.** $15\sqrt{5}$ **17.** $40\sqrt{2} - 60$ **18.** $\dfrac{x^2 + 2}{x^2 + 5x + 6}$ **19.** $\dfrac{m^2 - 2}{m^2 - 5m}$

20. $-2x(x - 3)(x - 1)$ **21.** $x^3(x + 7)(x - 2)$ **22.** $ax(x - 7)(x - 1)$

23. $py(x + 6)(x - 2)$ **24.** $-\dfrac{21}{13}$ **25.** -30 **26.** $\left(-\dfrac{8}{3}, \dfrac{2}{3}\right)$

27. $4x^4 + 2x^3 - 8x^2 + 12x + 8$ **28.** $\left(\dfrac{19}{7}, \dfrac{8}{7}\right)$ **29.** $7\sqrt{2}$ **30.** -2

problem set 33

1. 442 **2.** 2800 g **3.** 2025 g **4.** $N_B = 5, N_R = 50$

5. $R_M T_M + R_C T_C = 540, R_M = 40, R_C = 60, T_M + T_C = 11; 240 \text{ miles}$

6. $\dfrac{x^2 + 4}{xa}$ **7.** $\dfrac{(ay + mx + my)(a + m)}{xy(x + y)}$ **8.** 84 m^2 **9.** 24 m^2

10. $y = \dfrac{5}{3}x + \dfrac{26}{3}$ **11.** $\dfrac{11\sqrt{10}}{10}$ **12.** $\dfrac{4\sqrt{30}}{15}$ **13.** 56 m **14.** 18

15. 1.5×10^{-22} **16.** $\dfrac{x^2 + 4xy}{1 - xy}$ **17.** $12\sqrt{2}$ **18.** $1 + y$

19. $\dfrac{ax + b(x + y) + x^2(cx + 4)}{x^2(x + y)}$ **20.** $\dfrac{-2x - 6}{x^2 + 2x - 8}$ **21.** $-x^2(x - 5)(x + 1)$

22. $k^2(x - 5)(x - 2)$ **23.** $ap(x + 4)(x - 5)$ **24.** -23 **25.** $\dfrac{36}{5}$

26. $\left(-\dfrac{4}{3}, \dfrac{7}{3}\right)$ **27.** $2x^3 + 4x^2 + 8x + 15 + \dfrac{30}{x - 2}$ **28.** 36

29. 12 ft **30.** 8 ft^2

problem set 34

1. $R_B T_B + 40 = R_M T_M, R_M = 50, R_B = 60, T_M = T_B + 4; 2 \text{ a.m.}$

2. $R_R T_R + 7 = R_S T_S, R_S = 6, R_R = 2\frac{1}{2}, T_S = T_R; 2 \text{ hr}$

3. $R_R T_R = R_P T_P, R_R = 10, R_P = 3, T_R + T_P = 13; 30 \text{ miles}$

4. 60 tons   5. 150 g   6. 6, 7, 8   7. $\dfrac{y - a^2b^2}{b - a^2}$ 8. $\dfrac{1 - bx}{x^2}$ 9. 500 ft^2

10. 121.5 ft^2 11. $y = \dfrac{1}{2}x$ 12. $-\dfrac{23\sqrt{2}}{6}$ 13. 0   14. 602.88

15. 1.4×10^{-16} 16. $\dfrac{4y^2 + x}{3y^2 - 1}$ 17. $28\sqrt{3}$ 18. $1 - 5y$

19. $\dfrac{ax^2 + bx^2 + cx(x + y)}{x^3(x + y)}$ 20. $\dfrac{x^2 - 9x + 13}{x^2 - 6x + 9}$ 21. $2x^2(x + 2)(x - 1)$

22. $ap(x - 4)(x + 2)$ 23. $y(x - 2)(x - 2)$ 24. $-\dfrac{13}{2}$ 25. $-\dfrac{9}{2}$

26. $\left(\dfrac{12}{7}, -\dfrac{10}{7}\right)$ 27. $x^4 + 3x^3 + 5x^2 + 3x$ 28. $\left(-\dfrac{2}{7}, -\dfrac{10}{7}\right)$

29. $4\sqrt{11}$ 30. -12

problem set 35

1. |1200 ———→| $R_L T_L = R_M T_M + 1200$, $R_L = 3R_M$, $T_L = 30$, $T_M = 30$; $R_M = 20$ yds/min, $R_L = 60$ yds/min

2. |———→| $R_T T_T + R_B T_B = 56$, $R_T = 4$, $R_B = 6$, $T_T + T_B = 12$; $T_T = 8$, $T_B = 4$

3. |———→| $R_D T_D = R_B T_B$, $T_D = 12$, $T_B = 4$, $R_B = R_D + 6$; 36 miles

4. $N_Y = 4$, $N_B = 10$ 5. 100 tons   6. 400 g   7. 725.6 in^3 8. $\dfrac{1}{2}$ 9. $\dfrac{1}{3}$

10. 64   11. $-\dfrac{1}{4}$ 12. 81   13. $\dfrac{31\sqrt{35}}{35}$ 14. $-\dfrac{41\sqrt{10}}{10}$ 15. $\dfrac{a - 4b}{xy}$

16. $\dfrac{7x + 6y}{4}$ 17. 46 ft   18. 52 ft^2 19. 19.76 ft^2 20. $y = -\dfrac{4}{3}x + \dfrac{1}{3}$

21. $\dfrac{2x^2 - 2x + 4}{x^2(x + y)}$ 22. $\dfrac{3x^2 + 7x}{x^2 + x - 6}$ 23. $-a(x + 7)(x - 5)$

24. $-x^2(x + 4)(x - 2)$ 25. $36 - 14\sqrt{3}$ 26. 1×10^{10} 27. $-\dfrac{76}{5}$ 28. 30

29. $\dfrac{9}{4}$ 30. $\sqrt{122}$

problem set 36

1. 1360   2. $N_L = 10$, $N_F = 20$ 3. $\dfrac{15}{25}$

4. |———→| $R_D T_D = R_H T_H$, $T_D = 10$, $T_H = 8$, $R_H = R_D + 10$; 400 miles

5. |———→ 80| $R_B T_B = R_J T_J + 80$, $T_B = 4$, $T_J = 4$, $R_J = 30$; $R_B = 50$ mph

6. 49, 56, 63, 70   **7.** $\dfrac{x-3}{x+2}$ **8.** $\dfrac{x-2}{x+3}$ **9.** $\dfrac{x-2}{x+5}$ **10.** -9

11. $-\dfrac{1}{9}$ **12.** 27   **13.** $\dfrac{1}{9}$ **14.** $\dfrac{y+4x}{3xy+1}$ **15.** $\dfrac{4-3x}{7+2x}$ **16.** $\dfrac{3\sqrt{15}}{5}$

17. 1×10^7 **18.** $\dfrac{5\sqrt{21}}{21}$ **19.** $24\sqrt{6}-96$ **20.** $5\sqrt{7}$ **21.** $\dfrac{-3x^2+6x+6}{x^2(x+2)(x+1)}$

22. $\dfrac{ax^2p+acx^2+a^2x+mx+b}{a^2x^4}$ **23.** $\dfrac{2}{3}$ **24.** 8   **25.** $y=2,\ y=\dfrac{1}{3}x-2$

26. $71.4\ \text{ft}^3$ **27.** $x^3-x^2+x-1-\dfrac{1}{x+1}$ **28.** $\sqrt{130}$ **29.** $3-\dfrac{3y^3}{x^5}$

30. -16

problem set 37

1. 3200 g   **2.** 420 g   **3.** 700 g   **4.** $D_B=100, D_T=200$ **5.** 200   **6.** $\dfrac{x+1}{x+4}$

7. $x-2$ **8.** $-\dfrac{1}{16}$ **9.** $\dfrac{1}{81}$ **10.** $-\dfrac{1}{81}$ **11.** 81   **12.** $128\ \text{ft}^2$

13. $128\ \text{ft}^3$ **14.** $\dfrac{x^3+1}{x^4-1}$ **15.** $\dfrac{ax+y}{ax-my}$ **16.** $-\dfrac{29\sqrt{14}}{14}$ **17.** $\dfrac{7\sqrt{33}}{33}$

18. $y=\dfrac{9}{4}x+11$ **19.** 72.56 m   **20.** $(1,3)$

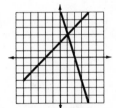

21. $\dfrac{x^2-4x}{x^2+4x-5}$ **22.** $\dfrac{6x+16}{x(x+2)}$ **23.** $\dfrac{41}{16}$ **24.** $\dfrac{30}{11}$ **25.** $30\sqrt{6}-12$

26. $24\sqrt{10}-80$ **27.** $\dfrac{y^7}{x^4}$ **28.** 3   **29.** 5×10^{20} **30.** $\dfrac{4}{y}-\dfrac{12}{p}$

problem set 38

1. 15   **2.** 320 kg   **3.** 64 g   **4.** $n=98, d=28$ **5.** 120 miles

6. $x^3+9x^2+27x+27$ **7.** x^3+3x^2+3x+1 **8.** $3, -2$ **9.** $0, 5, -7$

10. $5, -7$ **11.** $\dfrac{x+4}{x-3}$ **12.** $-\dfrac{1}{2}$ **13.** $-\dfrac{1}{8}$ **14.** 4   **15.** -2

16. 97.12   **17.** 768   **18.** $\dfrac{x^3-a}{a^2x-a}$ **19.** $\dfrac{mp^2-20}{15p^2}$ **20.** 3×10^{-10}

21. $\dfrac{61\sqrt{22}}{22}$ **22.** $-\dfrac{\sqrt{33}}{33}$ **23.** $72-108\sqrt{2}$ **24.** $y=-\dfrac{1}{5}x+\dfrac{23}{5}$ **25.** $\dfrac{92}{17}$

26. x^3-x^2+x-1 **27.** $\sqrt{26}$ **28.** $\dfrac{3x^2+12x+3}{x^2+5x+6}$ **29.** $\dfrac{y^{13}}{p^{20}}$ **30.** 40

problem set 39

1. 1820   **2.** 24,624   **3.** 144 g   **4.** 112 g   **5.** 15 miles   **6.** $m=8, y=6$

7. $M=4.2, A=5.6, B=2.4$ **8.** $3, -3$ **9.** $\dfrac{5}{9}, -\dfrac{5}{9}$ **10.** $0, -3, -8$

11. $x^3 - 3x^2 + 3x - 1$ 12. $\dfrac{x-2}{x-3}$ 13. -8 14. 9 15. $\dfrac{a^2x^2-4}{x^2+6a}$

16. $\dfrac{m^2p-6x}{m^2px-4}$ 17. 1×10^{-45} 18. $-\dfrac{59\sqrt{39}}{39}$ 19. $\dfrac{11\sqrt{6}}{6}$ 20. $15\sqrt{5}+2\sqrt{3}$

21. 8.71 22. 4647.2 23. 18 24. $y = -\dfrac{5}{6}x + \dfrac{17}{3}$ 25. $\dfrac{55}{36}$ 26. 3310

27. $\dfrac{2x^2+2x+5y+6xy}{y(x+1)^2}$ 28. $1+xy$ 29. $3+\dfrac{8}{z^4x^2y^6}$ 30. 24

problem set 40

1. $\$1,968,000$ 2. $18,240$ 3. 704 g 4. $N_S = 40, N_B = 120$ 5. 150 yd

6. $\dfrac{a^2}{ca-m}$ 7. $\dfrac{ax}{y-ac}$ 8. $\dfrac{6y}{m+yk+yax}$ 9. $\dfrac{mc}{a-bc}$

10. $y = 24, x = 10$ 11. $x^3 - 9x^2 + 27x - 27$ 12. $\pm\dfrac{7}{2}$ 13. $0, -3, 6$ 14. 1

15. 8 16. 8 17. $\dfrac{4x^2a^3+y^2}{2y^2a^2-2}$ 18. $\dfrac{xy^2-4p}{a^2yp-1}$ 19. 2×10^8 20. $30\sqrt{2}$

21. $\dfrac{23\sqrt{21}}{21}$ 22. 121.12 23. 32.56 yd 24. (a) $x = -1$ (b) $y = x - 2$

25. $-\dfrac{51}{2}$ 26. 510 27. $\dfrac{8x^2+11x+6}{x(x+2)(x+1)}$ 28. 5 29. $1 - \dfrac{3y^2}{p^2}$ 30. $-\dfrac{1}{24}$

problem set 41

1. $3, 5, 7, 9$ 2. 840 g 3. 400 4. $N_N = 60, N_D = 40$

5. $D_R = 64$ miles, $D_W = 12$ miles 6. $44 \text{ ft}^2 \times \dfrac{12 \text{ in}}{\text{ft}} \times \dfrac{12 \text{ in}}{\text{ft}} = (44)(12)(12) \text{ in}^2$

7. $42 \text{ yd}^2 \times \dfrac{3 \text{ ft}}{\text{yd}} \times \dfrac{3 \text{ ft}}{\text{yd}} \times \dfrac{12 \text{ in}}{\text{ft}} \times \dfrac{12 \text{ in}}{\text{ft}} = (42)(3)(3)(12)(12) \text{ in}^2$

8. $16 \text{ mi}^3 \times \dfrac{5280 \text{ ft}}{\text{mi}} \times \dfrac{5280 \text{ ft}}{\text{mi}} \times \dfrac{5280 \text{ ft}}{\text{mi}} \times \dfrac{12 \text{ in}}{\text{ft}} \times \dfrac{12 \text{ in}}{\text{ft}} \times \dfrac{12 \text{ in}}{\text{ft}}$

$= (16)(5280)(5280)(5280)(12)(12)(12) \text{ in}^3$

9. $\dfrac{mx}{cm+k}$ 10. $\dfrac{xyc}{cm+k}$ 11. $\dfrac{x^2km}{4pm-xy}$ 12. $p = 37.5, m = 25.5$

13. $2x^2 + 4x + 8 + \dfrac{15}{x-2}$ 14. $0, 2, 8$ 15. $0, \dfrac{9}{4}$ 16. 32 cm^2 17. 774.72 ft^3

18. $\dfrac{x+5}{x-2}$ 19. $\dfrac{1}{4}$ 20. $\dfrac{a^2x^2-a}{ax^2-4}$ 21. 7×10^{-22} 22. $\dfrac{-27\sqrt{35}}{35}$

23. $96 - 48\sqrt{3}$ 24. $y = 3x + 1$ 25. $-\dfrac{17}{2}$ 26. $\dfrac{1}{42}$

27. $\dfrac{-2x^3-2x^2+2x+4}{x^2(x-2)(x+2)}$ 28. $1+2x$ 29. $\dfrac{y^3}{x}$ 30. $\dfrac{5}{24}$

problem set 42

1. 396 2. 6000 3. 7200 g 4. 156 g 5. $N_S = 2, N_L = 10$

6. 3×10^{15} 7. 1×10^{-2} 8. 2×10^{51}

9. $40 \text{ yd}^3 \times \dfrac{3 \text{ ft}}{\text{yd}} \times \dfrac{3 \text{ ft}}{\text{yd}} \times \dfrac{3 \text{ ft}}{\text{yd}} \times \dfrac{12 \text{ in}}{\text{ft}} \times \dfrac{12 \text{ in}}{\text{ft}} \times \dfrac{12 \text{ in}}{\text{ft}} = (40)(3)(3)(3)(12)(12)(12) \text{ in}^3$

10. $\dfrac{m^2}{p + xm}$ 11. $\dfrac{axd}{pc - md}$ 12. $4x^2 - 8x + 16 - \dfrac{33}{x + 2}$ 13. $a = 45, b = 51$

14. $0, 4, 5$ 15. $\dfrac{5}{2}, -\dfrac{5}{2}$ 16. $\dfrac{x + 5}{x + 4}$ 17. $-\dfrac{1}{27}$ 18. $\dfrac{m - 4x}{6x - 1}$ 19. $\dfrac{-49\sqrt{33}}{33}$

20. $36 - 24\sqrt{3}$ 21. $5\sqrt{5}$ 22. 104.52 23. 3000 24. $y = -x - 5$

25. $-\dfrac{21}{4}$ 26. $\dfrac{5}{52}$ 27. $-\dfrac{16}{x^2 - 4}$ 28. $-1 + 3x^{-4}$ 29. $1 - x^2$ 30. $-\dfrac{27}{2}$

problem set 43

1. $400,000$ 2. 4000 3. 184 g 4. 2310 g 5. 40 miles

6. $\sin A = \dfrac{\text{opposite}}{\text{hypotenuse}}$, $\cos A = \dfrac{\text{adjacent}}{\text{hypotenuse}}$, $\tan A = \dfrac{\text{opposite}}{\text{adjacent}}$

7. (a) $.63$; (b) $.92$; (c) 1.59 8. (a) $.57$; (b) $.81$; (c) $.75$

9. $4 \text{ ft}^3 \times \dfrac{12 \text{ in}}{\text{ft}} \times \dfrac{12 \text{ in}}{\text{ft}} \times \dfrac{12 \text{ in}}{\text{ft}} = 4(12)(12)(12) \text{ in}^3$ 10. $\dfrac{az}{p + xz}$ 11. $\dfrac{xzc}{m + kc}$

12. $\dfrac{xcz}{zk + pc}$ 13. $b = \dfrac{45}{4}, a = \dfrac{17}{2}, c = \dfrac{17}{4}$ 14. $0, -5, -4$ 15. $\dfrac{9}{2}, -\dfrac{9}{2}$

16. $x^3 - 9x^2 + 27x - 27$ 17. 1 18. $\dfrac{1}{8}$ 19. $\dfrac{4x^2 + 1}{ay^2 - 4x}$ 20. $-\dfrac{25\sqrt{21}}{21}$

21. $30\sqrt{6} - 6$ 22. 923 cm^2 23. $y = -\dfrac{1}{7}x - \dfrac{37}{7}$ 24. $\dfrac{71}{19}$ 25. 6×10^{-7}

26. $\left(-\dfrac{3}{5}, \dfrac{13}{5}\right)$

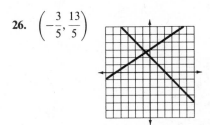

27. $\dfrac{6x^3 - x^2 - 2x + 2}{x^2(x^2 - 1)}$ 28. $8 - 16x^{-4}y^2$ 29. $\sqrt{37}$ 30. $-\dfrac{1}{8}$

problem set 44

1. 7 2. $-11, -9, -7, -5$ 3. 1080 tons 4. 55 g 5. 50 miles

6. $H = 10.44, y = 4.275, B = 66°$ 7. $m = 11.48, x = 7.98, B = 55°$

8. $x = \sqrt{13}, A = 31°, B = 59.0°$ 9. $4 \text{ mi}^2 \times \dfrac{5280 \text{ ft}}{\text{mi}} \times \dfrac{5280 \text{ ft}}{\text{mi}} = 4(5280)(5280) \text{ ft}^2$

10. $\dfrac{kd}{p + cd}$ 11. $\dfrac{3kpc}{Lp + cd}$ 12. $\dfrac{am}{x - dm}$ 13. $b = \dfrac{15}{2}, a = \dfrac{5\sqrt{13}}{2}$ 14. $0, 4, 6$

15. $+\dfrac{9}{5}, -\dfrac{9}{5}$ 16. $x^3 - 4x^2 + 16x - 64 + \dfrac{255}{x + 4}$ 17. $\dfrac{x + 5}{x - 3}$ 18. $-\dfrac{1}{16}$

19. 3208.32 in^3 20. $\dfrac{5x^3 + 1}{pm^2 + 5x}$ 21. $\dfrac{-79\sqrt{34}}{34}$ 22. $36 - 18\sqrt{2}$

23. (a) $y = 2$; (b) $y = \frac{1}{2}x - 4$ **24.** 7   **25.** $-\frac{3}{32}$

26. $\left(\frac{12}{5}, -\frac{1}{5}\right)$

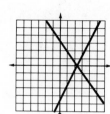

27. $\frac{x^2 - 2x - 2}{x^2 - 4}$ **28.** 4×10^{-32}

29. $\frac{1}{8}$ **30.** 8

problem set 45

1. 400   **2.** 272 g   **3.** 3024   **4.** $N_C = 3, N_R = 8$

5. $R_T = 70$ mph, $R_B = 40$ mph   **6.** $\pm\sqrt{3}$ **7.** $-2 \pm \sqrt{3}$ **8.** $-17 \pm \sqrt{2}$

9. $-\frac{2}{5} \pm \sqrt{3}$ **10.** 5.4   **11.** 6.5   **12.** $A = 48.2°, B = 41.8°$

13. $(100{,}000 \text{ mi}^2) \times \frac{5280 \text{ ft}}{\text{mi}} \times \frac{5280 \text{ ft}}{\text{mi}} \times \frac{12 \text{ in}}{\text{ft}} \times \frac{12 \text{ in}}{\text{ft}} = (100{,}000)(5280)(5280)(12)(12) \text{ in}^2$

14. 1×10^{-1} **15.** $\frac{13}{3}$ **16.** $\frac{4y}{4c - mx}$ **17.** $\frac{5xc}{m + kc}$ **18.** $0, 5, -10$

19. $\frac{5}{6}, -\frac{5}{6}$ **20.** $\frac{x + 5}{x + 2}$ **21.** $\frac{1}{243}$ **22.** 32 inches   **23.** $\frac{4x^2p - 1}{6px - p^2}$

24. $\frac{5\sqrt{10}}{2}$ **25.** $30\sqrt{3} - 20\sqrt{2}$ **26.** $y = -\frac{2}{3}x - \frac{19}{3}$ **27.** $\frac{43}{6}$

28. $\left(\frac{-45}{13}, \frac{43}{13}\right)$

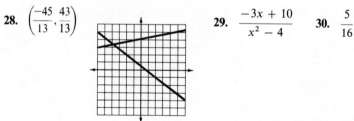

29. $\frac{-3x + 10}{x^2 - 4}$ **30.** $\frac{5}{16}$

problem set 46

1. 600   **2.** 6000 kg   **3.** 212 g   **4.** $N_E = 22, N_F = 77$ **5.** 20 mph

6. $-\frac{9\sqrt{10}}{10}$ **7.** $\frac{55\sqrt{14}}{14}$ **8.** $3^{3/4}$ **9.** $3^{1/2}$ **10.** $x^{11/6}y^{4/3}$ **11.** $x^{23/12}y^{9/4}$

12. $\frac{2}{5} \pm \sqrt{7}$ **13.** $\frac{1}{4} \pm \sqrt{5}$ **14.** $B = 66°, y = 7.69$ **15.** $A = 55.2°, C = \sqrt{33}$

16. 1×10^4 **17.** $\frac{kb}{xa - cb}$ **18.** $\frac{ybm}{x - pm}$ **19.** $y = \frac{12}{5}, x = \frac{16}{5}$

20. $0, -5, -8$ **21.** $\frac{x - 2}{x - 7}$ **22.** $\frac{1}{16}$ **23.** 339.36   **24.** $\frac{m^2 - x^2}{p^2 + 2x^2}$

25. $42\sqrt{2} - 21$ **26.** $y = \frac{9}{5}x + \frac{38}{5}$ **27.** $-\frac{29}{7}$ **28.** $-\frac{91}{10}$

29. $\left(-\dfrac{8}{3}, \dfrac{1}{3}\right)$

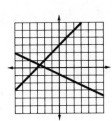

30. $\dfrac{3}{32}$

problem set 47

1. 4900   **2.** $-9, -7, -5, -3$ **3.** 2000   **4.** 328 g   **5.** $D_R = 40, D_W = 8$

6. $\dfrac{42\ \text{in}}{\text{sec}} \times \dfrac{1\ \text{ft}}{12\ \text{in}} \times \dfrac{1\ \text{mi}}{5280\ \text{ft}} \times \dfrac{60\ \text{sec}}{\text{min}} \times \dfrac{60\ \text{min}}{\text{hr}} = \dfrac{(42)(60)(60)\ \text{mi}}{(12)(5280)\ \text{hr}}$

7. $\dfrac{480\ \text{mi}}{\text{hr}} \times \dfrac{5280\ \text{ft}}{1\ \text{mi}} \times \dfrac{1\ \text{hr}}{60\ \text{min}} \times \dfrac{1\ \text{min}}{60\ \text{sec}} = \dfrac{(480)(5280)}{(60)(60)}\ \dfrac{\text{ft}}{\text{sec}}$

8. $\dfrac{15\ \text{in}}{\text{sec}} \times \dfrac{1\ \text{ft}}{12\ \text{in}} \times \dfrac{1\ \text{yd}}{3\ \text{ft}} \times \dfrac{60\ \text{sec}}{\text{min}} \times \dfrac{60\ \text{min}}{\text{hr}} = \dfrac{(15)(60)(60)}{(12)(3)}\ \dfrac{\text{yd}}{\text{hr}}$ **9.** $3^{7/6}$ **10.** $2^{5/12}$

11. $(x^2y)^{1/2}(y^2x)^{1/3} = xy^{1/2}y^{2/3}x^{1/3} = x^{4/3}y^{7/6}$ **12.** $(2 \cdot 2^{1/3})^{1/2} = 2^{1/2}2^{1/6} = 2^{2/3}$

13. $\dfrac{5\sqrt{6}}{2}$ **14.** $-\dfrac{104\sqrt{35}}{35}$ **15.** 5.61   **16.** 9.28   **17.** $3 \pm \sqrt{5}$ **18.** $0, -\dfrac{4}{7}$

19. 5×10^3 **20.** $\dfrac{acy}{m - pc}$ **21.** $\dfrac{cx}{a - bx}$ **22.** $2\dfrac{1}{2}$ **23.** 1872 in

24. $\dfrac{x^2p - m^2}{x - pm}$ **25.** $\dfrac{x - 2}{x - 5}$ **26.** $-\dfrac{1}{81}$ **27.** $\left(\dfrac{12}{5}, -\dfrac{7}{5}\right)$

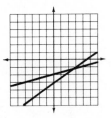

28. $y = \dfrac{2}{5}x + \dfrac{29}{5}$ **29.** $0, -2, -10$ **30.** $\dfrac{19}{40}$

problem set 48

1. 1600   **2.** 5600 g   **3.** 252 g   **4.** $N_G = 1, N_B = 4$ **5.** $R_F = 30, R_E = 60$

6. 68   **7.** No solution   **8.** 1

9. $\dfrac{60\ \text{mi}}{\text{hr}} \times \dfrac{5280\ \text{ft}}{\text{mi}} \times \dfrac{1\ \text{hr}}{60\ \text{min}} \times \dfrac{1\ \text{min}}{60\ \text{sec}} = \dfrac{(60)(5280)}{(60)(60)}\ \dfrac{\text{ft}}{\text{sec}}$

10. $(2 \cdot 2^{1/3})^{1/2} = 2^{1/2}2^{1/6} = 2^{2/3}$ **11.** $(m^2y)^{1/2}(m^4y)^{1/3} = my^{1/2}m^{4/3}y^{1/3} = m^{7/3}y^{5/6}$

12. $(2^3 \cdot 2^{1/3})^{1/2} = 2^{3/2}2^{1/6} = 2^{5/3}$ **13.** $\dfrac{41\sqrt{10}}{10}$ **14.** $\dfrac{311\sqrt{77}}{77}$ **15.** $-\dfrac{12}{7}, \dfrac{16}{7}$

16. $7, -1$ **17.** $c = 27°, b = 4.45$ **18.** $A = 48.2°$ **19.** 1×10^{24}

20. $\dfrac{cxy}{by + xyk - mx}$ **21.** $\dfrac{3\sqrt{65}}{2}$ **22.** $3x^2 - 3x + 3 - \dfrac{5}{x + 1}$ **23.** $\sqrt{41}$

24. $\dfrac{x + 7}{x + 3}$ **25.** $-\dfrac{1}{8}$ **26.** 283.52 in³   **27.** $\sqrt{145}$ **28.** $\dfrac{x^2 - x - 1}{-4x^2 + x + 2}$

29. $y = -\dfrac{8}{5}x + \dfrac{41}{5}$ **30.** $\dfrac{19}{3}$

problem set 49

1. -3 2. 420 3. 117.5 g 4. $N_Q = 5, N_R = 10$ 5. 220 yd/min

6. $y = -6x + 18$ 7. $N = 33.3$ 8. 68 9. 1

10. $\dfrac{200 \text{ in}}{\text{min}} \times \dfrac{1 \text{ ft}}{12 \text{ in}} \times \dfrac{1 \text{ min}}{60 \text{ sec}} = \dfrac{200}{(12)(60)} \dfrac{\text{ft}}{\text{sec}}$ 11. $2^{1/4} \cdot 2^{1/3} = 2^{7/12}$

12. $(3^2 \cdot 3^{1/2})^{1/3} = 3^{2/3} \cdot 3^{1/6} = 3^{5/6}$ 13. $(x^2y^3)^{1/2}(xy^4)^{1/2} = xy^{3/2}x^{1/2}y^2 = x^{3/2}y^{7/2}$

14. $(xy)^{1/3}(xy)^{1/2} = x^{1/3}y^{1/3}x^{1/2}y^{1/2} = x^{5/6}y^{5/6}$ 15. $-\dfrac{59\sqrt{14}}{14}$ 16. $\dfrac{91\sqrt{2}}{6}$

17. $\dfrac{5}{3} \pm \sqrt{5}$ 18. $2, -6$ 19. 2×10^{33} 20. $\dfrac{x^2}{m - xd - xc}$ 21. $\dfrac{mxy}{2y - 3x}$

22. $x^3 + 3x^2 + 3x + 1$ 23. $\dfrac{x - 3}{x - 2}$ 24. $\sqrt{82}$ 25. $y = -\dfrac{1}{9}x + \dfrac{22}{9}$

26. $(128.52)(12)(12)$ 27. $\dfrac{ay^2x - 3xy - 5a^2}{a^2y^4}$ 28. $\dfrac{-6x + 7}{x^2 - 9}$ 29. 12 30. $\dfrac{14}{3}$

problem set 50

1. $N_W = 2, N_T = 20$ 2. $D_S = 32, D_B = 72$ 3. 138 g 4. $-14, -12, -10$

5. 180 6. $-3 \pm \sqrt{13}$ 7. $-1 \pm \sqrt{6}$ 8. $\dfrac{5}{2} \pm \dfrac{\sqrt{5}}{2}$ 9. $y = 3x + 9$

10. $C = 73°, m = 3.1$ 11. 2 12. 31

13. $\dfrac{400 \text{ yd}}{\text{sec}} \times \dfrac{3 \text{ ft}}{1 \text{ yd}} \times \dfrac{1 \text{ mi}}{5280 \text{ ft}} \times \dfrac{60 \text{ sec}}{1 \text{ min}} \times \dfrac{60 \text{ min}}{1 \text{ hr}} = \dfrac{(400)(3)(60)(60)}{5280} \dfrac{\text{mi}}{\text{hr}}$

14. $(2 \cdot 2^{1/3})^{1/5} = 2^{1/5}2^{1/15} = 2^{4/15}$ 15. $(3^2 \cdot 3^{1/2})^{1/2} = 3 \cdot 3^{1/4} = 3^{5/4}$

16. $(m^3y^5)^{1/2}(m^2y^2)^{1/3} = m^{3/2}y^{5/2}m^{2/3}y^{2/3} = m^{13/6}y^{19/6}$ 17. $\dfrac{-62\sqrt{33}}{33}$ 18. 1×10^{24}

19. $\dfrac{2p + kpr}{mxr}$ 20. $\dfrac{3x^2m}{4m - xc}$ 21. $\dfrac{25}{4}$ 22. $2x^2 + 3x + 6 + \dfrac{23}{2x - 3}$

23. $0, 7, -4$ 24. $\dfrac{x + 2}{x + 5}$ 25. -343 26. 110.75 27. $y = -\dfrac{5}{3}x + \dfrac{5}{3}$

28. 2 29. 12 30. $4x^{-3} - \dfrac{3x^{-2}}{p}$

problem set 51

1. 408 2. 2400 3. 355 g 4. $N_D = 15, N_Q = 20$ 5. 400 miles

6. $2 + 9i$ 7. $-6 + 3i$ 8. $2 + 6i$ 9. $-3i$ 10. $-\dfrac{1}{2} \pm \dfrac{\sqrt{5}}{2}$ 11. $+1, -4$

12. $y = -4x + 10$ 13. 2.30 14. No solution

15. $\dfrac{20 \text{ in}}{\text{hr}} \times \dfrac{1 \text{ ft}}{12 \text{ in}} \times \dfrac{1 \text{ mi}}{5280 \text{ ft}} \times \dfrac{1 \text{ hr}}{60 \text{ min}} = \dfrac{20}{(12)(5280)(60)} \dfrac{\text{mi}}{\text{min}}$

16. $(3 \cdot 3^{1/2})^{1/5} = 3^{1/5}3^{1/10} = 3^{3/10}$ 17. $(3^2 \cdot 3^{1/3})^{1/2} = 3 \cdot 3^{1/6} = 3^{7/6}$

18. $(x^2y^2m)^{1/2}(xym^2)^{1/4} = xym^{1/2}x^{1/4}y^{1/4}m^{1/2} = x^{5/4}y^{5/4}m$ 19. $\dfrac{201\sqrt{26}}{26}$ 20. $-\dfrac{1}{243}$

21. 5×10^{22} **22.** $\dfrac{xm}{k + cm}$ **23.** $\dfrac{pR_1}{3R_1 - xp}$ **24.** 9936 **25.** $y = \dfrac{2}{3}x + \dfrac{2}{3}$

26. 3 **27.** $4p^2 - 2p$ **28.** $2x^2 - 4x + 6 - \dfrac{8}{x + 2}$ **29.** $-\dfrac{50}{7}$ **30.** $-\dfrac{10}{9}$

problem set ER-1

1. 8550 **2.** 400 **3.** 198 g **4.** $N_D = 40, N_N = 50$ **5.** 700 mi

6. -141.1043 **7.** -17.4403 **8.** -7.11458 **9.** $-10 - 4i$ **10.** $-1 + 8i$

11. $-\dfrac{1}{2} \pm \dfrac{\sqrt{5}}{2}$ **12.** $1, -5$ **13.** $y = -6x - 24$ **14.** $P = 21.43, M = 17.91$

15. 8 **16.** $\dfrac{2 \text{ ft}}{\text{sec}} \times \dfrac{1 \text{ mi}}{5280 \text{ ft}} \times \dfrac{60 \text{ sec}}{1 \text{ min}} \times \dfrac{60 \text{ min}}{\text{hr}} = \dfrac{2(60)(60)}{5280} \dfrac{\text{mi}}{\text{hr}}$

17. $(2 \cdot 2^{1/3})^{1/4} = 2^{1/4} \cdot 2^{1/12} = 2^{1/3}$ **18.** $(5^2 \cdot 5^{1/3})^{1/2} = 5 \cdot 5^{1/6} = 5^{7/6}$

19. $(x^3 y^2 m)^{1/2} (xy^2 m^2)^{1/5} = x^{3/2} y m^{1/2} \, x^{1/5} y^{2/5} m^{2/5} = x^{17/10} y^{7/5} m^{9/10}$ **20.** $-\dfrac{57\sqrt{26}}{26}$

21. $-\dfrac{1}{27}$ **22.** 7×10^{-25} **23.** $\dfrac{bm}{a + kb}$ **24.** $144(12)(12)$ **25.** $y = -\dfrac{4}{3}x - \dfrac{22}{3}$

26. $\sqrt{26}$ **27.** $3x^6 - \dfrac{12x^4}{y^5 p^3}$ **28.** $2x^2 - 5x + 10 - \dfrac{21}{x + 2}$ **29.** $-\dfrac{39}{5}$ **30.** $-\dfrac{18}{13}$

problem set 52

1. 75 ml 10%, 25 ml 50% **2.** 700 ml 10%, 700 ml 2% **3.** 460 g

4. $N_R = 6, N_Y = 10$ **5.** $D_L = 12, D_C = 16$ **6.** $-2 - 3i$ **7.** $5 - i$ **8.** $2 + 2i$

9. $2 + 6i$ **10.** $+1 \pm \sqrt{6}$ **11.** $\dfrac{5}{2} \pm \dfrac{\sqrt{33}}{2}$ **12.** $y = 2x - 8$ **13.** 26.92

14. 3 **15.** 67 **16.** $\dfrac{200 \text{ in}}{\text{hr}} \times \dfrac{1 \text{ ft}}{12 \text{ in}} \times \dfrac{1 \text{ mi}}{5280 \text{ ft}} \times \dfrac{1 \text{ hr}}{60 \text{ min}} = \dfrac{200}{(12)(5280)(60)} \dfrac{\text{mi}}{\text{min}}$

17. $2(2 \cdot 2^{1/4})^{1/2} = 2 \cdot 2^{1/2} 2^{1/8} = 2^{13/8}$ **18.** $3(3^2 \cdot 3^{1/4})^{1/2} = 3 \cdot 3 \cdot 3^{1/8} = 3^{17/8}$

19. $(4x^3 y^5)^{1/2} (8xy^2)^{1/3} = 2x^{3/2} y^{5/2} \cdot 2x^{1/3} y^{2/3} = 4x^{11/6} y^{19/6}$ **20.** $-\dfrac{7\sqrt{26}}{26}$ **21.** 1×10^9

22. $\dfrac{mpb + cpb - ap}{b}$ **23.** $\dfrac{x - 4}{x(x + 3)}$ **24.** -32 **25.** $\dfrac{x^2 y - p^3}{mp - 1}$ **26.** 1458 ft^3

27. $\dfrac{-x - 5}{x^2 - 9}$ **28.** $y = -\dfrac{2}{3}x - \dfrac{25}{3}$ **29.** $3x^2 + 12x + 47 + \dfrac{188}{x - 4}$ **30.** $\dfrac{2}{5}$

ER1. -418.76 **ER2.** -4.202

problem set 53

1. 29% **2.** 20 ml 20% solution, 80 ml 60% solution **3.** 50 ml 40% solution, 200 ml 80%

4. $N_B = 12, N_G = 68$ **5.** 100 mph

6. $7560 \text{ cm} \times \dfrac{1 \text{ m}}{100 \text{ cm}} \times \dfrac{1 \text{ km}}{1000 \text{ m}} = \dfrac{7560}{(100)(1000)} \text{ km}$

7. $32 \text{ yd} \times \dfrac{3 \text{ ft}}{1 \text{ yd}} \times \dfrac{12 \text{ in}}{1 \text{ ft}} \times \dfrac{2.54 \text{ cm}}{1 \text{ in}} \times \dfrac{1 \text{ m}}{100 \text{ cm}} = \dfrac{(32)(3)(12)(2.54)}{100} \text{ m}$

8. $15{,}740{,}000 \text{ cm}^2 \times \dfrac{1 \text{ m}}{100 \text{ cm}} \times \dfrac{1 \text{ m}}{100 \text{ cm}} \times \dfrac{1 \text{ km}}{1000 \text{ m}} \times \dfrac{1 \text{ km}}{1000 \text{ m}} = \dfrac{15{,}740{,}000}{1 \times 10^{10}} \text{ km}^2$

9. $.042 \text{ km}^2 \times \dfrac{1000 \text{ m}}{1 \text{ km}} \times \dfrac{1000 \text{ m}}{1 \text{ km}} \times \dfrac{100 \text{ cm}}{1 \text{ m}} \times \dfrac{100 \text{ cm}}{1 \text{ m}} \times \dfrac{1 \text{ in}}{2.54 \text{ cm}} \times \dfrac{1 \text{ in}}{2.54 \text{ cm}} \times \dfrac{1 \text{ ft}}{12 \text{ in}}$

$\times \dfrac{1 \text{ ft}}{12 \text{ in}} \times \dfrac{1 \text{ mi}}{5280 \text{ ft}} \times \dfrac{1 \text{ mi}}{5280 \text{ ft}} = \dfrac{(.042)(1000)(1000)(100)(100) \text{ mi}^2}{(2.54)(2.54)(12)(12)(5280)(5280)}$

10. 2   **11.** $2 + 8i$ **12.** $4i$ **13.** $-4 - 2i$ **14.** $\dfrac{5}{2} \pm \dfrac{3\sqrt{5}}{2}$ **15.** $3 \pm \sqrt{15}$

16. 20   **17.** 60   **18.** -1 **19.** $(2 \cdot 2^{1/3})^{1/5} = 2^{1/5} \cdot 2^{1/15} = 2^{4/15}$

20. $(3^4 \cdot 3^{1/4})^{1/2} = 3^2 \cdot 3^{1/8} = 3^{17/8}$ **21.** $(x^2 y)^{1/5}(xy^2)^{1/3} = x^{2/5}y^{1/5}x^{1/3}y^{2/3} = x^{11/15}y^{13/15}$

22. $-\dfrac{1}{32}$ **23.** $-12\sqrt{2}$ **24.** 1×10^4 **25.** $\dfrac{myc}{xc + ky}$ **26.** $\dfrac{p}{d - c}$

27. $B = \dfrac{5\sqrt{55}}{3}$ **28.** $x^2 - x - 1 + \dfrac{3}{x + 1}$ **29.** 23,978.88 in^2

30. $\left(-\dfrac{6}{7}, \dfrac{17}{7}\right)$

ER3. -8.107 **ER4.** -2.08162

problem set 54

1. 100 liters 20%, 300 liters 70%   **2.** 50 ml 10%, 100 ml 40%   **3.** 32%

4. $N_G = 20, N_D = 25$ **5.** 300 miles   **6.** $2.16R + 3.36U$ **7.** $-4.48R - 5.39U$

8. $39.48R - 14.28U$

9. $100 \text{ km}^2 \times \dfrac{1000 \text{ m}}{1 \text{ km}} \times \dfrac{1000 \text{ m}}{1 \text{ km}} \times \dfrac{100 \text{ cm}}{1 \text{ m}} \times \dfrac{100 \text{ cm}}{1 \text{ m}} = (100)(1000)(1000)(100)(100) \text{ cm}^2$

10. $100 \text{ ft}^2 \times \dfrac{12 \text{ in}}{1 \text{ ft}} \times \dfrac{12 \text{ in}}{1 \text{ ft}} \times \dfrac{2.54 \text{ cm}}{1 \text{ in}} \times \dfrac{2.54 \text{ cm}}{1 \text{ in}} = (100)(12)(12)(2.54)(2.54) \text{ cm}^2$

11. $\dfrac{60 \text{ mi}}{\text{hr}} \times \dfrac{5280 \text{ ft}}{\text{mi}} \times \dfrac{12 \text{ in}}{1 \text{ ft}} \times \dfrac{2.54 \text{ cm}}{\text{in}} \times \dfrac{1 \text{ m}}{100 \text{ cm}} \times \dfrac{1 \text{ km}}{1000 \text{ m}} \times \dfrac{1 \text{ hr}}{60 \text{ min}} \times \dfrac{1 \text{ min}}{60 \text{ sec}}$

$= \dfrac{(60)(5280)(12)(2.54)}{(100)(1000)(60)(60)} \dfrac{\text{km}}{\text{sec}}$

12. 4   **13.** $4 + 2i$ **14.** $-4 - 2i$ **15.** $-2 - 8i$ **16.** $\dfrac{3}{2} \pm \dfrac{\sqrt{37}}{2}$

17. $\dfrac{7}{2} \pm \dfrac{\sqrt{61}}{2}$ **18.** $y = \dfrac{3}{2}x - 9$ **19.** $\dfrac{59}{3}$ **20.** 6   **21.** 146   **22.** $\dfrac{xc}{p + c^2 m}$

23. $\dfrac{cp}{xp + yp - a}$ **24.** $(2^2 \cdot 2^{1/2})^{1/3} = 2^{2/3} \cdot 2^{1/6} = 2^{5/6}$

25. $(3^2 \cdot 3^{1/3})^{1/5} = 3^{2/5}3^{1/15} = 3^{7/15}$ **26.** $(xy)^{1/6}(xy^2)^{1/3} = x^{1/6}y^{1/6}x^{1/3}y^{2/3} = x^{1/2}y^{5/6}$

27. -3 **28.** $\dfrac{55\sqrt{21}}{21}$ **29.** $\dfrac{xp^3 - 1}{xy - p^2}$ **30.** $\dfrac{y^2}{4} - 8y^2 p^2$ **ER5.** 4.627

ER6. 57.67

problem set 55

1. 4, 5, 6 and $-2, -1, 0$ 2. 2, 4, 6   3. 48 ml 5%, 12 ml 40%

4. 108 g   5. 36, 39, 42   6. $\dfrac{yx}{kx - a - b}$ 7. $\dfrac{dc + ec}{dc - mp}$ 8. $\dfrac{pyb - mxb - dy}{mx - py}$

9. $32.8R - 22.8U$ 10. $-9.4R - 3.4U$

11. $4 \text{ yd}^3 \times \dfrac{3 \text{ ft}}{1 \text{ yd}} \times \dfrac{3 \text{ ft}}{1 \text{ yd}} \times \dfrac{3 \text{ ft}}{1 \text{ yd}} \times \dfrac{12 \text{ in}}{1 \text{ ft}} \times \dfrac{12 \text{ in}}{1 \text{ ft}} \times \dfrac{12 \text{ in}}{1 \text{ ft}} \times \dfrac{2.54 \text{ cm}}{1 \text{ in}} \times \dfrac{2.54 \text{ cm}}{1 \text{ in}} \times \dfrac{2.54 \text{ cm}}{1 \text{ in}}$

$\times \dfrac{1 \text{ m}}{100 \text{ cm}} \times \dfrac{1 \text{ m}}{100 \text{ cm}} \times \dfrac{1 \text{ m}}{100 \text{ cm}} = \dfrac{(4)(3)(3)(3)(12)(12)(12)(2.54)(2.54)(2.54)}{(100)(100)(100)} \text{ m}^3$

12. $1000 \dfrac{\text{ft}}{\text{sec}} \times \dfrac{12 \text{ in}}{1 \text{ ft}} \times \dfrac{2.54 \text{ cm}}{1 \text{ in}} \times \dfrac{1 \text{ m}}{100 \text{ cm}} \times \dfrac{1 \text{ km}}{1000 \text{ m}} \times \dfrac{60 \text{ sec}}{1 \text{ min}} = \dfrac{(1000)(12)(2.54)(60)}{(100)(1000)} \dfrac{\text{km}}{\text{min}}$

13. $-2 - 2i$ 14. $2 + i$ 15. $4 - 4i$ 16. $3 - 2i$ 17. $\dfrac{5}{2} \pm \dfrac{3\sqrt{5}}{2}$

18. $3 \pm \sqrt{15}$ 19. No solution   20. -5 21. $(2^4 \cdot 2^{1/2})^{1/2} = 2^2 \cdot 2^{1/4} = 2^{9/4}$

22. $(3^3 \cdot 3^{1/3})^{1/4} = 3^{3/4} \cdot 3^{1/12} = 3^{5/6}$ 23. $(x^2y)^{1/4}(x^5y^2)^{1/2} = x^{1/2}y^{1/4}x^{5/2}y = x^3y^{5/4}$

24. -243 25. $-\dfrac{117\sqrt{22}}{11}$ 26. 4×10^{-66} 27. 5   28. 144 ft^3 29. $\dfrac{x - 3}{x - 4}$

30. $8\sqrt{2}$ ER7. -215.222 ER8. -23.797

problem set ER-2

1. $-9, -7, -5$ 2. $-6, -4, -2, 0$ and $-10, -8, -6, -4$

3. 320 ml 50% and 480 ml 20%   4. 120 g   5. 16, 20, 24   6. 1.4451795

7. -6.5773403 8. $\dfrac{ab}{fb - m - x}$ 9. $\dfrac{dc + de}{cy - dk}$ 10. $\dfrac{mc - yfc + yab}{fc - ab}$

11. $-6.8R - 18.8U$ 12. $18.8R - 6.8U$

13. $\dfrac{500 \text{ m}}{\text{sec}} \times \dfrac{100 \text{ cm}}{1 \text{ m}} \times \dfrac{1 \text{ in}}{2.54 \text{ cm}} \times \dfrac{1 \text{ ft}}{12 \text{ in}} \times \dfrac{1 \text{ yd}}{3 \text{ ft}} \times \dfrac{60 \text{ sec}}{1 \text{ min}} = \dfrac{500(100)(60)}{(2.54)(12)(3)} \dfrac{\text{yds}}{\text{min}}$

14. $-2 + 7i$ 15. $3 + 3i$ 16. $6 - 14i$ 17. $3 \pm \sqrt{15}$ 18. $1 \pm \sqrt{3}$

19. no solution   20. 7   21. $(2^3 \cdot 2^{1/5})^{1/5} = 2^{3/5} \cdot 2^{1/25} = 2^{16/25}$

22. $(2^4 \cdot 2^{1/2})^{1/4} = 2 \cdot 2^{1/8} = 2^{9/8}$ 23. $(x^2y)^{1/6}(x^3y^2)^{1/2} = x^{1/3}y^{1/6}x^{3/2}y = x^{11/6}y^{7/6}$

24. -243 25. $-\dfrac{52\sqrt{33}}{11}$ 26. 5×10^{32} 27. $\dfrac{x + 3}{x - 4}$ 28. $\dfrac{26}{3}$ 29. 2355 ft^3

30. $\sqrt{145}$

problem set 56

1. $-8, -6, -4$ and $-4, -2, 0$ 2. 348 g, 92%   3. 40 miles

4. 400 gal 20%, 600 gal 80%   5. 400   6. 88 in^2 7. 540 ft^2 8. 4144.8 m^2

9. $\dfrac{bmp + bkc - amk}{mk}$ 10. $\dfrac{m^2 - ax - pma}{pm + x}$ 11. $-8.7R - 5U$ 12. $10R + 17.4U$

13. $60 \dfrac{\text{km}}{\text{hr}} \times \dfrac{1000 \text{ m}}{1 \text{ km}} \times \dfrac{100 \text{ cm}}{1 \text{ m}} \times \dfrac{1 \text{ in}}{2.54 \text{ cm}} \times \dfrac{1 \text{ hr}}{60 \text{ min}} \times \dfrac{1 \text{ min}}{60 \text{ sec}} = \dfrac{(60)(1000)(100)}{(2.54)(60)(60)} \dfrac{\text{in}}{\text{sec}}$

14. $400 \text{ yd}^3 \times \dfrac{3 \text{ ft}}{1 \text{ yd}} \times \dfrac{3 \text{ ft}}{1 \text{ yd}} \times \dfrac{3 \text{ ft}}{1 \text{ yd}} \times \dfrac{12 \text{ in}}{1 \text{ ft}} \times \dfrac{12 \text{ in}}{1 \text{ ft}} \times \dfrac{12 \text{ in}}{1 \text{ ft}} \times \dfrac{2.54 \text{ cm}}{1 \text{ in}} \times \dfrac{2.54 \text{ cm}}{1 \text{ in}}$

$\times \dfrac{2.54 \text{ cm}}{1 \text{ in}} = (400)(3)(3)(3)(12)(12)(12)(2.54)(2.54)(2.54) \text{ cm}^3$

15. $3 + 13i$ **16.** $6 + 5i$ **17.** $(2^2 \cdot 2^{1/5})^{1/6} = 2^{1/3} \cdot 2^{1/30} = 2^{11/30}$

18. $(y^4)^{1/2}(xy^2)^{1/2} = y^2 x^{1/2} y = y^3 x^{1/2}$ **19.** $(5^2 \cdot 5^{1/3})^{1/2} = 5 \cdot 5^{1/6} = 5^{7/6}$ **20.** $-\dfrac{16\sqrt{3}}{3}$

21. $\dfrac{7}{2} \pm \dfrac{\sqrt{77}}{2}$ **22.** $4 \pm 2\sqrt{6}$ **23.** -1 **24.** 5

25. $4x^3 + 12x^2 + 36x + 108 + \dfrac{323}{x-3}$ **26.** $\dfrac{ax^2 - y^3 p}{y^2 p - x}$ **27.** $y = -8x - 45$

28. $\dfrac{56}{3}$ **29.** $\dfrac{17}{3}$ **30.** $\dfrac{x(x^2 - 1) - (3x + 2)(x + 1) - 4a^2 y}{a^2 y(x^2 - 1)}$

ER9. 460.3984   **ER10.** 9.6574

problem set 57

1. 30 lb/in^2 **2.** 29.4 atm   **3.** 300 torr   **4.** 240 gal 20% and 760 gal 40%

5. 5, 7, 9, 11 or $-5, -3, -1, 1$ **6.** 1004.8 m^2 **7.** $\dfrac{xmp + xc - ap}{mp + c}$

8. $\dfrac{xd - pk - pdc}{d}$ **9.** $3.08R + 2.56U$ **10.** $34.8R - 20U$

11. $40\,\dfrac{\text{cm}}{\text{sec}} \times \dfrac{1\ \text{in}}{2.54\ \text{cm}} \times \dfrac{1\ \text{ft}}{12\ \text{in}} \times \dfrac{1\ \text{mi}}{5280\ \text{ft}} \times \dfrac{60\ \text{sec}}{1\ \text{min}} \times \dfrac{60\ \text{min}}{1\ \text{hr}} = \dfrac{(40)(60)(60)}{(2.54)(12)(5280)}\,\dfrac{\text{mi}}{\text{hr}}$

12. $1000\ \text{cm}^3 \times \dfrac{1\ \text{in}}{2.54\ \text{cm}} \times \dfrac{1\ \text{in}}{2.54\ \text{cm}} \times \dfrac{1\ \text{in}}{2.54\ \text{cm}} \times \dfrac{1\ \text{ft}}{12\ \text{in}} \times \dfrac{1\ \text{ft}}{12\ \text{in}} \times \dfrac{1\ \text{ft}}{12\ \text{in}}$

$= \dfrac{1000}{(2.54)(2.54)(2.54)(12)(12)(12)}\ \text{ft}^3$

13. $-2 - 3i$ **14.** $1 - 4i$ **15.** $6, -1$ **16.** $5, 1$ **17.** $y = \dfrac{11}{5}x + \dfrac{41}{5}$ **18.** 3

19. 4 **20.** -66 **21.** $-\dfrac{14}{5}$ **22.** $(3^2 \cdot 3^{1/3})^{1/2} = 3 \cdot 3^{1/6} = 3^{7/6}$

23. $[x^4(x^2 y)^{1/3}]^{1/2} = x^2 x^{1/3} y^{1/6} = x^{7/3} y^{1/6}$ **24.** $(2^2)^{1/3}(2)^{1/5} = 2^{2/3} \cdot 2^{1/5} = 2^{13/15}$

25. $\dfrac{\sqrt{10}}{10}$ **26.** $\dfrac{1}{32}$ **27.** $\dfrac{x^2 y - p^5 z}{xz - 4p^5}$ **28.** $-\dfrac{4}{7}$ **29.** $0, 7, -4$ **30.** $\dfrac{2x - 2a}{a^2(a + 2)}$

ER11. 4.9809   **ER12.** -743.9665

problem set 58

1. 600 K   **2.** 645 g   **3.** 27%   **4.** 40 ml 10%, 10 ml 40%   **5.** 45 miles

6. $-\dfrac{3}{8} \pm \dfrac{\sqrt{57}}{8}$ **7.** $\dfrac{1}{10} \pm \dfrac{\sqrt{41}}{10}$ **8.** $\dfrac{-1 \pm \sqrt{13}}{3}$ **9.** 3456 in^2 **10.** $\dfrac{xk - ak}{y + ck}$

11. $\dfrac{xy + xkc - ak}{y + kc}$ **12.** $-3.08R - 2.56U$ **13.** $7.1R - 7.1U$

14. $\dfrac{(70)(100)(60)(60)}{(2.54)(12)(5280)}\,\dfrac{\text{mi}}{\text{hr}}$ **15.** $(40)(12)(12)(12)(2.54)(2.54)(2.54)\ \text{cm}^3$

16. $(3^2)^{1/4} \cdot 3^{1/5} = 3^{1/2} \cdot 3^{1/5} = 3^{7/10}$ **17.** $(2^2 \cdot 2^{1/2})^{1/7} = 2^{2/7} \cdot 2^{1/14} = 2^{5/14}$

18. $(x^2 y^3)^{1/2}(xy^5)^{1/3} = xy^{3/2} x^{1/3} y^{5/3} = x^{4/3} y^{19/6}$ **19.** $-\dfrac{1}{8}$ **20.** $\dfrac{333\sqrt{55}}{55}$

21. $-2 + 5i$ **22.** 1×10^{34} **23.** $\dfrac{x + 2}{x - 5}$ **24.** 850.56 **25.** $\dfrac{x^2 a^2 p^3 - 4m}{xa - 5mp^3}$

26. $\left(\dfrac{20}{19}, -\dfrac{34}{19}\right)$ **27.** $4x^2 - 8x + 16 - \dfrac{37}{x + 2}$ **28.** -3

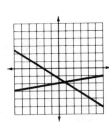

29. $\dfrac{15}{7}$ **30.** $\dfrac{ab + cx}{a^2(y + 1)}$ **ER13.** 7.6604 **ER14.** -5.23214

problem set 59

1. $-7, -6, -5, -4$ or $-6, -5, -4, -3$ **2.** 160 g   **3.** $D_W = 9, D_R = 32$

4. 40 liters 90%, 60 liters 70% **5.** 7 liters   **6.** $B = 3.5C - 12.5$

7. $N = .18S - .8$ **8.** $(4, -4)$ **9.** $\sqrt{34} \, \underline{/211^\circ}$ **10.** $26.1R - 15U$

11. $\dfrac{1}{4} \pm \dfrac{\sqrt{41}}{4}$ **12.** $\dfrac{5}{3}, -1$ **13.** $\dfrac{4}{3}, -1$ **14.** 9504 in^2 **15.** $\dfrac{4p - 4s}{c - 4x}$

16. $\dfrac{mx - sc + 4xs}{c - 4x}$ **17.** $\dfrac{(40)(2.54)(60)(60)}{100} \dfrac{\text{meters}}{\text{hr}}$ **18.** $\dfrac{18,000}{(2.54)(2.54)(2.54)(12)(12)(12)}$ ft^3

19. $6 + 5i$ **20.** $-2 + 2i$ **21.** 3 **22.** No solution

23. $(3 \cdot 3^{1/2})^{1/5} = 3^{1/5} \cdot 3^{1/10} = 3^{3/10}$ **24.** $(2^2 \cdot 2^{1/2})^{1/4} = 2^{1/2} \cdot 2^{1/8} = 2^{5/8}$ **25.** $-\dfrac{1}{8}$

26. $(x^2 y)^{1/2}(y^5 x^4)^{1/3} = xy^{1/2} y^{5/3} x^{4/3} = x^{7/3} y^{13/6}$ **27.** $\dfrac{92\sqrt{26}}{13}$ **28.** 39

29. $\dfrac{xya(a + y) + (x^2 + 2)(a + y) + yax^3}{y^2 a(a + y)}$ **30.** $\dfrac{y^4 x^4}{p^{11}}$ **ER15.** 23.7598

ER16. -9.8522

problem set 60

1. 20 **2.** $133\dfrac{1}{3}$ **3.** 24 **4.** 800 torr   **5.** $N_P = 90, N_T = 40$

6. $-2.84R + 2.84U$ **7.** $2\sqrt{5} \, \underline{/116.6^\circ}$ **8.** $(10, 20)$ **9.** $1, -\dfrac{3}{4}$ **10.** $1, -\dfrac{1}{3}$

11. $Na = .037C + 3.9$ **12.** 64 in^2 **13.** $\dfrac{ad - ycd - ym}{cd + m}$ **14.** $\dfrac{mk - mdz - xyz}{z}$

15. $-3 + 4i$ **16.** $1 - 15i$ **17.** $\dfrac{(60)(60)(60)}{(2.54)(12)(5280)} \dfrac{\text{mi}}{\text{hr}}$

18. $\dfrac{(1,400,000)}{(2.54)(2.54)(2.54)(12)(12)(12)(3)(3)(3)}$ yd^3 **19.** $(3^3 \cdot 3^{1/2})^{1/3} = 3 \cdot 3^{1/6} = 3^{7/6}$

20. $3 \cdot 3^{1/3} = 3^{4/3}$ **21.** $\dfrac{1}{27}$ **22.** $(m^2 p)^{1/2}(m^5 p^4)^{1/3} = mp^{1/2} m^{5/3} p^{4/3} = m^{8/3} p^{11/6}$

23. $-\dfrac{23\sqrt{2}}{6}$ **24.** 3×10^4 **25.** $\dfrac{32}{3}$ **26.** $12,458.88$ **27.** $\left(\dfrac{21}{8}, \dfrac{11}{4}\right)$

28. $0, 5, 10$ **29.** $y = \dfrac{2}{5}x + 18$ **30.** $6x^{-5} - 3y^{-2} p^{-2} x^{-11}$ **ER17.** -597.1918

problem set 61

1. .0003 kg/sec   **2.** 500 torr   **3.** 75 gal   **4.** 58.7 lb   **5.** 300 kg

6. $-19.4R + 5.2U$ **7.** $2\sqrt{10}\ \underline{/341.6°}$ **8.** (6, 20)   **9.** $\dfrac{2}{5} \pm \dfrac{2\sqrt{6}}{5}$ **10.** $\dfrac{1}{4} \pm \dfrac{\sqrt{57}}{4}$

11. $Pb = 42.6Sb - 31.9$ **12.** 175.84 cm²   **13.** $\dfrac{pkd - ak - dbx - dbm}{db}$

14. $\dfrac{4yc - xdc + xm}{2dc - 2m}$ **15.** $-2 - 7i$ **16.** $2 - 16i$ **17.** $\dfrac{(40)(60)(60)}{(2.54)(2.54)(2.54)}\ \dfrac{\text{in}^3}{\text{hr}}$

18. $4(12)(12)(12)(2.54)(2.54)(2.54)$ cm³   **19.** $(2^5 \cdot 2^{1/2})^{1/2} = 2^{5/2} \cdot 2^{1/4} = 2^{11/4}$

20. $(2^3 \cdot 2^{1/3})^{1/3} = 2 \cdot 2^{1/9} = 2^{10/9}$ **21.** $-\dfrac{1}{32}$

22. $(x^5 y)^{1/2}(y^2 x)^{1/4} = x^{5/2}y^{1/2}y^{1/2}x^{1/4} = x^{11/4}y$ **23.** $15\sqrt{6}$ **24.** 2×10^3

25. $C = \dfrac{52}{5}$ **26.** $(401.28)(12)(12)(12)$ **27.** $\left(-\dfrac{8}{3}, \dfrac{5}{3}\right)$ **28.** $0, 5, -3$

29. $y = \dfrac{3}{2}x + 7$ **30.** $-\dfrac{5}{4}$ **ER18.** 2.4763

problem set 62

1. 2   **2.** 400 K   **3.** 20 liters   **4.** 32 miles   **5.** 46.7%   **6.** $18.8R - 6.8U$

7. $\sqrt{29}\ \underline{/111.8°}$ **8.** (9, 6)   **9.** $\dfrac{1}{6} \pm \dfrac{\sqrt{59}}{6}i$ **10.** $\dfrac{1}{5} \pm \dfrac{\sqrt{14}}{5}i$

11. $Bi = 16Hg - 72.6$ **12.** 6631.68 in²   **13.** $\dfrac{cp + krc - arm}{rm}$

14. $\dfrac{3xr - 2pa - 2acr}{p + cr}$ **15.** $-2 - 5i$ **16.** $-7 + 5i$

17. $\dfrac{600}{(2.54)(2.54)(2.54)(12)(12)(12)(60)}\ \dfrac{\text{ft}^3}{\text{sec}}$ **18.** $(20)(3)(3)(3)(12)(12)(12)(2.54)(2.54)(2.54)$ cm³

19. $(2^4 2^{1/2})^{1/2} = 2^2 \cdot 2^{1/4} = 2^{9/4}$ **20.** $(2^2)^{1/4}2^{1/5} = 2^{1/2} \cdot 2^{1/5} = 2^{7/10}$ **21.** -32

22. $(xy^7)^{1/2}(x^5 y)^{1/3} = x^{1/2}y^{7/2}x^{5/3}y^{1/3} = x^{13/6}y^{23/6}$ **23.** $-\dfrac{79\sqrt{10}}{10}$ **24.** 1×10^8

25. $A = \dfrac{105}{12}, B = \dfrac{25}{4}, C = 25$ **26.** $3500(12)(12)(12)$ **27.** $(-3, 1)$

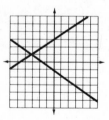

28. 0, 5, 10   **29.** $y = -\dfrac{2}{7}x - \dfrac{59}{7}$ **30.** $\dfrac{39}{7}$ **ER19.** 2.5984

problem set 63

1. 3   **2.** 480 K   **3.** 150 gal   **4.** 8000 ft   **5.** 104 g   **6.** $-16R - 13.2U$

7. $7.36R + 26.31U$ **8.** $27.33\ \underline{/74.51°}$ **9.** (10, 20)   **10.** $\dfrac{1}{4} \pm \dfrac{\sqrt{39}}{4}i$

11. $\dfrac{1}{3} \pm \dfrac{\sqrt{11}}{3}i$ **12.** $Mo = -7.3Zr + 1409$ **13.** 3.44×10^6 cm²

14. $\dfrac{ma + mcb - pb}{yb}$ **15.** $\dfrac{bp + zby}{a + bc}$ **16.** $-2 - 9i$ **17.** $-i$

18. $\dfrac{400}{(2.54)(12)(3)(60)}\dfrac{\text{yd}}{\text{sec}}$ **19.** $\dfrac{4(5280)(5280)(5280)(12)(12)(12)(2.54)(2.54)(2.54)}{(100)(100)(100)(1000)(1000)(1000)}\text{km}^3$

20. $(3 \cdot 3^{1/4})^{1/5} = 3^{1/5} \cdot 3^{1/20} = 3^{1/4}$ **21.** $(2^2 \cdot 2^{1/4})^{1/5} = 2^{2/5} \cdot 2^{1/20} = 2^{9/20}$ **22.** 36

23. $(xy^2)^{1/4}(x^3y)^{1/2} = x^{1/4}y^{1/2}x^{3/2}y^{1/2} = x^{7/4}y$ **24.** $-\dfrac{85\sqrt{14}}{14}$ **25.** 8×10^{40}

26. $\dfrac{10}{3}$ **27.** 200,000 **28.** $y = -\dfrac{5}{4}x - \dfrac{9}{2}$ **29.** 2 **30.** $\dfrac{7}{25}$ **ER20.** -24.2172

problem set 64

1. 1200 g **2.** 1600 K **3.** 40 liters 30%, 10 liters 80% **4.** 3200 km **5.** 89.4%

6. $\dfrac{a + abx + b}{1 + bx}$ **7.** $\dfrac{ax + ab + 4x^2}{x(x + b)}$ **8.** $\dfrac{2cm + m^2 + 8c^2}{c(2c + m)}$ **9.** $-5 + 2i$

10. $-2 + 6i$ **11.** $4 + 2\sqrt{6}$ **12.** $-3 + 13i$ **13.** $16 - 2i$ **14.** $-23 + 14i$

15. -10 **16.** $-16.13R + 16.74U$ **17.** $2\sqrt{13}\ \underline{/236.31°}$ **18.** $(20, 10)$

19. $-\dfrac{1}{3} \pm \dfrac{\sqrt{14}}{3}i$ **20.** $\dfrac{3}{4} \pm \dfrac{\sqrt{47}}{4}i$ **21.** $W = -.13Ir + 295$ **22.** 100.48 cm^2

23. $\dfrac{(400)(60)(60)}{(2.54)(2.54)(2.54)}\dfrac{\text{in}^3}{\text{hr}}$ **24.** $(3^2 \cdot 3^{1/3})^{1/3} = 3^{2/3} \cdot 3^{1/9} = 3^{7/9}$ **25.** $\dfrac{139\sqrt{35}}{35}$

26. $\dfrac{abf + acf}{d + mf}$ **27.** $\dfrac{xd + xmf - baf}{af}$ **28.** $\left(\dfrac{15}{11}, \dfrac{27}{11}\right)$

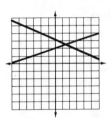

29. $0, -7, -8$ **30.** 1×10^{-54} **ER21.** -20.2783

problem set 65

1. 48 kg **2.** 16,000 torr **3.** 100 ml **4.** $N_E = 100, N_I = 500$ **5.** 80 kg

6. $R_W = 4, R_B = 12, T_W = \dfrac{3}{2}, T_B = \dfrac{1}{2}$ **7.** $R_C = 33, R_P = 99, T_P = 7, T_C = 5$

8. $R_2 = 30, R_1 = 60, T_1 = 2, T_2 = 4$ **9.** $\dfrac{acx + bx + c}{ac + b}$ **10.** $\dfrac{4ab + bc + 4}{c(ab + 1)}$

11. $\dfrac{xa + x + a^2}{a + 1}$ **12.** $-1 + 4i$ **13.** $-4 - 19i$ **14.** $-14 + 2i$

15. $5.48R + 9.14U$ **16.** $\sqrt{34}\ \underline{/300.96°}$ **17.** $\left(3, -\dfrac{6}{7}\right)$ **18.** $-\dfrac{7}{6} \pm \dfrac{\sqrt{73}}{6}$

19. $-\dfrac{5}{4} \pm \dfrac{\sqrt{57}}{4}$ **20.** $K = .15Ra - 150$ **21.** 460 ft^2 **22.** $\dfrac{(600)(12)(12)(12)}{60}\dfrac{\text{in}^3}{\text{min}}$

23. $[x^2(y^3)^{1/2}]^{1/2} = (x^2y^{3/2})^{1/2} = xy^{3/4}$ **24.** $(2^2 \cdot 2^{1/5})^{1/6} = 2^{1/3} \cdot 2^{1/30} = 2^{11/30}$

25. $\dfrac{-85\sqrt{2}}{6}$

26. $\dfrac{dab + dac + mabf + macf}{f}$ 27. $\dfrac{xf - cad - camf}{da + maf}$ 28. $\dfrac{x + 9}{x + 7}$ 29. 4×10^{10}

30. $\dfrac{4x - 11}{x^2 - 4}$ ER22. -21.8868

problem set 66

1. 50 g 2. 4000 K 3. 120 liters 20%, 80 liters 60% 4. 5 hr 5. 800 g

6. $\dfrac{1 + 7a}{x - 3}$ 7. $\dfrac{2x + 8}{x - 3}$ 8. $\dfrac{4 - 2x^2 - 4x}{x^2 - 4}$ 9. 200.96 ft^2 10. 135.28 in^2

11. $R_T = 50, R_P = 400, T_T = 6, T_P = 3$ 12. $R_P = 208, R_T = 52, T_P = 3, T_T = 7$

13. $\dfrac{a^2x^2 + ax + x}{ax + 1}$ 14. $\dfrac{mx^2 + m + x^2}{x(x^2 + 1)}$ 15. $(3 + \sqrt{6}) + 4i$ 16. $-26 + 2i$

17. $11.72R + 7.61U$ 18. $4\sqrt{5} \; \underline{/26.57^\circ}$ 19. $(8, 20)$ 20. $x = -\dfrac{3}{2} \pm \dfrac{\sqrt{15}}{2}$

21. $-\dfrac{1}{2} \pm \dfrac{\sqrt{105}}{10}$ 22. $(678.24)(100)(100) \text{ cm}^2$ 23. $Mg = 66.67Ca$

24. $\dfrac{10(2.54)(2.54)(2.54) \text{ cm}^3}{60} \dfrac{}{\text{min}}$ 25. $x^{1/2}$ 26. $x^{3/20}$ 27. $-\dfrac{79\sqrt{10}}{10}$

28. $\dfrac{pxd - cmd - mk}{d}$ 29. $\dfrac{cdy + ky - dm}{kp + cdp}$ 30. 3×10^{-22} ER23. -4.0274

problem set 67

1. 18,600 2. 200 cm^3 3. 100 lb 4. $N_D = 230, N_N = 305$ 5. 160 g

6. $\dfrac{-4 - \sqrt{3}}{13}$ 7. $\dfrac{6\sqrt{3} - 3\sqrt{2}}{10}$ 8. $\dfrac{\sqrt{5} + 1}{6}$ 9. $\dfrac{4x + 5}{x - 2}$ 10. $\dfrac{2x + 10}{x^2 - 9}$

11. $R_T = 59, R_P = 354, T_P = 3, T_T = 5$ 12. $\dfrac{4x^2b + 4xa + ab}{xb + a}$ 13. $\dfrac{amx + a + x^2}{mx + 1}$

14. $32 + 4i$ 15. $4 + i$ 16. $-13.44R - 11.87U$ 17. $5 \; \underline{/216.87^\circ}$

18. $(14, 10)$ 19. $\dfrac{1}{8} \pm \dfrac{\sqrt{15}}{8}i$ 20. $-\dfrac{1}{3} \pm \dfrac{\sqrt{14}}{3}i$ 21. $(2570)(12)(12)(12)$

22. $\dfrac{400}{(60)(2.54)(2.54)(2.54)} \dfrac{\text{in}^3}{\text{min}}$ 23. $Ag = -.2Au + 26$ 24. $2^{2/3}$

25. $x^{11/15}y^{13/15}p^{1/5}$ 26. $\dfrac{-164\sqrt{33}}{33}$ 27. $\dfrac{aby + acy}{py + m}$ 28. $\dfrac{mx - yab - pmab}{pma + ya}$

29. $y = \dfrac{3}{7}x + \dfrac{34}{7}$ 30. $\sqrt{58}$ ER24. -2.6434

problem set ER-3

1. 2000 2. 150 ml 3. 200 lb 4. $N_E = 8, N_F = 30$ 5. 8400

6. 12.47 7. 23.59 8. 3.78 9. 6.68 10. 38.95 11. $\dfrac{-2 + \sqrt{2}}{2}$

12. $\dfrac{12\sqrt{2} + 4}{17}$ 13. $3\sqrt{3} + 5$ 14. $\dfrac{5x + 21}{x^2 - 9}$ 15. $\dfrac{-11 - 2x}{x^2 - 16}$

16. $R_G = 10, R_B = 20, T_G = 14, T_B = 7$ 17. $\dfrac{4rx^2 + 4mr + xm}{x^2 + m}$

18. $\dfrac{3a^3 + 3ax + a^2x}{a^2 + x}$ **19.** $-23 + 14i$ **20.** $9 - i$ **21.** $37.42R - 24.22U$

22. $(20, 15)$ **23.** $\dfrac{1}{4} \pm \dfrac{\sqrt{3}}{4}i$ **24.** $\dfrac{5}{3}, -1$ **25.** $\dfrac{700(60)}{(2.54)(2.54)(2.54)}\dfrac{\text{in}^3}{\text{hr}}$ **26.** $3^{4/21}$

27. $x^{17/6}y^{13/6}$ **28.** $\dfrac{5\sqrt{6}}{2}$ **29.** $\dfrac{ym - pxa - cxya}{cxy + px}$ **30.** $y = -\dfrac{9}{13}x - \dfrac{46}{13}$

problem set 68

1. 3.5×10^6 K **2.** 3×10^7 lb/in^2 **3.** 80 tons **4.** 800 ml 40%, 1200 ml 80%

5. 5 p.m. **6.** $\dfrac{-2 + 2x}{x - 5}$ **7.** $\dfrac{6}{x - 2}$ **8.** $\dfrac{3x^2 + 20x + 27}{x^2 - 25}$ **9.** $\dfrac{12 + 4\sqrt{2}}{7}$

10. $\dfrac{10 + 6\sqrt{2}}{7}$ **11.** $\dfrac{-6 - 8\sqrt{2}}{23}$ **12.** $R_g = 19, R_R = 57, T_g = 9, T_R = 3$

13. $\dfrac{m^3 + 2m}{m^2 + 1}$ **14.** $\dfrac{ax^2 + x + a^2}{ax + 1}$ **15.** $-20i$ **16.** $3 - 2i$ **17.** $-34.64R - 20U$

18. $\sqrt{13}\ \underline{/\ 303.69°}$ **19.** $(30, 20)$ **20.** $\dfrac{5}{6} \pm \dfrac{\sqrt{71}}{6}i$ **21.** $-\dfrac{5}{6} \pm \dfrac{\sqrt{71}}{6}i$ **22.** 152 ft^2

23. $(4)(12)(12)(12)(60)\dfrac{\text{in}^3}{\text{hr}}$ **24.** $Cr = 37V - 2083$ **25.** $3^{13/8}$ **26.** $x^{7/12}m^{13/6}$

27. $\dfrac{-61\sqrt{2}}{3}$ **28.** $\dfrac{mc + zkm - yza}{za}$ **29.** $-\dfrac{17}{11}$ **ER25.** 19.7403 **ER26.** 1.3193

ER27. 46.5689

problem set 69

1. 2×10^{13} K **2.** 100 ml **3.** $N_F = 40, N_T = 20$ **4.** 160 g

5. $-8, -6, -4$ **6.** $\dfrac{mpa}{xa - p}$ **7.** $\dfrac{mbp + ma - xp^2}{xp^2 - ma}$ **8.** $\dfrac{6x - 2}{x - 2}$

9. $\dfrac{4x^2 + 19x - 26}{x^2 + 8x + 12}$ **10.** $\dfrac{-4 - \sqrt{2}}{7}$ **11.** $\dfrac{3\sqrt{3} + 1}{26}$ **12.** $\dfrac{\sqrt{3} + 1}{2}$

13. $R_B = 13, R_X = 26, T_B = 5, T_X = 4$ **14.** $\dfrac{xyb + xya + ab}{b + a}$ **15.** $\dfrac{m(ay + 1) + xy^2}{y(ay + 1)}$

16. $-8 - 27i$ **17.** $\sqrt{6} + i$ **18.** $8.36R - 5.22U$ **19.** $\sqrt{41}\ \underline{/\ 128.66°}$

20. $(10, 16)$ **21.** $1, \dfrac{1}{3}$ **22.** $\dfrac{2}{3} \pm \dfrac{\sqrt{17}}{3}i$ **23.** 401.92 in^2 **24.** $Co = -1.5Ni + 255$

25. $2^{13/8}$ **26.** $x^{13/10}y^{17/20}$ **27.** $\dfrac{-73\sqrt{14}}{14}$ **28.** $4x^2 + 4x + 4 + \dfrac{2}{x - 1}$

29. $y = \dfrac{3}{4}x + \dfrac{13}{2}$ **30.** 198 **ER28.** 2.4561 **ER29.** 283.3739 **ER30.** 23.2480

problem set 70

1. 20 torr **2.** 12,000 **3.** 130 ml of 10% and 70 ml of 30% **4.** 48 miles

5. 3.2% **6.** See text **7.** $\dfrac{1}{3} \pm \dfrac{\sqrt{14}}{3}i$ **8.** $7, -4$ **9.** $-\dfrac{1}{4} \pm \dfrac{\sqrt{31}}{4}i$

10. $\dfrac{my - ryc + 3mc}{ry - 3m}$ **11.** $\dfrac{4x + 9}{x - 2}$ **12.** $\dfrac{-1 - 3\sqrt{3}}{13}$

13. $R_M = 40, R_P = 80, T_M = 4, T_P = 5$　14. $\dfrac{2x^2 + x}{x + 1}$　15. $\dfrac{a^2b + a^2 + b^2}{ab + a}$

16. $4i$　17. $-5 + i$　18. $14.5R + 25.22U$　19. $4\sqrt{2}\,\underline{/\,315°}$　20. $(16, 8)$

21. $\dfrac{1}{3} \pm \dfrac{\sqrt{14}}{3}i$　22. 125.6 m^2　23. $Pb = 37B - 144$　24. $3^{9/4}$　25. $x^{13/6}y^{7/2}$

26. $\dfrac{-99\sqrt{10}}{10}$　27. $x^3 - 6x^2 + 12x - 8$　28. 8　29. $\dfrac{x^2y^2 - 1}{x^2 - 6y}$　30. 3.5

ER31. 2.9481　**ER32.** 101.4536　**ER33.** -3.4091

problem set 71

1. 100 ml　2. $10,200$　3. 8200　4. $N_F = 200, N_J = 40$　5. $2.5 \times 10^5 \text{ torr}$

6. $S = .075C - 2$　7. $-3.46R + 2U$　8. $20\sqrt{10}\,\underline{/\,198.43°}$　9. See lesson 70

10. $-\dfrac{1}{5} \pm \dfrac{\sqrt{34}}{5}i$　11. $-\dfrac{7}{6} \pm \dfrac{\sqrt{13}}{6}$　12. $\dfrac{xm_2}{am_2 - xy}$　13. $\dfrac{3x^2 + 6x - 1}{(x - 4)(x + 2)}$

14. $\dfrac{-3 + 9\sqrt{5}}{88}$　15. $R_C = 33, R_P = 99, T_C = 5, T_P = 7$　16. $\dfrac{ab^3 + ab + b^2}{b^2 + 1}$

17. $\dfrac{x^3y^2 + x^2 + xy^2}{xy^2 + 1}$　18. $-22 - 20i$　19. $6 + 7i$　20. $(12, 5)$

21. $x = \dfrac{1}{4} \pm \dfrac{\sqrt{31}}{4}i$　22. $1,360,000$　23. $2^{1/2}$　24. 16　25. $a^{7/4}y^{9/4}$

26. $\dfrac{-64\sqrt{15}}{15}$　27. $\dfrac{4\sqrt{13}}{3}$　28. $-\dfrac{11}{2}$　29. $y = -\dfrac{4}{3}x - \dfrac{28}{3}$　30. -1×10^{27}

ER34. 1.5714　**ER35.** 60.8594　**ER36.** -61.1369

problem set 72

1. 218 grams　2. $N_D = 8, N_P = 20$　3. 96 miles　4. 100　5. $1.4 \times 10^{10} \text{ K}$

6. $\dfrac{-17 - 14\sqrt{3}}{23}$　7. $-11 + 8\sqrt{2}$　8. $\dfrac{36 + 8\sqrt{6}}{19}$　9. $H = -9C + 1022$

10. $12.11R - 6.59U$　11. $\sqrt{97}\,\underline{/\,66.04°}$　12. See Lesson 70

13. $-\dfrac{5}{4} \pm \dfrac{\sqrt{15}}{4}i$　14. $-\dfrac{5}{4} \pm \dfrac{\sqrt{57}}{4}$　15. $\dfrac{pm_2}{ckm_2 + pk}$　16. $\dfrac{7x - 5}{x - 2}$

17. $R_A = 40, R_B = 80, T_A = 4, T_B = 3$　18. $\dfrac{a^3x^3 - ax^2 - a^2x}{a^2x - 1}$　19. $-11 - 13i$

20. $\left(\dfrac{24}{7}, -\dfrac{2}{7}\right)$

21. $\dfrac{3}{2} \pm \dfrac{\sqrt{15}}{2}i$　22. $52(100)(100) \text{ cm}^2$

23. $3^{9/4}$　24. 32　25. $ax^{1/4}y^{3/2}$　26. $\dfrac{-23\sqrt{10}}{10}$　27. $\dfrac{x + 5}{x + 3}$　28. $y = -\dfrac{2}{5}x$

29. 2×10^7　30. $\dfrac{51}{7}$　**ER37.** 3.4468　**ER38.** 157.8085　**ER39.** -3.1388

problem set 73

1. $R_G = 4$ mph, $R_B = 8$ mph, $T_G = 6$ hr, $T_B = 3$hr

2. $R_C = 4$ mph, $R_A = 16$ mph, $T_A = 5$ hr, $T_C = 7$ hr   3. 1.5×10^4 K

4. 1200 liters   5. 5, 7, 9 or $-1, 1, 3$ 6. $-7 + 4\sqrt{3}$ 7. $\dfrac{6 - 5\sqrt{2}}{7}$

8. $\dfrac{-19 + 11\sqrt{2}}{34}$ 9. $N = 70F - 2050$ 10. $-8.47R - 4.88U$ 11. $5\sqrt{2} \,\underline{/225°}$

12. See Lesson 70   13. $-\dfrac{1}{6} \pm \dfrac{\sqrt{47}}{6}i$ 14. $1, -\dfrac{4}{3}$ 15. $\dfrac{kmpc}{m - apc}$ 16. $\dfrac{cmx_2}{ax_2 - bmc}$

17. $\dfrac{a^2x^2 - ax^2 - a^2}{ax - x}$ 18. $7 - 9i$ 19. $\dfrac{2x - 4}{x^2 + 7x + 10}$ 20. $(4, 10)$

21. $\dfrac{1}{6} \pm \dfrac{\sqrt{83}}{6}i$ 22. $(27,632)(100)(100)$ cm^2 23. $2^{9/4}$ 24. 27   25. $4x^{4/3}y^{25/6}$

26. $\dfrac{-31\sqrt{42}}{14}$ 27. $y = 4x + 8$ 28. 86   29. 20   **ER40.** 3.2919

ER41. 102.6302   **ER42.** -7.4815

problem set 74

1. $R_D = 60$, $R_E = 120$, $T_D = 2$, $T_E = 3$ 2. 2.88×10^{-8} torr   3. 25 days

4. 5 hr   5. $N_L = 18$, $N_P = 20$ 6. $\dfrac{4x + 10}{(x + 1)(x + 1)}$ 7. $\dfrac{-x^2 - 4x - 9}{(x - 5)(x + 3)}$

8. $\dfrac{8 - 5\sqrt{2}}{7}$ 9. $\dfrac{15 - 7\sqrt{3}}{13}$ 10. $\dfrac{30 + \sqrt{3}}{39}$ 11. $V = -1.25K + 146$

12. See Lesson 70   13. $18.06R + 1.58U$ 14. $2\sqrt{17} \,\underline{/255.96°}$ 15. $\dfrac{1}{5} \pm \dfrac{\sqrt{19}}{5}i$

16. $\dfrac{bcx}{1 - ca}$ 17. $\dfrac{-cbf}{pf + cba}$ 18. $\dfrac{a^2y^2 + a^2y + ya}{y + 1}$ 19. $-9 - 6i$ 20. $(20, 30)$

21. $-\dfrac{1}{6} \pm \dfrac{\sqrt{23}}{6}i$ 22. 41.72 in^2 23. $2^{8/3}$ 24. -8 25. $m^{11/12}p^{31/12}$

26. $\dfrac{-75\sqrt{14}}{28}$ 27. 6×10^3 28. $y = x + 2$ 29. $\dfrac{86}{3}$ 30. $\dfrac{3}{2}$ **ER43.** 3.0505

ER44. 53.9186   **ER45.** -1388.8035

problem set ER-4

1. $R_H = 30$ mph, $R_M = 60$ mph, $T_H = 8$ hr, $T_M = 6$hr   2. 2400 torr   3. 160

4. 10 p.m.   5. $N_H = 20$, $N_T = 18$ 6. -8.352×10^{-7} 7. -6.4975×10^{-17}

8. $\dfrac{4x + 6}{x^2 + 2x + 1}$ 9. $\dfrac{-x^2 + 8x - 3}{x^2 - 3x - 10}$ 10. $\dfrac{-6 + \sqrt{3}}{11}$ 11. $\dfrac{19 - 11\sqrt{2}}{7}$

12. $\dfrac{25 - 7\sqrt{3}}{239}$ 13. $25.17R + 11.12U$ 14. $\sqrt{29} \,\underline{/-21.8°}$ 15. $\dfrac{-3}{10} \pm \dfrac{\sqrt{51}}{10}i$

16. $\dfrac{dyb}{ymb - a}$ 17. $\dfrac{cbf}{-cab - af}$ 18. $\dfrac{m^2p^3 + m^3p + pm^2}{p^2 + m}$ 19. $-7 + 5i$

20. $\dfrac{1}{2} \pm \dfrac{3}{2}i$ 21. $(8, 5)$ 22. $248(12)(12)$ in^2 23. $2^{5/3}$ 24. -32

25. $m^{11/15}p^{8/5}$ **26.** $\dfrac{48\sqrt{35}}{35}$ **27.** 6×10^{-3} **28.** $y = -9x + 15$ **29.** 16

30. $-\dfrac{12}{13}$

problem set 75

1. $R_C = 4$ mph, $R_T = 12$ mph, $T_C = 8$ hr, $T_T = 6$ hr   **2.** 2×10^{-1} liters   **3.** 200 ml

4. 312 g   **5.** $N_R = 2, N_B = 5$ **6.** $(4, 2, 3)$ **7.** $(6, 1, 2)$ **8.** $-7.29R + 4.90U$

9. $\sqrt{17} \,\underline{/\,194.04°}$ **10.** $\dfrac{2x^2 + 15x + 2}{(x + 5)(x - 2)}$ **11.** $-3 - \sqrt{2}$ **12.** $\dfrac{-5 + 4\sqrt{2}}{4}$

13. $Al = 10.5B + 138$ **14.** See Lesson 70   **15.** $\dfrac{rta - tcm - mrc}{tm + mr}$

16. $\dfrac{mrc + mrx}{ra - cm - xm}$ **17.** $\dfrac{1}{14} \pm \dfrac{\sqrt{29}}{14}$ **18.** $\dfrac{x^2 + 2x}{x + 1}$ **19.** $7 - 8i$

20. $(15, 39)$ **21.** 3720 **22.** $5^{9/4}$ **23.** -32 **24.** $m^{11/12}y^{3/4}$

25. $\dfrac{-19\sqrt{6}}{2}$ **26.** 9×10^2 **27.** $y = \dfrac{11}{7}x + \dfrac{6}{7}$ **28.** 3 **29.** $\dfrac{29}{54}$

ER46. -1.1568×10^{24} **ER47.** 5.3220

problem set 76

1. $T_S = 4$ hr, $T_H = 8$ hr, $R_S = 60$ mph, $R_H = 40$ mph   **2.** 5000 **3.** 450 liters

4. $49, 56, 63$ **5.** 8 mph   **6.** -1 **7.** 9 **8.** No solution   **9.** $(1, 2, 4)$

10. $(2, 1, -2)$ **11.** $-9.1R + 4.05U$ **12.** $\sqrt{53} \,\underline{/\,254.05°}$ **13.** $\dfrac{9x + 8}{x^2 - 2x - 15}$

14. $\dfrac{3\sqrt{2} - 2}{7}$ **15.** $\dfrac{-13 - 5\sqrt{5}}{4}$ **16.** $-2 \pm \dfrac{\sqrt{14}}{2}$ **17.** $\dfrac{maxR_2}{aR_2 - bxm}$

18. $\dfrac{mbxR_1}{aR_1 - max}$ **19.** $\dfrac{ax + a + x}{x + 1}$ **20.** 0 **21.** $(14, 10)$ **22.** $-\dfrac{7}{4} \pm \dfrac{\sqrt{41}}{4}$

23. $15{,}700{,}000$ cm^2 **24.** $\dfrac{(4000)(60)}{(2.54)(2.54)(2.54)(12)(12)(12)} \dfrac{\text{ft}^3}{\text{min}}$ **25.** $2^{7/12}$ **26.** 27

27. $95 + 48\sqrt{5}$ **28.** $y = \dfrac{4}{5}x + \dfrac{28}{5}$ **29.** $2\sqrt{41}$ **30.** $\dfrac{15}{17}$

ER48. 1973.3134 **ER49.** -1.7117×10^{16}

problem set 77

1. $R_R = 45$ mph, $R_J = 135$ mph, $T_R = 3$ hr, $T_J = 7$ hr

2. $-4, -2, 0, 2$ and $-10, -8, -6, -4$ **3.** 4000 g   **4.** 400

5. 20 and 10 or -20 and -10 **6.** $\dfrac{10}{2}$ **7.** 9 **8.** 9 **9.** 25

10. $S = .95I - 88$ **11.** $(1, 3, 4)$ **12.** $8.91R + 12.54U = 15.38 \,\underline{/\,54.61°}$

13. $\dfrac{-3x^2 - 20x - 3}{x^2 + 5x - 14}$ **14.** $5\sqrt{5} - 11$ **15.** $\dfrac{-6 - 4\sqrt{2}}{3}$ **16.** $2 - \sqrt{2}$

17. $-\dfrac{1}{2} \pm \dfrac{\sqrt{3}}{2}i$ **18.** $\dfrac{ax}{mcx - a}$ **19.** $\dfrac{aR_1 + ax}{mxR_1}$ **20.** $\dfrac{a^3 + ax + a^2}{a^2 + x}$

21. $-3 + 2i$ **22.** $720{,}000 \text{ cm}^2$ **23.** $400(2.54)(2.54)(2.54)(60)\dfrac{\text{cm}^3}{\text{min}}$ **24.** $2^{43/30}$

25. -2 **26.** 14 **27.** $\dfrac{19\sqrt{10}}{5}$ **28.** 5×10^{-1} **29.** 6.25×10^{-27}

ER50. 1.4545 **ER51.** 5.8011×10^{13}

problem set 78

1. $\$7140$ **2.** $T_E = 6 \text{ hr}, T_D = 30 \text{ hr}, R_E = 20 \text{ kph}, R_D = 30 \text{ kph}$ **3.** 27.3 percent

4. 20 ml 60% and 30 ml 90%   **5.** 80 swords and 280 spears   **6.** $\dfrac{14{,}000}{1000}$ liters

7. $\dfrac{(4)(12)(12)(12)(2.54)(2.54)(2.54)}{1000}$ liters   **8.** 0 **9.** 1 **10.** $(1, 3, 2)$

11. $2.24R - .58U = 2.31 \,\underline{/\!-14.52°}$ **12.** $\dfrac{2x^2 + 10x + 3}{x^2 - 9}$ **13.** $\dfrac{5 - 3\sqrt{2}}{7}$

14. $\dfrac{-1 - \sqrt{3}}{4}$ **15.** $\dfrac{3 + 4\sqrt{2}}{23}$ **16.** See Lesson 70   **17.** $\dfrac{mbxR_1}{R_1R_2 - mxR_2}$

18. $\dfrac{aR_1R_2 - maxR_2}{mxR_1}$ **19.** $\dfrac{ax^2 - ax - x^2}{x - 1}$ **20.** $5 + 2i$ **21.** $(20, 40)$ **22.** $1, -\dfrac{1}{3}$

23. 48 ft^2 **24.** $\dfrac{40(60)(60)}{(2.54)(2.54)(2.54)}\dfrac{\text{in}^3}{\text{hr}}$ **25.** $3^{13/4}$ **26.** $\dfrac{-32\sqrt{14}}{7}$ **27.** $28 - \sqrt{2}$

28. 3×10^3 **29.** $y = \dfrac{1}{5}x + 6$ **30.** 4 **ER52.** 3.2701×10^2

ER53. 1.7237×10^4

problem set 79

1. $\$273$ **2.** 4 **3.** $N_B = 800, N_P = 1000$ **4.** 200 **5.** $N_D = 20, N_C = 10$

6. 2 **7.** 16 **8.** 100 **9.** $(2, 2, 4)$ **10.** $7.64R + 1.41U$ **11.** $2\sqrt{10} \,\underline{/\,108.43°}$

12. $\dfrac{x - 4}{(x - 8)(x + 2)}$ **13.** $7 - 4\sqrt{2}$ **14.** $-6 - 5\sqrt{2}$ **15.** 5 **16.** $\dfrac{1}{2} \pm \dfrac{\sqrt{33}}{6}$

17. $\dfrac{xmpc}{ca - xmq}$ **18.** $\dfrac{ayc}{xpc + xqy}$ **19.** $\dfrac{3a^3 - 12a}{a^2 - 3}$ **20.** $-6 + 2i$ **21.** $(12, 8)$

22. $\dfrac{1}{2} \pm \dfrac{\sqrt{21}}{2}$ **23.** $\dfrac{600(60)}{(2.54)(2.54)(2.54)(12)(12)(12)}\dfrac{\text{ft}^3}{\text{hr}}$ **24.** 8138.88 **25.** $x^{7/4}y^{7/4}$

26. 8 **27.** $\left(\dfrac{3}{7}, -\dfrac{13}{7}\right)$ **28.** $\dfrac{3a(x + a) + 3x(x + a) + 7a^2x^2}{a^2x(x + a)}$

29. $0, 8, -4$ **30.** $y = x - 1$ **ER54.** 1.8713 **ER55.** -4.3039×10^2

problem set 80

1. 625 **2.** 100 **3.** $R_R = 75 \text{ mph}, R_J = 25 \text{ mph}, T_R = 5 \text{ hr}, T_J = 15 \text{ hr}$

4. 140 liters 10%, 60 liters 40%   **5.** 2800 **6.** $\dfrac{1}{10} + \dfrac{4}{5}i$ **7.** $-\dfrac{16}{13} - \dfrac{2}{13}i$

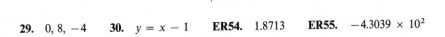

8. $\dfrac{1}{29} - \dfrac{17}{29}i$ **9.** 3   **10.** 64   **11.** (2, 3, 1)

12. $-2.6R + 12.56U = 12.83 \,\underline{/\,101.70°}$ **13.** $\dfrac{-3x^2 - 2x - 2}{x^2 - 9}$ **14.** $\dfrac{8 + 3\sqrt{2}}{2}$

15. $\dfrac{\sqrt{3} - 17}{26}$ **16.** $\dfrac{5 + 4\sqrt{2}}{7}$ **17.** $\dfrac{20\sqrt{29}}{29}$ **18.** See Lesson 70   **19.** $\dfrac{2a^4 - a^2}{a^2 + 1}$

20. $4 + 8i$ **21.** $(6, -6)$ **22.** $-\dfrac{3}{4} \pm \dfrac{\sqrt{7}}{4}i$ **23.** $4x^2 + 16x + 63 + \dfrac{254}{x - 4}$

24. $244{,}920{,}000 \text{ cm}^2$ **25.** $3^{17/8}$ **26.** $\dfrac{17\sqrt{5}}{5}$ **27.** $\dfrac{x^3y + 3(x + y) + 2x(x + y)}{x^2y(x + y)}$

28. 18   **29.** $\sqrt{170}$ **ER56.** 1.8290   **ER57.** 1.3118×10^5

problem set 81

1. 75   **2.** $R_R = 80$ mph, $R_P = 240$ mph, $T_P = 8$ hr, $T_R = 4$ hr   **3.** 4320 K

4. $D_R = 40$ mi, $D_W = 20$ mi   **5.** 276 g, 32.4%   **6.** $\dfrac{a(x + 1)}{x + 1 + xa}$ **7.** $\dfrac{b(xy + 1)}{axy + a + cy}$

8. $\dfrac{m(xy + a)}{x^2y + ax + x}$ **9.** $\dfrac{5}{17} - \dfrac{14}{17}i$ **10.** $1 + i$ **11.** 2   **12.** 25   **13.** $(1, -6, -2)$

14. $-26.48R + 14.96U$ **15.** $\sqrt{65} \,\underline{/\,119.74°}$ **16.** $\dfrac{-x^2 - 2}{x^2 - 9}$ **17.** $7 - 4\sqrt{2}$

18. $\dfrac{-3 - \sqrt{5}}{8}$ **19.** $\dfrac{16}{5}$ **20.** $\dfrac{pxm}{z - pym}$ **21.** $\dfrac{za}{xm + mya}$ **22.** $2 - 4i$

23. $(5, 9)$ **24.** See Lesson 70   **25.** $-1, -\dfrac{2}{3}$ **26.** $x^{19/6}y^{13/6}$ **27.** $\dfrac{-29\sqrt{2}}{2}$

28. 32   **29.** $\left(\dfrac{3}{8}, \dfrac{3}{2}\right)$

30. $\dfrac{400(60)}{(2.54)(2.54)(2.54)(12)(12)(12)} \dfrac{\text{ft}^3}{\text{min}}$

ER58. 1.3328   **ER59.** 1.2797

problem set 82

1. 400   **2.** $T_T = 3$ hr, $T_M = 6$ hr, $R_T = 90$ kph, $R_M = 70$ kph   **3.** 4 yd/sec

4. $N_S = 20, N_I = 60$ **5.** 640 ml of 5% and 160 ml of 20%   **6.** $x^{3a/2}y^{11x/4}$

7. $x^{a/2-2}y^{-a+4}$ **8.** x^{2a+4} **9.** $x^{ab+a}y^{ab+2a}$ **10.** $\dfrac{a(ax + 1)}{ax^2 + 2x}$ **11.** $\dfrac{b(ab + a)}{a^2b + a^2 + b^2}$

12. $-1 - 3i$ **13.** $-\dfrac{1}{2} + 2i$ **14.** 4   **15.** $(-2, 1, 2)$

16. $-3.76R + 3.63U = 5.23 \,\underline{/\,136.01°}$ **17.** $\dfrac{3x^3 + 11x^2 + 6x - 20}{x(x^2 - 4)}$ **18.** $\dfrac{-3\sqrt{2} - 2}{7}$

19. $-5\sqrt{3} - 9$ **20.** $\dfrac{cax}{a - px + cxy}$ **21.** $\dfrac{pmx - am}{xym - ax}$ **22.** $-8 - i$

23. $(7, -10)$ **24.** $\dfrac{5}{4} \pm \dfrac{\sqrt{65}}{4}$ **25.** $-\dfrac{195\sqrt{7}}{14}$ **26.** -9 **27.** $x^{8/3}y^{4/3}$

28. $4(1000)(60)(60)\,\dfrac{cm^3}{hr}$ **29.** $\sqrt{145}$ **30.** $y = -\dfrac{9}{8}x - \dfrac{11}{8}$ **ER60.** 1599.1549

ER61. 3.1415×10^6

problem set 83

1. 160 **2.** $373°C$ **3.** 824 g, 6.8% **4.** 240 ml   **5.** $350{,}000$ tons   **6.** $y^{4b/3}x^{c-p}$

7. $y^{2b/3}x^{c+p+1}$ **8.** x^{ab+b+4} **9.** y^{a+2} **10.** $\dfrac{p(am+1)}{a^2m + a + am^2}$

11. $\dfrac{bx-1}{pbx - p - bx}$ **12.** $-\dfrac{1}{2} - \dfrac{5}{2}i$ **13.** $-\dfrac{1}{6} + \dfrac{7}{6}i$ **14.** 49 **15.** $(1, 3, 3)$

16. $8.95R + 4.37U$ **17.** $\sqrt{73}\ \underline{/\,69.44°}$ **18.** $\dfrac{-5x^3 + 6x - 12}{x^2(x-3)}$ **19.** $\dfrac{34 - 23\sqrt{2}}{14}$

20. $-442 - 6\sqrt{5}$ **21.** $\dfrac{pcy}{my + acy - pbc}$ **22.** $\dfrac{mxy}{py + pbx - x^2y}$ **23.** $-5 - i$

24. $-\dfrac{1}{4} \pm \dfrac{\sqrt{23}}{4}i$ **25.** $B = -50Y + 5050$ **26.** 1130.40 ft^2 **27.** $-\dfrac{33\sqrt{15}}{5}$

28. $\dfrac{40(60)(60)}{(2.54)(12)(5280)}\,\dfrac{mi}{hr}$ **29.** 10 **30.** 114 **ER62.** 9.3737 **ER63.** 5.7509×10^{11}

problem set 84

1. 3 **2.** $T_B = 2$ hr, $T_P = 4$ hr, $R_B = 140$ mph, $R_P = 160$ mph   **3.** $N_D = 10, N_H = 20$

4. 160 ft^3 of 10% and 240 ft^3 of 50% **5.** $-5, -3, -1$, and $-9, -7, -5$

6. $B = 20, t_D = 1$ **7.** $(0, -3), \left(\dfrac{12}{5}, \dfrac{9}{5}\right)$ **8.** $a^{-x}b^{yx}$ **9.** $a^{x/2}b^{x/3 - 2}$

10. $\dfrac{x(cm + x)}{acm + ax + mb}$ **11.** $\dfrac{a(c^2 + b)}{3c^2 + 2b}$ **12.** $\dfrac{8}{17} + \dfrac{2}{17}i$ **13.** $\dfrac{10}{11} - \dfrac{7\sqrt{2}}{11}i$

14. 36 **15.** $(2, 1, 1)$ **16.** $3.39R + 3.88U = 5.15\ \underline{/\,48.86°}$ **17.** $\dfrac{2x + 7}{x - a}$

18. $\dfrac{11 - 7\sqrt{2}}{23}$ **19.** $\dfrac{-7 - 5\sqrt{3}}{4}$ **20.** $\dfrac{md - c^2}{cm}$ **21.** $\dfrac{c^2}{d - pc}$ **22.** -2

23. $(9, 6)$ **24.** See Lesson 70   **25.** $Na = -5Mg + 360$

26. $\left(\dfrac{16}{13}, \dfrac{30}{13}\right)$ **27.** 144 m^2

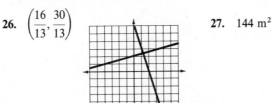

28. $100(12)(12)(12)(2.54)(2.54)(2.54)(60)$ cm^3/min   **29.** $\dfrac{137\sqrt{15}}{10}$ **30.** 32

ER64. 3.4854×10^{-7} **ER65.** -5.8808×10^{-17}

problem set 85

1. 800 1   2. 40 miles   3. 900 liters   4. 72 g   5. 16,800

6.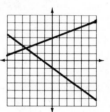
(number line: points at 1 and 2, marks at 0, 1, 2)

7. (number line: points at −3, −2, −1)

8. (number line: point at 1, arrow right, marks at 1, 2)

9. (number line: point at −1, marks at −2, −1, 0)

10. $B = 12,\ t_D = 4$

11. $\left(-\dfrac{1}{2} + \dfrac{\sqrt{17}}{2},\ \dfrac{1}{2} + \dfrac{\sqrt{17}}{2}\right)\left(-\dfrac{1}{2} - \dfrac{\sqrt{17}}{2},\ \dfrac{1}{2} - \dfrac{\sqrt{17}}{2}\right)$ 12. $x^{7a/3}y^{5ba/3}$ 13. x^{3a+2}

14. $\dfrac{ax + 1}{ax^2 + x + a^2}$ 15. $\dfrac{a^2 m - mx}{a^3 - 2ax}$ 16. $\dfrac{1}{2} + \dfrac{1}{2}i$ 17. $-\dfrac{1}{2} + \dfrac{1}{2}i$ 18. 1

19. $(-6, -3, -12)$ 20. $20\ \underline{/\,140°}$ 21. $\sqrt{109}\ \underline{/\,253.3°}$ 22. $\dfrac{21 + 11\sqrt{3}}{26}$

23. $\dfrac{aym}{x - cy - xym + 2}$ 24. $-3 - 7i$ 25. $\dfrac{-25\sqrt{14}}{14}$ 26. $\dfrac{-3 \pm \sqrt{6}}{3}$

27. $\dfrac{(40)(60)}{(12)(12)(3)(3)}\dfrac{\text{yd}^2}{\text{hr}}$ 28. $\dfrac{-3 \pm \sqrt{7}}{2}$ 29. $\left(-\dfrac{80}{23}, \dfrac{37}{23}\right)$

(graph of two intersecting lines)

30. $y = -\dfrac{3}{2}x + \dfrac{25}{2}$ **ER66.** -2.1131×10^{-9} **ER67.** -4.4940×10^{-13}

problem set 86

1. $R_B = 60$ mph, $R_J = 240$ mph, $T_B = 2$ hr, $T_J = 6$ hr   2. 20 hr

3. 120 in³ 80% and 480 in³ 30%   4. $N_H = 3,\ N_P = 132$ 5. 39.3%, 1464 g

6. See text   7. $\dfrac{43}{28}$ 8. (number line: points at −4, −3, −2, arrow)   9. (number line: points at 5, 6, 7, arrow)

10. $T_D = 3,\ B = 8$ 11. $\left(\dfrac{1}{5} + \dfrac{2\sqrt{19}}{5},\ -\dfrac{2}{5} + \dfrac{\sqrt{19}}{5}\right)\left(\dfrac{1}{5} - \dfrac{2\sqrt{19}}{5},\ -\dfrac{2}{5} - \dfrac{\sqrt{19}}{5}\right)$

12. $m^{3a+2}y^{-2}$ 13. $m^{2a}y^{-b}$ 14. $\dfrac{m(mx + 1)}{mx^2 + x + x^2}$ 15. $\dfrac{a^2 + b}{a^2 + b + a}$ 16. $-\dfrac{3}{2} - \dfrac{7}{2}i$

17. $\dfrac{4}{5} - \dfrac{7}{5}i$ 18. 49   19. $(1, 2, -2)$ 20. $6.45R + 4.76U = 8.02\ \underline{/\,36.43°}$

21. $7 - 4\sqrt{3}$ 22. $\dfrac{3 - \sqrt{3}}{4}$ 23. $\sqrt{6} + 3i$ 24. $\dfrac{-74\sqrt{21}}{7}$ 25. $\dfrac{bmp}{by + 4b - apm}$

26. $-\dfrac{1}{3} \pm \dfrac{\sqrt{11}}{3}i$ 27. $\dfrac{(10)(1000)(100)}{(2.54)(60)(60)}\dfrac{\text{in}}{\text{sec}}$ 28. See Lesson 70   29. $3^{17/8}$ 30. $\dfrac{1}{3}$

ER68. -25.0963 **ER69.** 5.9146×10^{-8}

problem set 87

1. 0, 2, 4 and −10, −8, −6   2. 1175 g

3. $R_D = 50$ mph, $R_G = 70$ mph, $T_D = 8$ hr, $T_G = 16$ hr   4. 20 liters

5. $\dfrac{(1)(8)}{(.0821)(273)}$ moles   **6.** See paragraph 87.A   **7.** $\sqrt{106}$

8.

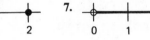

9.

10. $T_D = 1, B = 16$ **11.** $\left(\dfrac{1}{2} + \dfrac{\sqrt{3}}{2}, -\dfrac{1}{2} + \dfrac{\sqrt{3}}{2}\right)\left(\dfrac{1}{2} - \dfrac{\sqrt{3}}{2}, -\dfrac{1}{2} - \dfrac{\sqrt{3}}{2}\right)$ **12.** $x^{3b}y^{3a/4}$

13. $y^{7a/2+4}b^{1-2a}$ **14.** $\dfrac{r(1 + mr)}{m + m^2 r + r^2}$ **15.** $\dfrac{c(p^2 c - 1)}{p^2 c - 1 - c^3}$ **16.** $\dfrac{14}{13} - \dfrac{5}{13}i$

17. $\dfrac{27}{29} - \dfrac{5}{29}i$ **18.** 36   **19.** $(4, 2, -2)$ **20.** $44.77R - 21.84U$

21. $4\sqrt{10}\,\underline{/\,-71.57°}$ **22.** $-4 - 3i$ **23.** $-10\sqrt{3}$ **24.** $-1 - \sqrt{2}$

25. $-\dfrac{7}{2} - \dfrac{5\sqrt{3}}{2}$ **26.** $\dfrac{ycx}{yc - x^2 - 2yx}$ **27.** $-1 \pm i$ **28.** $\dfrac{10(60)}{(2.54)(2.54)(2.54)}\dfrac{in^3}{min}$

29. $59.44(12)(12)(12)$ in^3 **30.** $-\dfrac{40}{13}$ **ER70.** 2.5499   **ER71.** 1.6808×10^{13}

problem set 88

1. 2100   **2.** $D_R = 40$ km, $D_W = 25$ km   **3.** 560 gal   **4.** $N_S = 320, N_F = 50$

5. 80   **6.**

7.

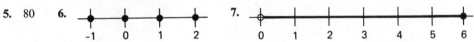

8.

9.

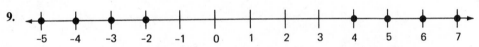

10. $T_D = 3, B = 14$ **11.** $\left(1 + \dfrac{\sqrt{2}}{2}, -1 + \dfrac{\sqrt{2}}{2}\right)\left(1 - \dfrac{\sqrt{2}}{2}, -1 - \dfrac{\sqrt{2}}{2}\right)$ **12.** $x^{3a/2+1}y^{-b}$

13. y^{2a^2+a-2} **14.** $\dfrac{k(am + 1)}{am^2 + m + m^2}$ **15.** $\dfrac{m(cb + d)}{acb + ad + cx}$ **16.** $-\dfrac{14}{17} + \dfrac{5}{17}i$

17. $-\dfrac{14}{17} - \dfrac{5}{17}i$ **18.** 25   **19.** $(1, 4, -2)$ **20.** $3.13R - 7.47U = 8.10\,\underline{/\,-67.27°}$

21. $\dfrac{-3 - 8\sqrt{2}}{7}$ **22.** $\dfrac{26 + 16\sqrt{3}}{23}$ **23.** $\dfrac{amx + cmx}{x - mca - mc^2}$ **24.** $\dfrac{xy - mc^2 y - mcx}{mx + mcy}$

25. 0   **26.** $\dfrac{-7\sqrt{6}}{12}$ **27.** $\dfrac{1 \pm \sqrt{141}}{14}$ **28.** $10(12)(12)(12)(60)\dfrac{in^3}{min}$ **29.** $3\sqrt{3}$

30. $\sqrt{101}$ **ER72.** 2.0398×10^3 **ER73.** 1.7775×10^3

problem set 89

1. $T_R = 2$ hr, $T_P = 10$ hr, $R_R = 80$ mph, $R_P = 30$ mph   **2.** 9.11 liters

3. $N_B = 5, N_R = 10$ **4.** 720 liters of 10% and 80 liters of 40%   **5.** 1536

6. $(1, 1, 1)$ **7.** $(1, 2, 3)$ **8.**

9.  **10.** $T_D = 2, B = 11$

11. $(-2 + \sqrt{5}, 2 + \sqrt{5}), (-2 - \sqrt{5}, 2 - \sqrt{5})$ **12.** $a^{2 - X/2}$ **13.** $m^{-3b/2} y^c$

14. $\dfrac{mx - 1}{xm^2 - 2m}$ **15.** $\dfrac{p(x - p)}{x^2 - xp - x^2 p}$ **16.** $\dfrac{7}{26} + \dfrac{9}{26} i$ **17.** $-\dfrac{7}{26} - \dfrac{9}{26} i$

18. 16 **19.** $6.06R + 7.77U$ **20.** $2\sqrt{29} \,\underline{/\,248.2°}$ **21.** $\dfrac{5 + 3\sqrt{3}}{4}$

22. $\dfrac{-18 - 11\sqrt{3}}{3}$ **23.** $5 + 2i$ **24.** $\dfrac{pryb}{bx - bmy + py}$ **25.** $\dfrac{xcb}{prb - pc + mcb}$

26. $3^{3/4}$ **27.** $-\dfrac{1}{12} \pm \dfrac{\sqrt{119}}{12} i$ **28.** $42(12)(12)(12)(2.54)(2.54)(2.54)$ cm³

29. $\dfrac{95}{13}$ **30.** $y = 5x + 17$ **ER74.** 3.2708 **ER75.** 1.7568×10^{-12}

problem set 90

1. 2960 mm **2.** 120 ml 60%, 180 ml 30% **3.** 75.8%

4. $R_S = 200$ mph, $R_L = 600$ mph, $T_S = 4$ hr, $T_L = 5$ hr **5.** 49

6.  **7.** 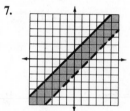 **8.** (2, 1, 3)

9. **10.** 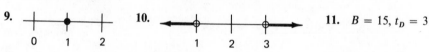 **11.** $B = 15, t_D = 3$

12. $(2 + \sqrt{2}, 2 - \sqrt{2}), (2 - \sqrt{2}, 2 + \sqrt{2})$ **13.** $x^{2b - ab/2}$ **14.** $\dfrac{a(cx^2 + 1)}{bcx^2 + b + x}$

15. $\dfrac{m(a^2 + m)}{a^3 + ma + ma^2}$ **16.** $\dfrac{16}{5} - \dfrac{3}{5} i$ **17.** $-\dfrac{12}{5} - \dfrac{9}{5} i$ **18.** $\dfrac{+6 - \sqrt{2}}{34}$

19. $-3 - 2\sqrt{2}$ **20.** $\dfrac{xa}{drm + dbma}$ **21.** $\dfrac{drmy}{x - dbmy}$

22. $2.94R + 2.57U = 3.90 \,\underline{/\,41.16°}$ **23.** $O_P = 7I_P - 110$ **24.** -3 **25.** $2 - 6i$

26. $\dfrac{56\sqrt{5}}{5}$ **27.** See Lesson 70 **28.** $\dfrac{400(60)}{(2.54)(2.54)(2.54)} \dfrac{\text{in}^3}{\text{min}}$ **29.** $6908(100)(100)$ cm²

30. $3x^2 - 6x + 10 - \dfrac{18}{x + 2}$ **ER76.** 698.8273 **ER77.** -4.4529×10^{10}

problem set 91

1. $W = 4$ mph, $T = 5$ hr **2.** $B = 11$ mph, $W = 3$ mph **3.** 20 kph **4.** 70

5. 25 sec **6.** **7.** 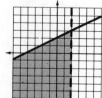 **8.** (3, 2, 1)

9. $(1, 3), (3, 1)$ **10.**

(number line with filled dot at -2, solid line to open circle at 4; marks $-2, -1, 0, 1, 2, 3, 4, 5$)

11. (number line with filled dots at $-3, -2, -1$, and $3, 4, 5$; arrow to left; marks $-3, -2, -1, 0, 1, 2, 3, 4, 5$)   **12.** $x^{-b/3}y^{3a+4}$

13. $\dfrac{x(1 + xy)}{y + xy^2 + yx}$ **14.** $\dfrac{m(m + a)}{a^2 + 2am}$ **15.** $-\dfrac{18}{25} + \dfrac{1}{25}i$ **16.** $-i$ **17.** 36

18. $-11.63R - 5.71U$ **19.** $5\sqrt{17}\,\underline{/\,104.04°}$ **20.** $\dfrac{-4 - 7\sqrt{2}}{2}$ **21.** $\dfrac{7 + 3\sqrt{2}}{2}$

22. $x^{17/15}y^{19/15}$ **23.** $\dfrac{xbyR_2}{aR_2 - xby}$ **24.** $\dfrac{aR_1R_2}{xbR_2 + xbR_1}$ **25.** $2i$ **26.** $\dfrac{89\sqrt{15}}{30}$

27. $-\dfrac{1}{6} \pm \dfrac{\sqrt{59}}{6}i$ **28.** $\dfrac{15(60)}{(2.54)(12)(3)}$ yd/min   **29.** $720{,}000$ cm^2 **30.** $\dfrac{-3x^2 - 9x + 4}{x^2 - 9}$

ER78. 4.2129 **ER79.** 2.6105×10^{-12}

problem set 92

1. 10 mph   **2.** $B = 10$ mph, $W = 2$ mph   **3.** 2 mph

4. $R_C = 4$ mph, $R_R = 8$ mph, $T_R = 6$ hr, $T_C = 10$ hr   **5.** 800 K

6. Two real number solutions   **7.** Two complex number solutions

8. (graph: shaded region with dashed and solid lines)   **9.** (number line with filled dots at $2, 3, 4, 5$)   **10.** $(4, -3, 2)$

11. $(0, -1)$ and $\left(\dfrac{4}{5}, -\dfrac{3}{5}\right)$ **12.** $y^{7a/3+6}$ **13.** $\dfrac{x(bc + 1)}{bc + 1 + ca}$ **14.** $\dfrac{m(3m + 1)}{6mx + 2x + 3m}$

15. $\dfrac{19}{25} - \dfrac{8}{25}i$ **16.** $\dfrac{14}{13} - \dfrac{8}{13}i$ **17.** 36 **18.** $4.82R - 6.36U = 7.98\,\underline{/\,-52.84°}$

19. $\dfrac{36 + 25\sqrt{2}}{2}$ **20.** $3 - \sqrt{5}$ **21.** $2^{11/15}$ **22.** $x^{7/4}y^4$ **23.** $\dfrac{m^2d + m^2bx - xdc}{xd}$

24. $\dfrac{m^2d}{ad + cd - bm^2}$ **25.** $-4i$ **26.** $\dfrac{-13\sqrt{2}}{2}$ **27.** $-\dfrac{1}{6} \pm \dfrac{\sqrt{23}}{6}i$

28. $\dfrac{10(100)(60)}{(2.54)(12)}\dfrac{\text{ft}}{\text{min}}$ **29.** $102.5(100)(100)$ cm^2 **30.** $x^2 - 6x + 18 - \dfrac{56}{x + 3}$

ER80. 8.5903×10^{-5} **ER81.** -7.8639×10^{-2}

problem set 93

1. .515   **2.** 60 gal   **3.** $W = 2$ mph   **4.** 350 ml   **5.** 16   **6.** $a, b,$ and d

7. 18   **8.** Discriminant $= 0$ so one real number solution   **9.**

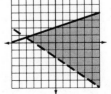

10.

A number line with points marked at 3, 4, 5, 9, 10, 11.

11. $(1, 1, 1)$

12. $\left(1 + \dfrac{\sqrt{2}}{2}, -1 + \dfrac{\sqrt{2}}{2}\right), \left(1 - \dfrac{\sqrt{2}}{2}, -1 - \dfrac{\sqrt{2}}{2}\right)$ **13.** $x^{-7a/6}y^{2-2a}$

14. $\dfrac{p(mp^2 - 1)}{m^2p^2 - m + m^2p}$ **15.** $\dfrac{x(ab^2 - 1)}{a^2b^2 - a + b^2}$ **16.** $-\dfrac{2}{3} + \dfrac{7}{3}i$ **17.** $-i$ **18.** 25

19. 0 **20.** $4\sqrt{5}\,\underline{/\,116.57°}$ **21.** $\dfrac{-13 - 2\sqrt{2}}{7}$ **22.** $-18 - 13\sqrt{2}$

23. $\dfrac{xR_2}{aR_2 + bR_2 - cx}$ **24.** $\dfrac{aR_1R_2 + bR_1R_2 - xR_2}{xR_1}$ **25.** $-3 + i$ **26.** $-\dfrac{92\sqrt{15}}{15}$

27. $-\dfrac{1}{4} \pm \dfrac{\sqrt{57}}{4}$ **28.** $\dfrac{20(60)(1000)}{(2.54)(2.54)(2.54)}\dfrac{\text{in}^3}{\text{min}}$ **29.** $N = -1.2R + 210$

30. $\dfrac{4a^3x + 6a^2x^4 + 3}{a^4x^3}$ **ER82.** 101.0966 **ER83.** 2.0715×10^{33}

problem set 94

1. 48 km   **2.** 800 ml   **3.** 1200 ml

4. $T_R = 4$ hr, $T_J = 16$ hr, $R_R = 90$ mph, $R_J = 30$ mph   **5.** $B = 12$ mph, $W = 3$ mph

6. $\left(\dfrac{4}{3}, 3\right)$ and $\left(-\dfrac{1}{2}, -8\right)$ **7.** $(1, \pm 2\sqrt{2})$ and $(-1, \pm 2\sqrt{2})$ **8.** $(1, -2, -7)$

9. a, d **10.** Discriminant $= -59$ so two complex roots   **11.**

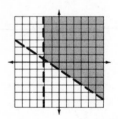

12.

A number line with a point marked at -1, showing -2, -1, 0.

13. $a^{-x/2+1}x^a$ **14.** $\dfrac{x(4 + x)}{32 + 4x}$ **15.** $\dfrac{5(2x + 2)}{5x + 4}$

16. $-\dfrac{15}{17} - \dfrac{8}{17}i$ **17.** $-1 + 4i$ **18.** 3 **19.** $23.83R - 6.88U = 24.8\,\underline{/-16.10°}$

20. $-23 + 16\sqrt{2}$ **21.** $\dfrac{2 - \sqrt{2}}{2}$ **22.** $\dfrac{mapq + mbpq - xq}{xp}$ **23.** $\dfrac{xq}{maq + mbq - xr}$

24. $4 + 3i$ **25.** $\dfrac{-79\sqrt{21}}{21}$ **26.** $1 \pm \sqrt{5}$ **27.** 13 liters   **28.** $y = -\dfrac{3}{2}x + 8$

29. $(10{,}048)(12)(12)\ \text{in}^2$ **30.** $\dfrac{30(5280)}{(60)(60)}\dfrac{\text{ft}}{\text{sec}}$ **ER84.** -335.4590

ER85. 1.4972×10^{-17}

problem set 95

1. 250   **2.** 4   **3.** 10 kph   **4.** $R_G = 30$ mph, $R_B = 60$ mph, $T_G = 10$ hr, $T_B = 5$ hr

5. 350 ml of 20% and 150 ml of 60%   **6.** $\left(\dfrac{3}{2} + \dfrac{\sqrt{57}}{2}, -\dfrac{3}{4} + \dfrac{\sqrt{57}}{4}\right), \left(\dfrac{3}{2} - \dfrac{\sqrt{57}}{2}, -\dfrac{3}{4} - \dfrac{\sqrt{57}}{4}\right)$

7. (0, 2) and (0, −2) **8.** (1, 1, 2) **9.** a, b **10.** 0 **11.**

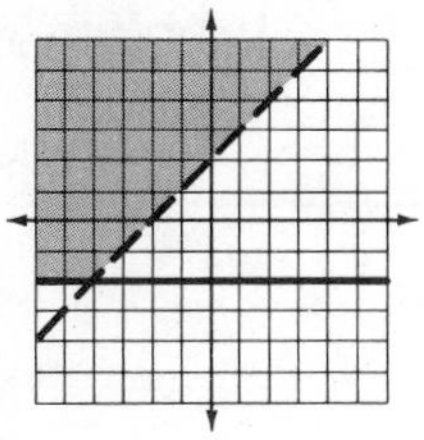

12.

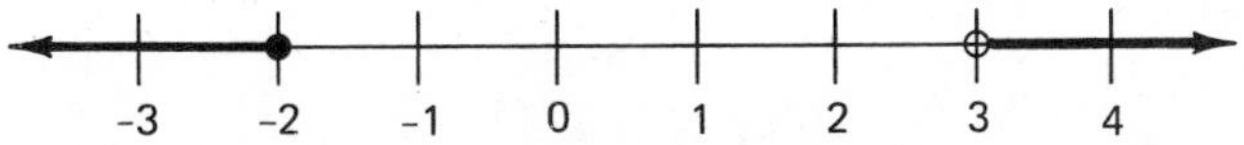

13. $a^{(x+1)/2}b^x$

14. $\dfrac{x(a^3b + 1)}{a^4b + a + ab^2}$ **15.** $2^{8/3}$ **16.** $-\dfrac{11}{13} - \dfrac{10}{13}i$ **17.** $1 - i$ **18.** 64

19. $2.34R + 3.52U = 4.23\,\underline{/56.39^\circ}$ **20.** $\dfrac{-7 - \sqrt{5}}{2}$ **21.** $\dfrac{-3 + \sqrt{7}}{2}$ **22.** $\dfrac{ma + md}{c + mr}$

23. $\dfrac{pc + pmr - md}{m}$ **24.** $7 - 4i$ **25.** 1.81 **26.** $-\dfrac{1}{8} \pm \dfrac{\sqrt{95}}{8}i$

27. $\left(\dfrac{10}{9}, \dfrac{22}{9}\right)$

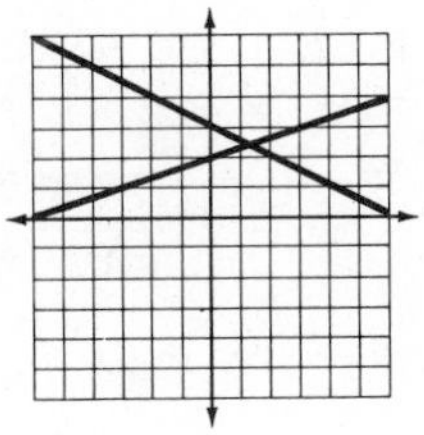

28. $\dfrac{40}{(12)(60)}\dfrac{\text{ft}}{\text{min}}$ **29.** $3\sqrt{7}$

30. $24(2.54)(2.54)(2.54)\text{ cm}^3$ **ER86.** 47.1219 **ER87.** -2.6988×10^8

problem set 96

1. $N_T = 96, N_U = 60$ **2.** 200 ml **3.** 1500 K **4.** 6 **5.** 2500 **6.** (1, −3)

7. (3, 2) **8.** $\left(-\dfrac{3}{4} - \dfrac{\sqrt{41}}{4}, \dfrac{3}{2} - \dfrac{\sqrt{41}}{2}\right), \left(-\dfrac{3}{4} + \dfrac{\sqrt{41}}{4}, \dfrac{3}{2} + \dfrac{\sqrt{41}}{2}\right)$

9. $(\sqrt{3}, \pm 2\sqrt{2}), (-\sqrt{3}, \pm 2\sqrt{2})$ **10.** (1, 2, −3) **11.** 17 **12.**

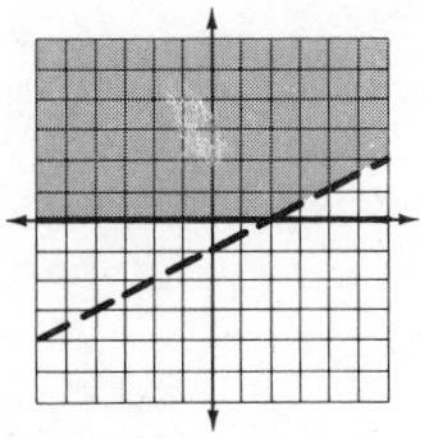

13.

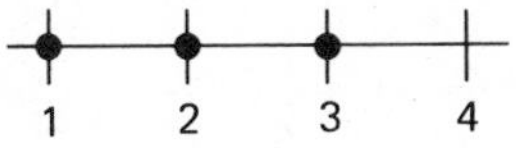

14. $x^{2a+1}y^{3b/2}$ **15.** $\dfrac{p(mx^2 + 1)}{m^2x^3 + mx - m^2x}$

16. $-\dfrac{7}{5} - \dfrac{9}{5}i$ **17.** $2^{7/12}$ **18.** $-5 + 2i$ **19.** 49

20. $-.68R + 2.24U = 2.34\,\underline{/106.89^\circ}$ **21.** $\dfrac{10 - 7\sqrt{2}}{4}$ **22.** $\dfrac{7 - 3\sqrt{7}}{7}$ **23.** $4 - 10i$

24. $-\dfrac{40\sqrt{3}}{3}$ **25.** $\dfrac{ar^2 - cxr^2 - x}{1 + cr^2}$ **26.** $\dfrac{mycx}{pc - myx}$ **27.** $\dfrac{10}{3}$ **28.** See Lesson 70

29. $\dfrac{400(60)(60)}{(2.54)(2.54)(2.54)}\dfrac{\text{in}^3}{\text{hr}}$ **30.** 42 **ER88.** -3.7552×10^3 **ER89.** 3.4125×10^{26}

problem set 97

1. $D_R = 48$ mi, $D_W = 48$ mi   2. 540 ml 10%, 60 ml 60%

3. $T_M = 6$ hr, $T_C = 24$ hr, $R_M = 60$ mph, $R_C = 50$ mph   4. $W = 1$ mph, $B = 6$ mph

5. 2   6. $\dfrac{29}{110}$ 7. $\dfrac{119}{36}$ 8. $\left(2, -\dfrac{1}{2}\right)$

9. $\left(-\dfrac{5}{6} + \dfrac{\sqrt{97}}{6}, \dfrac{5}{2} + \dfrac{\sqrt{97}}{2}\right), \left(-\dfrac{5}{6} - \dfrac{\sqrt{97}}{6}, \dfrac{5}{2} - \dfrac{\sqrt{97}}{2}\right)$ 10. $(2, \pm 2\sqrt{3}), (-2, \pm 2\sqrt{3})$

11. a, b, c 12. 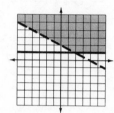   13.

14. $x^{13a/6}$ 15. $\dfrac{y(xy + 1)}{xy + 1 + y^2}$ 16. $2^{1/3}$ 17. $-2 + 3i$ 18. $-\dfrac{1}{2} + i$

19. $S = -50P + 380$ 20. $16.38R - 8.99U = 18.68 \,\underline{/-28.76°}$ 21. $-\dfrac{\sqrt{2}}{2}$

22. $-1 - \sqrt{3}$ 23. $\dfrac{xzn + yzn}{an - xz - yz}$ 24. $\dfrac{amn - xzn - xzm}{zn + zm}$ 25. $3 - i$

26. $\dfrac{211\sqrt{21}}{21}$ 27. $-\dfrac{1}{10} \pm \dfrac{3\sqrt{11}}{10}i$ 28. $(60)(2.54)(60)\,\dfrac{\text{cm}}{\text{min}}$ 29. $\dfrac{ya - x^2}{4y^2 - 3}$

30. 1×10^{42} **ER90.** 2.3155   **ER91.** 1.0575×10^{29}

problem set 98

1. 87.1%   2. $N_{pw} = 18, N_M = 2$ 3. 20 mph   4. 2000

5. 480 ml 5%, 720 ml 10%   6.

7.

8. No solution   9. $\dfrac{43}{42}$ 10. $(1, 2)$ 11. $(-4, -2), \left(6, \dfrac{4}{3}\right)$

12. $(\sqrt{5}, \pm\sqrt{3}), (-\sqrt{5}, \pm\sqrt{3})$ 13. $(2, -2, 1)$ 14. $\emptyset$ because $4 \notin \{$Negative integers$\}$

15.   16.   17. $x^{4 - 7a/4}$

18. $\dfrac{ab(b^2 + a)}{ab^2 + a^2 + b^3}$ 19. $2^{28/15}$ 20. $\dfrac{3}{2} + i$ 21. -3 22. $O_P = 4I_P - 180$

23. $1.69R + 9.63U = 9.78\,\underline{/80.05°}$ 24. $\dfrac{11 - 6\sqrt{2}}{7}$ 25. $\dfrac{6 - 4\sqrt{3}}{3}$ 26. $\dfrac{mab}{pb + pa}$

27. $4i$ **28.** $\dfrac{-23\sqrt{30}}{30}$ **29.** .116 liter   **30.** $\dfrac{(40)(60)}{(2.54)(12)}\dfrac{\text{ft}}{\text{min}}$ **ER92.** 125.7600

ER93. 1.2557×10^1

problem   **1.** $R_S = 40$ mph, $R_B = 80$ mph, $T_S = 6$ hr, $T_B = 4$ hr   **2.** 360 ml

set 99   **3.** $N_C = 60$, $N_S = 40$ **4.** 200   **5.** 10, 11, 12 and $-4, -3, -2$

6, 7, and **8.** See Lesson 99   **9.**   **10.** $\dfrac{179}{66}$

11. $\left(\dfrac{1}{2}, 1\right)$ **12.** $\left(\dfrac{5}{2} + \dfrac{\sqrt{33}}{2}, -\dfrac{5}{2} + \dfrac{\sqrt{33}}{2}\right)\left(\dfrac{5}{2} - \dfrac{\sqrt{33}}{2}, -\dfrac{5}{2} - \dfrac{\sqrt{33}}{2}\right)$

13. $(\sqrt{3}, \pm\sqrt{2}), (-\sqrt{3}, \pm\sqrt{2})$ **14.** $(2, 2, 8)$ **15.** None   **16.**

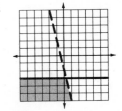

17. 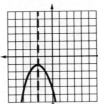   **18.** a^{2x+3+b}

19. $\dfrac{xy + 1}{x^2 y^2 + xy - y}$ **20.** $3^{9/4}$ **21.** $-1 + 2i$ **22.** i

23. $-4.25R + 9.05U = 10\,\underline{/115.16°}$ **24.** $\dfrac{5\sqrt{2} - 4}{2}$ **25.** $\dfrac{-17 - \sqrt{5}}{4}$ **26.** $\dfrac{-63\sqrt{26}}{26}$

27. $5 + 5i$ **28.** $\dfrac{mxk + mxc}{a - mkd - cdm}$ **29.** $\dfrac{400(12)(12)(12)}{60}\dfrac{\text{in}^3}{\text{sec}}$ **ER94.** .4804

ER95. 3.2975×10^{16}

problem   **1.** $P_P = \$40$, $M_U = \$8$ **2.** \$3000   **3.** \$9600   **4.** 600 liters   **5.** 6 mph

set 100   **6.** $y = -(x + 2)^2 - 1$

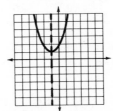

7. $y = (x + 1)^2 + 1$

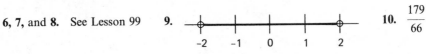

8. $y = -(x + 1)^2 - 2$

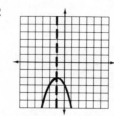

9.

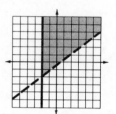

10. $\dfrac{59}{15}$ **11.** $\left(-\dfrac{1}{8}, \dfrac{7}{4}\right)$ **12.** $\left(\dfrac{2}{3} + \dfrac{\sqrt{19}}{3}, -2 + \sqrt{19}\right), \left(\dfrac{2}{3} - \dfrac{\sqrt{19}}{3}, -2 - \sqrt{19}\right)$

13. $(\sqrt{6}, \pm\sqrt{10}), (-\sqrt{6}, \pm\sqrt{10})$ **14.** $(2, -1, 2)$ **15.** $-\dfrac{15}{4}$ **16.**

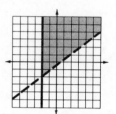

17.

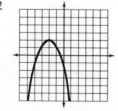

18. $y^{1-2a}a^{x/2+1}$ **19.** $\dfrac{my - 1}{my^2 - 2y}$

20. $7^{13/6}$ **21.** $-\dfrac{12}{13} - \dfrac{5}{13}i$ **22.** $\dfrac{2}{13} - \dfrac{3}{13}i$ **23.** $9.26R + 7.52U = 11.93\ \underline{/\ 39.08°}$

24. $\dfrac{2\sqrt{3}}{3} - \dfrac{\sqrt{2}}{2}$ **25.** $\dfrac{8 - 5\sqrt{2}}{2}$ **26.** $\dfrac{12\sqrt{10}}{5}$ **27.** $7 + 4i$ **28.** $\dfrac{xR_2}{mR_2 - cxR_2 - x}$

29. $-\dfrac{1}{4} \pm \dfrac{\sqrt{31}}{4}i$ **30.** $\dfrac{2p^4m}{y} - \dfrac{6}{xy^2}$ **ER96.** -64.6503 **ER97.** 7.8858×10^{23}

**problem
set 101**

1. 460 miles   **2.** 3 mph   **3.** 9   **4.** $980   **5.** $1200   **6.** 3   **7.** -10

8. No answer   **9.** $y = (x + 2)^2 - 2$

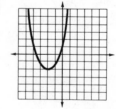

10. $y = -(x + 2)^2 + 2$

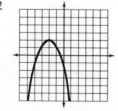

11.

12. $(3, -3)$

13. $\left(\dfrac{3}{2} - \dfrac{\sqrt{17}}{2}, -3 - \sqrt{17}\right), \left(\dfrac{3}{2} + \dfrac{\sqrt{17}}{2}, -3 + \sqrt{17}\right)$ **14.** $(2, \pm2\sqrt{2}), (-2, \pm2\sqrt{2})$

15. $(1, 3, 4)$ **16.**

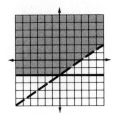

17.

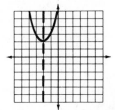

18. $\dfrac{7}{15}$ **19.** $x^{-5a/2}y^{8a/3}$

20. $k - k^2x$ **21.** $x^{23/12}y^{5/6}$ **22.** $\dfrac{1}{2} + \dfrac{7}{2}i$ **23.** $-2 + i$ **24.** $2\sqrt{29}\ \underline{/\ 68.2°}$

25. $\dfrac{am + xb - amx^2}{ax^2 - a}$ **26.** $\dfrac{\sqrt{2} - 4}{7}$ **27.** $2 + 3i$ **28.** $\dfrac{-62\sqrt{15}}{15}$

29. See Lesson 70 **30.** $(1013.8)(12)(12)(12)$ in^3 **ER98.** 115.1314

ER99. -8.8687×10^{-1}

problem set 102

1. \$890 **2.** 4 mph **3.** $T_M = 4$ hr, $T_D = 8$ hr, $R_M = 50$ mph, $R_D = 90$ mph

4. 450 K **5.** 22, 33, 44 **6.** $x^2 - xy + y^2$ **7.** $x^2 + xy + y^2$ **8.** $\emptyset$

9. $y = (x + 2)^2 + 2$

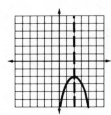

10. $y = -(x - 2)^2 - 2$

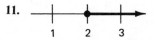

11.

12. $\dfrac{7}{4}$ **13.** $(2, 3)$ **14.** $(1 + \sqrt{2}, -1 + \sqrt{2}), (1 - \sqrt{2}, -1 - \sqrt{2})$

15. $\left(\dfrac{5}{2}, \pm\dfrac{\sqrt{15}}{2}\right), \left(-\dfrac{5}{2}, \pm\dfrac{\sqrt{15}}{2}\right)$ **16.** $(2, -1, -1)$ **17.**

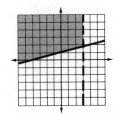

18.

19. $x^{2ab-5b/2}$ **20.** $\dfrac{m^3 + 1}{m^4 + 2m}$ **21.** $x^{13/20}y^{17/20}$ **22.** $-\dfrac{3}{5} - \dfrac{4}{5}i$ **23.** $\dfrac{21}{29} - \dfrac{20}{29}i$

24. $\dfrac{35 + 16\sqrt{5}}{5}$ **25.** $-5.25R + 1.09U = 5.36 \;\underline{/\; 168.27°}$ **26.** $\dfrac{pac}{bap - mc}$

27. 16   **28.** $9i$ **29.** $21\sqrt{3}$ **ER100.** 2.7848   **ER101.** -5.2789×10^{-5}

---

problem set 103

1. $N_D = 45, N_G = 4$ **2.** 2880 miles   **3.** 360 ml of 10% and 40 ml of 20%

4. $B = 10$ mph, $W = 4$ mph   **5.** 500% of cost, 83.3% of selling price

6. $\dfrac{623}{100,000,000}$ **7.** $\dfrac{1607}{99,000}$ **8.** $\dfrac{10,021,512}{9,990,000}$ **9.** $\dfrac{11711}{900}$ **10.** $m^2 + mp + p^2$

11. a, c **12.** $y = -(x - 2)^2 + 2$

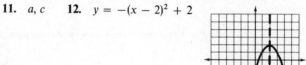

13.

14. $\dfrac{41}{28}$ **15.** $(9, 20)$

16. $\left(\dfrac{1}{4} + \dfrac{\sqrt{13}}{4}, -1 + \sqrt{13}\right), \left(\dfrac{1}{4} - \dfrac{\sqrt{13}}{4}, -1 - \sqrt{13}\right)$ **17.** $(2, 4, 6)$

18.   **19.**   **20.** $x^{-2a-4}y^{-5a/2}$

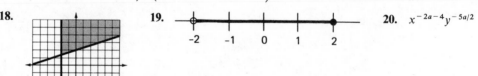

21. $\dfrac{k^2x - 1}{k^3x^2 - kx - k}$ **22.** $2^{7/15}$ **23.** $\dfrac{9}{10} + \dfrac{17}{10}i$ **24.** $-13 - 9\sqrt{2}$ **25.** $2 - 6i$

26. $\dfrac{-61\sqrt{10}}{10}$ **27.** $40.01R - 28.05U = 48.86 \;\underline{/\; -35.03°}$ **28.** $\dfrac{xyb}{myb + mxb - axy}$

29. $1, -\dfrac{5}{4}$ **30.** $\dfrac{4000(60)}{(2.54)(2.54)(2.54)(12)(12)(12)} \dfrac{\text{ft}^3}{\text{min}}$ **ER102.** 301.0300

ER103. 8.8426×10^{-16}

---

problem set 104

1. 1284 g   **2.** $D_W = 24$ mi, $D_T = 40$ mi   **3.** 60 ml   **4.** 6000

5. 25% of selling price, 33.3% of cost   **6.** $5, -\dfrac{3}{2}$ **7.** $-\dfrac{2}{3}, \dfrac{3}{2}$ **8.** $\dfrac{1233}{9,990,000}$

9. $\dfrac{165}{9990}$ **10.** $m^2 - mp + p^2$ **11.** $x^2 - xy + y^2$

12. $y = (x + 3)^2 - 1$

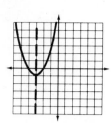

13.

-1   0   1

14. $\dfrac{18}{5}$

15. $(14, 40)$ **16.** $\left(\dfrac{1}{5} + \dfrac{\sqrt{31}}{5}, -1 + \sqrt{31}\right), \left(\dfrac{1}{5} - \dfrac{\sqrt{31}}{5}, -1 - \sqrt{31}\right)$

17. $(1 + \sqrt{3}, -1 + \sqrt{3}), (1 - \sqrt{3}, -1 - \sqrt{3})$ **18.** $(1, 4, -2)$ **19.** $(-2, 2)$

20.

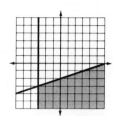

21. $5, -\dfrac{3}{2}$ **22.** $5, -\dfrac{3}{2}$ **23.** $0, 2, -\dfrac{1}{3}$ **24.** $-\dfrac{1}{3}, 2$

25. $-\dfrac{2}{3}, -2$ **26.** $0, -\dfrac{2}{3}, -2$ **27.** $\dfrac{5}{2}, -1$ **28.** $-\dfrac{1}{2}, -4$ **29.** $x^{11a/6 + 8}y^{2b/3}$

30. $-\dfrac{5}{4} - \dfrac{1}{4}i$ **31.** $\dfrac{-25 - 11\sqrt{5}}{10}$ **ER104.** 2.9218 **ER105.** 8.8400×10^4

problem set 105

1. $\$550$ **2.** $R_C = 50$ mph, $R_P = 300$ mph, $T_C = 5$ hr, $T_P = 4$ hr   **3.** 12 mph

4. 350 torr   **5.** $(1, -2, 2)$ **6.** $(2, -2, 3)$ **7.** $(2, -1, -3)$ **8.** $\dfrac{7006}{9,990,000}$

9. $\dfrac{40,616}{9900}$ **10.** $m^2 + mp + p^2$ **11.** $y = -(x + 1)^2 - 2$

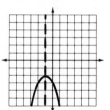

12. All integers   **13.** $\dfrac{73}{20}$ **14.** $(20, 28)$

15. $\left(\dfrac{3}{10} + \dfrac{\sqrt{89}}{10}, -\dfrac{3}{2} + \dfrac{\sqrt{89}}{2}\right), \left(\dfrac{3}{10} - \dfrac{\sqrt{89}}{10}, -\dfrac{3}{2} - \dfrac{\sqrt{89}}{2}\right)$

16. $\left(\dfrac{4}{5} + \dfrac{\sqrt{31}}{5}, -\dfrac{2}{5} + \dfrac{2\sqrt{31}}{5}\right), \left(\dfrac{4}{5} - \dfrac{\sqrt{31}}{5}, -\dfrac{2}{5} - \dfrac{2\sqrt{31}}{5}\right)$ **17.** $(3, -4)$

18.

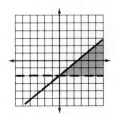

19. $\dfrac{5}{2} - \dfrac{1}{2}i$ **20.** $\dfrac{1}{6} \pm \dfrac{\sqrt{85}}{6}$

21. $8.17R - 6.29U = 10.31 \, \underline{/ -37.59°}$ **22.** $-\dfrac{1}{3}, -2$ **23.** $\dfrac{2}{3}, -1$ **24.** $-\dfrac{3}{2}, -5$

25. $0, -3, -\dfrac{5}{2}$ **26.** $-\dfrac{2}{3}, 5$ **27.** $3, \dfrac{5}{2}$ **28.** $0, -\dfrac{3}{2}, -2$ **29.** $-1, -\dfrac{1}{3}$

ER106. 367.7454 **ER107.** -4.9744×10^{17}

problem set 106

1. 320 ml 30%, 80 ml 60%   **2.** 84.2%   **3.** 47   **4.** 72   **5.** $P_P = \$420, M_U = \$980$

6. $(-2, 2, 3)$ **7.** $(1, 1, 2)$ **8.** $\dfrac{1212}{999{,}000}$ **9.** $x^2 - xy + y^2$

10. $y = -(x - 1)^2 + 2$ 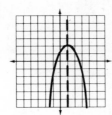   **11.** 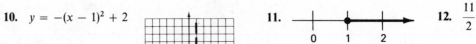   **12.** $\dfrac{11}{2}$

13. $(12, 9)$ **14.** $\left(-1 + \sqrt{21}, \dfrac{1}{4} + \dfrac{\sqrt{21}}{4}\right), \left(-1 - \sqrt{21}, \dfrac{1}{4} - \dfrac{\sqrt{21}}{4}\right)$

15. $\left(\dfrac{1}{5} + \dfrac{2\sqrt{19}}{5}, -\dfrac{2}{5} + \dfrac{\sqrt{19}}{5}\right), \left(\dfrac{1}{5} - \dfrac{2\sqrt{19}}{5}, -\dfrac{2}{5} - \dfrac{\sqrt{19}}{5}\right)$ **16.** $(4, -4)$

17. 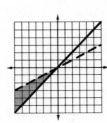   **18.** $-\dfrac{6}{25} - \dfrac{17}{25}i$ **19.** $x^{23/12}y^{17/12}$ **20.** 49

21. $4\sqrt{5} \, \underline{/ 116.57°}$ **22.** $y = -\dfrac{3}{5}x + \dfrac{9}{5}$ **23.** $\dfrac{p}{m^2xp - 1}$ **24.** $\dfrac{800(1000)}{(2.54)^3(12)^3(60)} \dfrac{\text{ft}^3}{\text{sec}}$

25. 2411.52 in^2 **26.** $0, -3, \dfrac{5}{2}$ **27.** $-1, -\dfrac{1}{2}$ **28.** $0, -2, -\dfrac{1}{5}$ **29.** $0, -1, -\dfrac{5}{3}$

30. $0, -\dfrac{3}{2}, 1$ **ER108.** 3.6674 **ER109.** 1.0083×10^{14}

problem set 107

1. \$3840   **2.** 42   **3.** 96   **4.** 5000   **5.** $t = 4$ hr, $B = 20$ mph

6. $(2my^2 + x)(4m^2y^4 - 2my^2x + x^2)$ **7.** $(a^4 + b^4)(a^8 - a^4b^4 + b^8)$

8. $(xy - p)(x^2y^2 + xyp + p^2)$ **9.** $(2x^4y^2 - my^3)(4x^8y^4 + 2x^4y^5m + m^2y^6)$

10. $\dfrac{3711}{900}$ **11.** $y = (x - 1)^2 - 2$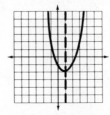

12.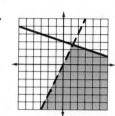

13. $(10, 8)$

14. $\left(-\dfrac{4}{5} + \dfrac{\sqrt{21}}{5}, \dfrac{2}{5} + \dfrac{2\sqrt{21}}{5}\right), \left(-\dfrac{4}{5} - \dfrac{\sqrt{21}}{5}, \dfrac{2}{5} - \dfrac{2\sqrt{21}}{5}\right)$ **15.** $(1, 2, 4)$

16. $(3, 2, 1)$ **17.**

18. $-\dfrac{4}{5} + \dfrac{3}{5}i$ **19.** $7 - 2\sqrt{5}$

20. $3^{7/18}$ **21.** $-\dfrac{\sqrt{15}}{3}$ **22.** $14.56 \,\underline{/-74.05°}$ **23.** $\dfrac{40(60)(60)}{5280} \dfrac{\text{mi}}{\text{hr}}$ **24.** $4\sqrt{5}$

25. $W = 15E - 500$ **26.** $\dfrac{b(a^3 + 1)}{a^4 + a + b^2}$ **27.** See Lesson 70   **28.** $-\dfrac{3}{2}, -3$

29. $0, -\dfrac{1}{3}, -3$ **30.** $3, -\dfrac{2}{3}$ **ER110.** 2.1911×10^4 **ER111.** 1.0382×10^{23}

problem set 108

1. 72 **2.** 74 **3.** $R_L = 3$ mph, $R_C = 6$ mph, $T_L = 21$ hr, $T_C = 10$ hr

4. 2400 torr   **5.** 60 ml   **6.** $8x^{3/2}y^{3/4}z^3$ **7.** $x + 2x^{1/2}y^{1/2} + y$

8. $x + 2x^{1/2}y^{-1/2} + y^{-1}$ **9.** $(xy^2 - 3m)(x^2y^4 + 3mxy^2 + 9m^2)$

10. $(4x^3y^2 + p^4z)(16x^6y^4 - 4x^3y^2p^4z + p^8z^2)$ **11.** $\dfrac{101,319}{99,000}$

12. $y = -(x - 2)^2 + 3$

13.

14. $(10, 9)$ **15.** $(-1, -3), (6, \tfrac{1}{2})$ **16.** $(1, 3, -3)$ **17.** $(3, 3, 3)$

18.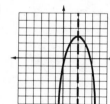

19. $\dfrac{6}{13} - \dfrac{9}{13}i$ **20.** $\dfrac{26 + 19\sqrt{2}}{46}$ **21.** $3^{4/3}$ **22.** $x^{7/2}y$

23. $x^{-5a/3 - 2/3}y^{3b/2}$ **24.** $\dfrac{-29\sqrt{6}}{3}$ **25.** $\dfrac{ka(k^2a - 1)}{k^3a - k - a^2}$

26. $9.77R + 10.97U = 14.69 \,\underline{/48.31°}$ **27.** $\dfrac{1000(1000)}{60} \dfrac{\text{ml}}{\text{sec}}$ **28.** $\dfrac{5}{2}, -2$ **29.** $4, -\dfrac{1}{2}, 0$

30. $-\dfrac{2}{3}, 3$ **ER112.** 6.8252 **ER113.** 5.0267×10^{23}

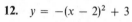

**problem
set 109**

1. 63   2. 16   3. 765 g   4. $B = 11$ mph, $W = 4$ mph

5. 33.3% of selling price and 50% of cost

6.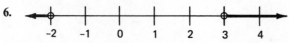

7. 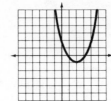   8. $4x^6y^4z^6$

9. $x - 2x^{1/2}y^{1/4} + y^{1/2}$ 10. $x - y$ 11. $(2x^3 - y^2p)(4x^6 + 2x^3y^2p + y^4p^2)$

12. $(3x^4y^3 + p^2m^5)(9x^8y^6 - 3x^4y^3p^2m^5 + p^4m^{10})$ 13. $\dfrac{1361}{99,900}$

14. $y = (x - 2)^2 - 1$ 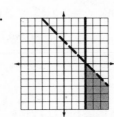   15.

16. $(14, 10)$ 17. $(2, 3, -4)$ 18. $\left(\dfrac{1}{2} + \dfrac{\sqrt{11}}{2}, -\dfrac{1}{2} + \dfrac{\sqrt{11}}{2}\right), \left(\dfrac{1}{2} - \dfrac{\sqrt{11}}{2}, -\dfrac{1}{2} - \dfrac{\sqrt{11}}{2}\right)$

19. $(1, 2, -2)$ 20.   21. $-\dfrac{1}{3} - \dfrac{1}{3}i$ 22. $\dfrac{-23 - 7\sqrt{5}}{4}$

23. $a^{4 - 5b/2}x^{2 - b/2}$ 24. $2^{7/8}$ 25. $\dfrac{23\sqrt{6}}{6}$ 26. 49   27. $\sqrt{241}\,\underline{/-75.07°}$

28. $-\dfrac{5}{3}, -2$ 29. $0, \dfrac{1}{4}, -1$ 30. $0, -\dfrac{3}{2}, -4$ **ER114.** 11.0895

ER115. 6.4972×10

**problem
set 110**

1. $N_N = 10, N_D = 15, N_Q = 1$ 2. $N_B = 2, N_G = 2, N_Y = 3$ 3. 12

4. 8, 9, 10 and $-5, -4, -3$ 5. 400

6.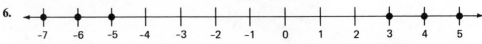

7.   8. $x + 2x^{1/2}y^{1/4} + y^{1/2}$

9. $x - 2x^{1/2}y^{-1/2} + y^{-1}$ 10. $\dfrac{x}{y}$ 11. $(x - m^2y^2)(x^2 + xm^2y^2 + m^4y^4)$

12. $(2x^2y - 3mp^4)(4x^4y^2 + 6x^2ymp^4 + 9m^2p^8)$ 13. $\dfrac{10,111}{9900}$

14. $y = -(x + 2)^2 + 3$

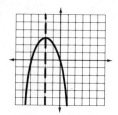

15.

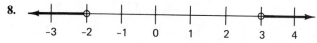

16. 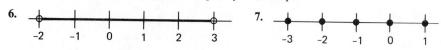   **17.** $(-10, 10)$ **18.** $(2, 4, 6)$

19. $\left(\dfrac{1}{2} + \dfrac{\sqrt{7}}{2}, -\dfrac{1}{2} + \dfrac{\sqrt{7}}{2}\right), \left(\dfrac{1}{2} - \dfrac{\sqrt{7}}{2}, -\dfrac{1}{2} - \dfrac{\sqrt{7}}{2}\right)$ **20.** $\dfrac{5}{2}, -1$ **21.** $\dfrac{40(60)(60)}{12} \dfrac{\text{ft}}{\text{hr}}$

22. $\dfrac{3}{7} - \dfrac{2}{7}i$ **23.** $\dfrac{6 + 4\sqrt{3}}{3}$ **24.** $a^{-5x/2}y^{1+3x/2}$ **25.** $x^{3/2}y$ **26.** $\dfrac{65\sqrt{14}}{14}$

27.

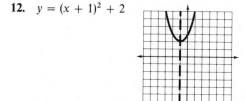

28. $8.14R + 14.14U = 16.32 \,\underline{/\,60.07°}$ **29.** $\dfrac{5}{2}, -2$

30. $0, -\dfrac{3}{2}, 5$ **ER116.** 3.8388 **ER117.** 4.0477×10^{24}

problem set 111

1. $N_N = 5, N_D = 5, N_Q = 10$ **2.** 43

3. $\$1440$ **4.** $B = 20$ mph, $T_D = 10$ hr, $T_U = 5$ hr

5. $T_H = 3$ hr, $T_F = 9$ hr, $R_H = 400$ mph, $R_F = 200$ mph

6.

7.

8.

9. $x - x^{1/2}y^{-1/4} + x^{1/2}y^{1/2} - y^{1/4}$ **10.** $(p^2x^2 - k)(p^4x^4 + kp^2x^2 + k^2)$ **11.** $\dfrac{39,742}{9900}$

12. $y = (x + 1)^2 + 2$

13.

14.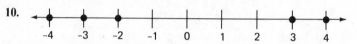

15. a, b **16.** $\dfrac{37}{12}$

17. $-\dfrac{1}{10} \pm \dfrac{\sqrt{79}}{10}i$ **18.** $y = 3x - 22$ **19.** $(8, 12)$ **20.** $(4, -3, 1)$

21. $(4, 1), (-\frac{1}{2}, -8)$ **22.** $(2, 4, -5)$ **23.** $-\frac{5}{6} + \frac{1}{6}i$ **24.** $a^{-1}y^{2-x/2}$ **25.** $x^{25/6}y^{4/3}$

26. $\dfrac{-75\sqrt{14}}{14}$ **27.** $\dfrac{36 - 5\sqrt{2}}{89}$ **28.** $0, -\dfrac{2}{3}, -1$ **29.** $2, -\dfrac{1}{2}$ **30.** $-2, -\dfrac{2}{3}$

ER118. 1.3719×10^5 **ER119.** 3.1599×10^{15}

problem set 112

1. $N_N = 10, N_D = 5, N_Q = 4$ **2.** 120   **3.** 140 ml 60%, 60 ml 70%

4. $R_C = 60$ mph, $R_D = 40$ mph, $D = 480$ miles   **5.** 3400 K

6. $10^{.7664}$, $\log 5.8437912 = .7664$ **7.** $10^{9.7664}$, $\log (584,312 \times 10^4) = 9.7664$

8. $10^{-8.2336}$, $\log (.00058423 \times 10^{-5}) = -8.2336$ **9.**

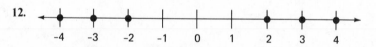

10.

11.

12.

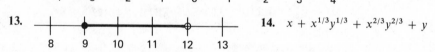

13.

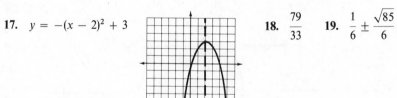

14. $x + x^{1/3}y^{1/3} + x^{2/3}y^{2/3} + y$

15. $(2p^2k^5 - xm^2)(4p^4k^{10} + 2p^2k^5xm^2 + x^2m^4)$ **16.** $\dfrac{313}{99,000}$

17. $y = -(x - 2)^2 + 3$ **18.** $\dfrac{79}{33}$ **19.** $\dfrac{1}{6} \pm \dfrac{\sqrt{85}}{6}$

20. $(20, 9)$ **21.** $(0, -2), (\frac{6}{5}, \frac{8}{5})$ **22.** $(5, 5, 5)$ **23.** $(2, 3, 4)$ **24.** $-\frac{1}{2}i$

25. $\dfrac{8 + 3\sqrt{2}}{4}$ **26.** $x^{1/2}y^{7/6}$ **27.** $-\dfrac{31\sqrt{3}}{3}$ **28.** $0, 1, -\dfrac{3}{2}$ **29.** $1, -\dfrac{2}{3}$

30. $0, -\frac{1}{3}, -2$ **ER120.** 3.2125   **ER121.** 7.9636×10^{-2}

problem set 113

1. 8, 10, 12 and $-4, -2, 0$ **2.** 160 ml

3. $R_S = 400$ mph, $R_F = 800$ mph, $T_S = 5$ hr, $T_F = 6$ hr   **4.** $B = 13$ mph

5. 17   **6.** A **7.** C **8.** B **9.** $10^{9.6794}$, $\log (47,830 \times 10^5) = 9.6794$

10. $10^{-7.1457}$, $\log (.000715 \times 10^{-4}) = -7.1457$

11. $10^{-10.4202}$, $\log (38,000 \times 10^{-15}) = -10.4202$

12. $x - 2x^{1/2}y^{-1/2} + y^{-1}$ **13.** $(mp + 3x^2y^3)(m^2p^2 - 3mpx^2y^3 + 9x^4y^6)$

14.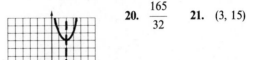

15.

16.   **17.**

18. $\dfrac{1046}{990,000}$ **19.** $y = (x - 2)^2 + 3$ **20.** $\dfrac{165}{32}$ **21.** $(3, 15)$

22. $(2, 4, -2)$ **23.** $(3 + \sqrt{10}, 3 - \sqrt{10}), (3 - \sqrt{10}, 3 + \sqrt{10})$ **24.** $(4, 3, 2)$

25. $O_P = 22.5I_P - 1512$ **26.** $\dfrac{1}{5} - \dfrac{3}{5}i$ **27.** $-\dfrac{99\sqrt{10}}{20}$ **28.** 16 **29.** $0, -\dfrac{2}{5}, -1$

30. $\dfrac{2}{5}, -1$ **ER122.** 3.9921×10^3 **ER123.** 1.3274×10^{-21}

problem set 114

1. 320 g   **2.** $D_S = 180$ mi, $D_B = 100$ mi

3. $R_B = 120$ mph, $R_H = 240$ mph, $T_B = 4$ hr, $T_H = 5$ hr   **4.** $N_Q = 1, N_D = 3, N_N = 10$

5. 24 **6.** $10^{-3.1451}$ **7.** $10^{9.6839}$ **8.** $10^{-4.1549}$ **9.** 5.74 **10.** 5.76×10^{-6}

11. 8.87×10^{-8} **12.** 1.34×10^{-7} **13.** 1.06×10^3 **14.** c

15.   **16.**

17.   **18.**

19. $x^{1/2} - 2x^{1/4}y^{-1/4} + y^{-1/2}$ **20.** $(xm^5 - 2py^2)(x^2m^{10} + 2xm^5py^2 + 4p^2y^4)$

21. $\dfrac{1037}{990}$ **22.** $y = (x + 1)^2$ **23.** $\dfrac{23}{4}$ **24.** $(12, 12)$

25. $(2, 4, -4)$ **26.** $(4, 1, 1)$ **27.** $-8 - 3i$ **28.** $\dfrac{-25 - 7\sqrt{2}}{17}$ **29.** $\dfrac{1}{6} \pm \dfrac{\sqrt{73}}{6}$

30. $-2, \dfrac{4}{3}$ **ER124.** 7.9474 **ER125.** 8.2801×10^{26}

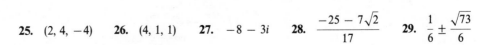

problem set 115

1. $-48, -42, -36$ 2. 400 K   3. $B = 10$ mph, $W = 3$ mph   4. 2000

5. 29   6. 2.6904   7. 2.8477   8. 2.4949   9. 8.91×10^{-8}

10. 3.98×10^{-4} 11. 8.71×10^{-2} 12. A 13. $(390.72)(12)(12)(12)$ in^3 14. $\dfrac{51}{5}$

15.

16.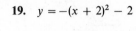

17.    18. $\dfrac{10,372}{9900}$

19. $y = -(x + 2)^2 - 2$

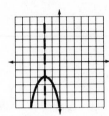

20. $x - 2x^{1/2}y^{3/4} + y^{3/2}$

21. $(3m^3p - x^4y)(9m^6p^2 + 3m^3px^4y + x^8y^2)$ 22. $\dfrac{35}{36}$ 23. $(15, 21)$

24. $(1, -3, -3)$ 25. $(-3, -3, -3)$ 26. $\dfrac{5}{17} - \dfrac{14}{17}i$ 27. $\dfrac{6 + \sqrt{6}}{2}$

28. $a^{x/2-4}m^{x/2}$ 29. 49   30. $0, 1, -\dfrac{2}{3}$ **ER126.** -277.836

ER127. 3.1439×10^{17}

problem set 116

1. $N_N = 8, N_D = 6, N_Q = 2$ 2. 25% of P_P and 20% of S_P

3. $T_G = 5$ hr, $T_M = 10$ hr, $R_G = 40$ mph, $R_M = 65$ mph   4. 420 ml of 30% and 180 ml of 60%

5. $864 paid, $3456 markup   6. 2.58   7. 3.9318   8. 2.3372

9. 4.57×10^{-6} 10. 9.64×10^{-3} 11. 9.07×10^{-5} 12. E

13. $(114.2)(100)(100)(100)$ cm^3 14. $\dfrac{4\sqrt{2}}{3}$ 15.

16. 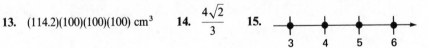   17.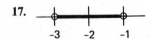

18. 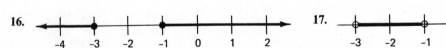   19. 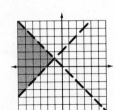   20. $\dfrac{20,221}{9900}$

21. $y = -(x - 2)^2 + 2$

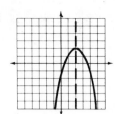

22. $x - 2x^{1/2}y^{-1/2} + y^{-1}$

23. $(m - 2p^2k^3)(m^2 + 2mp^2k^3 + 4p^4k^6)$ **24.** $\dfrac{43}{45}$ **25.** $(16, 9)$ **26.** $(2, -4, 4)$

27. $-\dfrac{1}{5} + \dfrac{8}{5}i$ **28.** $\dfrac{-6 + 2\sqrt{2}}{7}$ **29.** $x^{29/12}y^{1/4}$ **30.** $-\dfrac{1}{2}, +6$ **ER128.** 2.7482

ER129. 1.3281×10^{-1}

problem set 117

1. 4977 **2.** 36 g **3.** $N_Q = 9, N_N^* = 9$ **4.** $B = 15$ mph $T_U = 4$ hr, $T_D = 8$ hr

5. 41 **6.** 1.87×10^{-5} **7.** 2.13×10^2 **8.** 8.57×10^{-2} **9.** 2.21×10^{-2}

10. 2.5031 **11.** 6.9957 **12.** 2.4318 **13.** 9.08×10^{-6} **14.** 3.72×10^{-3}

15. 1.12×10^{-9} **16.** $184(12)(12)$ in^2 **17.** $\dfrac{162}{99,000}$

18. $y = -(x + 3)^2 - 1$

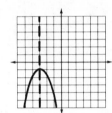

19. All integers

20.

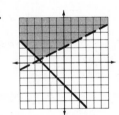

21.

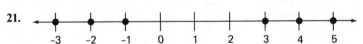

22.

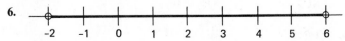

23. $-3 \pm i$ **24.** $x^{1/2} - 2x^{1/4}y^{1/4} + y^{1/2}$

25. $(x^2y - 3p^2m^3)(x^4y^2 + 3x^2yp^2m^3 + 9p^4m^6)$ **26.** $\dfrac{393}{80}$ **27.** $(3, 3, -3)$ **28.** 2

29. $3^{7/10}$ **30.** $-7\sqrt{2} - 11$ **ER130.** -425.4343 **ER131.** -1.4555×10^2

problem set 118

1. 1500 ml **2.** $T_G = 5$ hr, $T_T = 10$ hr, $R_G = 50$ mph, $R_T = 40$ mph **3.** 1000

4. $N_N = 5, N_D = 5, N_Q = 14$ **5.** 41

6.

7.

8.

9.   **10.**

11.   **12.**

13. 5.75×10^{13} **14.** 3.02×10^3 **15.** 3.29×10^1 **16.** 9.6840   **17.** 11.0214

18. 1.7696   **19.** 9.10×10^{-7} **20.** 5.89×10^{-3} **21.** 3.31×10^{-2}

22. $\dfrac{(5000)(1000)}{(60)(2.54)(2.54)(2.54)} \dfrac{\text{in}^3}{\text{sec}}$ **23.** $(3516.8)(100)(100) \text{ cm}^2$ **24.** $\dfrac{13}{990}$

25. $y = -(x + 1)^2 + 2$ **26.** See Lesson 70   **27.** $\dfrac{281}{75}$

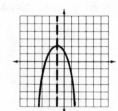

28. $\dfrac{-2}{13} - \dfrac{10}{13}i$ **29.** $a^{b/2}x^{b/2}$ **30.** $0, -1, 2$ **ER132.** 3.3783

ER133. 2.2208×10^{-8}

problem set 119

1. 210 liters   **2.** 29   **3.** $S_N = 2, M_N = 26$ **4.** $B_N = 25, L_N = 30$

5. $M_N = 32, B_N = 35$ **6.**

7.   **8.**

9.   **10.**

11.

12.   **13.** 3.28×10^{-13} **14.** 4.15×10^{-2}

15. 7.02×10^{-10} **16.** 11.4868   **17.** 6.0074   **18.** 1.2076   **19.** 1.20×10^{-6}

20. 7.41×10^{-4} **21.** 5.88×10^{-10} **22.** $\dfrac{40(12)(12)(12)}{60} \dfrac{\text{in}^3}{\text{sec}}$ **23.** $\dfrac{2161}{99,900}$

24. 1200 in^3

25. $y = -(x - 1)^2 - 2$

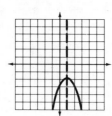

26. $-\dfrac{1}{5} \pm \dfrac{\sqrt{6}}{5}$ **27.**   (6, 4)

28. $2\sqrt{13}\,\underline{/-56.31°}$ **29.** $7.62R + 5.96U = 9.67\,\underline{/38.03°}$ **30.** $0, \frac{1}{5}, -2$

ER134. -2.5445×10^4 **ER135.** 9.1016×10^6

**problem
set 120**

1. 20 liters   **2.** 1200 K   **3.** $B = 10$ mph, $W = 4$ mph   **4.** $S_N = 1, G_N = 2$

5. $R_N = 10, Y_N = 5$ **6.**

7.

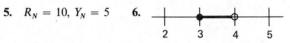

8.

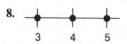

9.

10.

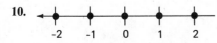

11. 5.94×10^{11}

12. 1.28×10^2 **13.** 8.4×10^{-2} **14.**   10.3788   **15.**   4.1403   **16.**   1.2757

17. 3.89×10^{-4} **18.** 5.62×10^{-8} **19.** 7.41×10^{-4} **20.** $\dfrac{4000}{(2.54)(12)(3)(60)}\dfrac{\text{yd}}{\text{sec}}$

21. $\dfrac{4234}{99,990}$ **22.**   720 ft³   **23.** $y = -(x - 2)^2 - 3$

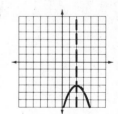

24. $-1.52R + 4.49U = 4.74\,\underline{/108.70°}$ **25.**   (2, 3, −2)   **26.**   (4, 9)   **27.**   (1, 2, 3)

28. $-13 + 6\sqrt{5}$ **29.** $0, \frac{7}{3}, -1$ **ER136.**   2.0657   **ER137.** 6.7513×10^4

**problem
set 121**

1. $N_N = 47, N_D = 22$ **2.** 128   **3.** 28   **4.** $Y_N = 10, M_N = 6$

5. $P_N = 10, D_N = 5$ **6, 7, 8.** See Lesson 121   **9.**

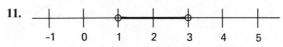

10.

11.

12.

13.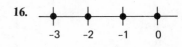

14. 7.24×10^{16} **15.** 3.74×10^1 **16.** 1.39×10^{-5} **17.** 8.4868 **18.** 4.1524

19. 2.7959 **20.** 9.55×10^{-5} **21.** 5.89×10^{-9} **22.** 7.41×10^{-11}

23. $(226.08)(12)(12)(3)(3)$ in^2 **24.** $\dfrac{(1000)(2.54)(60)(60)}{(100)(1000)}\dfrac{\text{km}}{\text{hr}}$ **25.** $\dfrac{167}{99,000}$

26. $-\dfrac{1}{10} \pm \dfrac{\sqrt{79}}{10}i$ **27.** $-1 - \dfrac{1}{2}i$ **28.** $-3 + \sqrt{2}$ **29.** $x^{3/4}y^{17/12}$

ER138. $-.5106$ **ER139.** 6.7706

problem set 122

1. $-1, 1, 3$ and $7, 9, 11$ **2.** $T_O = 5$ hr, $T_M = 10$ hr, $R_O = 130$ mph, $R_M = 80$ mph

3. 80 miles   **4.** 34 **5.** $S_N = 4, M_N = 24$ **6.** $\{2, 7, 13\}$ **7.** $\{1, 2, 3, 7, 8, 9\}$

8. $\{1, 2, 3, 4, 5, 7, 8, 9, 10, 13, 15\}$ **9, 10, 11.** See lesson 122   **12.**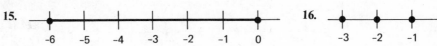

13. **14.** **15.** **16.**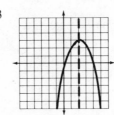

17. 2.32×10^{17} **18.** 4.21 **19.** 5.41×10^{-5} **20.** 10.8477 **21.** 3.1785

22. 1.6021 **23.** 1.07×10^{-4} **24.** 1.51×10^{-3} **25.** 9.31×10^{-10} **26.** $\dfrac{1234}{999,000}$

27. $y = -(x - 2)^2 + 3$ **28.** $-1, -\dfrac{2}{3}$ **29.** $3 - 4i$

ER140. 4.5160 **ER141.** 3.1921×10^{-5}